CHEMISTRY

FOR CCEA AS LEVEL

2nd EDITION

Dr Wingfield Glassey

ISBN: 978-1-78073-101-8

Second Edition
Second Impression 2018

Layout and design: April Sky Design
Printed by: GPS Colour Graphics Ltd, Belfast

The Author

Dr. Wingfield Glassey teaches Chemistry at Friends' School Lisburn, and is an examiner for GCE AS and A2 Chemistry. Dr. Glassey also maintains a professional interest in the teaching and learning of science, and has published scholarly articles in a number of peer-reviewed journals including *The Journal of Chemical Education*.

Copyright

Colourpoint Educational
An imprint of Colourpoint Creative Ltd
Colourpoint House
Jubilee Business Park
Jubilee Road
Newtownards
County Down
Northern Ireland
BT23 4YH

Tel: 028 9182 6339
Fax: 028 9182 1900
E-mail: sales@colourpoint.co.uk
Web site: www.colourpoint.co.uk

Publisher's Note: This book has been through a rigorous quality assurance process by an independent person experienced in the CCEA specification prior to publication. It has been written to help students preparing for the AS Chemistry specification from CCEA. While Colourpoint Educational, the author and the quality assurance person have taken every care in its production, we are not able to guarantee that the book is completely error-free. Additionally, while the book has been written to address the CCEA specification, it is the responsibility of each candidate to satisfy themselves that they have fully met the requirements of the CCEA specification prior to sitting an exam set by that body. For this reason, and because specifications and CCEA advice change with time, we strongly advise every candidate to avail of a qualified teacher and to check the contents of the most recent specification for themselves prior to the exam. Colourpoint Creative Ltd therefore cannot be held responsible for any errors or omissions in this book or any consequences thereof.

CONTENTS

Unit AS1: Basic Concepts in Physical and Inorganic Chemistry

Unit AS2: Further Physical and Inorganic Chemistry and an Introduction to Organic Chemistry

Unit AS 1:

Basic Concepts in Physical and Inorganic Chemistry

1.1 Formulas, Equations and Amounts of Substance

CONNECTIONS
- All matter is made from around 100 elements.
- Almost all life on earth is based on the element carbon.
- Hydrogen is the most abundant element in the universe.

Elements and Compounds

In this section we are learning to:

- Use the terms element and compound to classify pure substances.
- Recall examples of elements with metallic, molecular and giant structures.
- Distinguish between ionic compounds and nonmetal compounds.
- Recall examples of compounds with ionic, molecular and giant structures.
- Describe the structure of an ionic compound.
- Use the terms atom, molecule, monatomic and diatomic to describe the structure of molecular materials.
- Recall the effect of molecular size on the melting and boiling points of molecular materials.

Plastics, medicines, and the other materials in the world around us are made of matter. The matter within any material is a combination of elements and compounds. An element is the simplest type of pure substance. A compound is also a pure substance and contains two or more elements bonded together. Water and carbon dioxide are examples of compounds. Water is a compound of the elements hydrogen and oxygen, and carbon dioxide is a compound of the elements carbon and oxygen.

A mixture contains a combination of compounds and elements. Crude oil, paint and air are all mixtures. Approximately 99% of air is made up of the elements nitrogen and oxygen. The remaining 1% is principally a mixture of water vapour, carbon dioxide and the element argon. The relationship between elements, compounds and mixtures is summarised in Figure 1.

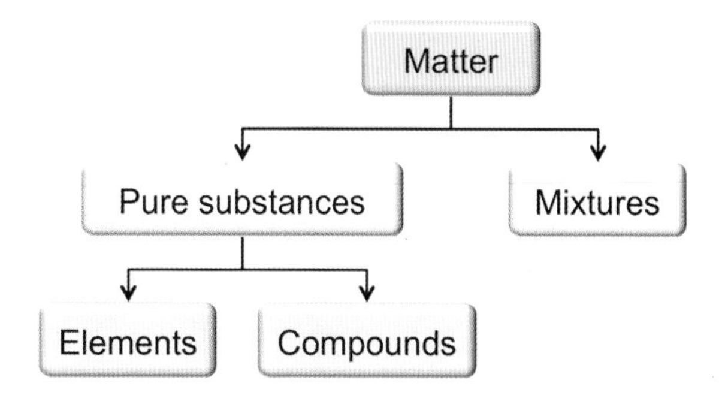

Figure 1: The relationship between elements, compounds and mixtures.

Elements

There are approximately 100 elements. The properties of one element can be quite different to those of another. Relationships between the properties and behaviour of different elements are best understood by constructing a Periodic Table. The form of the modern Periodic Table is shown in Figure 2. In the modern Periodic Table each element is represented by a chemical symbol and is assigned an atomic number. The atomic number of the elements increases from left to right across each row and down each column. When arranged in this way the properties of the elements are similar within each group (column), and vary in a predictable way from left to right across each period (row). For example, the distribution of metals and nonmetals in Figure 2 suggests that the elements behave less like metals and more like nonmetals from left to right across a period, and towards the top of each group.

Metals and nonmetals have very different properties. This results from differences in the arrangement of matter within each element. If we define an **atom** to be the smallest amount of an element, we can define an **element** to be a pure substance that contains only one type of atom. The atoms in a metal such as iron or silver are tightly packed together in an ordered arrangement. The resulting metallic structure is shown in Figure 3a. In contrast the atoms in nonmetals bond together to form molecules as in oxygen (Figure 3b),

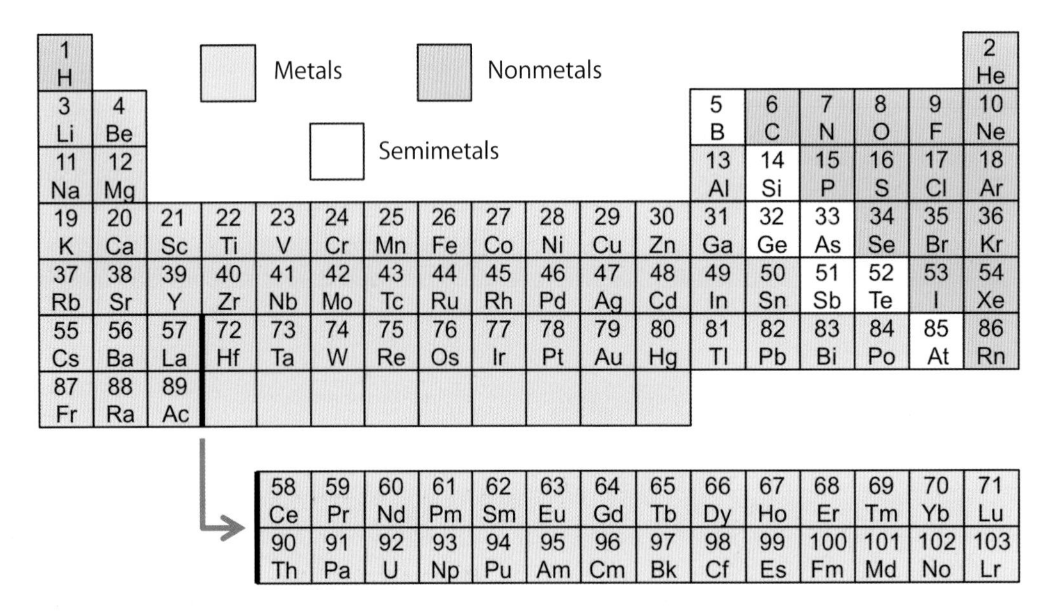

Figure 2: Classification of the elements as metals, nonmetals and semimetals. The term semimetal refers to elements that have properties in common with metals and nonmetals.

or to create giant structures such as the network of carbon atoms in diamond (Figure 3c).

The term **molecule** is used when referring to two or more atoms bonded together to form a single uncharged particle. Substances that are made of molecules are referred to as **molecular materials**. The elements oxygen and nitrogen are both molecular materials. They are also examples of **diatomic elements** as the molecules in oxygen and nitrogen each contain two atoms. Molecules containing two atoms are referred to as **diatomic molecules**.

Substances with giant structures such as diamond (Figure 3c) are held together by strong bonds between the atoms. As a result, they have high melting points and are solids under normal conditions. Graphite - the major component in pencil 'lead' - is a different form of carbon that also contains a network of carbon atoms. Like diamond, graphite has a high melting point and is a solid under normal conditions. Diamond and graphite are examples of allotropes where the term **allotrope** is used to refer to different physical forms of the same element.

In contrast, molecular materials may be solids, liquids or gases under normal laboratory conditions. Elements containing heavier atoms such as bromine and iodine are more likely to be liquids or solids. For example, under normal laboratory conditions oxygen is a gas, bromine is a liquid, and iodine is a solid.

Oxygen, bromine and iodine are diatomic elements. Elements containing larger molecules are even more likely to be liquids or solids. Rhombic sulfur and white phosphorus are examples of allotropes containing large molecules. Both are solids. Molecules of rhombic sulfur and white phosphorus are shown in Figure 4.

(a) Atoms in iron
Formula: Fe

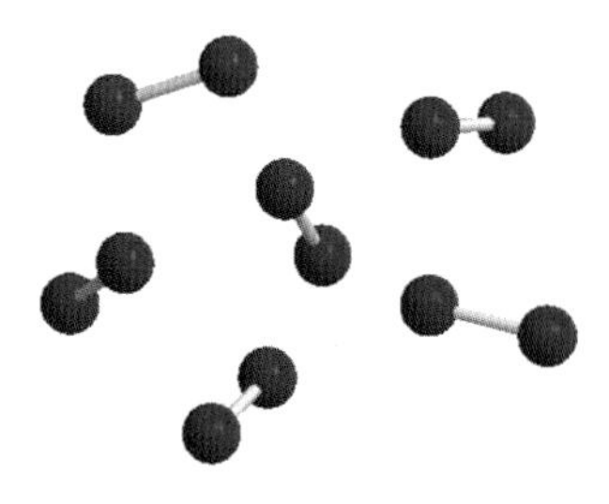

(b) Oxygen molecules
Formula: O_2

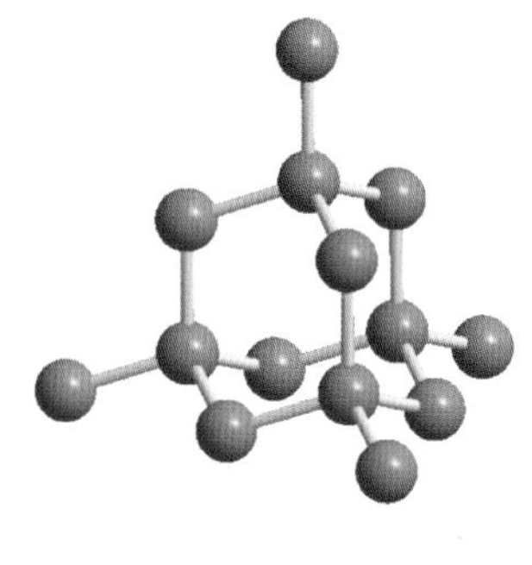

(c) Atoms in diamond
Formula: C

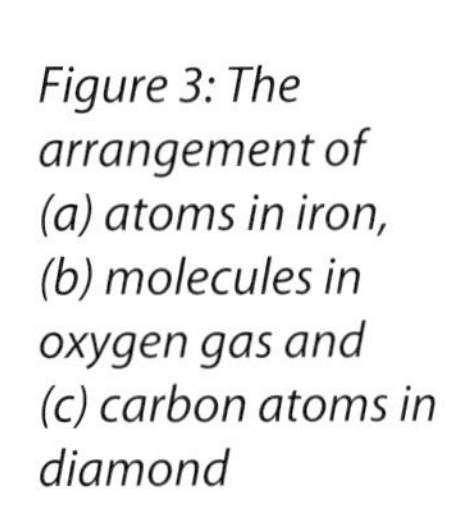

Figure 3: The arrangement of (a) atoms in iron, (b) molecules in oxygen gas and (c) carbon atoms in diamond

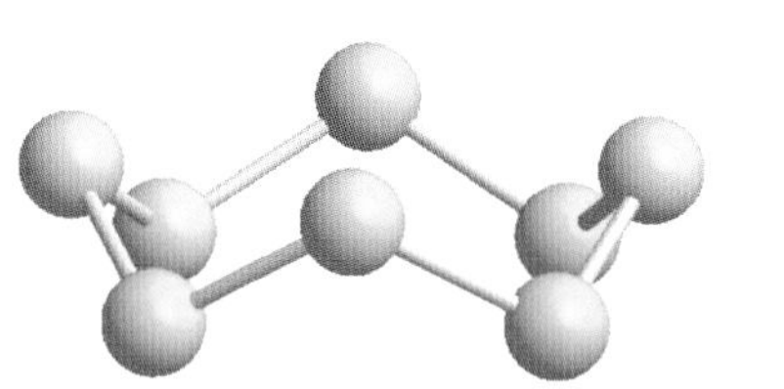

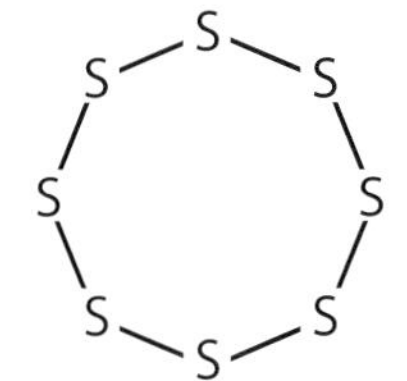

(a) Sulfur molecule
Formula: S_8

(b) Phosphorus molecule
Formula: P_4

Figure 4: A molecule of (a) rhombic sulfur and (b) white phosphorus.

The relationship between the size of the molecules in a substance and its melting point is further illustrated by the elements in Group VIII; a group of unreactive gases known as the noble gases. The noble gases are examples of **monatomic elements**. A substance is described as monatomic if the particles in the substance each contain a single atom. The fact that heavy noble gases such as krypton (Kr) and xenon (Xe) are gases, while much lighter elements such as sulfur are solids, demonstrates that the melting point of a molecular material is determined by the size of the molecules in the material.

Exercise 1.1A

1. Which of the following is (a) a diatomic element, (b) a gas, and (c) solid?

   nitrogen bromine silver sulfur iodine

Before moving to the next section, check that you are able to:

- Use the terms atom, molecule, monatomic and diatomic to describe elements.
- Recall examples of elements with metallic, molecular and giant structures.
- Recall the effect of molecular size on melting and boiling point.

Compounds

The term **compound** refers to a pure substance that contains two or more elements bonded together. A compound cannot be easily separated into its constituent elements. Water, carbon dioxide and sodium chloride (table salt) are all examples of compounds.

The compounds formed when metals combine with nonmetals are known as **ionic compounds** and are very different to the compounds formed when nonmetals combine. For example, sodium chloride is an ionic compound formed by the reaction between sodium metal (Na) and the nonmetal chlorine (Cl). Ionic compounds such as sodium chloride contain ions packed tightly together in a regular pattern known as a lattice. The lattice in sodium chloride is held together by attractive forces between sodium (Na^+) ions and chloride (Cl^-) ions. The packing of the ions in the sodium chloride lattice is shown in Figure 5a.

Ionic compounds are solids. The ions in the lattice separate and move independently when the solid dissolves or melts. Similarly, the molecules in nonmetal compounds separate and move independently when the compound dissolves or changes state. As in elements, compounds containing small molecules such as hydrogen chloride, HCl and sulfur dioxide,

sodium (Na^+) ion

chloride (Cl^-) ion

(a) Sodium chloride

silicon (Si) atom

oxygen (O) atom

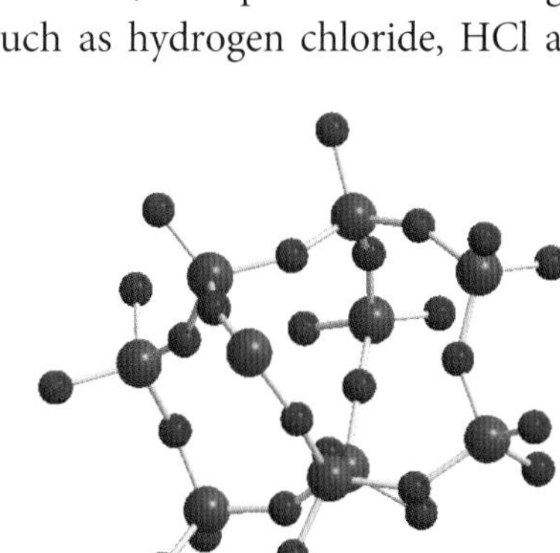

(b) Quartz

Figure 5: The structure of (a) sodium chloride and (b) quartz.

SO_2 are gases. Compounds made of larger molecules such as phosphorus(III) chloride, PCl_3 and dichloromethane, CH_2Cl_2 are liquids, and compounds containing even larger molecules such as phosphorus(V) chloride, PCl_5 and phosphorus(V) oxide, P_4O_{10} are solids.

Many nonmetal compounds such as the mineral quartz – an allotrope of silicon dioxide, SiO_2 – have a giant structure held together by strong bonds between the atoms. The structure of quartz is shown in Figure 5b. The strong bonding between atoms in giant structures such as quartz does not break easily and explains why substances with a giant structure have very high melting points, and tend to be insoluble in water and other solvents.

Exercise 1.1B

1. Which of the following (a) is an ionic compound, (b) is a molecular material, and (c) has a giant structure.

   NO_2 $NaNO_3$ $MgBr_2$ CCl_4 SiO_2

2. Which one of the following is a molecular substance?

   A CaO B CO C Cr_2O_3 D CuO

   (CCEA June 2013)

3. Which one of the following compounds is not molecular?

   A carbon dioxide
   B calcium chloride
   C hydrogen chloride
   D phosphorus trichloride

   (Adapted from CCEA June 2011)

Before moving to the next section, check that you are able to:

- Recall the definition of a compound.
- Distinguish between ionic compounds and nonmetal compounds.
- Describe the structure of an ionic compound.
- Recall examples of compounds with ionic, molecular and giant structures.

Chemical Formulas

In this section we are learning to:

- Use the formula for a substance to deduce its composition and structure.
- Write the formula for ionic compounds and nonmetal compounds, including those containing an element with variable valency.

Interpreting Formulas

The **chemical formula** or 'formula' of a substance describes the amount of each element in the substance. The formula of a substance also represents the smallest amount of the substance that can participate in a chemical reaction. For example, the formula of sodium chloride is NaCl and indicates that there is one sodium (Na^+) ion for every chloride (Cl^-) ion in the compound. Similarly the formula for magnesium chloride, $MgCl_2$ indicates that there are two chloride (Cl^-) ions for every magnesium (Mg^{2+}) ion in the compound.

In molecular materials such as water and carbon dioxide, the formula of the compound describes the number of atoms in each molecule of the substance. For example, the formula of water, H_2O indicates that each molecule of water contains two hydrogen atoms and one oxygen atom. Similarly, the formula for carbon dioxide, CO_2 indicates that each molecule of carbon dioxide contains one carbon atom and two oxygen atoms.

The formula of an element is determined by how the atoms in the element are organised. In a metal such as iron, the atoms are closely packed in an ordered array (Figure 3a) and are indistinguishable. As a result, a single metal atom can be used to represent any one atom in the structure, and the formula of a metal such as iron (formula: Fe) or gold (formula: Au) represents one atom of the element. Similarly, a single atom can be used to represent any atom in a monatomic element such as helium (formula: He) or neon (formula: Ne), and any atom in giant structures such as diamond (formula: C).

In elements with a molecular structure the formula of the element represents the atoms in one molecule of the element. For example, the formulas of hydrogen (formula: H_2) and chlorine (formula: Cl_2) represent two atoms of the element, indicating that the element is made up of diatomic molecules. Similarly, formulas can be used to indicate that the molecules in white phosphorus contain four phosphorus atoms (formula: P_4), and the molecules in rhombic sulfur contain eight sulfur atoms (formula: S_8).

Writing Formulas

The formula of a compound is determined by the **valency** or 'combining power' of the elements in the compound. The valency of many elements can be determined from their location in the Periodic Table. This relationship is illustrated in Figure 6.

When a metal combines with one or more nonmetals to form an ionic compound such as NaCl, the metal loses electrons to form a positive ion and the nonmetal gains electrons to form a negative ion. From Figure 6 we see that sodium (Na) belongs to Group I and has a valency of 1. The valency of sodium tells us that when sodium combines with nonmetals each sodium atom will lose one electron to form a sodium ion (Na^+). Similarly, chlorine belongs to Group VII and has a valency of 1. The valency of chlorine tells us that when chlorine reacts with a metal such as sodium, each chlorine atom will gain one electron to form a chloride ion (Cl^-). In this way the relationship between the location of an element in the Periodic Table and its valency can be used to deduce the charges on the ions in an ionic compound.

An ionic compound does not have a charge and, as a result, *the charges on the ions in one formula of an ionic compound add to zero*. This idea can be used to construct the formula for any ionic compound. For example, magnesium (Mg) belongs to Group II and has a valency of 2. The valency of magnesium tells us that magnesium forms Mg^{2+} ions when it combines with nonmetals to form an ionic compound. We have already seen that chlorine (Group VII) has a valency of 1 and forms chloride (Cl^-) ions when it combines with metals. If the charges on the ions in magnesium chloride add to zero the compound must contain two chloride (Cl^-) ions for each magnesium (Mg^{2+}) ion.

This explains why the formula of magnesium chloride: $MgCl_2$

represents one Mg^{2+} ion and two Cl^- ions

Worked Example 1.1i

Write the formula for (a) magnesium oxide, (b) sodium oxide and (c) aluminium bromide.

Strategy

- Use the valency of each element to calculate the charges on the ions.
- Deduce the number of positive and negative ions needed to balance the charge in one formula.

Solution

(a) Magnesium (Group II) has a valency of 2 and forms Mg^{2+} ions. Oxygen (Group VI) has a valency of 2 and forms O^{2-} ions. Magnesium oxide contains one Mg^{2+} ion for every O^{2-} ion. The formula of magnesium oxide is MgO.

(b) Sodium (Group I) has a valency of 1 and forms Na^+ ions. Oxygen (Group VI) has a valency of 2 and forms O^{2-} ions. Sodium oxide contains two Na^+

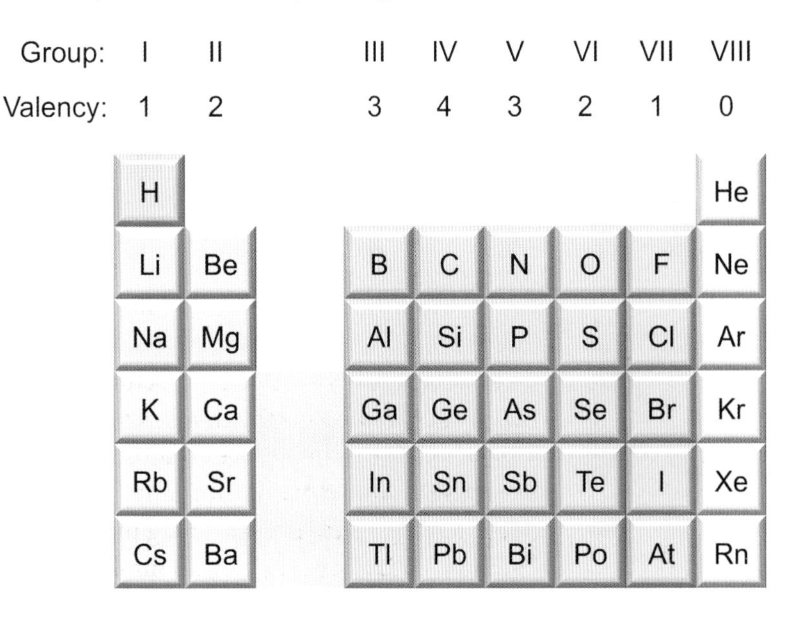

Figure 6: The relationship between the valency of an element and its location in the Periodic Table.

ions for every O^{2-} ion. The formula of sodium oxide is Na_2O.

(c) Aluminium (Group III) has a valency of 3 and forms Al^{3+} ions. Bromine (Group VII) has a valency of 1 and forms Br^- ions. Aluminium bromide contains three Br^- ions for every Al^{3+} ion. The formula of aluminium bromide is $AlBr_3$.

Exercise 1.1C

1. Write the chemical formula for (a) lithium fluoride, (b) potassium chloride, (c) magnesium bromide, (d) magnesium sulfide, and (e) aluminium oxide.

In nonmetal compounds such as carbon dioxide (formula: CO_2) and water (formula: H_2O), the atoms are held together by covalent bonds. The valency of an element tells us how many covalent bonds can be formed by an atom of the element when it combines with other elements to form a compound. For example, carbon belongs to Group IV and has a valency of 4. This tells us that a carbon atom can form up to four covalent bonds when bonding with other nonmetals. When carbon combines with oxygen (Group VI, valency 2) two oxygen atoms (total valency 2 + 2 = 4) are needed to equal the combining power of one carbon atom (valency 4). As a result, a molecule of carbon dioxide contains 2 oxygen atoms and 1 carbon atom.

This explains why the formula of carbon dioxide:

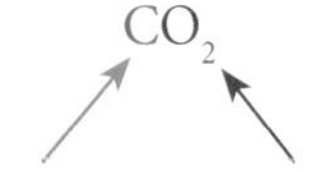

represents one carbon atom and two oxygen atoms

one molecule of carbon dioxide

Compounds of Hydrogen

Hydrogen, H is a nonmetal. It has a valency of 1 and combines with metals to form ionic compounds known as metal hydrides. Sodium hydride, NaH and calcium hydride, CaH_2 are examples of metal hydrides. Hydrogen also has a valency of 1 when it combines with nonmetals to form molecular compounds such as water, H_2O and ammonia, NH_3.

Worked Example 1.1ii

Deduce the formula of the compounds (a) H_xF, (b) H_xS, (c) NH_x and (d) CH_x.

Strategy

- Determine the valency of each element.
- Deduce the value of x needed to balance the combining power of each element.

Solution

(a) Hydrogen (H) and fluorine (F) both have a valency of 1. One hydrogen atom is needed to combine with one fluorine atom. Formula: HF

(b) Hydrogen (H) has a valency of 1 and sulfur (S) has a valency of 2. Two hydrogen atoms are needed to combine with one sulfur atom. Formula: H_2S

(c) Hydrogen (H) has a valency of 1 and nitrogen (N) has a valency of 3. Three hydrogen atoms are needed to combine with one nitrogen atom. Formula: NH_3

(d) Hydrogen (H) has a valency of 1 and carbon (C) has a valency of 4. Four hydrogen atoms are needed to combine with one carbon atom. Formula: CH_4

Elements with Variable Valency

A number of metals and nonmetals have the ability to vary their valency when forming compounds. For example, carbon can combine with oxygen to form carbon monoxide, CO in which the valency of carbon is 2 and carbon dioxide, CO_2 in which carbon has a valency of 4. Similarly iron can combine with chlorine to form iron(II) chloride, $FeCl_2$ in which iron has a valency of 2, and iron(III) chloride, $FeCl_3$ in which the valency of iron is 3.

Worked Example 1.1iii

Tin is a 'poor metal' and will behave in a similar way to carbon and other nonmetals in Group IV when bonding with other elements. Write the formula for (a) tin(II) chloride, (b) tin(IV) chloride, (c) copper(II) oxide and (d) copper(I) oxide.

Strategy

- Roman numerals are used to indicate the valency of tin and copper in each compound.

Solution

(a) Tin has a valency of 2 and chlorine has a valency of 1. Two chlorine atoms are needed to equal the combining power of tin. Formula: $SnCl_2$

(b) Tin has a valency of 4 and chlorine has a valency of 1. Four chlorine atoms are needed to equal the combining power of tin. Formula: $SnCl_4$

(c) The ions in copper(II) oxide are copper(II), Cu^{2+} and oxide, O^{2-}. One oxide ion is needed to balance the charge on one copper(II) ion. Formula: CuO

(d) The ions in copper(I) oxide are copper(I), Cu^+ and oxide, O^{2-}. Two copper(I) ions are needed to balance the charge on one oxide ion. Formula: Cu_2O

Exercise 1.1D

1. Write the chemical formula for
(a) phosphorus(III) bromide, (b) nitrogen(IV) oxide, (c) sulfur(IV) fluoride and
(d) mercury(II) chloride.

2. Write the chemical formula for (a) sulfur(IV) oxide, (b) phosphorus(V) chloride,
(c) manganese(IV) oxide and (d) xenon(II) fluoride.

Before moving to the next section, check that you are able to:

- Use formulas to deduce the composition and structure of compounds.
- Determine the valency of an element from its location in the Periodic Table.
- Write the formula for ionic and nonmetal compounds, including those containing an element with variable valency.

Naming Compounds

In this section we are learning to:

- Name ionic compounds, including those containing molecular ions.
- Use prefixes to name nonmetal compounds.
- Name ionic and nonmetal compounds, including those containing an element with the ability to vary its valency.

Ionic Compounds

When a metal combines with a nonmetal to form an ionic compound the metal loses electrons to form a **cation** and the nonmetal gains electrons to form an **anion**. For example, when sodium reacts with oxygen to form sodium oxide, Na_2O each atom of sodium forms a sodium cation, Na^+ and each oxygen gains electrons to form an oxide anion, O^{2-}. Ionic compounds are named by following the name of the cation with the name of the anion. In the case of sodium oxide, Na_2O the name indicates that the compound contains sodium cations and oxide anions.

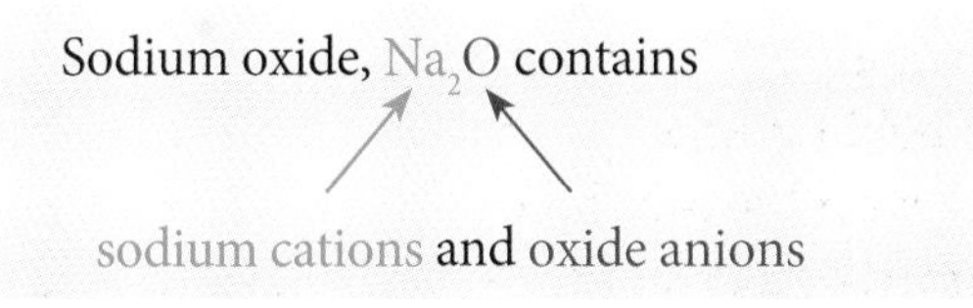

The names and formulas of common cations and anions are summarised in Table 1. Ions such as ammonium (NH_4^+), carbonate (CO_3^{2-}) and sulfate (SO_4^{2-}) are listed separately under the heading **molecular ions** as they are *positively or negatively charged ions that contain two or more atoms held together by covalent bonds*. The atoms in a molecular ion remain bonded together when the compound containing the molecular ion melts, reacts, or dissolves to form a solution.

Worked Example 1.1iv

Name the following ionic compounds.

(a) Na_2S (b) $MgBr_2$ (c) Na_2CO_3
(d) KNO_3 (e) $(NH_4)_2CO_3$ (f) K_2SO_4

Strategy

- Use the formula to identify the cations and anions in each compound.
- Use the names of the cation and anion to name the compound.

Solution

(a) Na_2S contains sodium cations (Na^+) and sulfide (S^{2-}) anions. Na_2S is the formula for sodium sulfide.

(b) $MgBr_2$ contains magnesium cations (Mg^{2+}) and bromide (Br^-) anions. $MgBr_2$ is the formula for magnesium bromide.

Cations	Anions	Molecular Ions
Hydrogen, H^+	Hydride, H^-	*Cations:*
		Ammonium, NH_4^+
Group I:	*Group V:*	*Anions:*
Lithium, Li^+	Nitride, N^{3-}	Peroxide, O_2^{2-}
Sodium, Na^+		Hydroxide, OH^-
Potassium, K^+	*Group VI:*	Nitrate, NO_3^-
	Oxide, O^{2-}	Nitrite, NO_2^-
Group II:	Sulfide, S^{2-}	Sulfate, SO_4^{2-}
Magnesium, Mg^{2+}		Thiosulfate, $S_2O_3^{2-}$
Calcium, Ca^{2+}	*Group VII:*	Hydrogensulfate, HSO_4^-
Barium, Ba^{2+}	Fluoride, F^-	Sulfite, SO_3^{2-}
	Chloride, Cl^-	Carbonate, CO_3^{2-}
Other metals:	Bromide, Br^-	Hydrogencarbonate, HCO_3^-
Aluminium, Al^{3+}	Iodide, I^-	Phosphate, PO_4^{3-}
Copper(II), Cu^{2+}		Chlorate, ClO_3^-
Iron(II), Fe^{2+}		Hypochlorite, ClO^-
Iron(III), Fe^{3+}		*Anions containing metals:*
Zinc, Zn^{2+}		Chromate, CrO_4^{2-}
Silver, Ag^+		Dichromate, $Cr_2O_7^{2-}$
		Permanganate, MnO_4^-

Table 1: Common cations and anions

(c) Na_2CO_3 contains sodium cations (Na^+) and carbonate (CO_3^{2-}) anions. Na_2CO_3 is the formula for sodium carbonate.

(d) KNO_3 contains potassium cations (K^+) and nitrate (NO_3^-) anions. KNO_3 is the formula for potassium nitrate.

(e) $(NH_4)_2CO_3$ contains ammonium (NH_4^+) cations and carbonate (CO_3^{2-}) anions. $(NH_4)_2CO_3$ is the formula for ammonium carbonate.

(f) K_2SO_4 contains potassium cations (K^+) and sulfate (SO_4^{2-}) anions. K_2SO_4 is the formula for potassium sulfate.

Exercise 1.1E

1. Name the following ionic compounds.
   (a) $CaCl_2$ (b) Al_2O_3 (c) AgBr
   (d) LiH (e) ZnS
2. Name the following ionic compounds.
   (a) $ZnCO_3$ (b) $ZnSO_4$ (c) $Mg(NO_3)_2$
   (d) Mg_3N_2 (e) $KMnO_4$
3. Name the following ionic compounds.
   (a) $AgNO_3$ (b) NH_4NO_3 (c) $NaNO_2$
   (d) $Al_2(SO_4)_3$ (e) Na_2SO_3
4. Name the following ionic compounds.
   (a) Na_2O (b) Na_2O_2 (c) $Ca(HCO_3)_2$
   (d) NaClO (e) K_2CrO_4

Nonmetal Compounds

When naming nonmetal compounds the amount of each element in the compound can be described by using a prefix such as *di* or *tri*. For instance, the prefixes *mon*, *di* and *tri* could be used to distinguish carbon monoxide, CO from carbon dioxide, CO_2 and sulfur dioxide, SO_2 from sulfur trioxide, SO_3. Similarly, the prefixes *tri* and *penta* could be used to distinguish phosphorus trichloride, PCl_3 from phosphorus pentachloride, PCl_5. The use of prefixes is further illustrated by the examples in Table 2.

Element Ratio in Compound	Prefix	Example
1:1	mon	nitrogen **mon**oxide, NO
1:2	di	nitrogen **di**oxide, NO_2
1:3	tri	boron **tri**fluoride, BF_3
1:4	tetra	carbon **tetra**chloride, CCl_4
1:5	penta	phosphorus **penta**chloride, PCl_5
1:6	hexa	sulfur **hexa**fluoride, SF_6

Table 2: Using prefixes to name nonmetal compounds

Worked Example 1.1v

Use prefixes to name the following compounds of nitrogen. (a) NO (b) NO_2 (c) N_2O (d) N_2O_4

Strategy

- If there is more than one atom of the first element include a prefix for every element in the formula.

Solution

(a) NO contains one oxygen atom (prefix: *mon*) for every nitrogen atom. The compound NO is called nitrogen monoxide.

(b) NO_2 contains two oxygen atoms (prefix: *di*) for every nitrogen atom. The compound NO_2 is called nitrogen dioxide.

(c) N_2O contains two nitrogen atoms (prefix: *di*) for every oxygen (prefix: *mon*). The compound N_2O is called dinitrogen monoxide.

(d) N_2O_4 contains four oxygen atoms (prefix: *tetra*) for every two nitrogen atoms (prefix: *di*). The compound N_2O_4 is called dinitrogen tetroxide.

Note: The 'a' at the end of the prefix is removed to avoid the combination of vowels 'ao' in tetraoxide.

Exercise 1.1F

1. Use prefixes to name the following compounds.

   (a) SF_4 (b) SO_3 (c) $SiCl_4$ (d) PBr_3 (e) XeF_4

Before moving to the next section, check that you are able to:

- Identify the ions in an ionic compound, including those containing molecular ions, and use the names of the ions to name the compound.
- Use prefixes to name nonmetal compounds.

Elements with Variable Valency

When naming an ionic compound containing an element that has the ability to vary its valency, the valency of the element is included in the name. For example, the name iron chloride is not sufficient to distinguish between the compounds iron(II) chloride, $FeCl_2$ and iron(III) chloride, $FeCl_3$. Several cations formed by metals with the ability to vary their valency are included in Table 1.

Worked Example 1.1vi

Name the following compounds of iron.

(a) FeS (b) $FeSO_4$ (c) $Fe(NO_3)_3$ (d) Fe_2O_3

Strategy

- Identify the anion.
- Use the charge on the anion to calculate the valency of iron.

Solution

(a) The compound FeS contains one sulfide (S^{2-}) anion for every iron cation. The cation must have a valency of two (Fe^{2+}) if the charges on the ions in the formula add to zero.

The compound FeS is iron(II) sulfide.

(b) The compound $FeSO_4$ contains one iron cation for every sulfate (SO_4^{2-}) anion. The iron cation must have a valency of two (Fe^{2+}) if the charges on the ions in the formula add to zero.

The compound $FeSO_4$ is iron(II) sulfate.

(c) The compound $Fe(NO_3)_3$ contains three nitrate (NO_3^-) anions for every iron cation. The cation must have a valency of three (Fe^{3+}) if the charges on the ions in the formula add to zero.

The compound $Fe(NO_3)_3$ is iron(III) nitrate.

(d) The compound Fe_2O_3 contains three oxide (O^{2-}) anions for every two iron cations. The cations must each have a valency of three (Fe^{3+}) if the charges on the ions in the formula add to zero.

The compound Fe_2O_3 is iron(III) oxide.

Exercise 1.1G

1. Name the following ionic compounds.

   (a) CuO (b) $CuSO_4$ (c) Cu_2O (d) $CuCl_2$

   (e) $HgCl_2$ (f) Hg_2Cl_2

A similar approach can be used when naming nonmetal compounds containing elements with the ability to vary their valency. For example, the name phosphorus chloride does not distinguish between phosphorus(III) chloride, PCl_3 and phosphorus(V) chloride, PCl_5 and it becomes necessary to specify the valency of phosphorus, or alternatively use a naming prefix, when naming the compounds.

Worked Example 1.1vii

The elements sulfur and nitrogen have the ability to vary their valency when forming compounds. Name the following compounds by specifying the valency of nitrogen and sulfur.

(a) NO (b) NO_2 (c) N_2O (d) SO_3 (e) SF_6

Strategy

- Determine the combined valency of the oxygen/fluorine atoms.
- The combined valency of the nitrogen/sulfur atoms is equal to the combined valency of the oxygen/fluorine atoms.

Solution

(a) The nitrogen atom in NO has the same valency as the oxygen atom (2). The compound NO is nitrogen(II) oxide.

(b) The valency of the nitrogen atom in NO_2 is the same as the combined valency of the two oxygen atoms (2 + 2 = 4). The compound NO_2 is nitrogen(IV) oxide.

(c) The combined valency of the nitrogen atoms in N_2O (1 + 1 = 2) is the same as the valency of the oxygen atom (2). The compound N_2O is nitrogen(I) oxide.

(d) The valency of the sulfur atom in SO_3 is equal to the combined valency of the oxygen atoms (2 + 2 + 2 = 6). The compound SO_3 is sulfur(VI) oxide.

(e) The valency of the sulfur atom in SF_6 is equal to the combined valency of the fluorine atoms (6 × 1 = 6). The compound SF_6 is sulfur(VI) fluoride.

Exercise 1.1H

1. Name the following compounds.

(a) SF_4 (b) $SnCl_2$ (c) $SiCl_4$
(d) PBr_3 (e) XeF_4 (f) PbO_2

Before moving to the next section, check that you are able to:

- Name ionic and nonmetal compounds using prefixes, and by specifying the valency of an element with the ability to vary its valency.

Chemical Equations

In this section we are learning to:

- Use the chemical equation for a reaction to relate the amount of reactants used and the amount of products formed by the reaction.
- Write chemical equations for reactions, including the use of state symbols to describe the physical state of the reactants and products.

Interpreting Chemical Equations

A **chemical equation** is used to summarise the change that occurs when a chemical reaction takes place. For example, when hydrogen burns in air it reacts with the oxygen in air to form water. The chemical change that occurs in the reaction is described by the chemical equation:

$$2H_2 + O_2 \rightarrow 2H_2O$$

The equation reads: two formulas of hydrogen ($2H_2$) react with one formula of oxygen (O_2) to form two formulas of water ($2H_2O$). In this example the reactants (to the left of the arrow) and the products (to the right of the arrow) are all molecular materials. This allows for the equation to be read as: two molecules of hydrogen ($2H_2$) react with one molecule of oxygen (O_2) to form two molecules of water ($2H_2O$).

A slightly different approach is needed when interpreting the chemical equation for a reaction involving non-molecular materials. For example, magnesium burns in air to form magnesium oxide. The equation for the reaction is:

$$2Mg + O_2 \rightarrow 2MgO$$

The equation reads: two formulas of magnesium (2Mg) react with one formula of oxygen (O_2) to produce two formulas of magnesium oxide (2MgO). Magnesium and magnesium oxide are not molecular materials. At this point it is helpful to recall that the formula Mg represents one atom of the metal, and the formula MgO represents the ions in one formula of magnesium oxide. In this way the equation for the reaction can also be read: two atoms of magnesium (2Mg) react with one molecule of oxygen (O_2) to produce two formulas of magnesium oxide (2MgO), each containing one magnesium ion (Mg^{2+}) and one oxide ion (O^{2-}).

Writing Chemical Equations

When learning to write chemical equations it is often helpful to begin by writing the chemical equation in words. For example, sodium chloride is formed when sodium metal reacts with chlorine gas. The so-called word equation for the reaction is:

sodium + chlorine → sodium chloride

The names of the reactants (to the left of the arrow) and products (to the right of the arrow) can then be replaced with a formula to produce the unbalanced chemical equation for the reaction.

Unbalanced equation: $Na + Cl_2 \rightarrow NaCl$

In any chemical reaction *the number of atoms of an element in the reactants must be the same as the number of atoms of the element in the products*. This is equivalent to stating that atoms cannot be formed or destroyed in the reaction. As a result, the number of sodium ions in the sodium chloride formed by the reaction must be same as the number of sodium atoms that reacted. Similarly, the number of chloride ions in the sodium chloride formed by the reaction must be the same as the number of atoms of chlorine gas that reacted. This balance is achieved by reacting two atoms of sodium (2Na) with one molecule of chlorine (Cl_2) to form two formulas of sodium chloride (2NaCl), each containing one sodium ion (Na^+) and one chloride ion (Cl^-).

Balanced equation: $2Na + Cl_2 \rightarrow 2NaCl$

Worked Example 1.1viii

Write the balanced chemical equation for the reaction that occurs when aluminium powder reacts with bromine to form aluminium bromide, $AlBr_3$.

Solution

Aluminium is a metal (formula: Al) and bromine is diatomic (formula: Br_2).

Unbalanced equation: $Al + Br_2 \rightarrow AlBr_3$

Balanced equation: $2Al + 3Br_2 \rightarrow 2AlBr_3$

Exercise 1.1I

1. Coal (mostly carbon) forms carbon dioxide when it burns in air. If the supply of oxygen is limited carbon monoxide is formed instead of carbon dioxide. Write an equation for the reaction that occurs when coal burns in air to form (a) carbon dioxide and (b) carbon monoxide.
2. The metal Beryllium was first isolated by Wohler in 1828 as a product of the reaction between potassium and beryllium chloride. (a) Write an equation for the reaction. (b) Beryllium chloride can be prepared by the reaction of beryllium with chlorine or hydrogen chloride. Write equations for both of these reactions.

   (Adapted from CCEA June 2015)
3. Magnesium burns in air to form magnesium oxide. A small amount of magnesium nitride is also formed as a result of the reaction between magnesium and nitrogen in the air. Write a balanced equation for the formation of magnesium nitride.
4. Silicon dioxide is used to make glass. Hydrofluoric acid, HF has to be stored in plastic containers as it reacts with the 'silicon dioxide' in glass to form silicon tetrafluoride and water. Write an equation for the reaction.

   (CCEA January 2013)
5. Phosphorus reacts with an excess of chlorine to form phosphorus(V) chloride. If the amount of chlorine is limited phosphorus(III) chloride is formed. Write a chemical equation for the reaction of phosphorus with (a) an excess of chlorine, and (b) a limited amount of chlorine.

State Symbols

The chemical equation for a reaction relates the amount of each reactant used, and the amount of products formed in the reaction. It can also be used to describe the physical states of the reactants and products by including a **state symbol** after the formula for each reactant and product. For example, if hydrogen reacts with oxygen at room temperature, the water formed in the reaction is formed as droplets of liquid water. This information is included in the chemical equation by adding the state symbol (l) or (g) after each formula to identify the individual reactants and products as liquids or gases.

$$2H_{2(g)} + O_{2(g)} \rightarrow 2H_2O_{(l)}$$

If the reaction occurred at temperatures above 100 °C the reaction would instead produce steam and the equation would be written:

$$2H_{2(g)} + O_{2(g)} \rightarrow 2H_2O_{(g)}$$

Clearly these two quite different outcomes could not be distinguished without the use of state symbols.

Worked Example 1.1ix

Write a chemical equation for the reaction that occurs when sulfur burns in air to form sulfur dioxide. Include state symbols.

Solution

The chemical equation for the reaction is:

$$S + O_2 \rightarrow SO_2$$

Note that sulfur can be represented by the formula S or S_8 when writing equations.

Sulfur is a solid. The state symbol for a solid is (s). Oxygen and sulfur dioxide are gases. The state symbol for a gas is (g).
Including state symbols gives the equation:

$$S_{(s)} + O_{2(g)} \rightarrow SO_{2(g)}$$

Exercise 1.1J

1. Write the chemical equation for the following reactions. Include state symbols.
   (a) sodium + chlorine → sodium chloride
   (b) hydrogen + chlorine → hydrogen chloride
   (c) calcium + oxygen → calcium oxide
   (d) magnesium + bromine
   → magnesium bromide
2. Write the chemical equation for the following reactions. Include state symbols.
   (a) iron + sulfur → iron(II) sulfide
   (b) calcium carbonate
   → calcium oxide + carbon dioxide
   (c) iron + bromine → iron(III) bromide
   (d) aluminium + iodine → aluminium iodide

Before moving to the next section, check that you are able to:

- Interpret the chemical equation for a reaction in terms of the number of formulas reacting or formed in the reaction.
- Write balanced chemical equations that include state symbols

Aqueous Solutions

In this section we are learning to:

- Use the term aqueous to describe a solution made by dissolving one or more substances in water.
- Identify the ions present in acids, alkalis and solutions of metal salts.
- Recall the solubility rules for compounds and use solubility to distinguish bases from alkalis.

An **aqueous solution** is formed by dissolving a substance in water. Compounds and ions in an aqueous solution are identified by the state symbol (aq). For example, the formula for hydrochloric acid is written $HCl_{(aq)}$ to distinguish it from hydrogen chloride gas, $HCl_{(g)}$. Similarly, the formula for sodium hydroxide solution is written $NaOH_{(aq)}$ to distinguish it from solid sodium hydroxide, $NaOH_{(s)}$.

Solutions Containing Salts

Ionic compounds such as metal chlorides, nitrates and sulfates are examples of **salts**, and are formed when an acid reacts with a base.

Reactions that form salts:

acid + base → salt + other products

Examples

acid + metal oxide → salt + water
acid + metal carbonate
→ salt + water + carbon dioxide
acid + metal → salt + hydrogen

When a salt dissolves in water it breaks apart into its constituent ions. The ions then become surrounded by water molecules and move freely throughout the solution. For example, dissolving magnesium chloride produces an aqueous solution of magnesium chloride, $MgCl_{2(aq)}$ that contains aqueous magnesium ions, $Mg^{2+}_{(aq)}$ and aqueous chloride ions, $Cl^-_{(aq)}$.

$MgCl_{2\,(s)} \rightarrow \underbrace{Mg^{2+}{}_{(aq)} + 2Cl^{-}{}_{(aq)}}$

aqueous magnesium chloride, $MgCl_{2\,(aq)}$

Similarly, sodium sulfate dissolves in water to form an aqueous solution of sodium sulfate, $Na_2SO_{4(aq)}$ that contains aqueous sodium ions, $Na^{+}{}_{(aq)}$ and aqueous sulfate ions, $SO_4^{2-}{}_{(aq)}$.

$Na_2SO_{4\,(s)} \rightarrow \underbrace{2Na^{+}{}_{(aq)} + SO_4^{2-}{}_{(aq)}}$

aqueous sodium sulfate, $Na_2SO_{4\,(aq)}$

It is possible to make some general statements about the solubility of salts and other compounds in water. The rules used to determine the solubility of compounds in water are summarised in Table 3.

The solubility of compounds in water

- All nitrate salts are soluble.
- Group I salts and ammonium salts are soluble.
- Chloride, bromide, and iodide salts are soluble.
  Exceptions: Silver and lead salts are insoluble.
- Carbonates, oxides, and hydroxides are insoluble.
  Exceptions: Barium hydroxide is soluble.
- Hydrogencarbonates are soluble.
- Sulfate salts are soluble.
  Exceptions: Silver, lead and barium salts are insoluble.
- Sulfide salts are insoluble.
  Exceptions: Group II sulfides are soluble.

Table 3: The solubility of compounds in water.

Worked Example 1.1x

Which of the following compounds is soluble in water? Explain your reasoning.

(a) Na_2O (b) $FeCl_3$ (c) $PbSO_4$ (d) Al_2O_3

(e) FeS (f) $Ba(NO_3)_2$ (g) $Cu(OH)_2$

Strategy

- Apply the solubility rules in Table 3.

Solution

(a) Sodium oxide, Na_2O is soluble because all Group I salts are soluble.

(b) Iron(III) chloride, $FeCl_3$ is soluble because all chlorides (except silver and lead) are soluble.

(c) Lead sulfate, $PbSO_4$ is NOT soluble because all sulfates (except silver, lead and barium) are soluble.

(d) Aluminium oxide, Al_2O_3 is NOT soluble because oxides (except Group I and ammonium) are insoluble.

(e) Iron(II) sulfide, FeS is NOT soluble because sulfides (except Group I, Group II and ammonium) are insoluble.

(f) Barium nitrate, $Ba(NO_3)_2$ is soluble because all nitrates are soluble.

(g) Copper(II) hydroxide, $Cu(OH)_2$ is NOT soluble because hydroxides (except Group I, ammonium and barium) are insoluble.

Exercise 1.1K

1. Which of the following compounds are soluble in water?
   (a) $MgBr_2$ (b) AgBr (c) $CuCl_2$ (d) $PbCl_2$
2. Which of the following compounds are soluble in water?
   (a) Na_2S (b) KOH (c) Na_2CO_3 (d) $(NH_4)_2CO_3$
3. Which of the following compounds are soluble in water?
   (a) $BaSO_4$ (b) $BaCl_2$ (c) $MgCO_3$ (d) $CuSO_4$
4. Which of the following compounds are soluble in water?
   (a) $Mg(OH)_2$ (b) CaO (c) $Ba(OH)_2$
   (d) $Ca(HCO_3)_2$

Before moving to the next section, check that you are able to:

- Identify the ions present in solutions of metal salts.
- Recall the solubility rules and use the rules to determine the solubility of compounds in water.

Acids and Alkalis

Many compounds dissolve in water to form aqueous solutions that behave as acids or alkalis. An **acid** is a substance that forms hydrogen (H^+) ions in solution. For example, hydrochloric acid $HCl_{(aq)}$ is an aqueous solution containing hydrogen ions, $H^{+}{}_{(aq)}$ and chloride ions, $Cl^-(aq)$. Hydrochloric acid is formed when hydrogen chloride, HCl(g) dissolves in water.

$$HCl_{(g)} \rightarrow \underbrace{H^+_{(aq)} + Cl^-_{(aq)}}_{\text{hydrochloric acid, } HCl_{(aq)}}$$

Similarly sulfuric acid, $H_2SO_4(aq)$ is an aqueous solution containing hydrogen ions, $H^+(aq)$ and hydrogensulfate ions, $HSO_4^-(aq)$. It is formed when sulfur trioxide, SO_3(g) reacts with water.

$$SO_{3\,(g)} + H_2O_{(l)} \rightarrow \underbrace{H^+_{(aq)} + HSO_4^-{}_{(aq)}}_{\text{sulfuric acid, } H_2SO_{4\,(aq)}}$$

An **alkali** is a solution containing hydroxide (OH^-) ions. Aqueous sodium hydroxide, NaOH(aq) is an alkali, and is formed when solid sodium hydroxide dissolves in water.

$$NaOH_{(s)} \rightarrow \underbrace{Na^+_{(aq)} + OH^-_{(aq)}}_{\text{aqueous sodium hydroxide, } NaOH_{(aq)}}$$

Aqueous ammonia, $NH_3(aq)$, also known as 'ammonia solution', is an alkali as a fraction of the ammonia molecules react with water to form ammonium ions, $NH_4^+(aq)$ and hydroxide ions, $OH^-(aq)$.

$$NH_{3\,(aq)} + H_2O_{(l)} \rightarrow \underbrace{NH_4^+{}_{(aq)} + OH_{(aq)}}_{\text{aqueous ammonium hydroxide, } NH_4OH_{(aq)}}$$

A **base** is a compound that reacts with an acid to form a salt. Metal oxides, hydroxides, carbonates and hydrogencarbonates are all bases. Oxides such as calcium oxide, CaO form an alkali by reacting with water to produce hydroxide ions.

$$CaO_{(s)} + H_2O_{(l)} \rightarrow Ca^{2+}_{(aq)} + 2OH^-_{(aq)}$$

Soluble carbonates such as sodium carbonate, $Na_2CO_3(s)$ also form an alkali when dissolved in water as aqueous carbonate ions react with water to form hydroxide ions.

$$CO_3^{2-}{}_{(sq)} + H_2O_{(l)} \rightarrow HCO_3^-{}_{(aq)} + OH^-_{(aq)}$$

Exercise 1.1L

1. Vinegar is an aqueous solution of ethanoic acid, $C_2H_4O_2$. Explain, with the help of an equation, why vinegar is an acid.
2. Phosphoric acid, H_3PO_4 is a weak acid found in fizzy drinks. (a) Explain, with the help of an equation, why phosphoric acid is an acid. (b) Suggest why phosphoric acid is a weak acid.
3. Beryllium chloride reacts vigorously with water:

   $BeCl_2 + 2H_2O \rightarrow Be(OH)_2 + 2HCl$

   Describe and explain what is observed when Universal Indicator is added to a solution of (a) beryllium chloride in water, and (b) sodium chloride in water.

   (Adapted from CCEA June 2015)
4. Explain how the reaction between zinc oxide and sulfuric acid can be used to demonstrate that zinc oxide is a base.

   $ZnO_{(s)} + H_2SO_{4(aq)} \rightarrow ZnSO_{4(aq)} + H_2O(l)$
5. Suggest, with the help of a chemical equation, why aqueous ammonium carbonate is an alkali.

Before moving to the next section, check that you are able to:

- Identify the ions present in common acids such as hydrochloric acid and sulfuric acid.
- Use the solubility of metal salts to distinguish alkalis from bases.
- Account for the presence of hydroxide ions in common alkalis such as aqueous sodium hydroxide and aqueous ammonia.

Ionic Equations

In this section we are learning to:

- Use the solubility rules for metal salts to account for the formation of a precipitate in aqueous solution.
- Recall that neutralisation occurs when hydrogen ions combine with hydroxide ions to form water.
- Write ionic equations for chemical reactions including neutralisation and precipitation.

In many chemical reactions, particularly those that take place in aqueous solution, some of the molecules and ions in the reaction mixture are not involved in the reaction. An **ionic equation** contains only those compounds and ions that react, or are formed, by the reaction. Ionic equations do not include compounds and ions that are unchanged by the reaction. As a result, the ionic equation for a reaction gives a clear picture of the chemical change that occurs during the reaction.

Precipitation Reactions

Adding a few drops of aqueous barium chloride, $BaCl_{2(aq)}$ to an aqueous solution containing sulfate ions, $SO_4^{2-}{}_{(aq)}$ produces a white precipitate of barium sulfate, $BaSO_{4(s)}$ where the term **precipitate** refers to an insoluble solid formed in solution by a chemical reaction.

Chemical equation:

$BaCl_{2(aq)} + SO_4^{2-}{}_{(aq)} \rightarrow BaSO_{4(s)} + 2Cl^-{}_{(aq)}$

The precipitate is formed when barium (Ba^{2+}) ions from the barium chloride solution combine with the sulfate ions already in the solution. The ionic equation reminds us that the chloride ions from the barium chloride solution do not get involved in the reaction and, for this reason, are known as **spectator ions**.

Ionic equation:

$Ba^{2+}{}_{(aq)} + SO_4^{2-}{}_{(aq)} \rightarrow BaSO_{4(s)}$

Precipitates formed by the addition of silver nitrate solution to solutions containing (from L to R) chloride, bromide and iodide ions.

Worked Example 1.1xi

Explain, in terms of the ions present, what happens when a few drops of aqueous barium chloride are added to the following solutions. Write an ionic equation for any reactions that occur.

(a) $Na_2SO_{4(aq)}$ (b) $NaCl_{(aq)}$ (c) $KOH_{(aq)}$ (d) $Na_2CO_{3(aq)}$

Strategy

- Use the solubility rules in Table 3 to determine if the cations and anions in the reaction mixture can combine to form an insoluble salt.

Solution

(a) Barium ion (Ba^{2+}) forms a precipitate with sulfate ion (SO_4^{2-}): $Ba^{2+}{}_{(aq)} + SO_4^{2-}{}_{(aq)} \rightarrow BaSO_{4(s)}$

(b) A precipitate will not form as all sodium salts are soluble and barium ion (Ba^{2+}) does not form a precipitate with chloride ion (Cl^-).

(c) A precipitate will not form as all potassium salts are soluble and barium ion (Ba^{2+}) will not form a precipitate with chloride ion (Cl^-) or hydroxide ion (OH^-).

(d) Barium ion (Ba^{2+}) forms a precipitate with carbonate ion (CO_3^{2-}):

$Ba^{2+}{}_{(aq)} + CO_3^{2-}{}_{(aq)} \rightarrow BaCO_{3(s)}$

Exercise 1.1M

1. Write the ionic equation for the reaction that occurs when silver nitrate solution is added to aqueous sodium iodide. Include state symbols.
2. Write the ionic equation for the reaction that occurs when aqueous calcium chloride is added to dilute sodium carbonate solution. Include state symbols.

Before moving to the next section, check that you are able to:

- Use the solubility rules for metal salts to account for the formation of a precipitate in aqueous solution.
- Write an ionic equation to describe the formation of a precipitate.

Neutralisation Reactions

In a **neutralisation reaction** an acid reacts with a base to form a salt and water. For example, adding sodium hydroxide solution to dilute hydrochloric acid produces a solution of sodium chloride in water.

Chemical equation:

$HCl_{(aq)} + NaOH_{(aq)} \rightarrow NaCl_{(aq)} + H_2O_{(l)}$

The acid is neutralised when hydrogen (H^+) ions from the acid react with hydroxide (OH^-) ions from the base to form water. The sodium ions and chloride ions in the reaction mixture are spectator ions and are not included when writing the ionic equation.

Ionic equation: $H^+_{(aq)} + OH^-_{(aq)} \rightarrow H_2O_{(l)}$

Worked Example 1.1xii

Write the chemical equation and ionic equation for the reaction that occurs when aqueous potassium hydroxide is added to dilute sulfuric acid.

Strategy

- Write a balanced chemical equation for the reaction.
- Re-write the chemical equation showing all of the ions in solution.
- Identify and remove spectator ions to form the ionic equation.

Solution

Chemical equation:

$H_2SO_{4(aq)} + 2KOH_{(aq)} \rightarrow K_2SO_{4(aq)} + 2H_2O_{(l)}$

Ions in solution:

$H^+_{(aq)} + HSO_4^-{}_{(aq)} + 2K^+_{(aq)} + 2OH^-_{(aq)}$

$\rightarrow 2K^+_{(aq)} + SO_4^{2-}{}_{(aq)} + 2H_2O_{(l)}$

The potassium (K^+) ions are spectator ions. They are not involved in the reaction and do not appear in the ionic equation.

Ionic equation:

$H^+_{(aq)} + HSO_4^-{}_{(aq)} + 2OH^-_{(aq)} \rightarrow SO_4^{2-}{}_{(aq)} + 2H_2O_{(l)}$

If an acid reacts with an insoluble base the ionic equation more closely resembles the chemical equation for the reaction but is still helpful. Consider, for example, the neutralisation reaction that occurs when dilute hydrochloric acid is added to calcium carbonate.

Chemical equation:

$CaCO_{3(s)} + 2HCl_{(aq)} \rightarrow CaCl_{2(aq)} + H_2O_{(l)} + CO_{2(g)}$

The chloride (Cl^-) ions in the hydrochloric acid are the only spectator ions as they remain in solution and are unchanged by the reaction. Removing the spectator ions gives the ionic equation for the reaction.

Ionic equation:

$CaCO_{3(s)} + 2H^+_{(aq)} \rightarrow Ca^{2+}_{(aq)} + H_2O_{(l)} + CO_{2(g)}$

Exercise 1.1N

1. Write the (a) chemical equation and (b) ionic equation for the reaction that occurs when copper(II) oxide is added to dilute hydrochloric acid.
2. Write the (a) chemical equation and (b) ionic equation for the reaction that occurs when aqueous sodium carbonate is added to dilute hydrochloric acid.

Before moving to the next section, check that you are able to:

- Recall that neutralisation occurs when hydrogen ions combine with hydroxide ions to form water.
- Identify spectator ions and write ionic equations for reactions involving solids, liquids and gases.

Amounts of Substance

In this section we are learning to:

- Define one mole of substance in terms of Avogadro's constant.
- Use Avogadro's constant to calculate the number of particles in a given amount of substance.
- Calculate the molar mass of a substance and use molar mass to relate amounts of a substance in grams and moles.

Previously we had begun to interpret chemical equations in terms of the amount of reactants used and the amount of products formed, for example:

$2Na_{(s)}$	+	$Cl_{2(g)}$	$\rightarrow$	$2NaCl_{(s)}$
2 formulas		1 formula		2 formulas

This interpretation becomes inconvenient if we are working with amounts of substance measured in grams and we must develop alternative measures for the amount of substance that allow us to easily relate quantities with masses that range from grams to tonnes (1 tonne = 1000 kg).

Avogadro's Constant and The Mole

One gram of a substance contains approximately 10^{23} atoms. As a result it becomes convenient to define an amount of substance that contains approximately 10^{23} atoms when working with amounts of substance measured in grams. For this reason amounts of substance are measured in units known as moles (Unit: mol) where one **mole** of a substance contains *the same number of atoms, molecules or ions as 6.02×10^{23} formulas of the substance*. The number 6.02×10^{23} is known as Avogadro's constant (Symbol: L).

Having defined a mole to be a number of formulas, chemical equations can be interpreted in terms of the moles of substance reacting and the moles of products formed. For example, the reaction between sodium and chlorine can be interpreted as the reaction between two moles of sodium (2Na) and one mole of chlorine (Cl_2) to form two moles of sodium chloride (2NaCl).

$2Na_{(s)}$	+	$Cl_{2\,(g)}$	→	$2NaCl_{(s)}$
2L formulas (2 moles)		L formulas (1 mole)		2L formulas (2 moles)

Molar Mass

The **molar mass** of a substance is defined to be *the mass of one mole of the substance in grams (Unit: g mol^{-1})*. For example, the molar mass of oxygen (O_2) is 32 g mol^{-1}. This means that one mole of oxygen gas (O_2) has a mass of 32 g, two moles have a mass of $2 \times 32 = 64$ g, three moles have a mass of $3 \times 32 = 96$ g and so on.

The molar mass of an element or compound can be obtained from the Periodic Table by adding the relative atomic mass (RAM) for each atom in the formula of the substance.

Worked Example 1.1xiii

Calculate the molar mass of (a) oxygen and (b) sodium chloride.

Strategy

- Add the RAMs for the atoms in one formula of the substance.
- The result has units of g mol^{-1}.

Solution

(a) The formula of oxygen (gas) is O_2
The RAM of O is 16
Molar mass of oxygen (O_2) = 16 + 16 = 32 g mol^{-1}

(b) The formula of sodium chloride is NaCl
The RAM of Na is 23 and the RAM of Cl is 35.5
Molar mass of sodium chloride = 23 + 35.5
= 58.5 g mol^{-1}

The relationship between the amount of substance in moles and its equivalent mass in grams is summarised by Equation 1.

$$\text{Moles} = \frac{\text{Mass}}{\text{Molar Mass}} \qquad \text{Equation 1}$$

Worked Example 1.1xiv

An electronic balance was used to measure 20 g of the following substances. Calculate the number of moles in each sample.
(a) iron (b) gold (c) water (d) sodium carbonate

Strategy

- Calculate the molar mass.
- Use the molar mass to calculate the number of moles in 20 g.

Solution

(a) Molar mass of iron (Fe) = 56 g mol^{-1}

Moles of iron

$$= \frac{\text{Mass}}{\text{Molar Mass}} = \frac{20\text{ g}}{56\text{ g mol}^{-1}} = 0.36\text{ mol}$$

The answer is rounded to two significant figures as its accuracy is limited by the mass, which has only two significant figures. For more on significant figures see the Appendix.

(b) Molar mass of gold (Au) = 197 g mol^{-1}

Moles of gold

$$= \frac{\text{Mass}}{\text{Molar Mass}} = \frac{20\text{ g}}{197\text{ g mol}^{-1}} = 0.10\text{ mol}$$

(c) Molar mass of water (H_2O) = 1 + 1 + 16 = 18 g mol^{-1}

Moles of water

$$= \frac{\text{Mass}}{\text{Molar Mass}} = \frac{20\text{ g}}{18\text{ g mol}^{-1}} = 1.1\text{ mol}$$

(d) Molar mass of Na_2CO_3 = (2 × 23) + 12 + (3 × 16)

= 106 g mol^{-1}

Moles of Na_2CO_3

$$= \frac{\text{Mass}}{\text{Molar Mass}} = \frac{20\text{ g}}{106\text{ g mol}^{-1}} = 0.19\text{ mol}$$

Exercise 1.10

1. Prozac tablets contain 20.0 mg of fluoxetine, $C_{17}H_{18}F_3NO$ in each tablet. The number of moles of fluoxetine in each tablet is:

   (a) 6.47×10^{-5} (b) 1.39×10^{-4}

   (c) 6.47×10^{-2} (d) 1.39×10^{-1}

   (Adapted from CCEA June 2013)

Before moving to the next section, check that you are able to:

- Define one mole of substance in terms of Avogadro's constant.
- Calculate the molar mass of a substance and use molar mass to relate amounts of a substance in grams and moles.

Using Avogadro's Constant

Avogadro's constant can be used to calculate the number of particles in a given amount of substance, or to calculate the amount of substance that contains a specified number of particles.

By definition, one mole of any substance contains Avogadro's constant (L) of formulas of the substance. In the case of a molecular material such as carbon dioxide (formula: CO_2), each formula represents one molecule. As a result, one mole of carbon dioxide contains L molecules, two moles contains 2L molecules, and so on. This relationship is summarised by Equation 2.

Number of molecules = Moles × L Equation 2

If the substance is a metal (formula: Fe, Au, ...), a monatomic gas (formula: He, Ne, ...), or an element with a giant structure such as diamond (formula: C), one formula represents one atom. The relationship between the number of atoms in the substance and the amount of substance in moles is summarised by Equation 3.

Number of atoms = Moles × L Equation 3

In the case of an ionic compound such as NaCl or $MgCl_2$, each formula contains the ions needed to describe the composition of the substance. If one formula contains N ions, one mole of formulas contains N moles of ions, and the total number of ions in a given number of moles of the compound is given by Equation 4.

Number of ions = Moles × N × L Equation 4

Worked Example 1.1xv

How many particles (ions, atoms or molecules) are in the following. Leave your answers in terms of Avogadro's number (L).

(a) 3 moles of nitrogen
(b) 2 moles of iron
(c) 2 moles of magnesium sulfate

Strategy

- Identify the particles in one formula.
- Multiply the number of particles in one formula by the number of moles.

Solution

(a) Nitrogen is a molecular material (formula: N_2). Each formula represents one molecule, therefore 1 mol contains L molecules and 3 mol contains 3 × L = 3L molecules.

(b) Iron is a metal (formula: Fe). Each formula represents one atom therefore 1 mol (L formulas) contains L atoms and 2 mol contains 2 × L = 2L atoms.

(c) Magnesium sulfate is an ionic compound (formula: $MgSO_4$). Each formula represents 1 magnesium (Mg^{2+}) ion and 1 sulfate (SO_4^{2-}) ion. As a result 1 mol (L formulas) contains 2L ions and 2 mol contains 2 × 2L = 4L ions.

Having demonstrated the use of Avogadro's constant to calculate the number of particles in a mole of substance, and the use of molar mass to relate the amount of substance in moles and grams, we can now combine these methods and use moles to determine the number of particles in a given mass of substance.

Worked Example 1.1xvi

Carbon, in the form of graphite, is a major component of the 'lead' in pencils. A single dot made by a pencil contains approximately 0.00010 g of carbon. How many atoms of carbon are in the dot?

Strategy

- Calculate the moles of carbon in the dot.
- Use Avogadro's constant to convert moles to atoms.

Solution

Moles of carbon in the dot

$$= \frac{\text{Mass}}{\text{Molar Mass}} = \frac{1.0 \times 10^{-4}\ \text{g}}{12\ \text{g mol}^{-1}} = 8.3 \times 10^{-6}\ \text{mol}$$

Carbon atoms in the dot =

Moles × L = $(8.3 \times 10^{-6})(6.02 \times 10^{23}) = 5.0 \times 10^{18}$

The use of significant figures in this example is discussed in the Appendix.

Exercise 1.1P

1. How many copper atoms are in a 1p coin, assuming the coin contains 1.00 g of copper?
2. What mass of sodium contains the same number of atoms as 1.00 g of lithium?

   (CCEA January 2003)
3. A female cockroach secretes a chemical with the formula $C_{11}H_{18}O_2$. It is reported that the male of the species will respond to as little as 60 molecules of the chemical. Calculate (a) the relative formula mass, (b) the mass of one mole, (c) the mass of one molecule, and (d) the mass of 60 molecules of the chemical.

   (CCEA June 2001)
4. X is an oxide of nitrogen. (a) 2.30 g of X contains 3.01×10^{22} molecules of X. Calculate the molar mass of X. (b) Deduce the formula of X.

   (CCEA June 2012)
5. (a) Write an equation for the reaction of sodium with oxygen to form sodium peroxide, Na_2O_2. (b) At higher temperatures a different oxide, Y is formed. One mole of Y contains Avogadro's constant of O^{2-} ions and 1.2×10^{24} Na^+ ions. Deduce the formula of Y.

   (Adapted from CCEA June 2015)
6. Which one of the following is the number of atoms present in 0.25 moles of $C_{12}H_{22}O_{11}$?

   A 6.8×10^{24} B 1.4×10^{25}
   C 2.7×10^{25} D 1.1×10^{26}

   (CCEA June 2013)

Before moving to the next section, check that you are able to:

- Use Avogadro's constant to calculate the number of particles in a given amount of substance.

Calculating Reacting Amounts

In this section we are learning to:

- Use chemical equations to relate the amounts of substances that react and the amounts of products formed in a chemical reaction.
- Identify a reactant as *limiting* if the amount of reactant present in the reaction mixture determines the amount of products formed.
- Calculate the amount of product formed in the presence of a limiting reactant.

The chemical equation for a reaction can be used to calculate the amount of products formed in a reaction and the amount of reactants needed to form a specified amount of product. Both are common problems faced by chemists and can be tackled by applying the following scheme:

Calculating Reacting Amounts

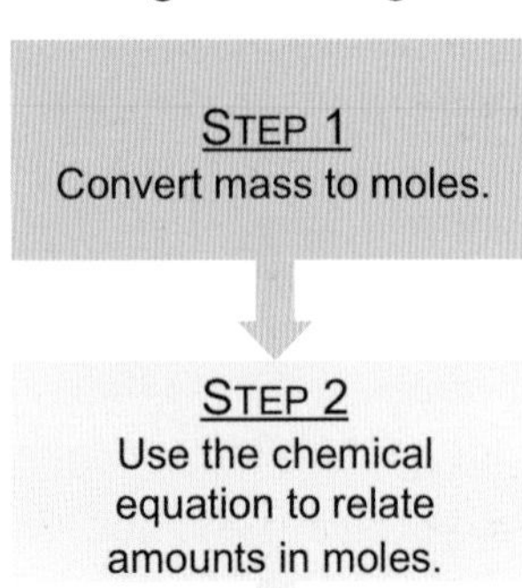

Worked Example 1.1xvii

Calculate the mass of magnesium oxide formed when 10 g of magnesium is completely burnt in oxygen. The chemical equation for the reaction is:

$$2Mg_{(s)} + O_{2(g)} \rightarrow 2MgO_{(s)}$$

Strategy

Apply the 3-step approach.

1. Calculate the number of moles in 10 g of magnesium.
2. Use the answer from step 1 to calculate the moles of MgO formed.
3. Use the answer from step 2 to calculate the mass of MgO formed.

Solution

1. Moles of Mg =

$$\frac{\text{Mass}}{\text{Molar Mass}} = \frac{10\text{ g}}{24\text{ g mol}^{-1}} = 0.42\text{ mol}$$

2. The chemical equation shows that 1 mol of MgO is formed for every mole of magnesium that reacts. Moles of MgO formed = Moles of Mg reacted = 0.42 mol
3. Mass of MgO formed = Moles × Molar mass = 0.42 mol × 40 g mol^{-1} = 17 g

Exercise 1.1Q

1. Aspirin, $C_9H_8O_4$ is prepared by reacting salicylic acid, $C_7H_6O_3$ with acetic anhydride, $C_4H_6O_3$. Calculate (a) the mass of anhydride needed to react with 500 g of salicylic acid, and (b) the mass of aspirin formed in the reaction.

   $C_7H_6O_3 + C_4H_6O_3 \rightarrow C_9H_8O_4 + C_2H_4O_2$

2. Lithium oxide, Li_2O is used in spacecraft to remove water from the air supply. Calculate the mass of water that can be removed by an 'air scrubber' containing 500 g of lithium oxide.

   $Li_2O_{(s)} + H_2O_{(g)} \rightarrow 2LiOH_{(s)}$

3. Mercury(II) oxide, HgO decomposes on heating. (a) Calculate the mass of mercury formed when 10 g of mercury(II) oxide decomposes. (b) Calculate the mass of mercury(II) oxide needed to produce 0.5 mol of oxygen.

   $2HgO_{(s)} \rightarrow 2Hg_{(l)} + O_{2(g)}$

4. Potassium chlorate, $KClO_3$ decomposes on heating. Calculate the mass of potassium chlorate needed to produce 0.50 mol of oxygen.

   $2KClO_{3(s)} \rightarrow 2KCl_{(s)} + 3O_{2(g)}$

5. Photosynthesis converts carbon dioxide into glucose, $C_6H_{12}O_6$ and oxygen. A fully grown tree will consume 300 kg of carbon dioxide each year and convert it to oxygen. Calculate the mass of oxygen produced by the tree each year.

   $6CO_{2(g)} + 6H_2O_{(l)} \rightarrow C_6H_{12}O_{6(s)} + 6O_{2(g)}$

   (CCEA June 2005)

6. In industry iron is produced by the reduction of iron(III) oxide, Fe_2O_3. Calculate the mass of carbon dioxide in kg released into the atmosphere for every 0.100 tonne of iron produced (1 tonne = 1000 kg).

   $Fe_2O_3 + 3CO \rightarrow 2Fe + 3CO_2$

7. Sodium carbonate is manufactured by the Solvay process. This is a two stage process.

   Stage 1
   Sodium hydrogencarbonate is formed

   $NaCl + NH_3 + CO_2 + H_2O \rightarrow NaHCO_3 + NH_4Cl$

Stage 2
Sodium hydrogencarbonate decomposes when heated. $2NaHCO_3 \rightarrow Na_2CO_3 + H_2O + CO_2$

(a) Calculate the number of moles of sodium hydrogencarbonate formed from 234 kg of sodium chloride. (b) Calculate the maximum mass of sodium carbonate formed in kg.

(Adapted from CCEA June 2015)

The three-step approach to calculating reacting amounts can also be used to determine the amount of an element in a given amount of a substance.

Worked Example 1.1xviii

In industry iron is obtained from iron(III) oxide, Fe_2O_3 using a Blast Furnace. How much iron can be obtained from 1 tonne (1 tonne = 1000 kg) of iron(III) oxide?

Strategy

Apply the three-step approach.

1. Calculate the moles of iron(III) oxide in 1 tonne of iron(III) oxide.
2. Use the answer from step 1 to calculate the number of moles of iron in 1 tonne of iron(III) oxide.
3. Use the answer to step 2 to calculate the mass of iron in 1 tonne of iron(III) oxide.

Solution

1. Moles of Fe_2O_3 in 1 tonne =

$$\frac{\text{Mass}}{\text{Molar Mass}} = \frac{1 \times 10^6 \text{ g}}{160 \text{ g mol}^{-1}} = 6.25 \times 10^3 \text{ mol}$$

The use of significant figures in this example is discussed in the Appendix.

2. Moles of Fe in 6.25×10^3 mol of Fe_2O_3
$= 2 \times 6.25 \times 10^3 = 1.25 \times 10^4$ mol
3. Mass of Fe in 1 tonne $= (1.25 \times 10^4 \text{ mol})(56 \text{ g mol}^{-1})$
$= 7.00 \times 10^5 \text{ g} = 700$ kg

Exercise 1.1R

1. One tonne (1000 kg) of sulfur costs £160. What is the cost (to the nearest £) of the sulfur needed to make 1 tonne of sulfuric acid, H_2SO_4.

(CCEA June 2006)

Before moving to the next section, check that you are able to:

- Use chemical equations to calculate the amount of reactants used and the amount of products formed in a chemical reaction.

Limiting Reactants

The compound iron(II) sulfide, FeS is formed by heating a mixture of iron and sulfur. If the amount of sulfur in the mixture is increased until the mixture contains just enough sulfur to react with the iron, the entire reaction mixture will be converted into iron(II) sulfide. If the amount of sulfur in the mixture is increased further, all of the iron will be converted to iron(II) sulfide and some sulfur will remain in the reaction mixture. The amount of iron(II) sulfide formed is limited by the amount of iron in the reaction mixture. In this way iron has become the **limiting reactant** and sulfur is described as being **in excess**.

The method used previously to calculate reacting amounts must be modified slightly to allow for the possibility of a limiting reactant.

Limiting Reactant Calculations

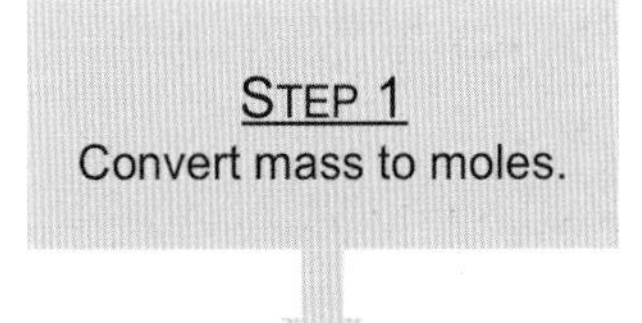

STEP 2
Use the chemical equation and the amount of limiting reactant to relate amounts in moles.

STEP 3
Convert moles to mass.

Worked Example 1.1ixx

How many moles of iron(II) sulfide are formed when a mixture containing 3.0 g of iron and 3.0 g of sulfur is heated?

$$Fe_{(s)} + S_{(s)} \rightarrow FeS_{(s)}$$

Strategy

Apply the three-step approach.

1. Calculate the moles of each reactant.
2. Use the moles of limiting reactant to calculate the moles of product formed.
3. Use the answer from step 2 to calculate the mass of product formed.

Solution

1. Moles of Fe $= \frac{\text{Mass of Fe}}{\text{Molar Mass of Fe}} = \frac{3.0 \text{ g}}{56 \text{ g mol}^{-1}}$
   $= 0.054$ mol

   Moles of S $= \frac{\text{Mass of S}}{\text{Molar Mass of S}} = \frac{3.0 \text{ g}}{32 \text{ g mol}^{-1}}$
   $= 0.094$ mol
2. The chemical equation shows that 1 mol of iron reacts with 1 mol of sulfur. Moles of S reacted = Moles of Fe reacted = 0.054 mol

   Not all of the sulfur reacts. Sulfur is in excess and iron is the limiting reactant.

   One mole of FeS is formed for every mole of iron that reacts. Moles of FeS formed = Moles of Fe reacted = 0.054 mol
3. Mass of FeS = Moles of FeS × Molar mass
   $= (0.054 \text{ mol})(88 \text{ g mol}^{-1}) = 4.8$ g

Exercise 1.1S

1. Calculate the amount of copper obtained by adding 0.42 g of iron to an excess of copper(II) sulfate solution.

   $Fe + CuSO_4 \rightarrow FeSO_4 + Cu$

   (Adapted from CCEA June 2002)
2. Iron(III) oxide can be reduced by carbon to form iron. Calculate the maximum mass of iron that can be produced by reacting 3.20 kg of iron(III) oxide with 0.72 kg of carbon.

   $2Fe_2O_3 + 3C \rightarrow 4Fe + 3CO_2$

   (CCEA January 2010)
3. The extraction and purification of uranium from its ore involves the following reaction between uranium(IV) fluoride and magnesium. Calculate the mass of uranium extracted by reacting 500 tonnes of uranium(IV) fluoride with 50 tonnes of magnesium.

   $2Mg + UF_4 \rightarrow U + 2MgF_2$

   (Adapted from CCEA June 2009)
4. Phosphoric acid, H_3PO_4 is manufactured by the reaction of sulfuric acid with calcium phosphate. Calculate the mass of phosphoric acid that would be obtained by reacting 60 kg of sulfuric acid with 60 kg of calcium phosphate.

   $3H_2SO_4 + Ca_3(PO_4)_2 \rightarrow 2H_3PO_4 + 3CaSO_4$

   (CCEA January 2009)
5. Phosphorus, P reacts with bromine at room temperature to form phosphorus tribromide, PBr_3, which is a liquid. It reacts with water immediately forming hydrogen bromide and phosphoric(III) acid, H_3PO_3. (a) Write the equation for the reaction of bromine with phosphorus. (b) Calculate the maximum mass of phosphorus tribromide formed when 6.2 g of phosphorus, which is an excess, reacts with 8.00 cm^3 of bromine, Br_2. The density of bromine is 3.10 g cm^{-3}. (c) Write the equation for the reaction of phosphorus tribromide with water.

   (Adapted from CCEA January 2012)
6. Titanium is extracted in a two-stage process. The first stage involves the conversion of titanium(IV) oxide to titanium(IV) chloride. In the second stage, the titanium(IV) chloride is reduced using magnesium. How much titanium would be obtained when 8.0 kg of titanium(IV) oxide is converted to titanium(IV) chloride and then reduced using 7.2 kg of magnesium?

   $TiO_2 + C + 2Cl_2 \rightarrow TiCl_4 + CO_2$

   $TiCl_4 + 2Mg \rightarrow Ti + 2MgCl_2$

   (CCEA June 2010)

Before moving to the next section, check that you are able to:

- Identify a limiting reactant as any reactant that limits the amount of product formed in a chemical reaction.
- Calculate the amount of reactants used and the amount of products formed in the presence of a limiting reactant.

Salts Containing Water

In this section we are learning to:

- Use the term water of crystallisation to refer to water that is chemically bonded within a salt.
- Refer to salts containing water of crystallisation as hydrated salts or 'hydrates'.
- Use the term anhydrous to refer to salts that do not contain water of crystallisation.
- Calculate the % water by mass in hydrated salts.
- Determine the formula of a hydrated salt by the technique of heating to constant mass.

Water of Crystallisation

Crystals of hydrated copper(II) sulfate, $CuSO_4.5H_2O$ have a characteristic blue colour due to water molecules ($5H_2O$) known as **water of crystallisation** that *are chemically bonded within the crystals and form part of the crystal structure.*

Hydrated copper(II) sulfate.

A salt is referred to as a **hydrated salt** or 'hydrate' if *the salt contains water of crystallisation.* The formula of hydrated copper(II) sulfate indicates that the salt contains 5 moles of water of crystallisation ($5H_2O$) for every mole of copper(II) sulfate in the crystal. The examples in Table 4 further illustrate how to describe the amount of water of crystallisation in a hydrate when naming the salt.

The water of crystallisation in a salt can be removed by heating. When blue crystals of hydrated copper(II) sulfate are heated the water of crystallisation evaporates leaving anhydrous copper(II) sulfate, $CuSO_4$, a white solid. A salt is referred to as an **anhydrous salt** if *the salt does not contain water of crystallisation.*

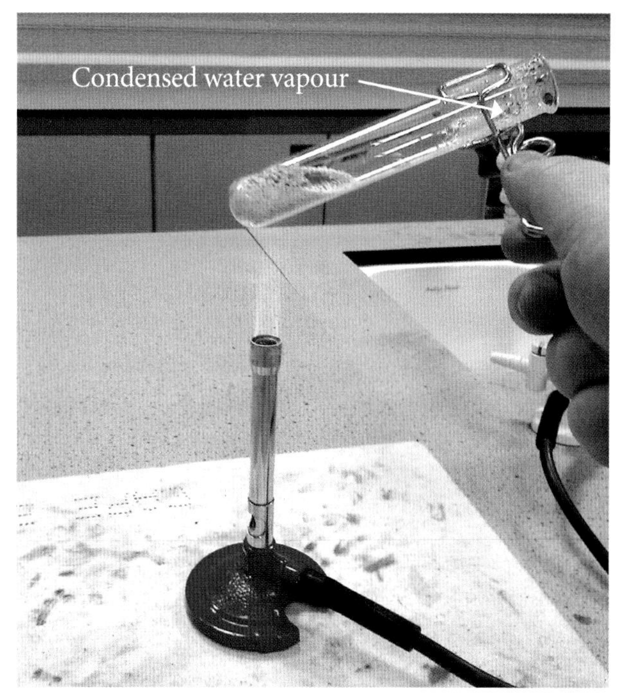

Forming anhydrous copper(II) sulfate, $CuSO_4$ by heating hydrated copper(II) sulfate, $CuSO_4.5H_2O$. The water removed from the salt condenses on the sides of the test tube.

Table 4: Examples of naming hydrated salts.

Formula	Name
$CoCl_2.2H_2O$	cobalt(II) chloride-2-water
$CuSO_4.5H_2O$	copper(II) sulfate-5-water
$Na_2CO_3.10H_2O$	sodium carbonate-10-water

Worked Example 1.1 xx

Calculate the percentage water by mass in copper(II) sulfate-5-water.

Solution

Molar mass of $CuSO_4.5H_2O$ = 250 g mol^{-1}

Molar mass of H_2O = 18 g mol^{-1}

Mass of H_2O in 1 mol of hydrate = 5 × 18 = 90 g

$$\% \text{ water by mass} = \frac{90\text{ g}}{250\text{ g}} \times 100 = 36\%$$

Exercise 1.1T

1. Calculate the percentage water by mass in crystals of washing soda, $Na_2CO_3.10H_2O$.
2. Carbon can be produced by heating cane sugar, $C_{12}H_{22}O_{11}$, with concentrated sulfuric acid. The reaction produces carbon, steam and diluted sulfuric acid. (a) Write an equation for the reaction. Do not include sulfuric acid in the equation. (b) Explain the meaning of the terms *hydrated* and *water of crystallisation*. (c) Explain whether the cane sugar is hydrated.

(Adapted from CCEA January 2012)

Before moving to the next section, check that you are able to:

- Use the terms hydrated and anhydrous to describe the amount of water of crystallisation in a salt.
- Name salts containing water of crystallisation.
- Calculate the percentage water by mass in a hydrated salt.

Determining the Formula of a Hydrate

The amount of water of crystallisation in a salt is determined by a technique known as **heating to constant mass**.

Method

- Weigh a sample of the hydrate in a pre-weighed crucible.
- Heat the sample in the crucible using a blue Bunsen flame.
- Allow the crucible to cool to room temperature before weighing the crucible.
- Repeat the process of heating and weighing until the mass remains constant.
- Calculate the mass of anhydrous salt formed.

Heating a solid compound to constant mass.

Worked Example 1.1xxi

Use the following information to determine the chemical formula for hydrated iron(II) sulfate, $FeSO_4.xH_2O$.

Mass of crucible	= 19.38 g
Mass of hydrate + crucible	= 22.32 g
Mass after heating (constant)	= 20.99 g

Strategy

Calculate:

1. The moles of water removed by heating.
2. The moles of anhydrous salt remaining after heating.
3. The number of moles of water per mole of anhydrous salt.

Solution

$$FeSO_4.xH_2O_{(s)} \rightarrow FeSO_{4(s)} + x\,H_2O_{(l)}$$

1. Mass of water removed from hydrate = 22.32 – 20.99 = 1.33 g

   Moles of water in hydrate

$$= \frac{\text{Mass}}{\text{Molar Mass}} = \frac{1.33\text{ g}}{18\text{ g mol}^{-1}} = 0.0739\text{ mol}$$

2. Mass of anhydrous salt = 20.99 – 19.38 = 1.61 g

   Moles of anhydrous salt ($FeSO_4$)

$$= \frac{\text{Mass}}{\text{Molar Mass}} = \frac{1.61\ \text{g}}{152\ \text{g mol}^{-1}} = 0.0106\ \text{mol}$$

3. $$x = \frac{\text{Moles of water}}{\text{Moles of anhydrous } FeSO_4} = \frac{0.0739}{0.0106} = 6.97$$

   Rounding gives x = 7 and the chemical formula of hydrated iron(II) sulfate becomes $FeSO_4.7H_2O$.

Exercise 1.1U

1. A sample of hydrated cobalt(II) chloride weighing 2.38 g was heated to constant mass. The sample weighed 1.20 g after heating. Calculate the formula of the hydrate.

2. The hydrated form of barium chloride has the formula, $BaCl_2.xH_2O$. Use the following information to find x.

   Mass of hydrate + crucible = 43.44 g
   Mass after heating = 43.26 g
   Mass of crucible = 42.22 g

Before moving to the next section, check that you are able to:

- Recall the procedure to determine the amount of water of crystallisation in a salt by heating to constant mass.
- Calculate the formula of a hydrated salt using data obtained by heating a sample of the hydrate to constant mass.

1.2 Atomic Structure and The Periodic Table

CONNECTIONS

- The amount of the isotope carbon-14 in an object can be used to date the object using a technique known as radiocarbon dating.
- Atoms are very small but can be 'seen' using a special kind of microscope known as a Scanning Tunnelling Microscope (STM).

Atoms Ions and Isotopes

In this section we are learning to:

- Describe the modern picture of the atom and recall the properties of the subatomic particles within an atom.
- Recall that elements are organised in order of increasing atomic number in the modern Periodic Table.
- Describe the formation of ions from atoms.
- Explain the existence of isotopes and use the mass number of an isotope to calculate the number of protons, electrons and neutrons in the isotope.
- Recall the definition of relative isotopic mass (RIM) and relative atomic mass (RAM) in terms of the carbon-12 mass standard.
- Calculate the relative atomic mass (RAM) of an element and the relative molecular mass (RMM) of a molecule from a mass spectrum.

The Modern Atom

The modern picture of the atom is based on the model proposed by Ernest Rutherford in 1911. In Rutherford's model, an atom is made up of negatively charged electrons orbiting a positively charged nucleus as shown in Figure 1a. We now know that the nucleus contains two types of particle: protons and neutrons. Protons and neutrons have approximately the same mass and are about 1840 times heavier than an electron. As a result, the nucleus of an atom contains almost all of the atoms' mass but occupies only a tiny fraction of the space inside the atom.

The particles that make up atoms are known as subatomic particles. The charges on subatomic particles such as protons and electrons are measured in units of electron charge (symbol: e). In these units each proton has a charge of 1+ and each electron has a charge of 1−. Neutrons do not have a charge. The properties of electrons, protons and neutrons are summarised in Figure 1b.

There are approximately 100 elements in the modern Periodic Table, each made from one type of atom. The elements are represented by their chemical symbols and are arranged in order of increasing atomic number as shown in Figure 2. The **atomic number** of an element is defined to be *the number of protons in the nucleus of one atom of the element*. For example, the element carbon (symbol: C) is element 6 in the Periodic Table. This tells us that the atomic number of

(a)

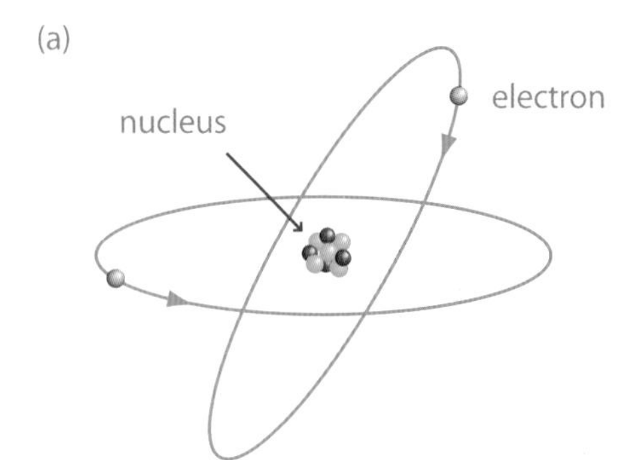

(b)

Subatomic Particle	Relative Mass	Relative Charge
proton	1	1+
neutron	1	0
electron	$\frac{1}{1840}$	1−

Figure 1: (a) Rutherford's model of the atom. (b) Properties of the subatomic particles in Rutherford's model.

Atomic number increasing →

1 H																	2 He
3 Li	4 Be											5 B	6 C	7 N	8 O	9 F	10 Ne
11 Na	12 Mg											13 Al	14 Si	15 P	16 S	17 Cl	18 Ar
19 K	20 Ca	21 Sc	22 Ti	23 V	24 Cr	25 Mn	26 Fe	27 Co	28 Ni	29 Cu	30 Zn	31 Ga	32 Ge	33 As	34 Se	35 Br	36 Kr
37 Rb	38 Sr	39 Y	40 Zr	41 Nb	42 Mo	43 Tc	44 Ru	45 Rh	46 Pd	47 Ag	48 Cd	49 In	50 Sn	51 Sb	52 Te	53 I	54 Xe
55 Cs	56 Ba	57 La	72 Hf	73 Ta	74 W	75 Re	76 Os	77 Ir	78 Pt	79 Au	80 Hg	81 Tl	82 Pb	83 Bi	84 Po	85 At	86 Rn
87 Fr	88 Ra	89 Ac															

58 Ce	59 Pr	60 Nd	61 Pm	62 Sm	63 Eu	64 Gd	65 Tb	66 Dy	67 Ho	68 Er	69 Tm	70 Yb	71 Lu
90 Th	91 Pa	92 U	93 Np	94 Pu	95 Am	96 Cm	97 Bk	98 Cf	99 Es	100 Fm	101 Md	102 No	103 Lr

Figure 2: The form of the modern Periodic Table. The atomic number of the elements increases from left to right across each row.

carbon is 6, and that every carbon atom has 6 protons in its nucleus.

Exercise 1.2A

1. Atoms consist of protons, neutrons and electrons. Complete the table giving the properties of a proton, a neutron and an electron.

	Relative mass	Relative charge
Proton		
Neutron		
Electron		

(CCEA June 2013)

Atoms and Ions

Atoms do not have a charge as they contain equal numbers of protons (1+ charge) and electrons (1– charge). When an electron is removed from an atom, the total charge on the protons is one greater than the total charge on the remaining electrons, and the atom becomes a cation with a single positive charge. In this way a sodium ion, Na^+ is formed by removing one electron from a sodium atom, and a copper(II) ion, Cu^{2+} is formed by removing two electrons from a copper atom.

Conversely, when an atom gains an electron the total charge on the protons is one less than the total charge on the electrons, and the atom becomes an anion with a single negative charge. In this way a chloride ion, Cl^- is formed when a chlorine atom gains an electron, and an oxide ion, O^{2-} is formed when an oxygen atom gains two electrons.

The relationship between the charge on an ion and the numbers of protons (p) and electrons (e) in the ion can be summarised in the form of an equation.

Charge on ion
= number of protons (p) – number of electrons (e)

Worked Example 1.2i

Identify the atoms or ions A, B and C.

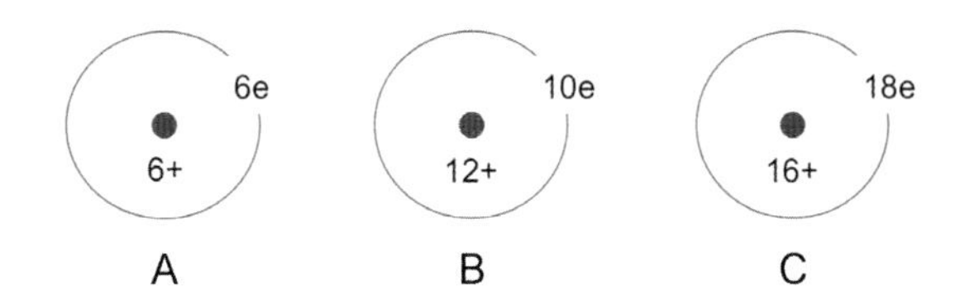

Strategy

- Use the charge on the nucleus to determine the atomic number.
- Calculate the charge on the ion (p – e).
- Recall that the charge on an atom (p – e) is zero.

Solution

(a) Atomic number = 6 (6+ charge on nucleus)

Charge on A = p − e = 6 − 6 = 0

A is a carbon atom

(b) Atomic number = 12

Charge on B = p − e = 12 − 10 = 2+

B is a magnesium ion, Mg^{2+}

(c) Atomic number = 16

Charge on C = p − e = 16 − 18 = 2−

C is a sulfide ion, S^{2-}

Exercise 1.2B

1. Which one of the following is the name of the species shown below?

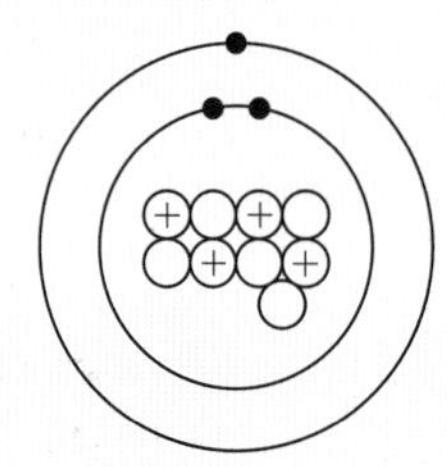

A beryllium atom
B beryllium ion
C lithium atom
D lithium ion

(CCEA January 2012)

2. Which one of the following elements contains the same number of electrons as a Mg^{2+} ion?

A calcium B fluorine C neon D sodium

(CCEA June 2011)

3. A caesium atom differs from a caesium ion because the atom has a greater

A atomic number
B mass number
C number of electrons
D number of protons

(CCEA June 2013)

Isotopes

Carbon has three isotopes: carbon-12 (^{12}C), carbon-13 (^{13}C) and carbon-14 (^{14}C). All three are present in carbon obtained from natural sources. The number of protons, neutrons and electrons in each isotope is summarised in Table 1.

The isotopes have the same electron configuration, and therefore behave in the same way when they react with other elements to form compounds. They are only distinguished by the mass numbers 12, 13 and 14, where the term **mass number** refers to the *total number of protons and neutrons in an atom.*

Mass Number
= number of protons (p) + number of neutrons (n)

By defining mass number in this way, the term **isotopes** is used when referring to *atoms with the same atomic number that have different mass numbers.*

Isotope	Symbol	Protons (p)	Neutrons (n)	Electrons (e)
carbon-12	^{12}C	6	6	6
carbon-13	^{13}C	6	7	6
carbon-14	^{14}C	6	8	6

Table 1: The number of protons, neutrons and electrons in the isotopes of carbon

Worked Example 1.2ii

Calculate the number of neutrons in an atom of the following isotopes.

(a) ^{31}P (b) ^{56}Fe (c) ^{197}Au (d) ^{238}U

Solution

(a) Phosphorus-31, ^{31}P is an isotope of phosphorus (p = 15)
Mass number = p + n = 31
therefore n = 31 − p = 16

(b) Iron-56, ^{56}Fe is an isotope of iron (p = 26)
Mass number = p + n = 56
therefore n = 56 − 26 = 30

(c) Gold-197, ^{197}Au is an isotope of gold (p = 79)
Mass number = p + n = 197
therefore n = 197 − 79 = 118

(d) Uranium-238, ^{238}U is an isotope of uranium (p = 92)
Mass number = p + n = 238
therefore n = 238 − 92 = 146

Exercise 1.2C

1. Complete the following table to show the number of subatomic particles in a ^{43}Ca atom.

	neutrons	electrons	protons
^{43}Ca			

(CCEA June 2014)

2. (a) Define what is meant by the *atomic number* of an atom. (b) Define what is meant by the *mass number* of an isotope.

(Adapted from CCEA June 2012)

3. Magnesium contains three isotopes: ^{24}Mg, ^{25}Mg and ^{26}Mg. Explain the meaning of the term *isotope*.

(Adapted from CCEA January 2013)

4. A sample of iron from a meteorite was found to contain the following isotopes: ^{54}Fe, ^{56}Fe and ^{57}Fe. (a) Complete the table to show the number of protons, neutrons and electrons present in each isotope. (b) Explain the difference, if any, in the chemical properties of the isotopes of iron.

isotope	protons	neutrons	electrons
^{54}Fe			
^{56}Fe			
^{57}Fe			

(CCEA June 2015)

5. (a) State the number of protons, electrons and neutrons in an atom of ^{23}Na. (b) Explain why ^{23}Na and ^{24}Na are regarded as isotopes.

(CCEA June 2006)

6. Magnesium reacts with chlorine to form magnesium chloride. Complete the table for the ions in magnesium chloride.

ion	protons	neutrons	electrons
$^{24}Mg^{2+}$			
$^{35}Cl^{-}$			

(Adapted from CCEA June 2012)

7. The isotope carbon-14 decomposes when a neutron in its nucleus changes into an electron and a proton forming a new element. (a) State the number of electrons, protons and neutrons in the new element. (b) Name the element produced when carbon-14 decomposes. (c) Explain why carbon-12 and carbon-14 are isotopes.

(Adapted from CCEA January 2012)

8. Francium was first synthesised according to the following equation. The symbol n represents a neutron. (a) State the relative mass of a neutron. (b) Show, by calculation, that the equation is balanced according to mass. (c) Explain why electrons are not used when balancing the equation according to mass.

$^{197}Au + {}^{18}O \rightarrow {}^{210}Fr + 5n$

(Adapted from CCEA January 2014)

The **relative isotopic mass (RIM)** of an isotope is *the mass of one atom of the isotope relative to one-twelfth of the mass of one atom of carbon-12*. On this scale protons and neutrons have a mass of one and the RIM of an isotope is equal to the mass number of the isotope. For example, the RIM of an atom of carbon-12 (6 protons + 6 neutrons) is 12 and the RIM of an atom of carbon-13 (6 protons + 7 neutrons) is 13. Relative masses do not have units.

Worked Example 1.2iii

Which one of the following statements about the isotope carbon-14 is incorrect?

(CCEA June 2002)

A It is used as a standard for measuring mass
B It has a relative isotopic mass of 14
C It has a mass number of 14
D It has eight neutrons

Solution

The standard for measuring mass is carbon-12.
Statement A is incorrect.

By choosing to use carbon–12 as a mass standard **Avogadro's constant** ($L = 6.02 \times 10^{23}$) can be interpreted as *the number of atoms in exactly 12 g of carbon-12.*

Exercise 1.2D

1. Avogadro's constant has the value 6.02×10^{23}. Define the term *Avogadro's constant.*

   (CCEA June 2012)

2. Explain the meaning of the term *carbon–12 standard.*

   (CCEA January 2012)

Atomic Mass

The **relative atomic mass (RAM)** of an element is *the average mass of one atom of the element relative to one-twelfth of the mass of an atom of carbon-12*. The RAM of an element is obtained directly from the Periodic Table. For example, the RAM of nitrogen (N) is 14 and the RAM of silicon (Si) is 28.

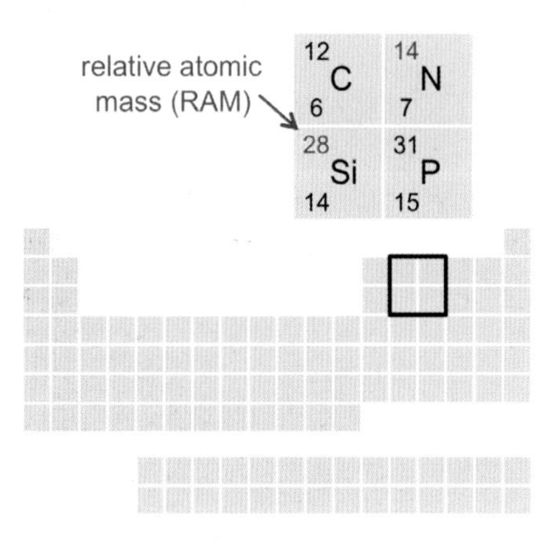

Before moving to the next section, check that you are able to:

- Describe the modern picture of the atom and recall the properties of the subatomic particles within an atom.
- Recall that elements are organised in order of increasing atomic number in the Periodic Table.
- Describe the formation of ions from atoms.
- Explain the existence of isotopes and use the mass number of an isotope to calculate the number of protons, electrons and neutrons in the isotope.
- Recall the definition of relative isotopic mass (RIM) and relative atomic mass (RAM) in terms of the mass of carbon–12.

Mass Spectrometry

Many elements contain significant amounts of two or more isotopes and, as a result, the atomic mass of the element can only be determined if the relative amount of each isotope in the element is known. This information is obtained using an experimental technique called mass spectrometry.

In a mass spectrometry experiment a small sample of the element is injected into a machine known as a mass spectrometer. Once in the spectrometer (Figure 3) the sample is vaporised and then bombarded with a beam of high energy electrons. The electrons in

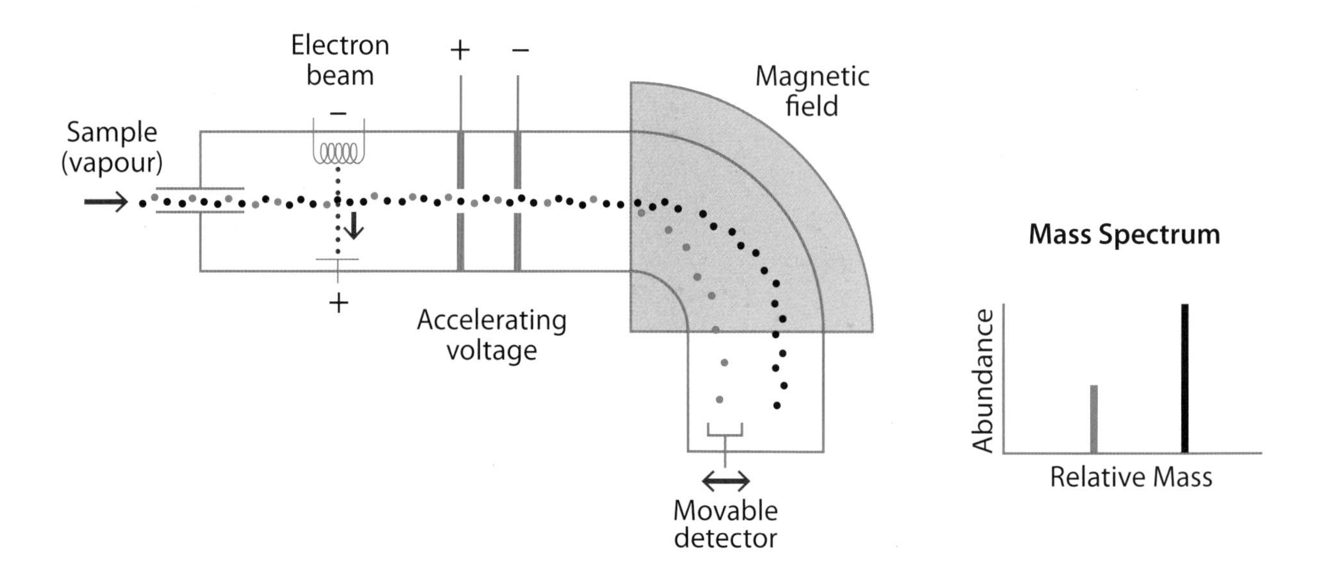

Figure 3: Separating ions with different masses using a mass spectrometer. The number of ions with each mass is recorded in the form of a mass spectrum.

the beam knock electrons from the particles in the vapour to form positively charged ions; a process known as ionisation. For instance, when a sample of neon gas is injected into a mass spectrometer the neon atoms are converted to neon ions: $Ne(g) + e^- \rightarrow Ne^+(g) + 2e^-$. The Neon ions ($Ne^+$) are then accelerated and passed through a magnetic field. When a charged particle such as an ion passes through a magnetic field it experiences a force and is deflected from its path. The size of the deflection depends on the mass of the ion, making it possible to calculate the mass of the ion by measuring the size of the deflection. The number of ions detected is recorded and used to construct the **mass spectrum** for the sample; a plot that shows the relative number of ions detected with each mass. The mass spectrum for neon gas is shown in Figure 4.

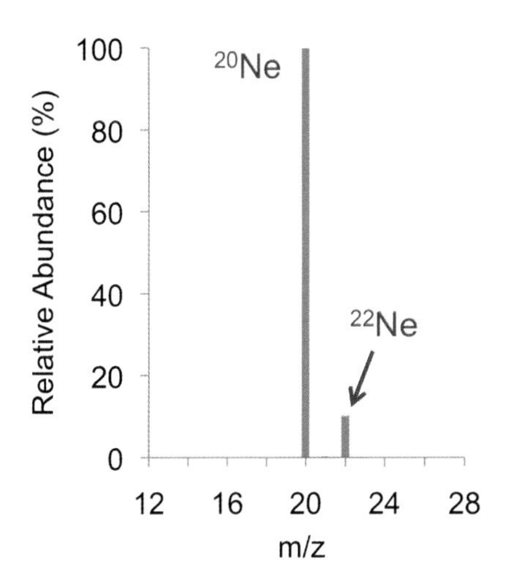

Figure 4: Mass spectrum for neon gas.

In reality a mass spectrometer detects the mass-to-charge ratio, m/z of an ion. In routine mass spectroscopy experiments the charge on each ion (z) is the same size as the charge on one electron (e), and the ratio m/z = m/e = m/1 is equal to the mass of the ion (m).

Calculating Atomic Mass

The mass spectrum for Neon in Figure 4 reveals that 90.48% of the neon atoms in a sample of neon gas are neon-20. Thus, for every 10,000 neon atoms in a sample of neon gas, 9048 (90.48%) are neon-20, and 10000 – 9048 = 952 are neon-22. This is equivalent to stating that: for every 100 neon atoms in a sample of neon gas, 90.48 are neon-20 (RIM = 20) and 100 – 90.48 = 9.52 are neon-22 (RIM = 22).

If we recall that the relative atomic mass (RAM) of an element is the average mass of one atom of the element, the RAM for neon can be calculated from the relative abundance of each isotope as follows.

RAM of Ne

$$= \frac{\text{Mass of } ^{20}\text{Ne atoms} + \text{Mass of } ^{22}\text{Ne atoms}}{\text{Total number of atoms}}$$

$$= \frac{(20 \times 90.48) + (22 \times 9.52)}{100}$$

$$= \frac{1810 + 209}{100} = 20.19$$

If we compare the calculated RAM for neon with the value given on the Periodic Table, we are reminded that the RAMs on the Periodic Table are often rounded to the nearest whole number. This is a reasonable approximation when most of the atoms in an element are the same isotope. Chlorine is an exception. The RAM of chlorine is 35.5 as chlorine contains significant amounts of two isotopes: ^{35}Cl (RIM = 35) and ^{37}Cl (RIM = 37).

Worked Example 1.2iv

Chlorine contains two isotopes: ^{35}Cl and ^{37}Cl. The relative abundance of the ^{35}Cl isotope is 75.78%. Calculate the atomic mass (RAM) of chlorine to one decimal place.

Solution

For every 100 atoms of chlorine in a sample of chlorine gas:

75.78 atoms (75.78%) are ^{35}Cl atoms and
(100 – 75.78) = 24.22 atoms are ^{37}Cl atoms.
RAM of Cl

$$= \frac{\text{Mass of } ^{35}\text{Cl atoms} + \text{Mass of } ^{37}\text{Cl atoms}}{\text{Total number of atoms}}$$

$$= \frac{(35 \times 75.78) + (37 \times 24.22)}{100}$$

$$= \frac{2652 + 896.1}{100} = 35.48$$

= 35.5 to 1 decimal place

Exercise 1.2E

1. Calculate the relative atomic mass of sodium in a sample containing 2.00% ^{24}Na and 98.00% ^{23}Na by mass to two decimal places.

   (CCEA June 2006)

2. Naturally occurring carbon contains 98.89% carbon-12 and 1.11% carbon-13. Calculate the relative atomic mass of carbon to two decimal places.

   (Adapted from CCEA January 2012)

3. Boron contains the isotopes ^{10}B and ^{11}B. The relative atomic mass of the element is 10.80. Which one of the following is the approximate ratio of the number of lighter atoms to heavier atoms?

   A 1:3 B 1:4 C 1:9 D 4:1

   (CCEA January 2014)

The following example demonstrates how to calculate the RAM for an element containing more than two isotopes.

Worked Example 1.2v

A sample of carbon was found to have the following composition. Calculate the relative atomic mass of the carbon sample to two decimal places.

Isotope:	^{12}C	^{13}C	^{14}C
% abundance:	98.50	1.25	0.25

(CCEA June 2005)

Solution

For every 100 atoms 98.50 atoms (98.50%) are ^{12}C atoms, 1.25 atoms (1.25%) are ^{13}C atoms and 0.25 atoms (0.25%) are ^{14}C atoms.

RAM of C =

$$\frac{\text{Mass of } ^{12}\text{C atoms} + \text{Mass of } ^{13}\text{C atoms} + \text{Mass of } ^{14}\text{C atoms}}{\text{Total number of atoms}}$$

$$= \frac{(12 \times 98.50) + (13 \times 1.25) + (14 \times 0.25)}{100}$$

$$= \frac{1182 + 16.3 + 3.5}{100} = 12.02$$

Exercise 1.2F

1. Several isotopes of iodine are produced in nuclear reactions. The percentage abundance of each isotope in a sample of radioactive dust is given in the table. Calculate the relative atomic mass of iodine to one decimal place.

Isotope:	^{127}I	^{129}I	^{131}I
% abundance:	95.91	2.49	1.60

(Adapted from CCEA June 2004)

2. A sample of iron from a meteorite was found to contain the following isotopes: ^{54}Fe, ^{56}Fe and ^{57}Fe. Use the relative abundance of each isotope to calculate the relative atomic mass of iron to one decimal place.

m/z ratio:	54	56	57
% abundance:	5.8	91.6	2.6

(CCEA June 2015)

3. (a) Explain what is meant by the term *isotope*.
   (b) Use the relative abundance of each calcium isotope in the table to calculate the relative atomic mass of calcium to **two** decimal places.

Isotope:	^{40}Ca	^{42}Ca	^{43}Ca	^{44}Ca
% abundance:	96.9	0.6	0.2	2.3

(Adapted from CCEA June 2014)

4. Xenon has a number of naturally occurring isotopes. The percentage abundance of each isotope is given in the table. Calculate the relative atomic mass of Xenon.

Relative Mass:	129	131	132	134	136
% abundance:	27	23	28	12	10

(CCEA January 2009)

Determining Molecular Mass

The technique of mass spectrometry can also be used to accurately determine the mass of individual molecules. The mass of a molecule is referred to as the **relative molecular mass (RMM)** of the molecule and is defined to be *the mass of the molecule relative to one-twelfth of the mass of an atom of carbon-12.*

When a molecule (M) enters the mass spectrometer it is vaporised before being ionised by collisions with a beam of high energy electrons.

$$M_{(g)} + e^-_{(g)} \rightarrow M^+_{(g)} + 2e^-_{(g)}$$

The molecular ions (M^+) formed by this process pass into the mass spectrometer and are detected. A molecular ion (M^+) has the same RMM as its parent molecule (M), and as a result, *the mass-to-charge ratio for the molecular ion peak is equal to the RMM of the molecule (M).*

The mass spectrum of naphthalene, $C_{10}H_8$ in Figure 5 contains a large molecular ion peak at an RMM of 128 that results from the ionisation and detection of molecules with the formula $^{12}C_{10}{}^{1}H_8$. The mass spectrum also contains a molecular ion peak at an RMM of 129 that results from the ionisation and detection of molecules with the formula $^{12}C_9{}^{13}C_1H_8$.

Worked Example 1.2vi

The mass spectrum of chlorine, Cl_2 is shown below. Which peak should not be present?

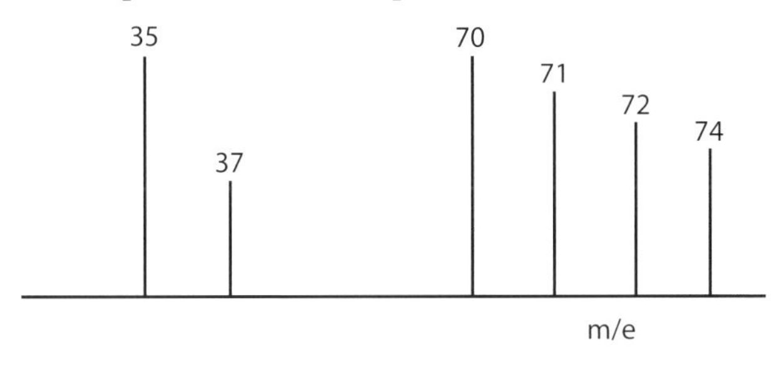

(Adapted from CCEA January 2011)

Solution

Peaks would be expected at the following RMMs.

Species:	^{35}Cl	^{37}Cl	$^{35}Cl^{35}Cl$	$^{35}Cl^{37}Cl$	$^{37}Cl^{37}Cl$
RMM:	35	37	70	72	74

The peak for an RMM of 71 should not appear in the mass spectrum.

Having defined RMM in this way, the **relative formula mass (RFM)** of a compound becomes *the average mass of one formula of the compound relative to one-twelfth of the mass of an atom of carbon-12*, and is obtained by adding the RAMs for the atoms in one formula. For example, the RAM of chlorine is 35.5, and the RFM of a chlorine molecule (Cl_2) is 35.5 + 35.5 = 71. The lack of a peak corresponding to an RMM of 71 in the mass spectrum of chlorine (See Worked Example 1.2vi) demonstrates that the RFM of a compound is an average value, and does not necessarily correspond to the mass of an actual molecule.

Exercise 1.2G

1. Chlorine exists as the isotopes, ^{35}Cl and ^{37}Cl with the result that the relative atomic mass of chlorine is 35.5. Which one of the following statements is correct?

   A The isotopes have different chemical properties.

   B The ^{37}Cl isotope has a natural abundance of 75%.

   C The mass spectrum of chlorine, Cl_2 includes peaks at 70, 72 and 74.

   D The nuclei of the isotopes have the same number of neutrons.

   (CCEA June 2008)

2. What is the purpose of mass spectrometry?

   (CCEA January 2012)

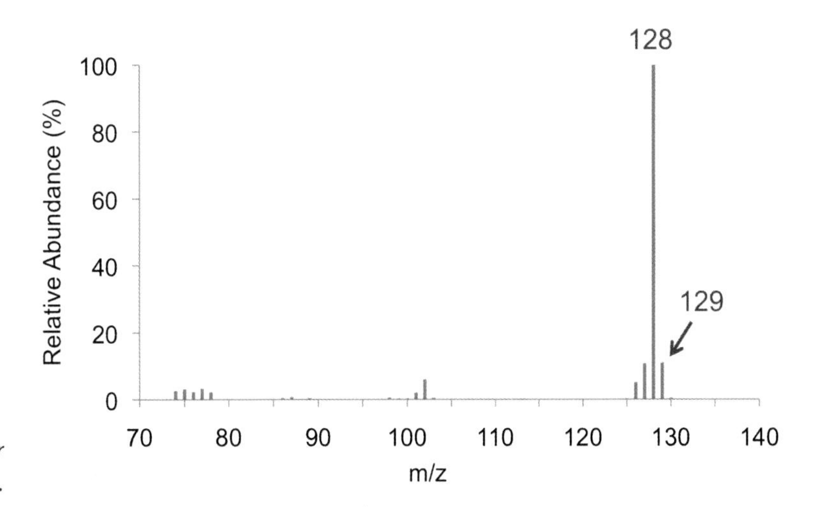

Figure 5: Mass spectrum for naphthalene, $C_{10}H_8$.

Before moving to the next section, check that you are able to:

- Calculate the relative atomic mass (RAM) of an element given the percentage abundance of each isotope in the element.
- Calculate the RMM of a molecule from its mass spectrum.
- Explain the terms *relative molecular mass (RMM)* and *relative formula mass (RFM)*.

Atomic Structure

In this section we are learning to:

- Describe the arrangement of electrons in atoms in terms of atomic orbitals.
- Use spd-notation to write electron configurations for atoms and ions.
- Explain how ionisation energies can be used to provide evidence for electron shells and subshells in atoms.

Atoms are much too small to be seen with instruments such as a light microscope. They must, instead, be visualised with much more sensitive instruments such as a Scanning Tunnelling Microscope (STM). STM images of atoms on the surface of pieces of gold and silicon are shown in Figure 6. In an STM image each atom appears as a cloud of electrons that results from electrons moving within regions of space known as **atomic orbitals.**

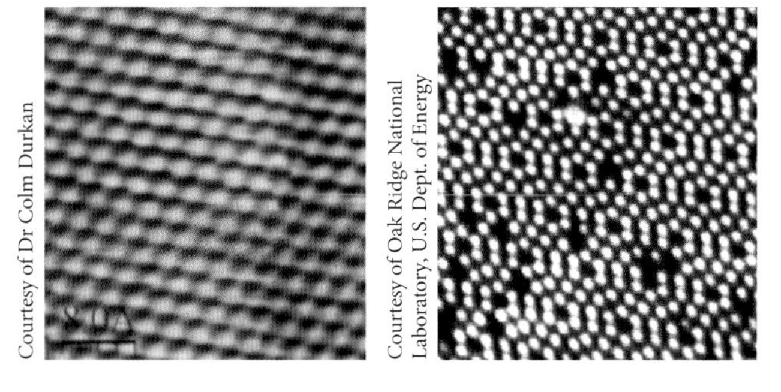

Figure 6: STM images of atoms on the surface of (a) gold and (b) silicon.

Atomic Orbitals

The electrons in an atom are arranged in atomic orbitals. There are four types of atomic orbital: s, p, d, and f, each with a characteristic shape. An s-orbital is a sphere (ball) centred on the nucleus of the atom. In contrast, a p-orbital has 2 lobes and is shaped like a

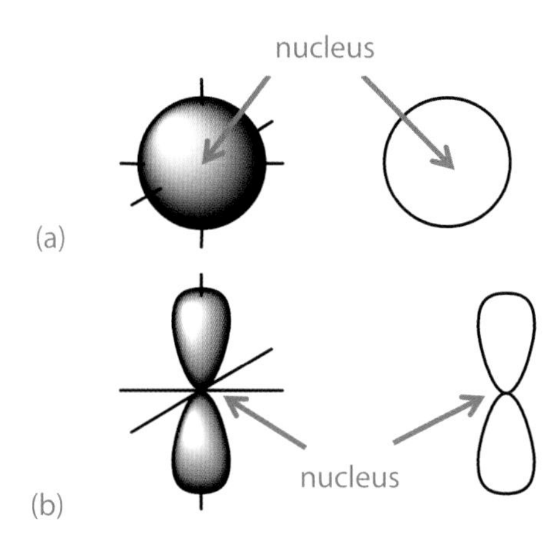

Figure 7: The shape and cross-section of (a) an s-orbital and (b) a p-orbital.

'dumb-bell' with the nucleus between the lobes. The shapes of the d and f orbitals are more complex. When drawing orbitals it is only necessary to draw the cross-section; the shape obtained by cutting through the orbital. The shapes and cross-sections of s and p-type atomic orbitals are shown in Figure 7.

The atomic orbitals in an atom are arranged in shells. Each shell is assigned a number (n) that is then used to identify the s, p, d and f orbitals belonging to the shell. For example, the first shell (n=1) contains one s-type atomic orbital. This orbital is referred to as the 1s orbital to indicate that it is an s-orbital belonging to the first shell (n=1). The second shell (n=2) contains one s-type orbital and a set of three p-type atomic orbitals. The s-orbital is referred to as the 2s orbital and the set of p-orbitals is referred to as the 2p orbitals to indicate that they belong to the second shell.

The sets of s, p, d and f orbitals belonging to each shell are known as subshells. We have already seen that the first shell contains the 1s subshell while the second shell contains the 2s and 2p subshells. The third shell (n=3) is bigger and consists of the 3s, 3p and 3d subshells. As with other s and p subshells, the 3s subshell contains a single s-type orbital and the 3p subshell contains a set of three p-type atomic orbitals. The 3d subshell consists of a set of 5 d-type atomic orbitals. Other d-type subshells such as the 4d and 5d subshells also contain a set of 5 d-type atomic orbitals. The fourth and fifth shells are larger still and contain sets of f-type atomic orbitals. The 4f and 5f subshells each contain 7 f-type atomic orbitals.

Each atomic orbital holds a maximum of two electrons. As a result, the single s-orbital in an s-subshell holds a maximum of 2 electrons, the set of 3 p-orbitals in a p-subshell holds a maximum of $3 \times 2 = 6$ electrons, and

Figure 8: The energies of the subshells associated with the first four shells in a hydrogen atom.

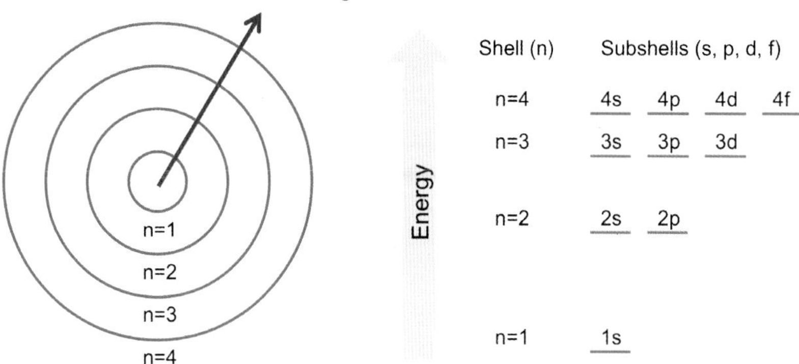

the set of 5 d-orbitals in a d-subshell holds up to 5 × 2 = 10 electrons. The sets of 7 f-orbitals in the 4f and 5f subshells each hold up to 7 × 2 = 14 electrons.

The energy level diagram in Figure 8 reveals that the subshells associated with each shell in a hydrogen atom have the same energy. As a result the electron in a hydrogen atom has the same energy when it occupies any one of the atomic orbitals within the subshells associated with a particular shell. This situation is unique to hydrogen and ions containing one electron such as He^{+}, Li^{2+}, and Be^{3+}. *In contrast, in atoms and ions with two or more electrons, the electrons have different energies when they occupy atomic orbitals associated with different subshells.*

Exercise 1.2H

1. Which one of the following orbitals is occupied by an electron with the energy level n=2?

   A A dumb-bell shaped orbital
   B A spherically shaped orbital
   C An s or d orbital
   D An s or p orbital

(CCEA January 2012)

Electron Configuration

The **electron configuration** of an atom *details how the electrons are distributed amongst the subshells*. For example, the electron configuration of a nitrogen atom: $1s^2\ 2s^2\ 2p^3$ indicates that there are 2 electrons in the 1s subshell ($1s^2$), 2 electrons in the 2s subshell ($2s^2$), and three electrons in the 2p subshell ($2p^3$).

The subshell filling order can be deduced from the energy level diagram in Figure 9a. According to the **Aufbau principle** *the lowest energy subshell fills first. The subshells then fill in order of increasing energy until all of the electrons have been placed in subshells.* The subshell filling order: 1s, 2s, 2p, 3s, 3p, 4s, 3d, 4p ... that results from applying the Aufbau principle can be remembered by 'following the arrows' in Figure 9b.

The electron configuration obtained using the Aufbau principle is referred to as the **ground state** of the atom as *the electrons occupy the lowest energy subshells available to produce the electron configuration with the lowest energy*.

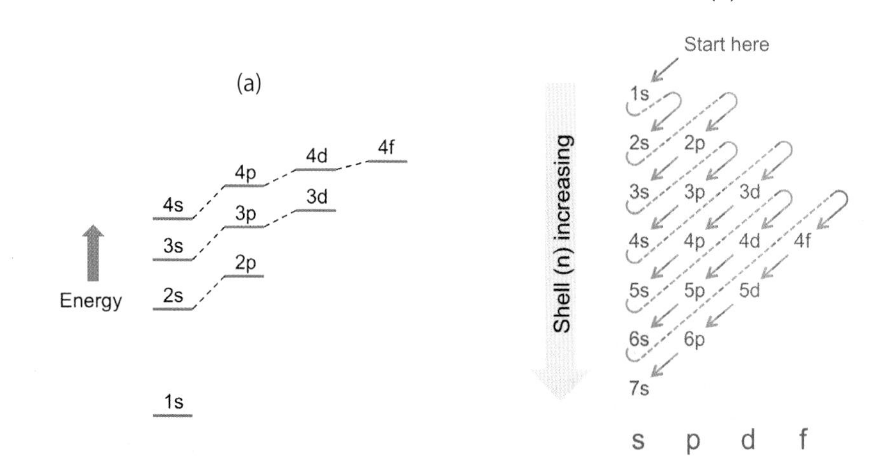

Figure 9: (a) The energies of the subshells in an atom with two or more electrons. (b) The subshell filling order is obtained by 'following the arrows'.

Worked Example 1.2vii

Write the ground state electron configuration for an atom of: (a) helium, (b) lithium, (c) oxygen, (d) aluminium and (e) iron.

Strategy

- Recall that subshells fill in the order: 1s 2s 2p 3s 3p 4s 3d ...
- Recall that an s-subshell holds up to 2 electrons, a p-subshell holds up to 6 electrons and a d-subshell holds up to 10 electrons.

Solution

(a) Helium has 2 electrons. Both electrons go in the 1s subshell. The electron configuration is $1s^2$.

(b) Lithium has 3 electrons. Two electrons fill the 1s subshell. The third electron goes in the 2s subshell. The electron configuration is $1s^2\ 2s^1$.

(c) Oxygen has 8 electrons. The first four electrons fill the 1s and 2s subshells. The remaining four electrons go in the 2p subshell. The electron configuration is $1s^2\ 2s^2\ 2p^4$.

(d) Aluminium has 13 electrons. Twelve electrons fill the 1s, 2s, 2p and 3s subshells. The remaining electron goes in the 3p subshell. The electron configuration is $1s^2\ 2s^2\ 2p^6\ 3s^2\ 3p^1$.

(e) Iron has 26 electrons. Twenty electrons fill the 1s, 2s, 2p, 3s, 3p and 4s subshells. The remaining six electrons go in the 3d subshell. The electron configuration is $1s^2\ 2s^2\ 2p^6\ 3s^2\ 3p^6\ 4s^2\ 3d^6$.

Electron Configurations for Ions

The ground state electron configuration for a negative ion is obtained by adding electrons to the electron configuration for the corresponding atom.

$1s^2\ 2s^2\ 2p^6\ 3s^2\ 3p^5 \rightarrow 1s^2\ 2s^2\ 2p^6\ 3s^2\ 3p^6$

chlorine atom (Cl) → chloride ion (Cl^-)

Conversely, the electron configuration for a positive ion is obtained by removing electrons from the electron configuration for the corresponding atom. When removing electrons the Aufbau principle is followed in reverse as the electrons in the highest energy orbitals are the easiest to remove.

$1s^2\ 2s^2\ 2p^6\ 3s^1 \rightarrow 1s^2\ 2s^2\ 2p^6$

sodium atom (Na) → sodium ion (Na^+)

Worked Example 1.2viii

Use spd-notation to write the ground state electron configuration for (a) Mg^{2+} (b) F^- and (c) S^{2-}.

Strategy

- Write the electron configuration for the atom.
- Add or remove electrons to form the electron configuration for the ion.

Solution

(a) The electron configuration for a magnesium atom is $1s^2\ 2s^2\ 2p^6\ 3s^2$. The electron configuration for a magnesium ion (Mg^{2+}) is obtained by removing two electrons. The electron configuration for a Mg^{2+} ion is $1s^2\ 2s^2\ 2p^6$.

(b) The electron configuration for a fluorine atom is $1s^2\ 2s^2\ 2p^5$. The electron configuration for a fluoride ion (F^-) is obtained by adding one electron. The electron configuration for a F^- ion is $1s^2\ 2s^2\ 2p^6$.

(c) The electron configuration for a sulfur atom is $1s^2\ 2s^2\ 2p^6\ 3s^2\ 3p^4$. The electron configuration for a sulfide ion (S^{2-}) is obtained by adding two electrons. The electron configuration for a S^{2-} ion is $1s^2\ 2s^2\ 2p^6\ 3s^2\ 3p^6$.

Exercise 1.2I

1. Calcium oxide is an ionic substance. Use spd-notation to show the formation of a calcium ion and an oxide ion from a calcium atom and an oxygen atom and state the charge on each ion.

   (CCEA January 2006)

2. (a) Write the ground state electron configuration for neon. (b) Which of the following ions have the same ground state electron configuration as neon?

   Li^+ Al^{3+} F^- Cl^- K^+

3. Which one of the following elements forms an ion with a double negative change that has the same electronic configuration as argon?

   A Calcium   B Chlorine
   C Selenium   D Sulfur

   (CCEA June 2015)

The situation is slightly different when removing electrons from metals in the 'd-block' between groups II and III of the Periodic Table. For example, when an atom of iron (...$3s^2$ $3p^6$ $4s^2$ $3d^6$) forms a positive ion electrons are removed from the outermost s-subshell, and then from the d-subshell, until the required number of electrons has been removed.

... $3s^2$ $3p^6$ $4s^2$ $3d^6$ ➜ ... $3s^2$ $3p^6$ $3d^6$
iron atom (Fe)   iron II ion (Fe^{2+})

➜ ... $3s^2$ $3p^6$ $3d^5$
iron III ion (Fe^{3+})

Worked Example 1.2ix

Use spd-notation to write the electron configuration for the ground state of a V^{2+} ion.

(Adapted from CCEA June 2008)

Strategy

- Write the ground state electron configuration for a vanadium atom.
- Remove two electrons from the outermost s-subshell.

Solution

Ground state for a V atom:
$1s^2$ $2s^2$ $2p^6$ $3s^2$ $3p^6$ $4s^2$ $3d^3$

Ground state for a V^{2+} ion:
$1s^2$ $2s^2$ $2p^6$ $3s^2$ $3p^6$ $3d^3$

Exercise 1.2J

1. Use spd-notation to write the ground state electron configuration for the following atoms and ions.

   (a) Mn   (b) Mn^{2+} (c) Zn^{2+} (d) Ni

2. Use spd-notation to write the ground state electron configuration for the following atoms and ions.

   (a) Sc   (b) V   (c) V^{3+} (d) Ca

Electron Spin

Electrons have a property called **spin**. Each atomic orbital holds a maximum of 2 electrons: one 'spin-up' (↑) the other 'spin-down' (↓). The spin of the electrons in each atomic orbital can be shown by drawing the electron configuration of an atom or ion in the form of 'electrons-in-boxes'. The electrons-in-boxes representations for the electrons in atoms of lithium, carbon and oxygen are shown in Figure 10.

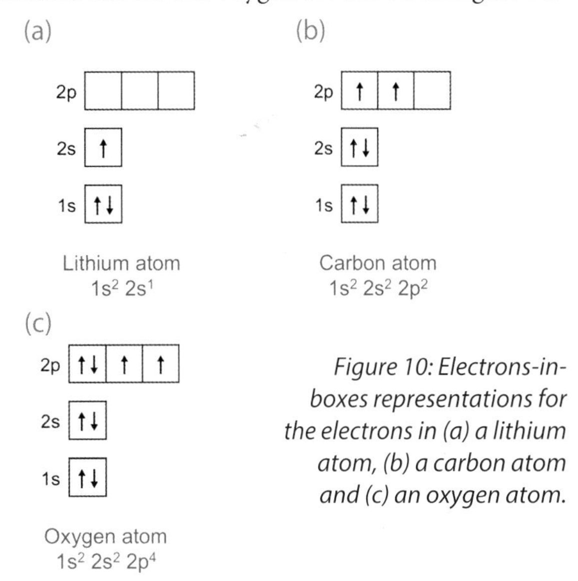

Figure 10: Electrons-in-boxes representations for the electrons in (a) a lithium atom, (b) a carbon atom and (c) an oxygen atom.

In an electrons-in-boxes representation each atomic orbital is represented by a box. The subshells are drawn in energy order with the orbitals in each subshell arranged side-by-side. The electrons-in-boxes configurations are obtained by applying the Aufbau principle while remembering to add one electron to each orbital in a subshell (↑) before pairing the electrons (↑↓). In lithium (Figure 10a) two electrons pair up in the 1s orbital: one spin-up (↑), the other spin-down (↓). The third electron then occupies the 2s orbital. In a carbon atom (Figure 10b) the electrons pair up in the 1s and 2s orbitals before the remaining electrons occupy orbitals in the 2p subshell. In oxygen (Figure10c), the 2p subshell contains additional electrons that begin to pair up (↑↓) once each orbital in the subshell contains a single electron (↑).

Exercise 1.2K

1. Which one of the following electron configurations contains two unpaired electrons?

   A $1s^2$ $2s^1$ B $1s^2$ $2s^2$ $2p^3$
   C $1s^2$ $2s^2$ $2p^4$ D $1s^2$ $2s^2$ $2p^6$ $3s^2$ $3p^5$

   (CCEA June 2006)

2. Which one of the following atoms contains one unpaired electron in its ground state?

   A oxygen B magnesium C chlorine D argon

   (Adapted from CCEA June 2011)

3. Magnesium reacts with chlorine to form magnesium chloride. Use the boxes below to complete the electron configuration for each type of ion in magnesium chloride.

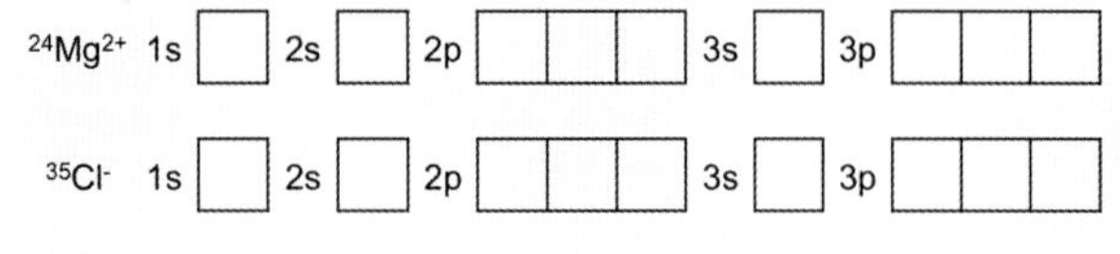

   (Adapted from CCEA June 2012)

4. Which one of the following diagrams represents the distribution of electrons amongst the 3d and 4s subshells in the ground state of an iron atom?

   A 3d | ↑↓ | ↑↓ | ↑↓ | | | 4s | ↑↓ |

   B 3d | ↑↓ | ↑↓ | ↑↓ | ↑ | ↑ | 4s | |

   C 3d | ↑↓ | ↑↓ | ↑↓ | ↑↓ | | 4s | |

   D 3d | ↑↓ | ↑ | ↑ | ↑ | ↑ | 4s | ↑↓ |

5. Which one of the following diagrams represents the distribution of electrons in the 3d and 4s subshells in the ground state of an iron(III) ion?

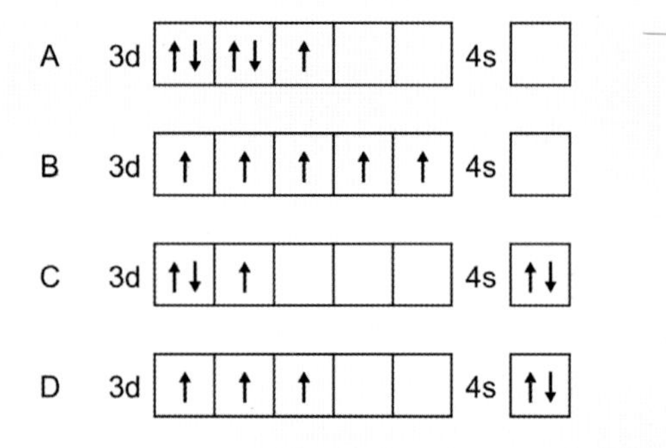

   (CCEA June 2014)

Exceptions to the Aufbau Principle

The Aufbau principle can be used to reliably predict the ground state electron configuration for most elements and ions. The ground state electron configurations for chromium and copper are amongst the few configurations not predicted correctly by the Aufbau principle. According to the Aufbau principle, the electron configurations for chromium and copper should be:

Cr [Ar] $4s^2$ $3d^4$ and Cu [Ar] $4s^2$ $3d^9$

where we have used the shorthand [Ar] to represent the electron configuration of an argon atom. The electron configurations are instead found to be:

Cr [Ar] $4s^1$ $3d^5$ and Cu [Ar] $4s^1$ $3d^{10}$

These exceptions demonstrate that *an electron configuration with a half-filled or completely filled d-subshell such as s^1d^5 or s^1d^{10} is more stable than a configuration with a part-filled d-subshell such as s^2d^4 or s^2d^9.*

Exercise 1.2L

1. (a) Write the electronic configuration of a Fe^{2+} ion. (b) With reference to spd-notation explain the stability of the Fe^{3+} ion relative to the Fe^{2+} ion.

   (CCEA June 2015)

Before moving to the next section, check that you are able to:

- Describe the shape of s and p orbitals and draw their cross-section.
- Use the Aufbau principle to construct the ground state electron configuration for atoms and ions of the elements in the first four periods (up to Kr).
- Use electrons-in-boxes notation to describe the filing of subshells.
- Explain exceptions to the Aufbau principle in terms of the stability of filled and half-filled subshells.

Evidence for Shells and Subshells

The **first ionisation energy (IE1)** of an element is *the energy needed to remove one mole of electrons from one mole of gas phase atoms to form one mole of gas phase ions with a 1+ charge.*

First ionisation energy (IE1): $E_{(g)} \rightarrow E^{+}_{(g)} + e^{-}_{(g)}$

The plot of IE1 against atomic number in Figure 11 reveals that IE1 generally increases from left to right across a period. Smaller fluctuations in IE1 across a period can be explained in terms of the arrangement of electrons in subshells and provide direct evidence for the existence of subshells in atoms. The plot of IE1 against atomic number in Figure 11 also reveals that IE1 decreases down a group. This is particularly evident for the noble gases (Group VIII) and the Alkali Metals (Group I).

Exercise 1.2M

1. The graph of first ionisation energy against atomic number for a series of ten consecutive elements in the Periodic Table is shown below. Which one of the following assignments is correct?

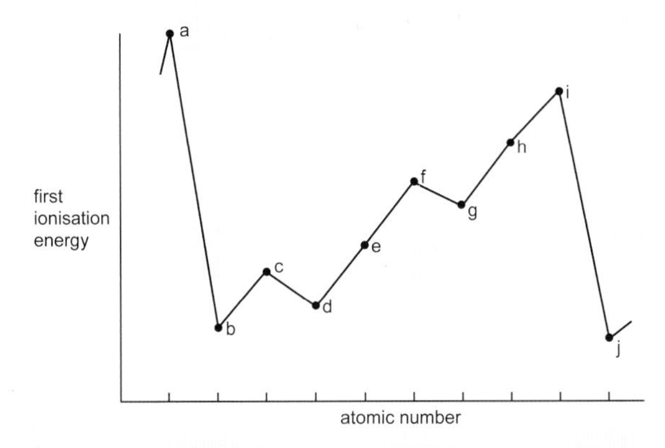

	Group II metal	Halogen
A	a	h
B	b	g
C	c	h
D	c	i

(Adapted from CCEA June 2014)

2. The first ionisation energy is shown against increasing atomic number. Which one of the following shows a Group I element together with a Group VII element?

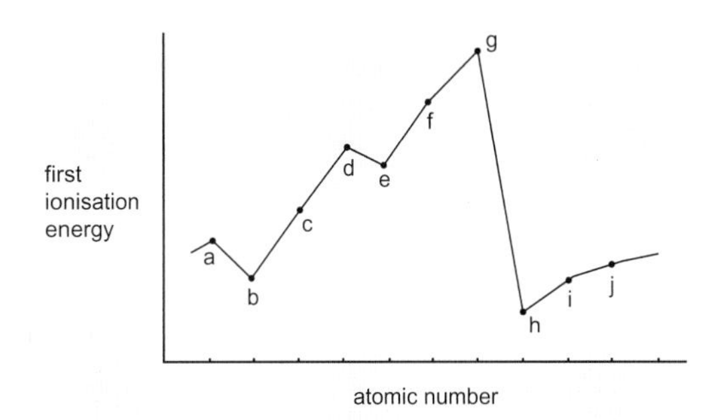

	Group I	Group VII
A	b	f
B	b	g
C	h	f
D	h	g

(CCEA January 2014)

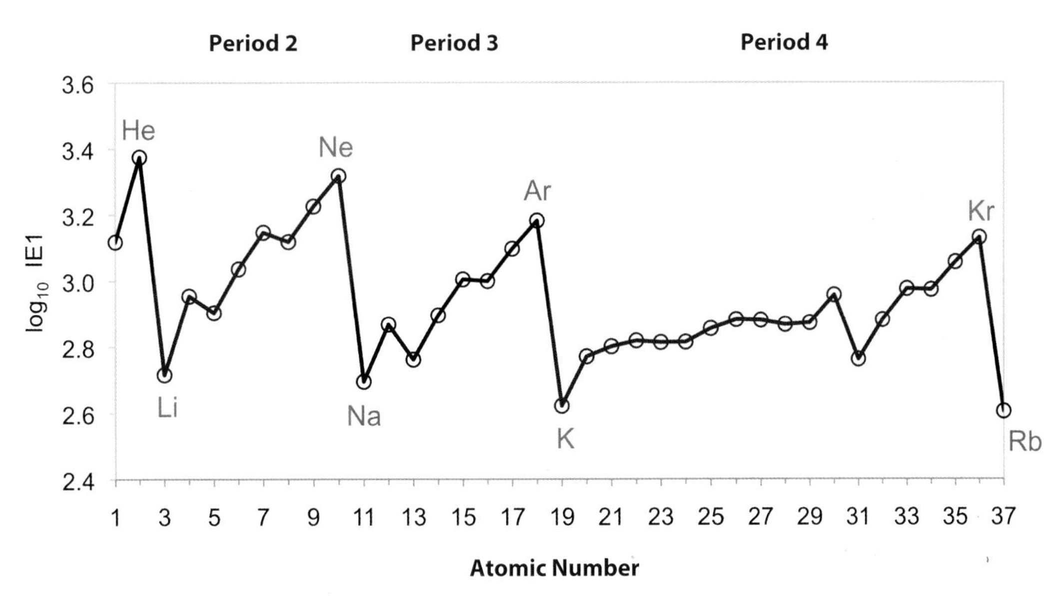

Figure 11: Graph of first ionisation energy (plotted as $\log_{10}$ IE1) against atomic number for the elements in Periods 1–4.

Trend in IE1 within a Group

The orbitals in a subshell get bigger as the energy of the subshell increases.

Subshell energy increasing

1s < 2s < 2p < 3s < 3p < 4s < 3d < 4p < ...

Orbital size increasing

The electrons in the outermost (biggest) subshell are attracted to the nucleus and, at the same time, repelled by electrons in the smaller subshells closer to the nucleus. This repulsion between electrons reduces the attraction felt by the electrons in the outermost subshell and is referred to as **shielding**. The shielding of outer electrons from the nucleus is illustrated in Figure 12. *As the number of filled subshells increases, electrons in the outermost subshell are further from the nucleus and better shielded by electrons in the subshells closer to the nucleus.* This allows the electrons in the outermost subshell to be more easily removed from the atom and explains why IE1 decreases down a group in the Periodic Table.

Exercise 1.2N

1. Which one of the following equations represents the first ionisation energy of fluorine?

   A $F_{2(g)} + 2e^- \rightarrow 2F^-_{(g)}$

   B $F_{(g)} + e^- \rightarrow F^-_{(g)}$

   C $F_{(g)} \rightarrow F^+_{(g)} + e^-$

   D $F_{2(g)} \rightarrow 2F^+_{(g)} + 2e^-$

   (CCEA June 2015)

2. State two reasons why the first ionisation energy of calcium is less than that of magnesium.

   (CCEA January 2009)

Trend in IE1 Across a Period

The plot of IE1 against atomic number in Figure 11 reveals that IE1 generally increases as atomic number increases from left to right across a period. Moving one element to the right across a period adds a proton to the nucleus and an electron to the outermost subshell. The resulting increase in shielding is not sufficient to offset the effect of adding a proton to the nucleus and, as a result, IE1 increases as the electrons in the outermost subshell are more strongly attracted to the nucleus and become harder to remove.

The plot of IE1 against atomic number in Figure 11 also reveals similar variations in IE1 across period 2, period 3, and those groups in period 4 in common

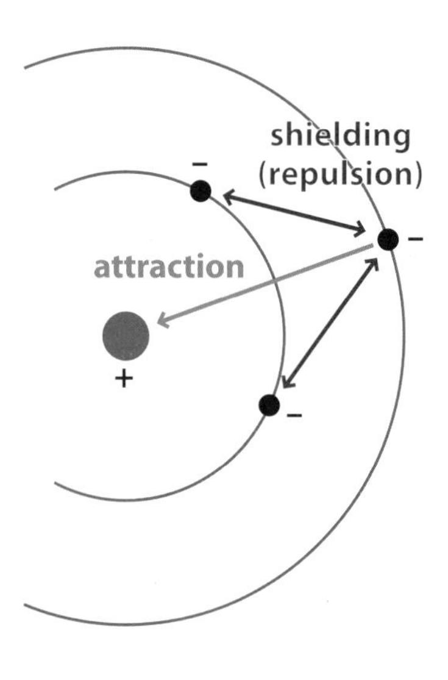

Figure 12: Shielding of the outer electrons in an atom.

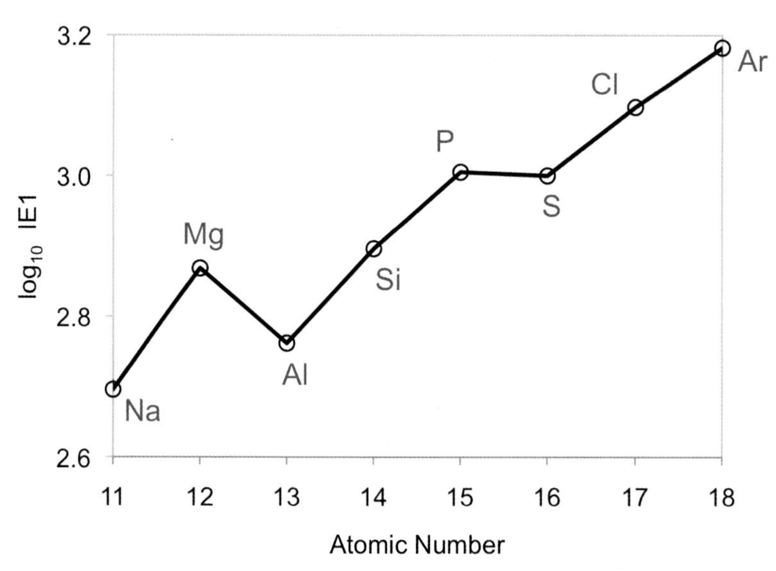

Figure 13: Graph of first ionisation energy (plotted as $\log_{10}$ IE1) against atomic number for the elements belonging to the third period.

with periods 2 and 3. The variations in IE1 across a period can be explained by the order in which the electrons occupy the outermost subshells. In this way, the variations in IE1 across a period can be used as evidence to support the existence of subshells, and the subshell filling order as defined by the Aufbau principle.

Consider, for example, the variations in IE1 across Period 3 (Figure 13). The electron configuration for a sodium atom is: $1s^2\ 2s^2\ 2p^6\ 3s^1$. The single 3s electron is the only electron in the n=3 shell and is effectively shielded by the electrons in the smaller 1s, 2s, and 2p subshells that lie between the 3s electron and the nucleus. As a result the 3s electron experiences a relatively small attraction to the nucleus, and is easy to remove.

Magnesium has an additional electron in the 3s subshell. The electron configuration for a magnesium atom is: $1s^2\ 2s^2\ 2p^6\ 3s^2$. Shielding of the 3s electrons in magnesium is similar to the shielding experienced by the 3s electron in sodium. As a result, the 3s electrons in magnesium experience a greater attraction to the nucleus and are harder to remove than the single 3s electron in sodium. This explains why IE1 for magnesium is greater than IE1 for sodium.

The electron configuration for an aluminium atom is: $1s^2\ 2s^2\ 2p^6\ 3s^2\ 3p^1$. The electron in the 3p subshell has a higher energy than the electrons in the 3s subshell and is therefore easier to remove from the atom. This explains why IE1 for aluminium is less than IE1 for magnesium.

IE1 then increases steadily from aluminium ($3p^1$) to phosphorus ($3p^3$) as electrons are added to the 3p subshell. Adding electrons to the 3p subshell does not increase the shielding of the electrons already in the 3p subshell with the result that the electrons in the 3p subshell are more strongly attracted to the nucleus and become harder to remove as the charge on the nucleus increases.

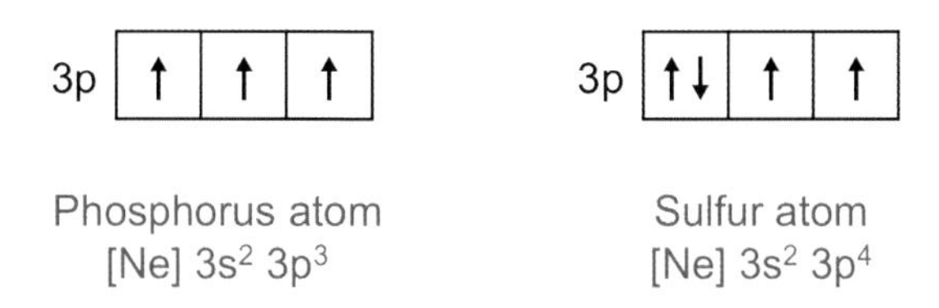

The electrons-in-boxes description of the 3p electrons in phosphorus ($3p^3$) and sulfur ($3p^4$) reveals that while the 3p electrons in phosphorus are equivalent, the 3p electrons in sulfur are not equivalent. In sulfur, the paired electrons in the 3p subshell repel (both are negatively charged) and have a higher energy than the unpaired electrons in the 3p subshell. As a result IE1 for sulfur, which involves removing a paired electron from the 3p subshell, is less than might otherwise be expected.

The unexpectedly low IE1 for sulfur can also be explained by noting that IE1 for phosphorus ($3p^3$) is, in fact, higher than might be expected on account of the stability associated with a half-filled subshell (p^3). Previously the stability associated with a half-filled subshell has been used to explain why an s^1d^5 configuration is preferred over an s^2d^4 configuration in chromium.

IE1 then increases from sulfur ($3p^4$) to argon ($3p^6$) as the atomic number increases and the electrons in the outermost 3p subshell become more strongly attracted to the nucleus. The increase in IE1 towards the right of the period can also be attributed to the stability associated with a full subshell of electrons ($3p^6$) in the same way as the stability associated with a full d-subshell was used to explain the preference for an s^1d^{10} configuration over an s^2d^9 configuration in copper.

Taken together, these explanations provide considerable evidence for the existence of subshells, and the subshell filling order defined by the Aufbau principle.

Worked Example 1.2x

The graph below shows the first ionisation energy (IE1) for the first twelve elements:

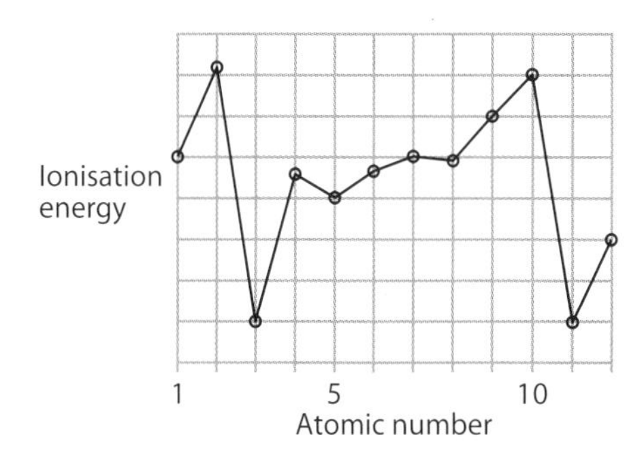

(a) Explain why oxygen has a lower IE1 than nitrogen.

(b) Explain why lithium and sodium have a low IE1.

(c) Explain why helium and neon have a high IE1.

(d) Write the equation defining IE1 for beryllium. Include state symbols.

(CCEA June 2007)

Solution

(a) The electron removed from the 2p subshell in oxygen ($2p^4$) is paired and is therefore easier to remove than an unpaired electron from the 2p subshell in nitrogen ($2p^3$).

Alternative answer:

The electron removed from the 2p subshell in nitrogen belongs to a half-filled subshell and is therefore more difficult to remove than an electron from the 2p subshell in oxygen.

(b) Lithium and sodium have one electron in their outermost shell that is well shielded from the nucleus and is therefore easily removed from the atom.

Alternative answer:

The electron in the outermost shell is easily removed to form a stable ion with a full outer shell of electrons.

(c) Helium and neon have full outer shells and are stable.

(d) $Be_{(g)} \rightarrow Be^{+}_{(g)} + e^{-}_{(g)}$

Exercise 1.20

1. (a) What is meant by the term *ground state*?
   (b) Use electrons-in-boxes notation to represent the ground state of a phosphorus atom.
   (c) Explain why phosphorus has an unusually high first ionisation energy.

(CCEA January 2003)

2. Which one of the following lists the first ionisation energies (in kJ mol^{-1}) for magnesium, aluminium, silicon, phosphorus and sulfur in this order?

A	496	736	577	786	1060
B	577	786	1060	1000	1260
C	736	577	786	1060	1000
D	786	1060	1000	1260	1520

(CCEA January 2011)

3. The graph represents the first ionisation energies of the elements sodium to argon.

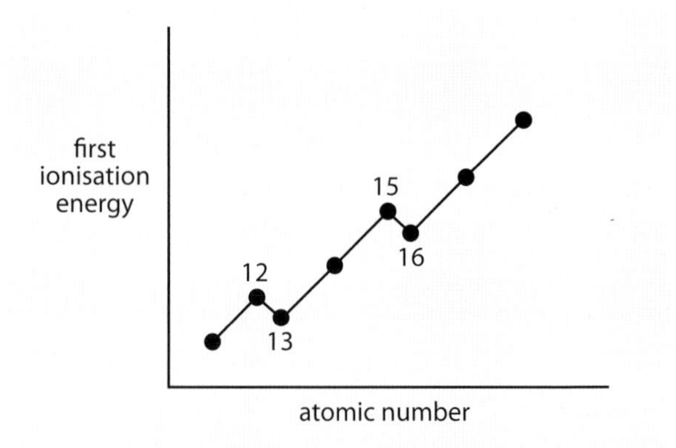

   (a) Explain the general rise in the first ionisation energy across the period.
   (b) Explain the decrease in ionisation energy from atomic number 12 to 13.
   (c) Explain the decrease in ionisation energy from atomic number 15 to 16.

(CCEA January 2008)

Successive Ionisation

Some of the most direct evidence for the existence of shells in atoms is revealed by plotting successive ionisation energies (IE1, IE2, IE3, ...) for an atom. The **second ionisation energy (IE2)** of an element is *the energy needed to remove one mole of electrons from one mole of gas phase ions with a 1+ charge to form one mole of gas phase ions with a 2+ charge.*

Second ionisation energy (IE2): $E^{+}_{(g)} \rightarrow E^{2+}_{(g)} + e^{-}_{(g)}$

Similarly, the **third ionisation energy (IE3)** of an element is *the energy needed to remove one mole of electrons from one mole of gas phase ions with a 2+ charge to form one mole of gas phase ions with a 3+ charge.*

Third ionisation energy (IE3): $E^{2+}_{(g)} \rightarrow E^{3+}_{(g)} + e^{-}_{(g)}$

The successive ionisation energies for magnesium: $1s^2\ 2s^2\ 2p^6\ 3s^2$ are plotted in Figure 14. The first and second ionisation energies (IE1 and IE2) are the energies needed to remove the first and then the second electron from the outermost 3s subshell. The ionisation energies IE3 to IE12 detail the energy needed to remove electrons from the 2p subshell (IE3 to IE8), and then from the 2s subshell (IE9 and IE10), before finally removing electrons from the 1s subshell (IE11 and IE12).

The electrons in the 3s subshell (n=3 shell) have low IE's as they are furthest from the nucleus and experience the greatest shielding. The electrons in the 2s and 2p subshells (n=2 shell) are closer to the nucleus and are less shielded than the electrons in the 3s subshell (n=3 shell). As a result, the electrons in the second (n=2) shell are more strongly attracted to the nucleus and are harder to remove than the electrons in the third (n=3) shell. This explains why the IE's for the electrons in the second shell (IE3 to IE10) are significantly greater than the ionisation energies for the electrons in the third shell (IE1 and IE2). The electrons in the 1s subshell (n=1) are even more difficult to remove as they are closest to the nucleus and are less shielded than the electrons in the second shell. As a result, the IE's for the 1s electrons (IE11 and IE12) are significantly greater than the IE's for the electrons in the second shell (IE3 to IE10). In this way *the successive ionisation energies for an atom can be used as evidence for the existence of electron shells in atoms.*

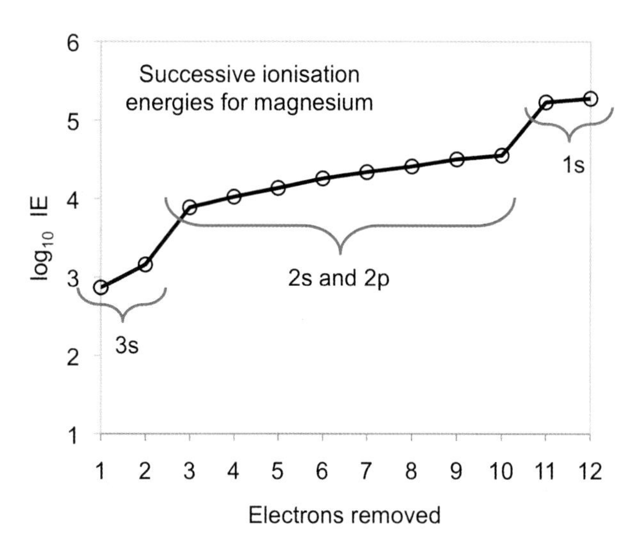

Figure 14: Successive ionisation energies for magnesium (plotted as $\log_{10}$ IE).

Worked Example 1.2xi

The first four ionisation energies of aluminium are 578, 1817, 2745 and 11578 kJ mol^{-1}. Label the subshells in the following diagram and use the electrons-in-boxes notation to show how the electrons are arranged in an Al^{2+} ion.

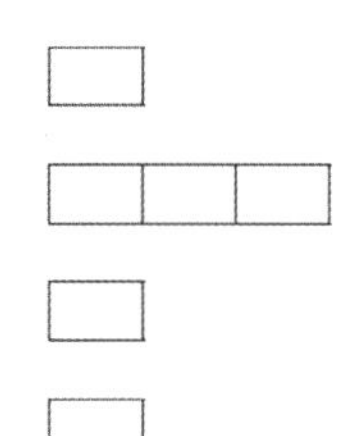

(a) Write the equation used to define the fourth ionisation energy of aluminium. Include state symbols.

(b) Explain why the third ionisation energy of aluminium is much smaller than the fourth ionisation energy.

(CCEA January 2010)

Solution

3p ☐☐☐

3s ↑

2p ↑↓ ↑↓ ↑↓

2s ↑↓

1s ↑↓

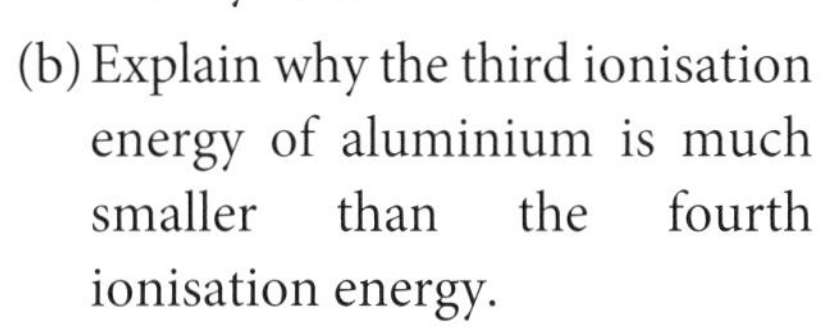

(a) $Al^{3+}{}_{(g)} \rightarrow Al^{4+}{}_{(g)} + e^-{}_{(g)}$

(b) The outermost (3s) electron in Al^{2+} is further from the nucleus and better shielded than the outermost (2p) electrons in Al^{3+}.

Exercise 1.2P

1. On the axes below sketch a graph to show the successive ionisation energies of boron.

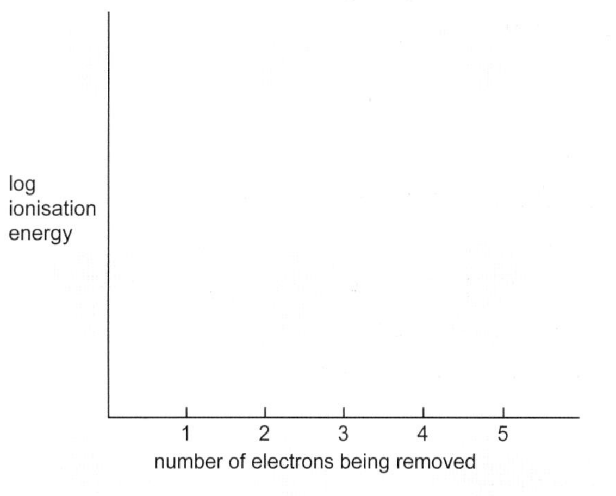

(CCEA June 2014)

2. Plot the successive ionisation energies of aluminium.

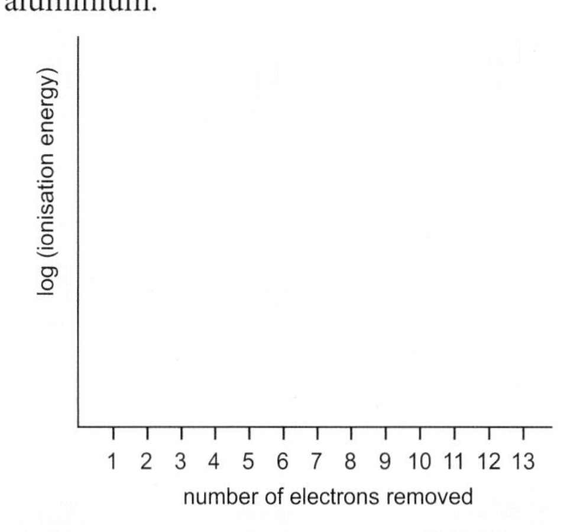

(CCEA June 2010)

3. The graph below shows the successive ionisation energies of an element when all its electrons are removed. (a) Name the element that gives rise to this graph. (b) Explain why the ionisation energies increase in section X. (c) Explain the large difference in ionisation energies in section Y.

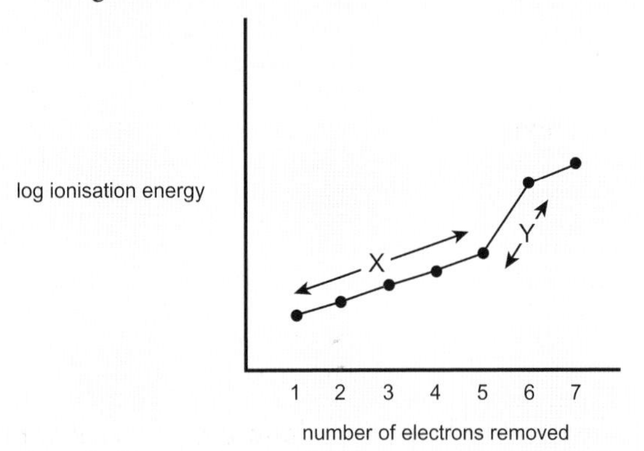

(CCEA January 2013)

4. (a) Give two reasons why potassium has a lower first ionisation energy than sodium. (b) Why is the second ionisation energy for a Group I metal much higher than the first ionisation energy?

(CCEA June 2011)

5. The first three ionisation energies of calcium are given in the table. (a) Write the equation for the second ionisation of calcium including state symbols. (b) Calculate the amount of energy, in kJ, required to form 8.0 g of $Ca^{2+}(g)$ ions from $Ca(g)$.

Ionisation Energy	first	second	third
Energy (kJ mol^{-1})	590	1145	4912

(Adapted from CCEA June 2014)

6. (a) Explain what is meant by the *second ionisation energy* of an element. (b) Write an equation, including state symbols, which represents the second ionisation energy of magnesium. (c) Give reasons why the third ionisation energy of magnesium is much larger than the second.

(Adapted from CCEA June 2012)

7. The Ca^{2+} ion has the same electron arrangement as an argon atom. (a) Write the electron arrangement for the Ca^{2+} ion. (b) The first ionisation energy of argon is 1520 kJ mol^{-1}. Explain why the third ionisation energy of calcium is much higher than the first ionisation energy of argon.

(CCEA June 2014)

8. The first six ionisation energies of an element Z are: 590, 1100, 4900, 6500, 8100 and 10500 kJ mol^{-1}. Which ion is formed when Z reacts with chlorine?

A Z^{+} B Z^{2+} C Z^{-} D Z^{2-}

(CCEA January 2004)

9. The first six ionisation energies of an element M are: 578, 1817, 2745, 11578, 14831 and 18378 kJ mol^{-1}. What is the formula of the oxide of M?

A MO B MO_2 C M_2O D M_2O_3

(Adapted from CCEA January 2006)

10. The graph shows how the **second** ionisation energy of the elements varies across a period. Which element is an alkali metal?

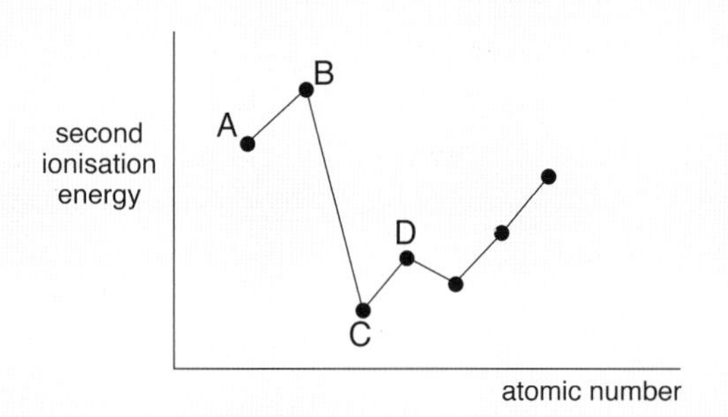

(CCEA June 2012)

Before moving to the next section, check that you are able to:

- Explain trends in IE1 within a group in terms of orbital size and the shielding of electrons.
- Explain trends in IE1 across a period in terms of the filling of subshells and the stability associated with filled and half-filled subshells.
- Define the successive ionisation energies of an atom and account for the size of successive ionisation energies in terms of the filling of subshells.

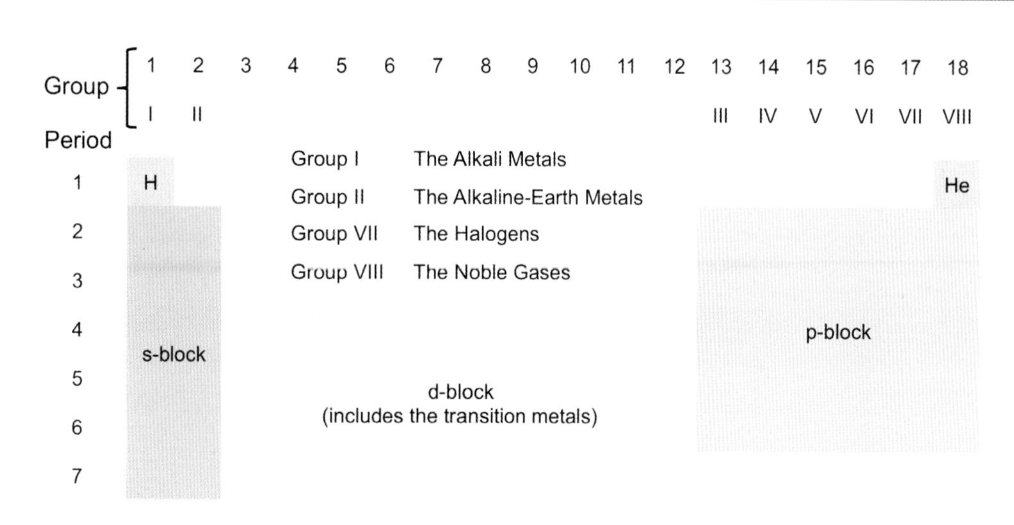

Figure 15: Organisation of the elements within the modern Periodic Table into periods, groups and blocks.

The Periodic Table

In this section we are learning to:

- Recall the organisation of elements within the Periodic Table in terms of atomic number, groups, periods and blocks.
- Determine the electron configuration of an element from its location in the Periodic Table.
- Explain trends in the melting point, first ionisation energy and atomic radius of the elements in the third period.

Organisation and Structure

In the modern Periodic Table the elements are arranged in order of increasing atomic number. Elements with similar chemical properties are arranged in columns known as **groups**, a number of which have common names and are referred to using roman numerals as shown in Figure 15. The rows of elements formed by arranging the elements in this way are known as **periods**. Many properties of the elements vary in a predictable way across periods and within groups. Trends in the properties of the elements within a period or group are referred to as **periodic trends**. Arranging the elements in this way also reveals properties that are shared by the elements in an entire region or 'block' of the Periodic Table. The scheme used to subdivide the Periodic Table into s, p, d and f blocks is shown in Figure 15.

The form of the modern Periodic Table also allows for the elements to be classified as metals and nonmetals based on their location in the Periodic Table. The distribution of metals and nonmetals within the Periodic Table is shown in Figure 16. A number of elements have been identified as **semimetals** on account of their having properties in common with both metals and nonmetals. The location of the semimetals reminds us that the transition from metallic behaviour to nonmetallic behaviour is gradual and is an example of a periodic trend. The trends in **metallic character** shown in Figure 16 remind us that the metals towards the bottom of the s-block have more metallic character, and are 'better' metals, than those in the d-block. Conversely p-block elements such as lead (Pb) and tin (Sn) are considered 'poor' metals, and the metallic character of the elements decreases to the extent that germanium (Ge) and silicon (Si) are classified as semimetals, and carbon (C) is considered a nonmetal.

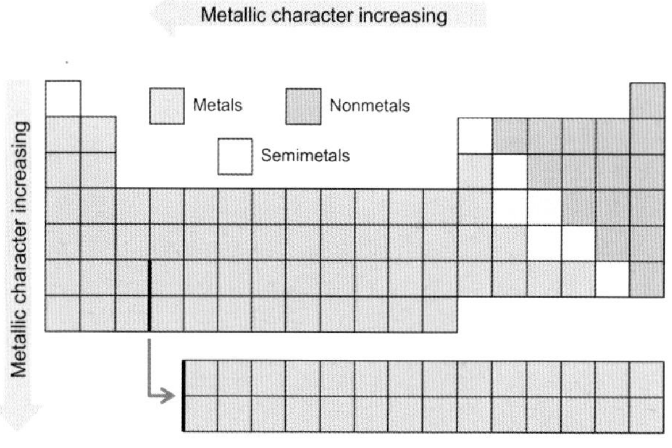

Figure 16: Periodic trends in metallic character.

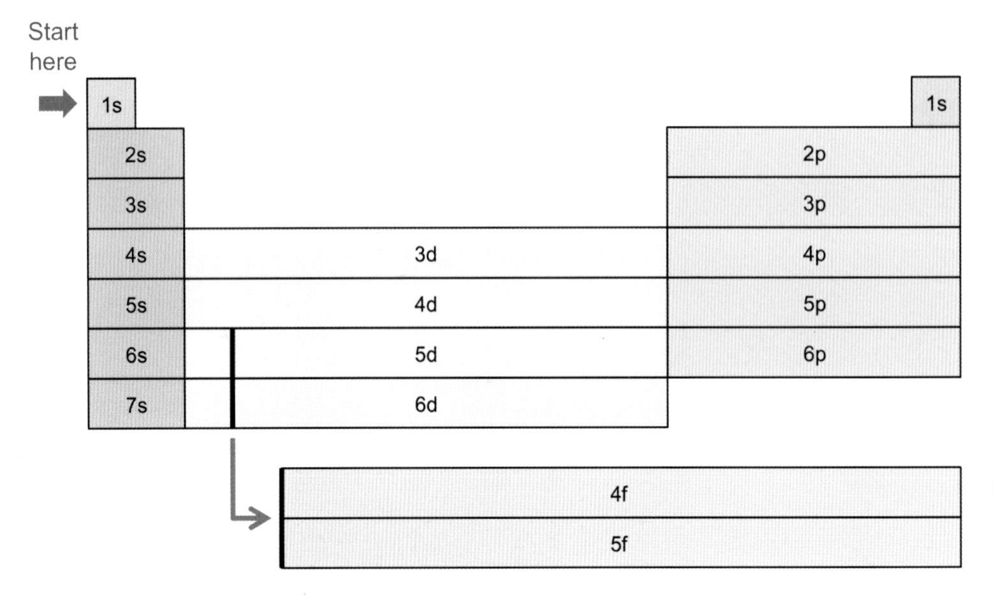

Figure 17: The subshell filling order for an atom superimposed on the Periodic Table.

Many properties of the elements are determined by their electron configuration. The electron configuration of an element can be determined from the Periodic Table by superimposing the subshell filling order on the Periodic Table as shown in Figure 17. The electron configuration for an element is obtained by filling the subshells in the order shown, adding an electron every time the atomic number increases by one, until the atomic number of the element is reached. For example, the electron configuration for an oxygen atom (atomic number = 8) is determined by first adding two electrons to the 1s subshell (H to He), followed by two electrons to the 2s subshell (Li to Be), and a further four electrons to the 2p subshell (Be to O).

In this way the 2s and 2p subshells belonging to the second shell (n=2) fill from left to right across the second period.

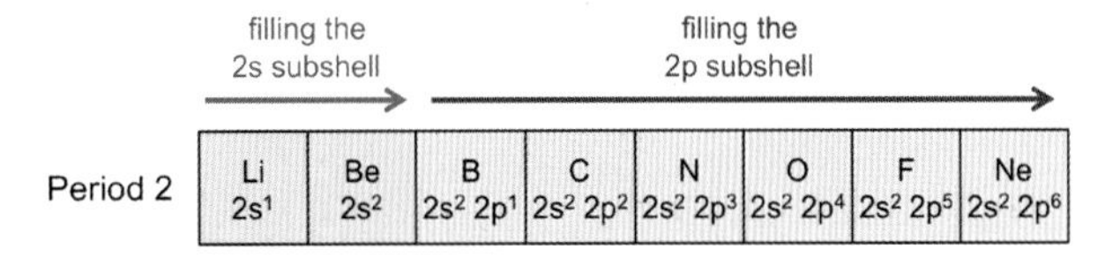

Similarly, the 3s and 3p subshells belonging to the third shell (n=3) fill from left to right across the third period, and the 4s, 3d and 4p subshells fill from left to right across the fourth period. The grouping of the 4s, 3d and 4p subshells in this way reminds us that *the outermost shell of an element in the fourth period contains the 4s, 3d and 4p subshells that together hold a total of eighteen electrons.*

The subshell filling order in Figure 17 also reveals that the outermost electrons in an s-block element are located in an s-subshell, the outermost electrons in a p-block element are located in a p-subshell, the outermost electrons in a d-block element are located in a d-subshell, and the outermost electrons in an f-block element are located in an f-subshell.

Further, as shown in Figure 18, *the arrangement of the electrons in the subshells belonging to the outermost shell is the same for each element in a group.*

Worked Example 1.2xii

Which one of the following represents the first five ionisation energies in kJ mol^{-1} of an s-block element?

	First	Second	Third	Fourth	Fifth
A	580	1800	2700	11600	14800
B	740	1500	7700	10500	13600
C	1000	2300	3400	4600	7000
D	14800	11600	2700	1800	580

(CCEA June 2009)

Strategy

- The electron configuration for an s-block element is s^1 or s^2.
- If the configuration is s^1 the second IE will be much larger than the first and the ionisation energies will then increase steadily as electrons are removed from a full p-subshell.
- If the configuration is s^2 the third IE will be much larger than the second and the ionisation energies will then increase steadily as electrons are removed from a full p-subshell.

Figure 18: The relationship between the arrangement of the outer shell electrons in s- and p-block elements and their location in the Periodic Table. Filled d-subshells have been omitted for clarity.

Group:	I	II	III	IV	V	VI	VII	VIII
Configuration:	s^1	s^2	s^2p^1	s^2p^2	s^2p^3	s^2p^4	s^2p^5	s^2p^6
	H $1s^1$							He $1s^2$
	Li $2s^1$	Be $2s^2$	B $2s^22p^1$	C $2s^22p^2$	N $2s^22p^3$	O $2s^22p^4$	F $2s^22p^5$	Ne $2s^22p^6$
	Na $3s^1$	Mg $3s^2$	Al $3s^23p^1$	Si $3s^23p^2$	P $3s^23p^3$	S $3s^23p^4$	Cl $3s^23p^5$	Ar $3s^23p^6$
	K $4s^1$	Ca $4s^2$	Ga $4s^24p^1$	Ge $4s^24p^2$	As $4s^24p^3$	Se $4s^24p^4$	Br $4s^24p^5$	Kr $4s^24p^6$
	Rb $5s^1$	Sr $5s^2$	In $5s^25p^1$	Sn $5s^25p^2$	Sb $5s^25p^3$	Te $5s^25p^4$	I $5s^25p^5$	Xe $5s^25p^6$
	Cs $6s^1$	Ba $6s^2$	Tl $6s^26p^1$	Pb $6s^26p^2$	Bi $6s^26p^3$	Po $6s^26p^4$	At $6s^26p^5$	Rn $6s^26p^6$

Solution

Answer B is consistent with an s^2 configuration.

Exercise 1.2Q

1. In which block of the Periodic Table is argon found? Explain your answer.

   (CCEA June 2013)

2. Which block in the Periodic Table contains silver?

   A d-block B f-block C p-block D s-block

   (CCEA June 2012)

3. Elements Q and R have ground state electron structures $1s^22s^22p^63s^2$ and $1s^22s^22p^5$ respectively. Write the formula of the compound formed when Q and R combine.

   (Adapted from CCEA January 2010)

4. (a) What property is used to order the elements in the Periodic Table? (b) Explain why transition metals are classified as d-block elements.

   (CCEA June 2009)

5. (a) Explain why all of the Group I elements are described as being s-block elements. (b) Explain why the ions of the Group I elements get bigger down the group.

   (Adapted from CCEA June 2011)

6. Use the following electron configurations to identify which block of the Periodic Table contains the elements X, Y and Z.

   X $1s^2 2s^2 2p^6 3s^2 3p^6 3d^3 4s^2$

   Y^{2-} $1s^2 2s^2 2p^6 3s^2 3p^6$

   Z^+ $1s^2 2s^2 2p^6 3s^2 3p^6$

   (Adapted from CCEA June 2008)

7. The successive ionisation energies of an element are given below in units of kJ mol^{-1}. Which group does the element belong to?

   590 1145 4912 6474 8144 10496 12320

   A Group I B Group II

   C Group III D Group IV

   (Adapted from CCEA January 2014)

8. The successive ionisation energies of an element in Period 2 are given in units of kJ mol^{-1}. Identify the element.

   1090 2350 4610 6220 37800 47000

   A Carbon B Fluorine C Nitrogen D Oxygen

   (Adapted from CCEA June 2015)

9. Which one of the following lists the first four ionisation energies of a Group II element?

   A 584, 1823, 2751, 11584

   B 744, 1457, 7739, 10547

   C 793, 1583, 3238, 4362

   D 1018, 1909, 2918, 4963

   (CCEA January 2013)

10. The successive ionisation energies (in kJ mol^{-1}) for elements X and Y are shown below. Which one of the following is the formula for a compound of X and Y?

	1st	2nd	3rd	4th	5th	6th	7th	8th
X	578	1817	2745	11577	14842	18379	23326	27465
Y	1314	3388	5301	7469	10990	13327	71330	84078

A XY_2 B X_2Y C X_2Y_3 D X_3Y_2

(Adapted from CCEA June 2013)

11. (a) Using outer electrons only draw diagrams to explain the formation of caesium chloride from caesium atoms and chlorine atoms. (b) Use spd-notation to write the electron configuration for the caesium and chloride ions in caesium chloride.

(CCEA June 2011)

12. An element X has 3 electrons in its outermost shell. Element Y has 6 electrons in its outermost shell. What is the empirical formula of the compound formed by X and Y?

(CCEA June 2008)

13. The element europium reacts with hydrogen to form europium hydride. Atoms of europium have their outer electrons in levels 5 and 6 i.e. $5s^2$ $5p^6$ $6s^2$. What is the formula of europium hydride?

(Adapted from CCEA June 2011)

Before moving to the next section, check that you are able to:

- Recall that the elements are organised in order of increasing atomic number within the Periodic Table.
- Describe the organisation of elements within the Periodic Table in terms of groups, periods and blocks.
- Determine the electron configuration of an atom from its location in the Periodic Table.
- Determine the location of an element in the Periodic Table from the arrangement of the electrons in its outer shell.

1.3 Chemical Bonding and Structure

CONNECTIONS

- Diamond and graphite are allotropes of carbon. Diamond is one of the hardest materials known but graphite is a soft brittle solid used in pencils.
- The formation of strong bonds between water molecules in ice can be used to explain why ice floats in water.

Ionic Compounds

In this section we are learning to:

- Explain the characteristic properties of ionic compounds in terms of their structure and the formation of ionic bonds in the compound.
- Use dot-and-cross diagrams to explain the formation of ionic bonds.

An ionic compound is formed when a metal combines with one or more nonmetals. Ionic compounds contain an ordered array of tightly packed ions known as an **ionic lattice**. Many ionic compounds form crystals with well-defined edges and shapes that reflect the order within the lattice. The ionic lattice in sodium chloride is illustrated in Figure 1.

In sodium chloride, NaCl the lattice consists of sodium ions (Na^+) surrounded by chloride ions (Cl^-) and vice versa. The ions are formed when sodium metal reacts with chlorine gas. During the reaction each sodium atom in the metal loses one electron to form a sodium ion and each chlorine atom in the gas gains one electron to form a chloride ion as shown in Figure 2. The ions in the lattice are held together by

Figure 1: The ionic lattice in sodium chloride.

ionic bonds that result from *strong electrostatic attraction between neighbouring positive and negative ions in the lattice*. Compounds held together by ionic bonds are referred to as ionic compounds.

A great deal of energy is needed to break the strong ionic bonds between ions in the lattice. As a result ionic compounds have high melting and boiling points. The melting point of sodium chloride is 801 °C.

The strength of the ionic bonds between neighbouring ions, and a lack of space due to the tight-packing of ions, prevents the ions from moving within the lattice. As a result there are no charged particles able to move within the lattice and the solid is unable to conduct electricity. The inability of the ions to move past each other in the lattice also explains why many ionic solids are hard materials.

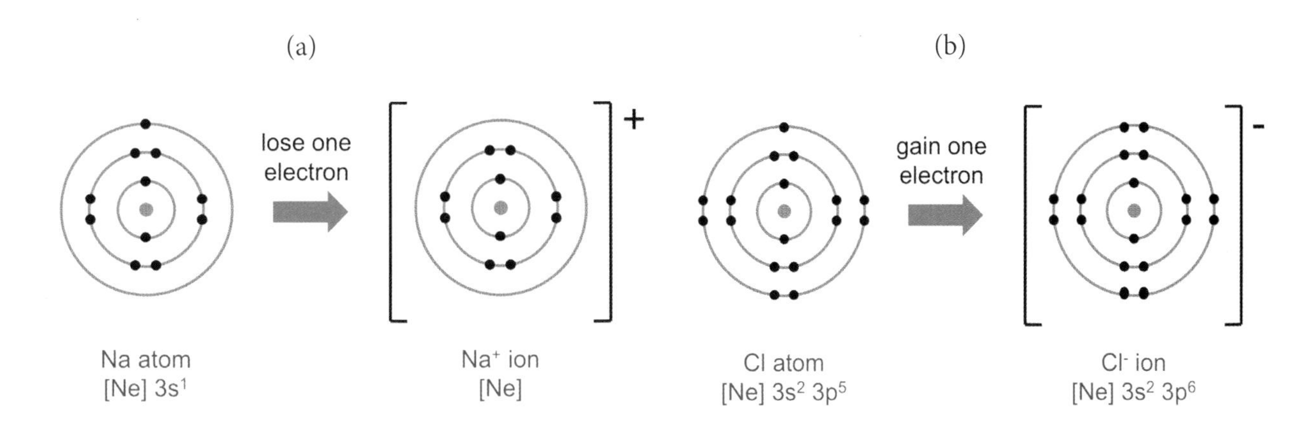

Figure 2: The transfer of one electron from sodium to chlorine to form (a) a sodium (Na^+) ion and (b) a chloride (Cl^-) ion.

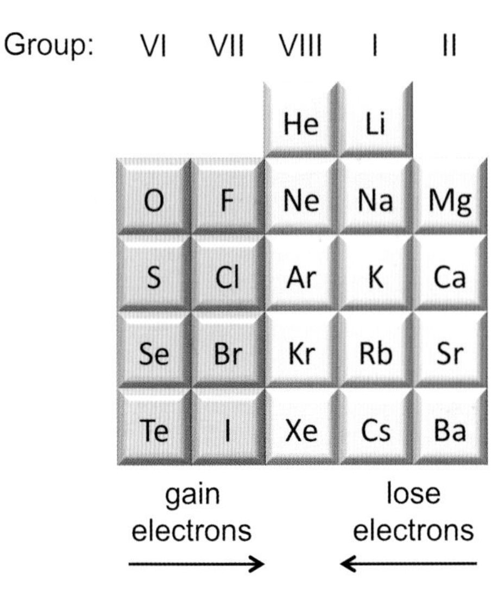

Figure 3: Achieving a noble gas configuration by losing or gaining electrons.

The ions in an ionic compound have the same electron configuration as a noble gas. As a result, when forming an ionic compound, *the metals towards the left side of the Periodic Table will lose electrons to achieve a noble gas configuration and the nonmetals towards the right side of the Periodic Table will gain electrons to achieve a noble gas configuration* as illustrated in Figure 3.

We can use the idea that metals react with nonmetals to form ions with full outer shells to determine the formula of an ionic compound. For example, when magnesium (Group II) combines with chlorine (Group VII) to form magnesium chloride, magnesium achieves a noble gas configuration by losing two electrons to form a magnesium (Mg^{2+}) ion. Similarly, chlorine (Group VII) achieves a noble gas configuration by gaining one electron to form a chloride (Cl^-) ion. The formation of ions in an ionic compound can be illustrated by drawing a 'dot-and-cross' diagram in which dots and crosses are used to distinguish outer shell electrons belonging to different atoms. The dot-and-cross diagram in Figure 4 describes the formation of ions when magnesium chloride is formed from atoms of magnesium and chlorine.

An ionic compound such as magnesium chloride does not have a charge as the charge on the positive ions in the lattice balances the charge on the negative ions in the lattice. As a result, the formula for magnesium chloride must be $MgCl_2$ as it represents an ionic lattice in which there are two chloride (Cl^-) ions for every magnesium ion (Mg^{2+}).

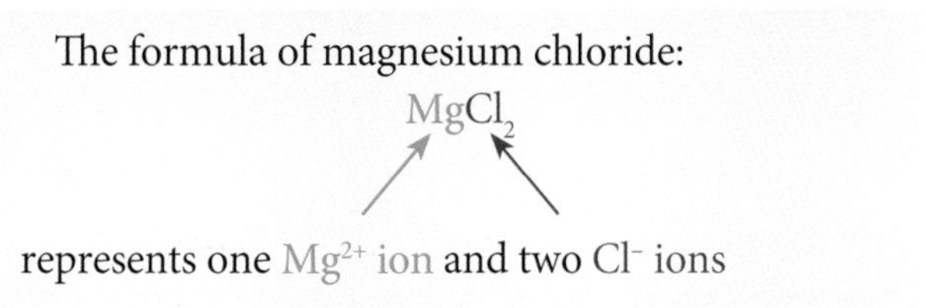

Worked Example 1.3i

Use dot-and-cross diagrams to explain how strontium atoms combine with fluorine atoms to form strontium fluoride. Show only outer shell electrons.

(CCEA June 2010)

Strategy

The diagram must clearly show:

- A strontium atom achieving a full outer shell by donating two electrons to form a strontium (Sr^{2+}) ion.
- Two fluorine atoms, each gaining one of the electrons donated by the strontium atom to form a fluoride (F^-) ion.

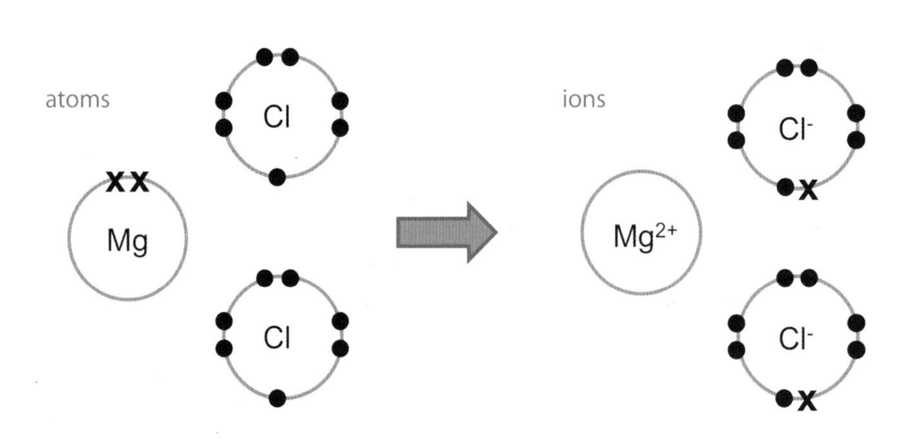

Figure 4: A dot-and-cross diagram showing the formation of ions when magnesium and chlorine combine to form magnesium chloride, $MgCl_2$. Only outer shell electrons are shown.

Solution

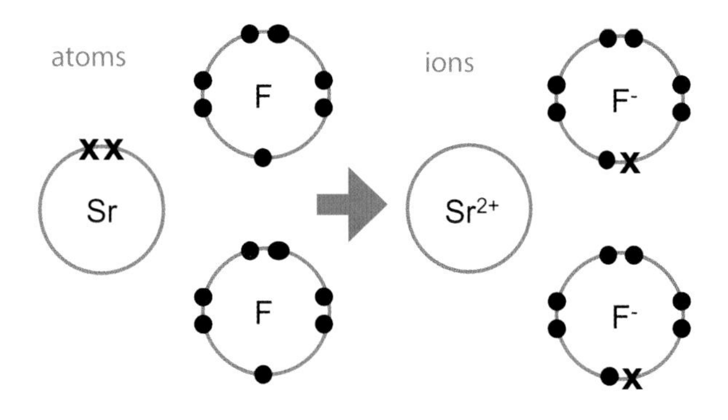

Exercise 1.3A

1. (a) Write an equation, including state symbols, for the formation of sodium fluoride from sodium and fluorine. (b) Draw dot-and-cross diagrams, using outer electrons only, to show the formation of sodium fluoride from sodium and fluorine atoms.

   (CCEA January 2008)

2. (a) Using outer electrons only draw diagrams to explain the formation of calcium bromide from calcium atoms and bromine atoms. (b) Use spd-notation to write the electron configuration for the calcium ions and bromide ions in calcium bromide.

3. Magnesium oxide is formed by the combustion of magnesium metal in oxygen. (a) Draw a dot-and-cross diagram, using outer shell electrons only, to show how magnesium oxide is formed from a magnesium atom and an oxygen atom. (b) State the type of bonding in magnesium oxide. (c) State **two** physical properties of magnesium oxide.

   (CCEA January 2013)

4. Sodium reacts with chlorine gas to form sodium chloride. (a) Draw dot-and-cross diagrams to show how sodium bonds with chlorine gas. Only outer shell electrons should be shown. (b) Name the type of bonding in sodium chloride. (c) The structure of sodium chloride is described as a lattice. Explain what is meant by the term *lattice*. (d) Apart from its appearance give **three** physical properties of sodium chloride.

   (CCEA June 2014)

5. Ammonium phosphate is an ionic compound consisting of ammonium and phosphate, PO_4^{3-} ions. (a) Write the formula for ammonium phosphate. (b) State and explain three physical properties you would expect ammonium phosphate to have.

   (CCEA January 2014)

Before moving to the next section, check that you are able to:

- Describe the ionic lattice and the nature of the bonding in an ionic compound.
- Explain the high melting point and lack of conductivity of ionic compounds in terms of the formation of strong ionic bonds.
- Use dot-and-cross diagrams to explain the formation of ions when an ionic compound is formed from its elements

Covalent Bonding

In this section we are learning to:

- Describe the formation of covalent bonds in molecules in terms of the sharing of electrons between atoms and the octet rule.
- Use structural formulas to represent covalent bonding in molecules and ions.
- Describe the formation and properties of coordinate bonds.
- Identify and explain exceptions to the octet rule.
- Describe the formation of bond dipoles in terms of the sharing of electrons between atoms with different electronegativity.
- Construct the permanent dipole for a molecule from bond dipoles and relate the size of the permanent dipole to the shape of the molecule.

The Octet Rule

As in ionic compounds, the atoms in nonmetal compounds such as water (H_2O) and carbon dioxide (CO_2) bond with each other in an attempt to achieve a full outer shell of electrons. The **octet rule** asserts that *atoms will attempt to gain, lose or share electrons when forming compounds in order to achieve a full outer shell containing eight electrons.* In a molecule of chlorine (Cl_2) two chlorine atoms achieve a full outer shell by sharing a pair of electrons as illustrated in Figure 5.

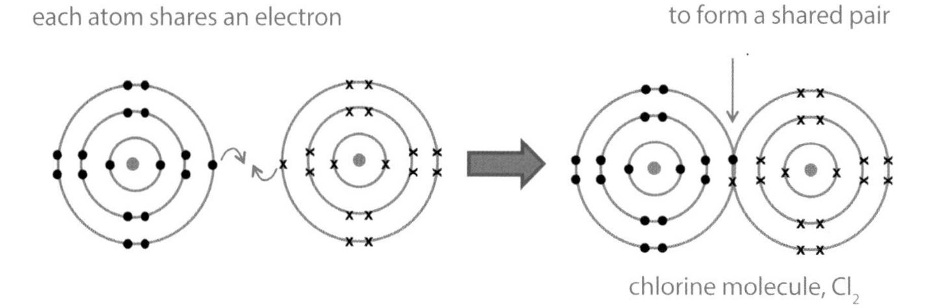

Figure 5: The formation of a covalent bond between two chlorine atoms to form a chlorine molecule, Cl_2.

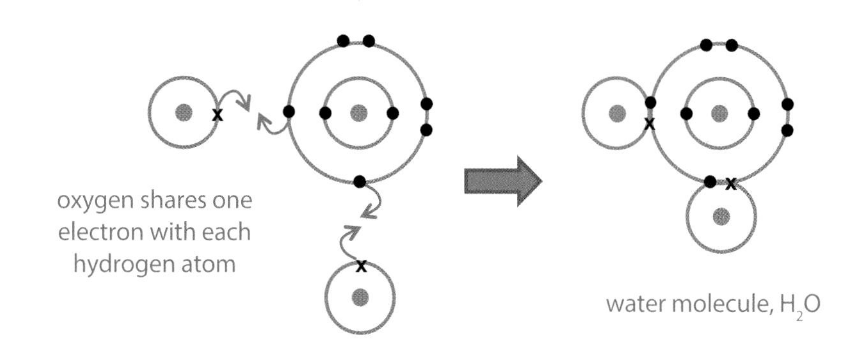

Figure 6: Covalent bonding in a water molecule, H_2O.

The shared electrons form a **covalent bond** between the chlorine atoms as *the negatively charged electrons in the shared pair are attracted to the positively charged nucleus in each atom*, effectively 'gluing' the atoms together.

The atoms in water are also held together by covalent bonds. A molecule of water (H_2O) contains two hydrogen atoms and one oxygen atom. Each hydrogen atom shares a pair of electrons with the oxygen atom to form a covalent bond as illustrated in Figure 6. The molecule is stable as all three atoms in the molecule have satisfied the Octet Rule by attaining a full outer shell.

Worked Example 1.3ii

(a) Write an equation for the formation of phosphorus trifluoride, PF_3 by the reaction of fluorine with phosphorus, P_4. (b) Draw a dot-and-cross diagram, using outer electrons only, to show the bonding in PF_3. (c) With reference to PF_3 explain the octet rule.

(CCEA June 2007)

Solution

(a) $P_4 + 6F_2 \rightarrow 4PF_3$

(b)

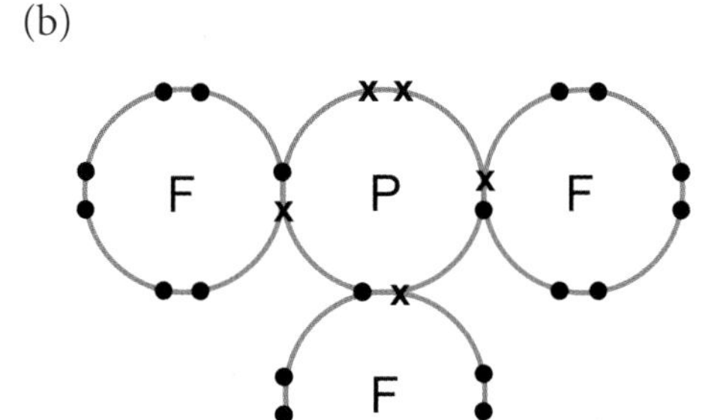

(c) All four atoms in PF_3 have satisfied the octet rule by sharing electrons to obtain a full outer shell containing eight electrons.

Exercise 1.3B

1. (a) Explain what is meant by the term *covalent bond*. (b) Explain what is meant by the term *octet rule*.

   (CCEA June 2013)

2. Phosphorus, P_4 reacts explosively with liquid bromine to form phosphorus tribromide. (a) Write an equation for the reaction. (b) Draw a dot-and-cross diagram, using outer electrons

only, to show the bonding in phosphorus tribromide. (c) State the *octet rule* and explain whether or not phosphorus obeys the octet rule in phosphorus tribromide.

(Adapted from CCEA June 2014)

3. The elements X and Y are in Groups VI and VII of the Periodic Table. Which one of the following describes the formula and bonding in the compound formed by X and Y?

	Formula	Bonding
A	XY_2	covalent
B	XY_2	ionic
C	X_2Y	covalent
D	X_2Y	ionic

(Adapted from CCEA January 2012)

Multiple Bonding

When two oxygen atoms combine to form a molecule of oxygen (O_2), the oxygen atoms achieve a full outer shell by using two electrons to create two shared pairs of electrons as illustrated in Figure 7. Each shared pair of electrons results in a covalent bond between the oxygen atoms. The pair of covalent bonds between the oxygen atoms is referred to as a **double bond**. Similarly, a **triple bond** is formed when two atoms each use three electrons to create three shared pairs of electrons. Molecules of nitrogen (N_2) and acetylene (C_2H_2) contain triple bonds.

The formation of two or more covalent bonds between a pair of atoms is known as **multiple bonding**. The double bond in a molecule of oxygen (O_2) and the triple bond in a nitrogen molecule (N_2) are examples of multiple bonds. Most covalent bonds are strong and a lot of energy is needed to break the bond. The additional shared pairs in a double or triple bond make the bond even stronger and more difficult to break than a single covalent bond.

Exercise 1.3C

1. Draw dot-and-cross diagrams, using outer electrons only, to show the formation of a carbon dioxide molecule from a carbon atom and an oxygen molecule.

(CCEA January 2012)

2. Which one of the following molecules has two double bonds?

A C_2H_4 B N_2 C CO_2 D $BeCl_2$

(Adapted from CCEA June 2009)

Structural Formula

The **structural formula** of a molecule shows how the atoms are connected as a result of forming covalent bonds between atoms. A line between two atoms represents a single covalent bond formed by sharing a pair of electrons. Unshared or 'lone' pairs of electrons in the outer shell of an atom are represented by a pair of dots.

Molecular formula	Structural formula
Cl_2	:C̈l̤—C̈l̤:
H_2O	H—Ö̤—H
O_2	:Ö=Ö:
N_2	:N≡N:

Exercise 1.3D

1. Draw structural formulas for ethane, C_2H_6 ethene, C_2H_4 and ethyne, C_2H_2.

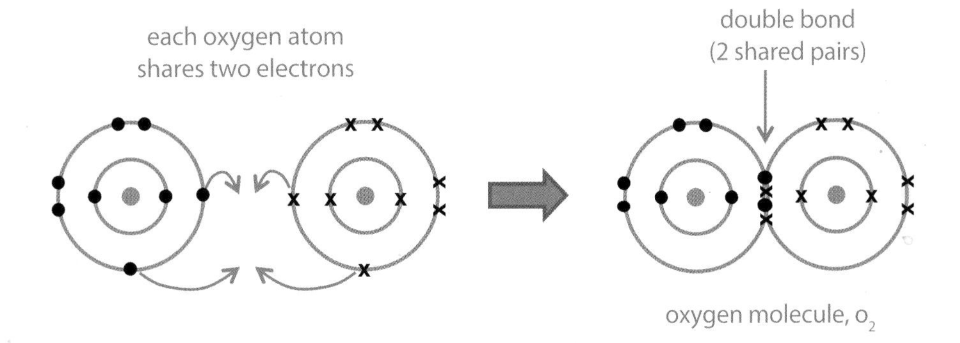

The Figure 7: The formation of a double bond (2 shared pairs) between two oxygen atoms to form a molecule of oxygen, O_2.

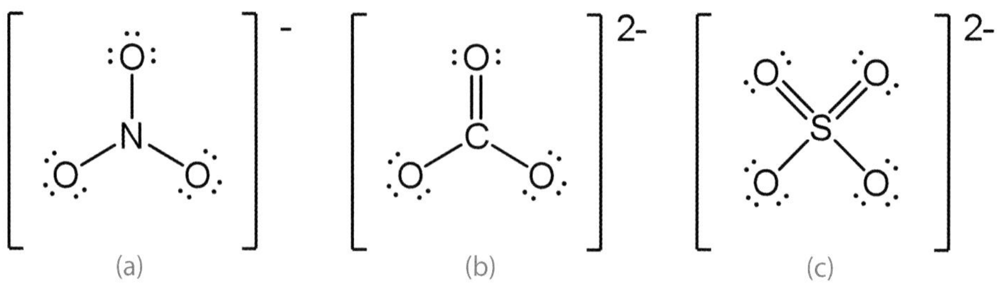

(a) (b) (c)

Figure 8: Structural formulas for (a) nitrate ion, NO_3^- (b) carbonate ion, CO_3^{2-} and (c) sulfate ion, SO_4^{2-}.

Molecular Ions

The structural formulas for nitrate ion (NO_3^-), carbonate ion (CO_3^{2-}) and sulfate ion (SO_4^{2-}) are shown in Figure 8. All three ions are **molecular ions** as they contain two or more atoms held together by covalent bonds. In compounds containing molecular ions such as calcium carbonate, $CaCO_3$ the ions in the lattice are tightly packed and are held together by attractive forces between the oppositely charged Ca^{2+} and CO_3^{2-} ions. As a result, compounds containing molecular ions have the same properties as other ionic compounds such as sodium chloride, NaCl.

Further, the strong covalent bonding within a molecular ion such as nitrate, NO_3^- or sulfate, SO_4^{2-} does not break when a salt containing the molecular ion melts or dissolves.

Melting magnesium sulfate:

$MgSO_{4(s)} \rightarrow Mg^{2+}{}_{(l)} + SO_4^{2-}{}_{(l)}$

Dissolving magnesium sulfate:

$MgSO_{4(s)} \rightarrow Mg^{2+}{}_{(aq)} + SO_4^{2-}{}_{(aq)}$

Coordinate Bonding

Most covalent bonds are formed when two atoms each use one electron to form a shared pair of electrons. For example, in a molecule of ammonia, NH_3 each hydrogen atom uses its electron to form a covalent bond with the nitrogen atom as shown in Figure 9.

Alternatively, two atoms may form a **coordinate bond** by *sharing a pair of electrons that has been donated by one of the atoms forming the bond.* Ammonium ion, NH_4^+ contains a coordinate bond. The coordinate bond in an ammonium ion is formed when the nitrogen atom in a molecule of ammonia shares the pair of electrons in its outer shell with a hydrogen ion (H^+) as shown in Figure 10. The coordinate bond in an ammonium ion is identical to the other N-H bonds in the ion; the only difference between the coordinate bond and the other covalent bonds in the ion is the way in which the coordinate bond was formed.

Formation of a coordinate bond in a molecule or ion is indicated by drawing an arrow between the atoms forming the bond. The direction of the arrow indicates the direction of electron donation in the bond. The use of an arrow to indicate the formation

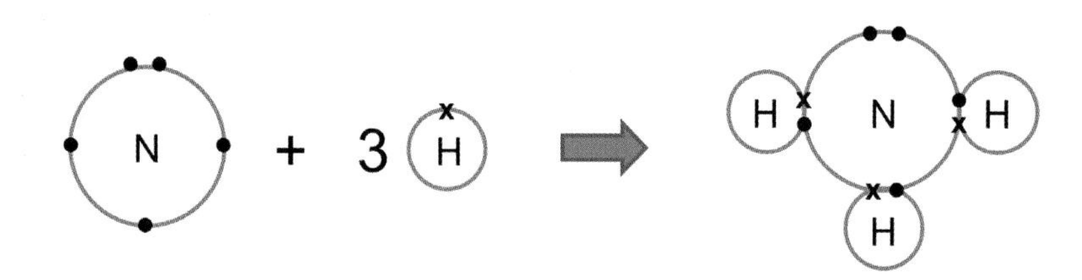

Figure 9: Covalent bonding in a molecule of ammonia, NH_3. Only outer shell electrons are shown.

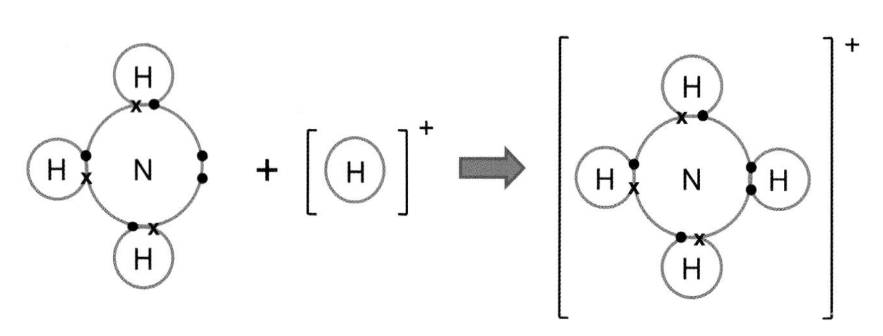

Figure 10: Formation of a coordinate bond in an ammonium ion, NH_4^+.

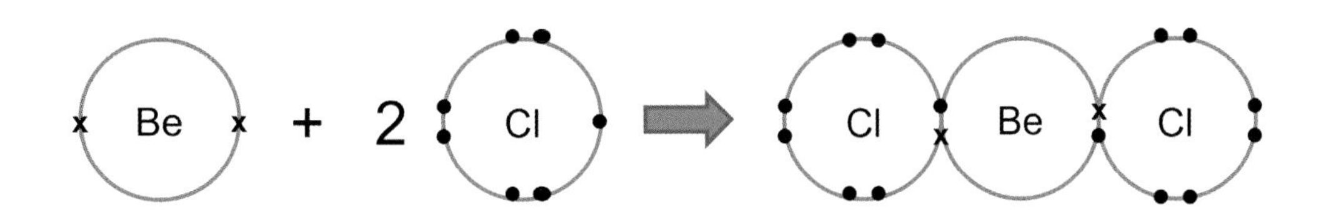

Figure 12: Covalent bonding in a molecule of beryllium chloride, $BeCl_2$.

of a coordinate bond in an ammonium ion, NH_4^+ is shown in Figure 11.

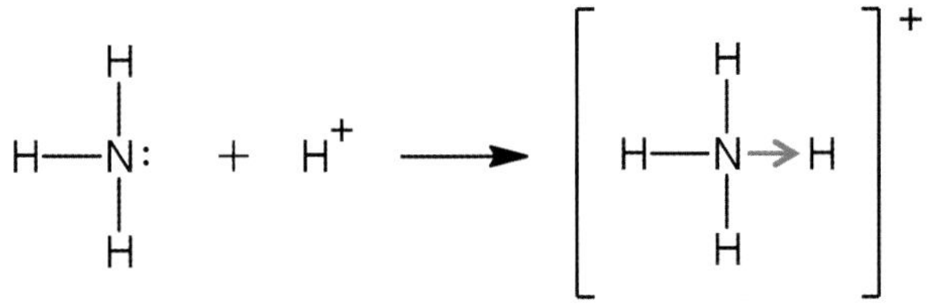

Figure 11: Using an arrow to indicate the formation of a coordinate bond in an ammonium ion, NH_4^+.

Exercise 1.3E

1. Boron trifluoride can combine with ammonia to form the following molecule. (a) Name the type of bond formed between the boron and nitrogen atoms. (b) Explain how this bond is formed.

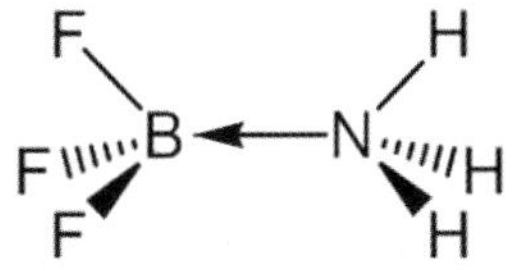

(CCEA June 2010)

2. Which one of the following substances has coordinate bonding in its structure?

   A Ammonia

   B Ammonium chloride

   C Carbon dioxide

   D Water

(CCEA January 2013)

Exceptions to the Octet Rule

As a rule, atoms will try to satisfy the octet rule when they combine to form compounds. However, in some cases a stable compound can still form when there are not enough electrons to form an octet of (eight) electrons around one or more of the atoms. For example, the dot-and-cross diagram for beryllium chloride, $BeCl_2$ (Figure 12) reveals that while each chlorine atom has satisfied the octet rule beryllium has not achieved a full outer shell and has not satisfied the octet rule. *Molecules in which one or more atoms have not satisfied the octet rule are described as exceptions to the octet rule.*

Boron trifluoride, BF_3 is also an exception to the octet rule. The dot-and-cross diagram for boron trifluoride (Figure 13) reveals that while each fluorine atom has attained a full outer shell and satisfied the octet rule, the boron atom has only six electrons in its outer shell and has not satisfied the octet rule.

It is also possible to form a stable compound that contains atoms with more than eight electrons in their

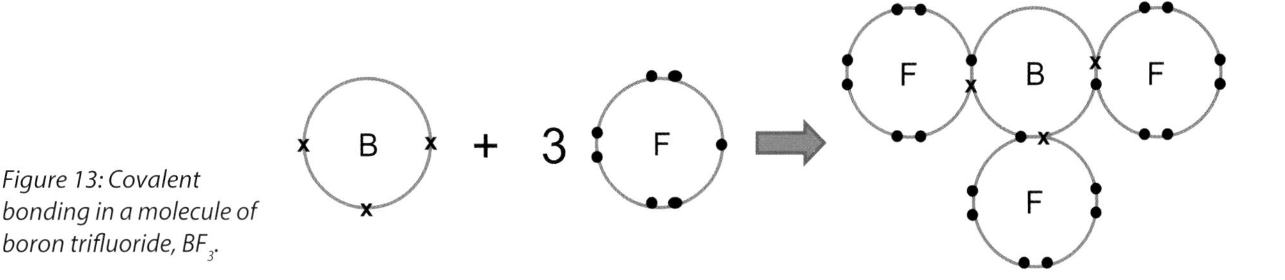

Figure 13: Covalent bonding in a molecule of boron trifluoride, BF_3.

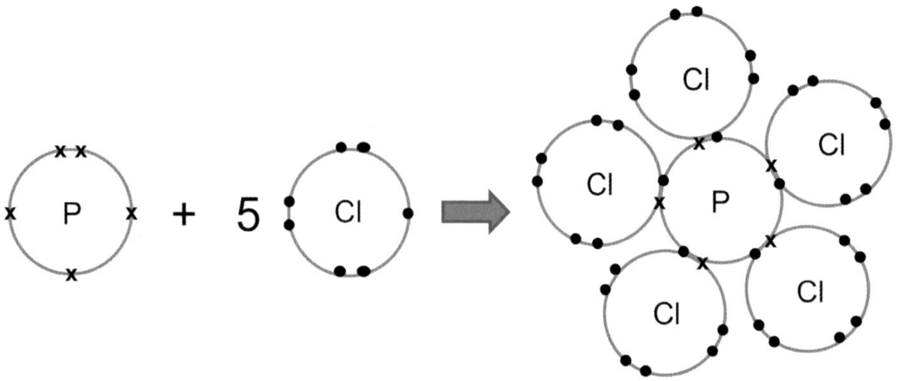

Figure 14: Covalent bonding in a molecule of phosphorus pentachloride, PCl_5.

outer shell. In the case of PCl_5 (Figure 14), each chlorine atom achieves a full outer shell by sharing an electron pair with phosphorus. As a result, phosphorus has ten electrons in its outer shell. *An outer shell that contains more than eight electrons is known as an expanded octet. Atoms with an expanded octet do not obey the octet rule and are exceptions to the octet rule.*

The sulfur atom in sulfur hexafluoride, SF_6 is another example of an atom with an expanded octet. In SF_6 each fluorine atom forms a single covalent bond with the sulfur atom. As a result, the outer shell of the sulfur atom contains twelve electrons; two from each of the six S-F bonds.

Exercise 1.3F

1. Beryllium chloride, $BeCl_2$ has a melting point of 400 °C. The high melting point of beryllium chloride is explained by the formation of a polymeric structure. Two units of the polymeric structure are shown below. (a) Explain whether beryllium, in the polymeric structure, obeys the octet rule. (b) Explain whether chlorine, in the polymeric structure, obeys the octet rule.

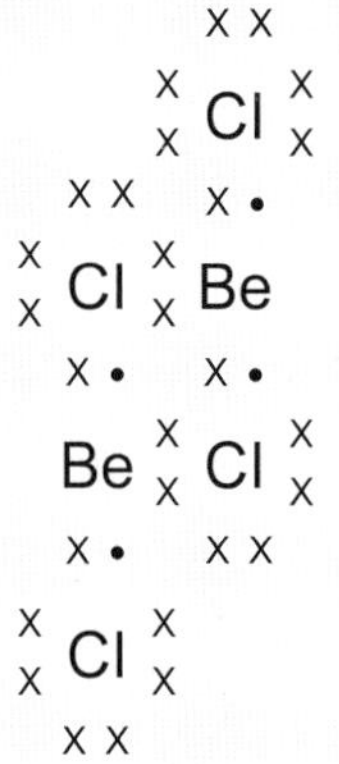

   (Adapted from CCEA June 2015)

2. Fluorine reacts with boron to form boron trifluoride. (a) Draw a dot-and-cross diagram, using outer shell electrons only, to show the bonding in boron trifluoride. (b) State the *octet rule*. (c) Explain whether or not the elements in boron trifluoride obey the octet rule.

   (CCEA January 2013)

3. In which of the following molecules does the underlined element (a) form two covalent bonds and (b) satisfy the octet rule?

   $\underline{Be}Cl_2$ $\quad \underline{C}H_4$ $\quad \underline{N}H_3$ $\quad \underline{O}H_2$

   (Adapted from CCEA June 2009)

4. Sulfur forms the fluorides SF_2 and SF_4. (a) Draw a dot-and-cross diagram for each compound, showing only outer electrons. (b) State the *octet rule*. (c) Explain whether sulfur is obeying the octet rule in each fluoride.

   (Adapted from CCEA January 2014)

5. (a) Draw a dot-and-cross diagram, using outer shell electrons only, to show the bonding in SF_6. (b) Explain whether SF_6 obeys the octet rule.

   (CCEA June 2010)

Before moving to the next section, check that you are able to:

- Explain the formation of covalent bonds in molecules and ions in terms of the sharing of electrons between atoms and the octet rule.
- Draw structural formulas to represent covalent bonding in molecules and ions.
- Describe the formation and properties of coordinate bonds.
- Identify and explain exceptions to the octet rule.

Polar Covalent Bonds

In molecules such as chlorine (Cl_2) covalent bonds are formed by the sharing of electrons between identical atoms. The electrons are shared equally by the atoms and the bond is described as a **nonpolar bond**. In contrast, the electrons in a covalent bond between atoms of different elements are not shared equally. For example, in a molecule of hydrogen chloride (HCl), the chlorine atom is more able to attract the shared electrons in the covalent bond. As a result, the chlorine atom acquires a small negative charge (δ-) and the hydrogen atom becomes slightly positive (δ+). The covalent bond in HCl is an example of a **polar bond** where we are using the term 'polar' to describe *a covalent bond in which the electrons forming the bond are not shared equally by the atoms*. The use of partial charges (δ+ and δ-) to describe the unequal sharing of electrons in polar covalent bonds is illustrated in Figure 15.

δ+ δ-
H—Cl:

δ+ δ- δ+
H—O—H

Figure 15: The use of partial charges to describe the unequal sharing of electrons in polar covalent bonds.

The **electronegativity** of an element can be used to quantify *the extent to which an atom of the element attracts the shared electrons in a covalent bond to itself when forming covalent bonds with other atoms.* In a polar covalent bond the more electronegative element acquires a small negative charge (δ–) and the less electronegative element acquires a small positive charge (δ+). The separation of charge (δ+ δ–) that results from the unequal sharing of electrons in a bond is known as a **bond dipole**. *The size of the dipole associated with a polar covalent bond becomes larger as the difference between the electronegativity values of the atoms forming the bond increases.*

The periodic trends in Figure 16 reveal that the electronegativity of the elements increases from left to right across a period and towards the top of each group. As a result metals on the far left of the Periodic Table have the lowest electronegativity values and nonmetals have the highest electronegativity values. This suggests that big differences in electronegativity between the metals and nonmetals in compounds such as NaCl and $MgCl_2$ give rise to ionic bonding, while smaller differences in compounds such as H_2O and CO_2 result in polar covalent bonding. In this way polar covalent bonding can be viewed as the link between ionic bonding and nonpolar covalent bonding.

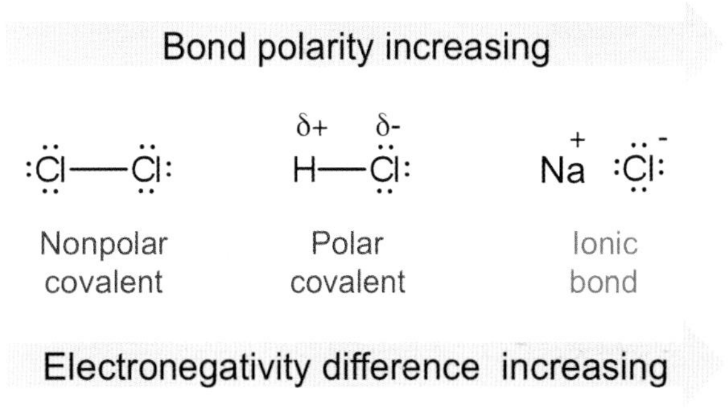

Bond polarity increasing

:Cl—Cl:	δ+ δ- H—Cl:	Na^+ $:Cl:^-$
Nonpolar covalent	Polar covalent	Ionic bond

Electronegativity difference increasing

Electronegativity increasing →

Electronegativity increasing ↑

H 2.1							He
Li 1.0	Be 1.5	B 2.0	C 2.5	N 3.0	O 3.5	F 4.0	Ne
Na 0.9	Mg 1.2	Al 1.5	Si 1.8	P 2.1	S 2.5	Cl 3.0	Ar
K 0.8	Ca 1.0	Ga 1.6	Ge 1.8	As 2.0	Se 2.4	Br 2.8	Kr
Rb 0.8	Sr 1.0	In 1.7	Sn 1.8	Sb 1.9	Te 2.1	I 2.5	Xe
Cs 0.7	Ba 0.9	Tl 1.8	Pb 1.9	Bi 1.9	Po 2.0	At 2.1	Rn

Figure 16: Periodic trends in electronegativity.

Exercise 1.3G

1. Francium is one of the least electronegative elements in the Periodic Table. (a) Explain the meaning of the term *electronegativity*. (b) State how electronegativity values change on going across a period.

   (CCEA January 2014)

2. The electronegativity values, not in order, for caesium, cobalt, fluorine and nitrogen are listed below. Which one is the value for the cobalt?

   A 0.70 B 1.80 C 3.00 D 4.00

   (CCEA January 2014)

3. (a) Explain the meaning of the term *electronegativity*. (b) Using electronegativity suggest why beryllium chloride is a covalent molecule and barium chloride is an ionic compound.

   (CCEA January 2011)

4. Which one of the following molecules does not contain a polar bond?

   A Fluorine

   B Hydrogen fluoride

   C Oxygen difluoride (OF_2)

   D Tetrafluoromethane (CF_4)

   (CCEA June 2014)

5. Which one of the following molecules contains the most polar bond?

   CH_4 NH_3 H_2O HF

   (CCEA January 2009)

6. Use the electronegativity values in Figure 16 to determine which one of the following bonds is the most polar.

Element:	B	C	N	O	F	I
Electronegativity:	2.0	2.5	3.0	3.5	4.0	2.5

   A B-F B N-F C C-I D O-I

   (Adapted from CCEA January 2012)

Permanent Dipoles

The presence of one or more polar covalent bonds in a molecule may result in the molecule having a positive end and a negative end. In the case of chloromethane, CH_3Cl the C-Cl bond dipole results in the chlorine end of the molecule having a small negative charge

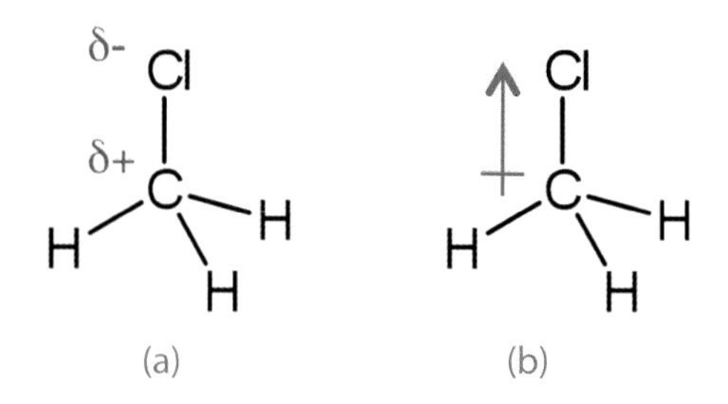

Figure 17: The use of (a) partial charges and (b) a dipole arrow to indicate the nature of the C-Cl bond dipole in chloromethane, CH_3Cl.

(δ-) and the methyl (CH_3) end of the molecule having a small positive charge (δ+). The direction of a bond dipole is indicated by a 'dipole arrow' which, by convention, points from the positive end of the dipole to the negative end of the dipole. The use of a dipole arrow to describe the C-Cl bond dipole in chloromethane, CH_3Cl is shown in Figure 17.

The separation of charge that gives rise to positive and negative ends in a molecule such as chloromethane is described as a **permanent dipole**. Molecules with a permanent dipole are referred to as **polar molecules**. The use of dipole arrows to describe individual bond dipoles makes it easier to visualise how the individual bond dipoles combine to produce a permanent dipole in the molecule. In the case of chloromethane (CH_3Cl) the C-H bond dipoles are very small and the permanent dipole for chloromethane is almost entirely due to the C-Cl bond dipole. In contrast, in water (H_2O) and ammonia (NH_3) several bond dipoles combine to produce a much larger permanent dipole. The process of combining bond dipoles to produce a molecular dipole is illustrated in Figure 18. The examples in Figure 18 demonstrate that the

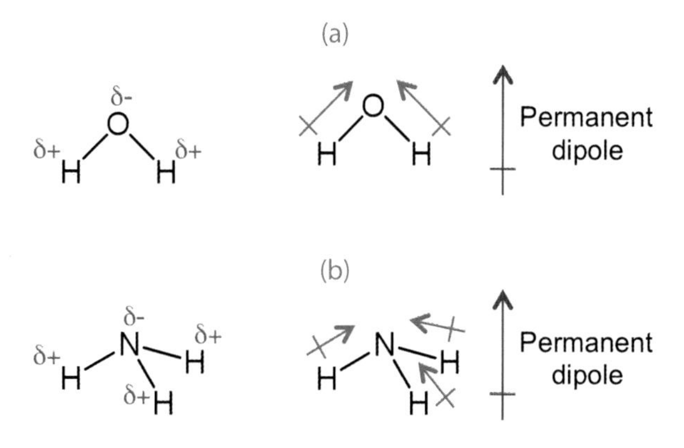

Figure 18: Combining bond dipoles to form a permanent dipole in (a) water, H_2O and (b) ammonia, NH_3.

permanent dipole for a molecule is obtained by adding the bond dipoles in a way that takes account of their size and direction.

The presence of bond dipoles in a molecule is, however, not always sufficient to produce a permanent dipole in the molecule. For instance, the bond dipoles in carbon dioxide (CO_2) and carbon tetrachloride (CCl_4) combine to produce molecules with no permanent dipole. The shapes of CO_2 and CCl_4 (Figure 19) are similar in that they represent the most symmetric way to arrange two and four identical atoms around a central atom. In contrast, the shapes of water and ammonia (Figure 18) are much less symmetric and give rise to permanent dipoles. Together these examples demonstrate that *highly symmetric molecules are less likely to have a permanent dipole than those with much less symmetry.*

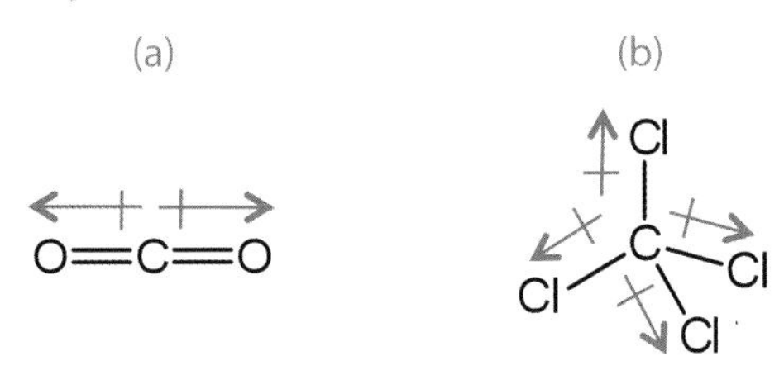

Figure 19: Combining bond dipoles in (a) carbon dioxide, CO_2 and (b) carbon tetrachloride, CCl_4 to produce molecules with no permanent dipole.

Exercise 1.3H

1. Water, H_2O and carbon dioxide, CO_2 both contain polar bonds. Explain why the carbon dioxide molecule is nonpolar.

(Adapted from CCEA June 2015)

Before moving to the next section, check that you are able to:

- Define electronegativity in terms of the sharing of electrons in covalent bonds.
- Explain the origin of bond dipoles in terms of electronegativity.
- Construct the permanent dipole for a molecule from bond dipoles.
- Relate the size of the permanent dipole for a molecule to its shape.

Covalently Bonded Materials

In this section we are learning to:

- Account for the physical properties of molecular materials in terms of their molecular structure and the weak attractive forces between the molecules.
- Recall that diamond and graphite are allotropes of carbon and have a giant covalent structure.
- Explain how the properties of diamond and graphite arise from the bonding within their giant structures.

Molecular Materials

Gases such as nitrogen (N_2), carbon dioxide (CO_2), and methane (CH_4) are molecular materials and are described as having a **molecular covalent structure**. The properties of a molecular material are determined by the nature of the covalent bonds in the molecules, and the much weaker forces of attraction between the molecules. The attractive forces between larger molecules, such as those in sulfur (S_8) and phosphorus (P_4), or molecules containing heavy atoms such as iodine (I_2), are significantly greater than the attractive forces between the molecules in gases such as nitrogen and carbon dioxide. As a result substances such as sulfur and iodine have higher melting points, and are more likely to be liquids or solids under normal laboratory conditions.

Molecular solids such as iodine, sulfur and ice are soft and brittle where a material is considered brittle if it powders easily when struck. The soft, brittle nature of molecular solids is due to the weak attractive forces between the molecules in the solid.

The solubility of a molecular material in a particular solvent is largely determined by the nature of the attractive forces between molecules in the solid and the nature of the attractive forces between molecules in the solvent. Molecular materials containing polar molecules such as ammonia (NH_3) will dissolve in polar solvents such as water. Conversely materials containing less polar molecules such as sulfur (S_8) and iodine (I_2) will dissolve in less polar solvents such as hexane.

The electrons in a molecular material are associated with individual molecules and are not free to move and carry a charge through the material. As a result molecular materials such as iodine and sulfur do not conduct electricity. Further, molecular materials do not contain ions and therefore cannot conduct an electric current when molten or dissolved in a solution.

Giant Covalent Materials

Many nonmetals have molecular covalent structures. Others prefer to form 'giant' structures and do not contain molecules. Diamond and graphite are different physical forms (allotropes) of the element carbon. In diamond each carbon atom forms a covalent bond with four neighbouring carbon atoms to form a three-dimensional network of atoms known as a **giant covalent structure**. The structure of diamond is shown in Figure 20.

The network of carbon atoms in diamond is extremely strong with each atom held tightly in place by four strong covalent bonds in a tetrahedral arrangement. The strength of the bonding between neighbouring carbon atoms can be used to explain why diamond is one of the hardest naturally occurring substances, has a very high melting point, and does not dissolve in water. The rigidity and organisation of the atoms in the giant structure also makes it possible to cut diamond into shapes with smooth faces and well-defined edges. This property helps make diamond ideal for use in jewellery and in drills for the mining and oil industries.

The carbon atoms in diamond use all four of their outer shell electrons to form covalent bonds with neighbouring carbon atoms. As a result, all of the outer shell electrons are held tightly in covalent bonds and there are no electrons free to move and carry a charge through the solid. This explains why diamond is unable to conduct electricity and is an electrical insulator.

In contrast, graphite is a soft powdery solid that conducts electricity. It has a slippery feel and is used as a solid lubricant. The carbon atoms in graphite form two-dimensional layers that are stacked as shown in Figure 20. Within each layer the carbon atoms use three of their four outer shell electrons to form covalent bonds with neighbouring carbon atoms. The strength of the covalent bonding between the carbon atoms in each layer can be used to explain why graphite has a very high melting point. The fourth outer shell electron on each atom is shared with atoms from neighbouring layers to form weak bonds between the layers. The electrons shared between the layers are described as **delocalised electrons** as they *are outer shell electrons that are not associated with an atom or ion, and are able to move freely*. The delocalised electrons in graphite can carry a charge through the solid making graphite an electrical conductor.

The weak bonds between layers can be easily overcome with the result that the layers can slide over each other and can be readily pulled apart. The weak bonding between layers explains why graphite is soft and the ability of the layers to slide over each other explains why graphite feels slippery; a property that makes graphite ideal for use in pencil 'lead' and graphite grease.

Exercise 1.3I

1. With reference to structure and bonding explain why graphite conducts electricity and why it may be used as a lubricant.

   (CCEA June 2003)

2. (a) Explain what is meant by the term *covalent*. (b) Describe the structures of diamond and graphite. (c) Explain why graphite conducts electricity. (d) Explain why diamond is exceptionally hard.

   (CCEA June 2009)

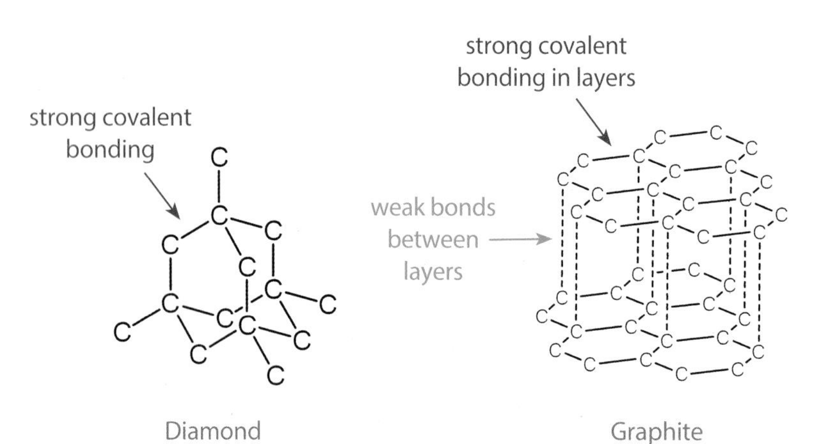

Figure 20: The giant covalent structures formed by carbon atoms in diamond and graphite.

3. All of the atoms in the giant covalent structure shown below are atoms of the same element.

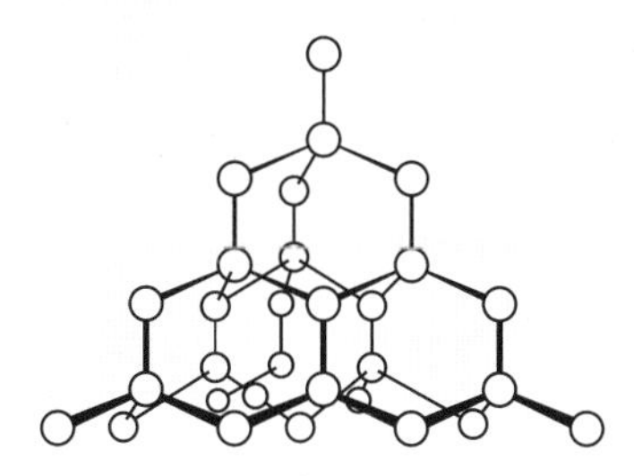

(a) Name the substance. (b) Explain whether the substance is hard or soft. (c) Explain whether the substance conducts electricity or not.

(CCEA June 2011)

4. A solid melts sharply at 100–101 °C, does not conduct electricity when molten, and dissolves in hydrocarbon solvents. The structure of the solid is:

A atomic B giant covalent

C ionic D molecular covalent

(CCEA January 2011)

5. Explain why diamond has a high melting point.

(Adapted from CCEA January 2013)

Before moving to the next section, check that you are able to:

- Account for the low melting point, solubility, and poor electrical conductivity of molecular covalent materials in terms of weak attractive forces between molecules and the absence of charged particles that can move and carry charge.
- Describe the giant covalent structures of diamond and graphite.
- Explain the high melting point, hardness, and poor electrical conductivity of diamond in terms of the formation of strong covalent bonds.
- Use a combination of strong bonding within layers and weak bonding between layers to explain the high melting point, electrical conductivity, and lubricating properties of graphite.

Metals

In this section we are learning to:

- Explain the physical properties of metals in terms of the formation of metallic bonds between the metal atoms.
- Use the model of metallic bonding in a metal to explain trends in the strength of metallic bonding within groups and periods.

In a metal the outer shell electrons of each metal atom become detached from the atom and are able to move freely through the metal. As a result, metals contain an array of positively charged metal ions in a 'sea' of delocalised electrons as shown in Figure 21. The metal ions are bonded together by forces of attraction between the metal ions and the delocalised electrons; a type of bonding referred to as **metallic bonding**.

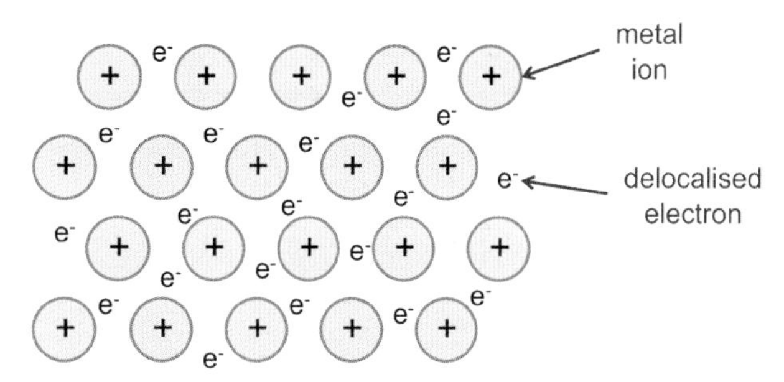

Figure 21: The model of metallic bonding in a Group I metal (one delocalised electron per ion).

The atoms in a Group I metal such as sodium have a single electron in their outer shell. As a result, the atoms in a Group I metal each lose one electron to form an ion with a 1+ charge surrounded by delocalised electrons. As the metal ions get bigger down the group the delocalised electrons are further from the positive nucleus of each metal ion and are therefore less attracted to the metal ions. This explains why the strength of metallic bonding in the Group I metals decreases down the group.

In a Group II metal such as magnesium, each metal atom loses two electrons to form an ion with a 2+ charge. The attraction between the ions and the delocalised electrons in a Group II metal is significantly greater than in a Group I metal and, as a result, the bonding in a Group II metal is significantly stronger than the bonding in a Group I metal. This explains why Group I metals are soft and can be cut with knife while Group II metals are much harder and have higher melting points than Group I metals.

Metallic bonds are similar to covalent bonds. Both types of bonding involve the sharing of electrons and can result in very strong bonds. The metallic bonding in most metals is strong and can be used to explain why the metal ions in a metal are tightly packed to produce a hard, dense solid with a high melting point. In contrast, the relatively weak metallic bonding in Group 1 metals such as sodium can be used to explain why the metal ions are less tightly packed to produce soft metals with low densities.

The ability of the outer shell electrons to move freely allows the metal ions to move without disturbing the bonding in the metal. The ability of metal ions to move past each other without disturbing the metallic bonding in the metal explains why metals can be deformed when struck with a harder object (metals are **malleable**) and why metals can be drawn into wires (metals are **ductile**).

The ability of the outer shell electrons to move freely through the metal can also be used to explain why metals are good **electrical conductors**. When a power supply (battery) is attached to a metal the mobile electrons move towards the positive terminal of the power supply resulting in an electrical current within the metal.

Worked Example 1.3iii

(a) Using a labelled diagram explain the bonding in sodium metal. (b) Metals are good conductors of electricity. Explain why the electrical conductivity of aluminium is greater than that of sodium.

(CCEA June 2014)

Solution

(a) Metallic bonding in sodium results from the attraction between a lattice of positively charged sodium ions and the delocalised outer shell electrons.

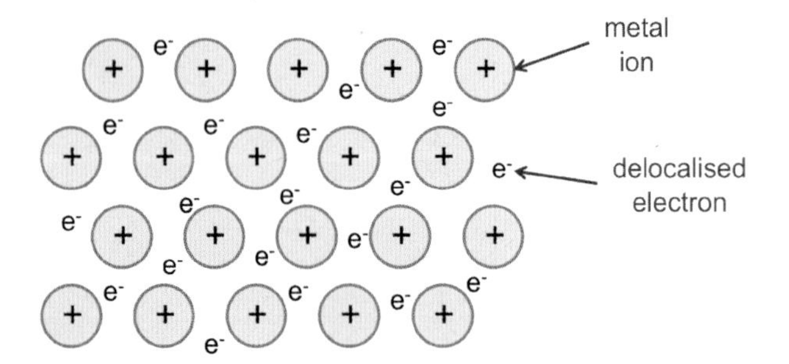

(b) Each atom in aluminium produces three delocalised electrons that can move and carry charge. In contrast, each atom in sodium can only produce one delocalised electron.

Exercise 1.3J

1. (a) Explain, with the help of a labelled diagram, what is meant by the term metallic bonding. (b) Suggest why the melting points of the Group I metals decrease from sodium to caesium. (c) Use the concept of metallic bonding to suggest why calcium should be a better electrical conductor than potassium.

   (CCEA June 2011)

2. Francium has a melting point of 27 °C and has the highest electrical conductivity of the alkali metals. (a) Explain, in terms of metallic bonding, why francium has a low melting point. (b) Explain, in terms of metallic bonding, why francium has the highest electrical conductivity.

   (CCEA January 2014)

3. Explain why sodium (a) is malleable and (b) conducts electricity.

   (CCEA January 2008)

4. Describe the bonding in a metal and use this to explain the ductility and electrical conductivity of a typical metal.

   (CCEA January 2006)

5. Compare the electrical conductivity of solid strontium metal with that of solid strontium fluoride. Explain your answer.

   (CCEA June 2010)

Before moving to the next section, check that you are able to:

- Describe the nature of metallic bonding within a metal and explain periodic trends in the strength of metallic bonding.
- Explain the high melting point, malleability, ductility, and electrical conductivity of metals in terms of the nature of the metallic bonding within the metal.

1.4 Shapes of Molecules and Ions

CONNECTIONS

- The characteristic shapes formed when atoms bond with each other make it possible for scientists to design materials with specific properties, including medicines to treat specific diseases.

VSEPR Theory

In this section we are learning to:

- Explain the shapes of molecules and ions in terms of the repulsion between groups of electrons in the outer shell of a central atom.
- Recall the shapes of molecules and ions that result from repulsion between four groups of electrons.

The shape of a molecule or ion is determined by the arrangement of the outer shell electrons about each of the central atoms. The structural formulas in Figure 1 reveal that the C, N and O atoms in methane (CH_4), ammonia (NH_3) and water (H_2O), each have four pairs of electrons in their outer shell.

If we use the term **bonding pair** when referring to *a pair of electrons in the outer shell of an atom that is shared to form a covalent bond*, and the term **lone pair** when referring to *a pair of electrons in the outer shell of an atom that is not involved in bonding*, we can say that the carbon atom in methane has four bonding pairs in its outer shell, the nitrogen atom in ammonia has three bonding pairs and one lone pair in its outer shell, and the oxygen atom in water has two bonding pairs and two lone pairs in its outer shell.

Valence Shell Electron Pair Repulsion (VSEPR) theory asserts that *a molecule or ion will always try to minimise the repulsion between outer shell electrons by arranging the outer shell electrons as far away from each other as possible.* In a molecule or ion with four pairs of electrons around the central atom, the repulsion between the electron pairs is minimised when the electron pairs point towards the corners of a tetrahedron. The tetrahedral arrangement of hydrogen atoms in methane (CH_4) is shown in Figure 2a. The angle formed by the atoms in adjacent C-H bonds is referred to as the **bond angle** in methane, and has the value 109.5°.

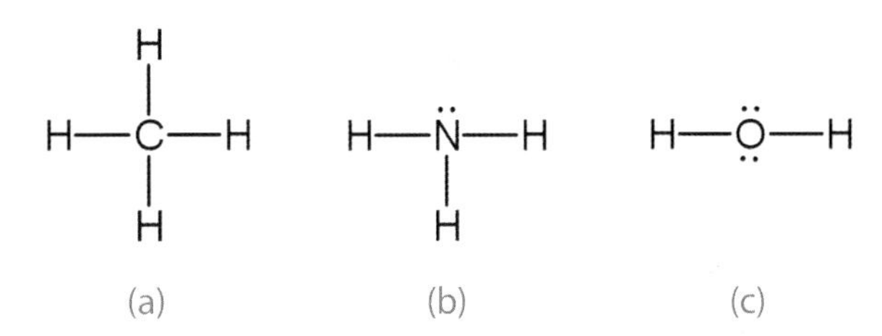

Figure 1: Structural formulas for (a) methane, (b) ammonia and (c) water.

The arrangement of the bonding pairs and lone pairs about the central atom in ammonia and water is also shown in Figure 2. The structures in Figure 2 reveal that the bond angle decreases as the number of lone pairs on the central atom increases. This occurs because *the repulsion between a bonding pair and a lone pair is greater than the repulsion between two bonding pairs*, and forces the bonding pairs further from the lone pairs.

Figure 2: The arrangement of electron pairs about the central atom in (a) methane, (b) ammonia and (c) water.

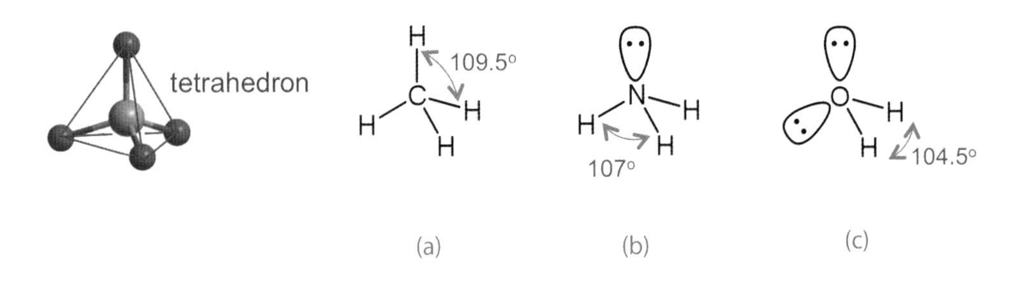

Repulsion increasing

Repulsion between lone pairs

Repulsion between a lone pair and bonding pair

Repulsion between bonding pairs

104.5°

The **molecular shapes** resulting from the arrangement of electron pairs in Figure 2 are shown in Figure 3. The structural formulas and names used to describe the shape of methane, ammonia and water are also given in Figure 3, and demonstrate that while all three molecules have a tetrahedral arrangement of electron pairs about the central atom, only methane has a tetrahedral shape.

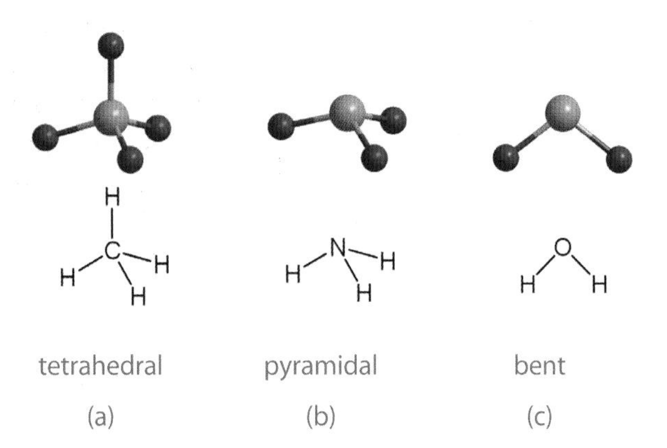

Figure 3: Structural formulas and names used to describe the molecular shapes of (a) methane, (b ammonia and (c) water.

Exercise 1.4A

1. Draw and explain the shape of an ammonia molecule.

   (CCEA June 2009)

2. **(a)** Draw a dot-and-cross diagram to show the bonding in phosphine, PH_3. (b) Draw and name the shape of a phosphine molecule.

   (CCEA January 2010)

3. (a) State the bond angle in a water molecule. (b) State the shape of a water molecule and explain why it adopts this shape. (c) Explain why the bond angle of water is different to the bond angle in methane.

   (Adapted from CCEA June 2012)

4. Sulfur difluoride has the same shape as a water molecule but the bond angle is 6° smaller. Draw and name the shape of sulfur difluoride, stating its bond angle.

   (CCEA January 2014)

5. State and explain the shapes of the following compounds, and give the values of the bond angles in each compound.

   methane CH_4

   hydrogen sulfide H_2S

   ammonia NH_3

   (Adapted from CCEA January 2012)

6. (a) Draw a dot-and-cross diagram, using outer shell electrons only, to show the bonding in oxygen difluoride, OF_2. (b) Suggest and explain the shape of the oxygen difluoride molecule.

   (CCEA June 2005)

7. (a) Draw the structure of chloroform, $CHCl_3$ showing all the bonds present. (b) Draw a dot-and-cross diagram using outer shell electrons only, to show the bonding in chloroform. (c) State and explain the shape of the chloroform molecule.

   (CCEA January 2006)

8. The structural formula for hydrogen peroxide is H-O-O-H. (a) Draw a dot-and-cross diagram for a molecule of hydrogen peroxide showing outer shell electrons only. (b) Suggest why a hydrogen peroxide molecule is not linear.

   (Adapted from CCEA June 2004)

9. Ammonium phosphate is an ionic compound consisting of ammonium and phosphate, PO_4^{3-} ions. (a) Write the formula for an ammonium ion. (b) Draw and name the shape of an ammonium ion. (c) State the angle between the bonds in an ammonium ion. (d) Write the formula for ammonium phosphate.

   (Adapted from CCEA January 2014)

10. Boron trifluoride can react with a fluoride ion to form the BF_4^- ion. (a) Draw a dot-and-cross diagram for the BF_4^- ion and use it to suggest the shape of the ion and its bond angle. (b) Name the type of bond formed between the fluoride ion and boron.

(Adapted from CCEA June 2014)

11. Sodium reacts with ammonia to form sodium amide, $NaNH_2$ and hydrogen. (a) Write an equation for the reaction. (b) Draw the shape of an amide ion, NH_2^-, showing any lone pairs of electrons. (c) State the shape of the amide ion. (d) Explain, in terms of electron pair repulsion, why the bond angle in an amide ion is smaller than the bond angle in an ammonia molecule.

(Adapted from CCEA June 2015)

Before moving to the next section, check that you are able to:

- Recall examples of molecules and ions with tetrahedral, pyramidal and bent shapes.
- Explain how the repulsion between bonding electrons and lone pairs can be used to explain the shapes of molecules and ions with four groups of electrons around a central atom.
- Recall the bond angles in CH_4, NH_3 and H_2O and use these angles to predict the bond angles around atoms with four groups of electrons in their outermost shell.

Shapes Based on Two or Three Repulsions

In this section we are learning to:

- Explain the shapes of molecules and ions in terms of the repulsion between groups of electrons in the outer shell of a central atom.
- Recall the shapes of molecules and ions that result from repulsion between two or three groups of electrons.

In carbon dioxide, CO_2 the central carbon atom forms a double bond with each oxygen atom (O=C=O). The repulsion between the electrons in one double bond and the electrons in the other double bond is minimised when the bonds are 180° apart and all three atoms lie in a straight line as shown in Figure 4. This arrangement of atoms is described as linear. Similarly, in hydrogen cyanide, HCN the central carbon atom forms a single covalent bond with hydrogen and a triple bond with the nitrogen atom (H-C≡N). The repulsion between the electrons in the single bond and the electrons in the triple bond is minimised when the bonds are 180° apart and the arrangement of the atoms about the carbon atom is linear.

In this way the molecules in Figure 4 demonstrate that *a linear arrangement of atoms results from the repulsion between two groups of electrons around a central atom.*

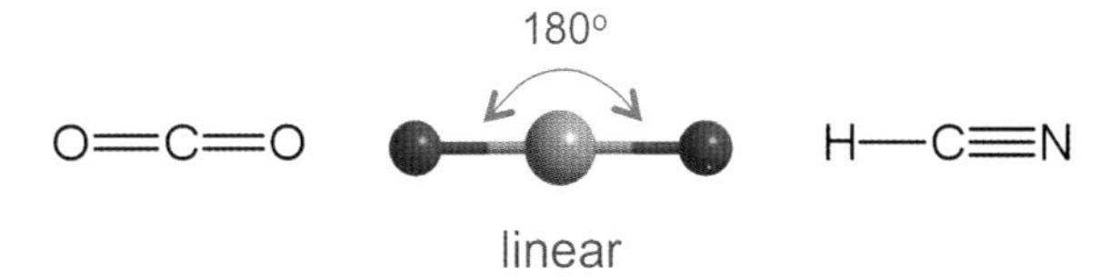

Figure 4: The linear shape of CO_2 and HCN that results from the linear arrangement of two groups of electrons about the central atom.

The atoms in a molecule of beryllium chloride, $BeCl_2$ also have a linear arrangement. Unlike the central carbon atom in CO_2 and HCN, the central beryllium atom in $BeCl_2$ does not have a full outer shell. The beryllium atom uses the two electrons in its outer shell to form a covalent bond with each chlorine atom. The repulsion between the electrons in one Be-Cl bond and the electrons in the other Be-Cl bond is minimised when the Be-Cl bonds are 180° apart giving the molecule a linear shape.

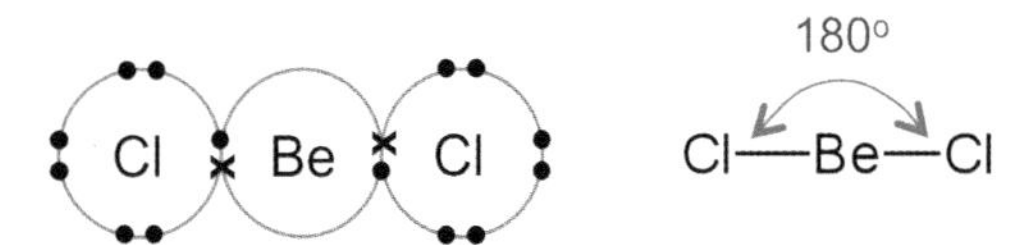

Boron trifluoride, BF_3 is an example of a molecule with three groups of electrons around a central atom. In BF_3 the central boron atom forms a single covalent bond with each fluorine atom. The repulsion between the electron pairs in the B-F bonds is minimised when the bonds are directed towards the corners of a triangle as shown in Figure 5. The shape of boron trifluoride is described as trigonal planar to reflect the trigonal arrangement of fluorine atoms with all four atoms in the same plane.

Formaldehyde, H_2CO also has a trigonal planar shape (Figure 5) in which the central carbon atom forms a covalent bond with each hydrogen atom, and a double bond with the oxygen atom. The angle between

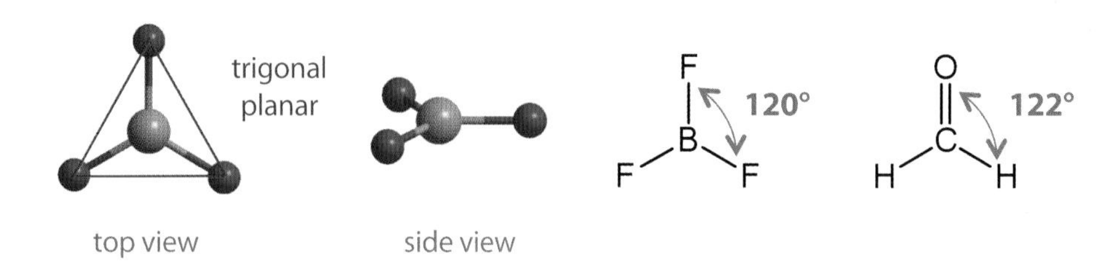

Figure 5: The trigonal planar arrangement of atoms in boron trifluoride, BF_3 and formaldehyde, H_2CO resulting from the trigonal planar arrangement of three groups of electrons about the central atom.

the C=O bond and each of the C-H bonds is slightly greater than 120° on account of greater repulsion between the two pairs of electrons in the C=O double bond and the pair of electrons in a C-H bond.

In this way the molecules in Figure 5 demonstrate that *a trigonal planar arrangement of atoms results from the repulsion between three groups of electrons around a central atom.*

Exercise 1.4B

1. Beryllium chloride can be prepared by the action of chlorine or hydrogen chloride on the metal. (a) Write the equation for the reaction of beryllium with hydrogen chloride. (b) Draw a dot-and-cross diagram to show the formation of beryllium chloride from beryllium and chlorine atoms showing only outer shell electrons. (c) Beryllium chloride can be said to obey and at the same time not obey the octet rule. Explain this contradiction. (d) Draw and name the shape of a beryllium chloride molecule. (e) Explain the shape of a beryllium chloride molecule.

   (CCEA January 2011)

2. (a) Using outer electrons only, draw the dot-and-cross diagram for boron trifluoride. (b) State the octet rule and explain why boron trifluoride does not obey the octet rule. (c) Draw and explain the shape of boron trifluoride.

   (CCEA January 2004)

3. Aluminium chloride, $AlCl_3$ reacts with a chloride ion to form the $AlCl_4^-$ ion. (a) Draw dot-and-cross diagrams, using outer electrons only, to show the bonding in $AlCl_3$ and $AlCl_4^-$. (b) State the type of bond formed between $AlCl_3$ and the Cl^- ion. (c) Draw and name the shapes of $AlCl_3$ and $AlCl_4^-$.

   (Adapted from CCEA June 2013)

4. Which one of the following molecules is the most polar?

   A BF_3 B CO_2 C F_2 D NH_3

   (CCEA June 2012)

Before moving to the next section, check that you are able to:

- Recall examples of molecules and ions with linear and trigonal planar arrangements of electrons about a central atom.
- Explain how the repulsion between bonding electrons and lone pairs can be used to explain the shapes of molecules and ions with two or three groups of electrons around a central atom.

Shapes Based on Five or Six Repulsions

In this section we are learning to:

- Explain the shapes of molecules and ions in terms of the repulsion between groups of electrons in the outer shell of a central atom.
- Recall the shapes of molecules and ions that result from repulsion between five or six groups of electrons in the outer shell of a central atom.

The shape of molecules such as PCl_5 and SF_4 is determined by the repulsion between the five pairs of electrons in the outer shell of the central atom. In PCl_5 the central phosphorus atom forms a single covalent bond with each chlorine atom. The repulsion between the bonding pairs of electrons in the P-Cl bonds is minimised when the chlorine atoms are located at the corners of a shape known as a *trigonal bipyramid* as shown in Figure 6.

In PCl_5 three of the chlorine atoms are arranged in trigonal planar arrangement about the central phosphorus atom. The remaining two chlorine atoms

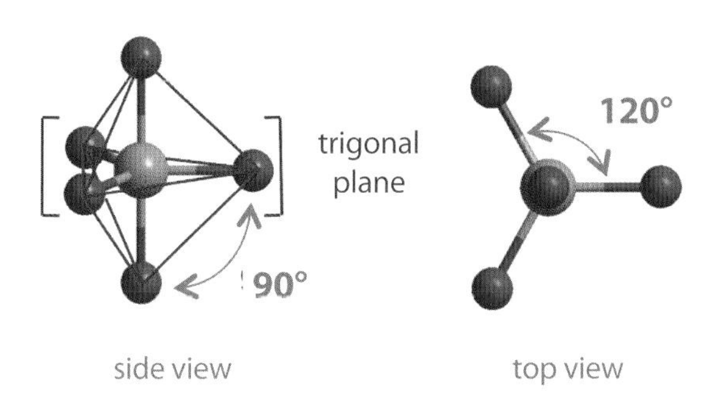

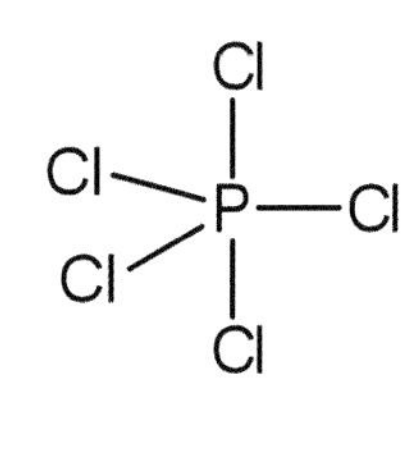

trigonal bipyramidal

Figure 6: The trigonal bipyramidal arrangement of chlorine atoms in PCl_5 resulting from the trigonal bipyramidal arrangement of five electron pairs about the central phosphorus atom.

lie above and below the trigonal plane. The angle between a bond in the trigonal plane and a bond above or below the trigonal plane is 90°. This angle is significantly smaller than the bond angles in molecules based on a tetrahedral shape (105-109°) and results in much greater repulsion.

When one of the electron pairs around the central atom is a lone pair, as in SF_4, *the repulsion between electron pairs is greatly reduced if the lone pair occupies one of the three sites in the trigonal plane.* The preference for lone pairs to occupy sites in the trigonal plane can also be used to explain the shapes of molecules and ions such as ClF_3 and I_3^- shown in Figure 7, all of which have five pairs of electrons arranged around the central atom.

The shapes of the molecules and ions in Figure 7 demonstrate that *the repulsion between five groups of electrons around a central atom may give rise to a trigonal pyramidal, see-saw, T-shaped, or linear arrangement of atoms.*

In sulfur hexafluoride, SF_6 the outer shell of the sulfur atom contains six pairs of electrons; one from each of the six S-F bonds. The repulsion between these electron pairs is minimised when the electron pairs point towards the corners of an octahedron as illustrated in Figure 8. This arrangement gives the molecule an *octahedral* shape in which the bond angle formed by neighbouring S-F bonds is 90°.

The outer shell of the xenon atom in xenon(IV) fluoride, XeF_4 also contains six pairs of electrons, two of which are lone pairs. The lone pairs are placed 180° apart, giving the molecule a *square planar* shape, to avoid the very great repulsion that would result from placing the lone pairs at 90° to each other.

The shapes of the molecules in Figure 8 demonstrate that *the repulsion between six groups of electrons around a central atom may give rise to an octahedral or square planar arrangement of atoms.*

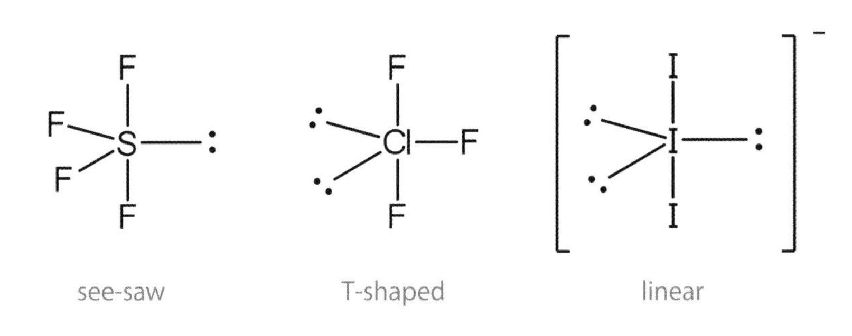

Figure 7: Structural formulas and names used to describe the shapes of SF_4, ClF_3 and I_3^-.

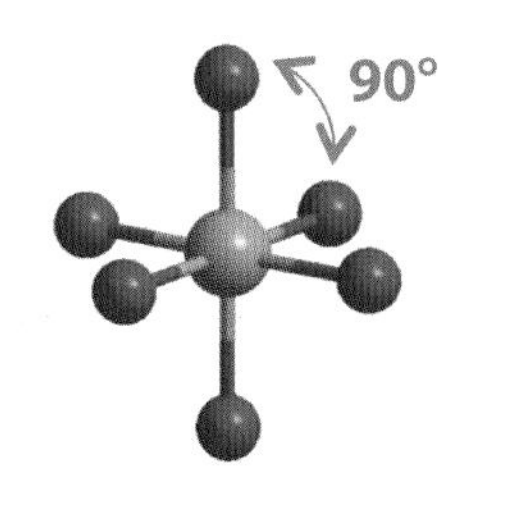

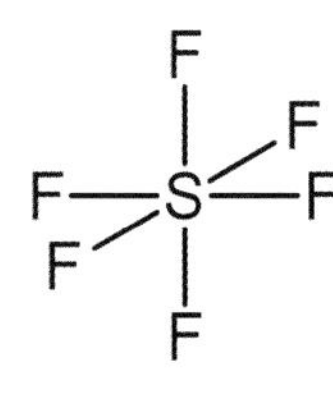

octahedral

F—Xe—F (square planar)

square planar

Figure 8: Structural formulas and names used to describe the shapes of SF_6 and XeF_4.

Exercise 1.4C

1. Explain why a molecule of silicon tetrachloride, $SiCl_4$ has a tetrahedral shape and a molecule of sulfur tetrafluoride, SF_4 has a see-saw shape.
2. Explain why a molecule of BF_3 is trigonal planar, a molecule of PF_3 is pyramidal and a molecule of ClF_3 is T-shaped.
3. Which one of the following molecules contains the smallest bond angle?

   A $BeCl_2$

   B BF_3

   C CH_4

   D SF_6

   (CCEA June 2012)

Before moving to the next section, check that you are able to:

- Recall examples of molecules and ions with trigonal bipyramidal, see-saw, T-shaped and linear shapes and explain how these arise from the repulsion between five groups of electrons in the outer shell of a central atom.
- Recall examples of molecules and ions with octahedral and square planar shapes and explain how these result from the repulsion between six groups of electrons on a central atom.

Nonpolar Molecules

In this section we are learning to:

- Explain the size of the permanent dipole for a molecule or ion in terms of the formation of polar bonds and their arrangement in the molecule.

The unequal sharing of the electrons in a polar covalent bond gives rise to partial charges (δ+ and δ-) on the atoms forming the bond. If the *average position of the positive (δ+) charges in a molecule coincides with the average position of the negative (δ-) charges in the molecule, the molecule will not have a permanent dipole.*

A molecule that does not have a permanent dipole is referred to as a **nonpolar molecule**. Molecules in which the bond dipoles are arranged more symmetrically about the central atom(s) will have smaller permanent dipoles than molecules in which the bond dipoles are less symmetrically arranged. The molecular shapes shown in Figure 9 are highly symmetric and will always give rise to nonpolar molecules.

Exercise 1.4D

1. Which one of the following compounds is nonpolar?

   A HCl B CCl_4 C CH_3Cl D $CHCl_3$

   (Adapted from CCEA June 2006)

2. (a) Using outer shell electrons only, draw the dot-and-cross structure of carbon dioxide. (b) Draw and name the shape of a carbon dioxide molecule. (c) Explain why carbon dioxide has this shape. (d) Explain why a carbon dioxide molecule does not have a permanent dipole.

   (CCEA June 2011)

3. Explain why (a) hydrogen fluoride is polar and (b) boron trifluoride is nonpolar.

   (CCEA June 2007)

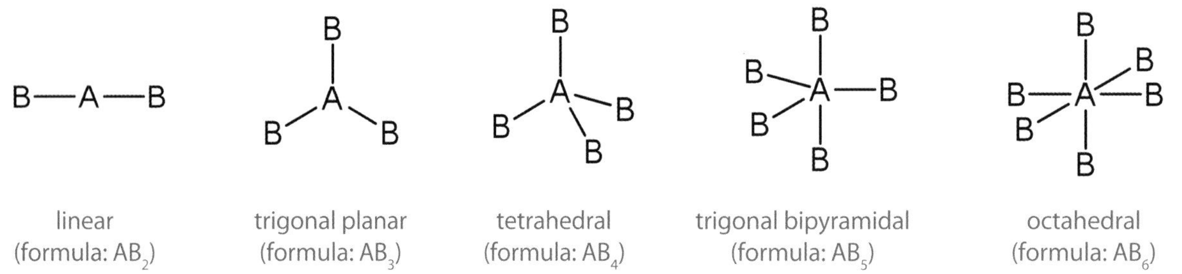

linear (formula: AB_2)

trigonal planar (formula: AB_3)

tetrahedral (formula: AB_4)

trigonal bipyramidal (formula: AB_5)

octahedral (formula: AB_6)

Figure 9: Molecular shapes that give rise to nonpolar molecules.

4. Which one of the following molecules is polar?

   A BF_3 B CF_4 C OF_2 D F_2

   (CCEA January 2010)

5. (a) Draw, name and explain the shape of the SF_6 molecule. (b) Suggest why SF_6 is a nonpolar molecule even though it contains polar bonds.

   (CCEA June 2010)

Before moving to the next section, check that you are able to:

- Recall that a permanent (molecular) dipole results from combining individual bond dipoles.
- Identify molecular shapes that give rise to nonpolar molecules.

1.5 Intermolecular Forces

CONNECTIONS

- Ice floats in water because the water molecules in a block of ice are held apart by strong forces.
- Oil and water don't mix because the forces between the molecules in oil and the forces between the molecules in water are different.

The term **intermolecular force** refers to *any type of bonding interaction between molecules*. The physical properties of many substances are determined by the nature of the intermolecular forces between molecules in the substance. For example, intermolecular forces can be used to explain why ice floats on water, and why some liquids mix while others don't. Variations in the strength of the intermolecular forces between molecules can also be used to explain why heavier elements such as iodine tend to be solids, while lighter elements such as hydrogen and oxygen have lower melting and boiling points and are gases.

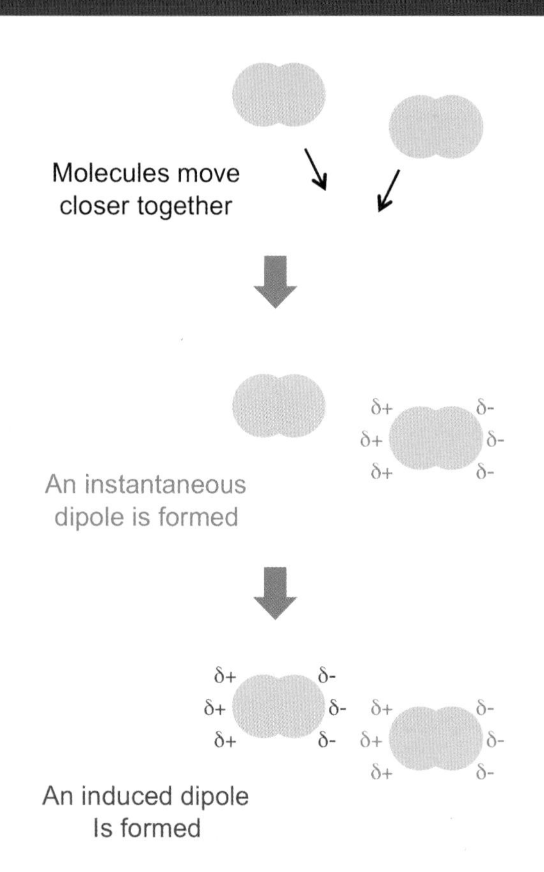

Figure 1: A van der Waals force resulting from the attraction between the oppositely charged ends of an instantaneous dipole and an induced dipole.

Van der Waals Forces

In this section we are learning to:

- Explain the nature of van der Waals forces of attraction between molecules.
- Recall the factors that affect the size of the van der Waals attraction between molecules and their effect on the melting and boiling point of a substance.

When two molecules approach each other, the electrons in one molecule repel the electrons in the other, and an instantaneous dipole (δ+ δ-) is created within one of the molecules as shown in Figure 1. The instantaneous dipole then induces a dipole in the neighbouring molecule, and the molecules experience a **van der Waals force** that results from *the attraction between the instantaneous and induced dipoles on neighbouring molecules.*

The size of the attraction between the instantaneous and induced dipoles on neighbouring molecules increases as the dipoles get bigger. *Molecules with more electrons will have larger dipoles and will experience greater van der Waals attraction*. For instance, the melting and boiling points of the noble gases (Group VIII) increase down the group as the number of electrons in each noble gas atom increases. Similarly, an increase in van der Waals attraction

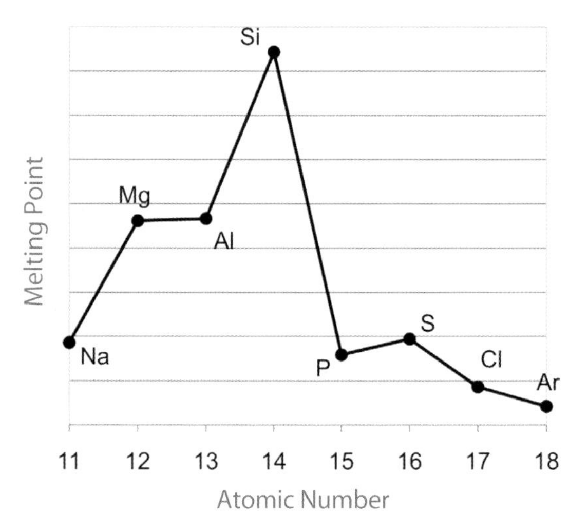

Figure 2: Melting points of the elements in the third period (Na-Ar).

Figure 3: Factors governing the magnitude of the van der Waals attraction between molecules.

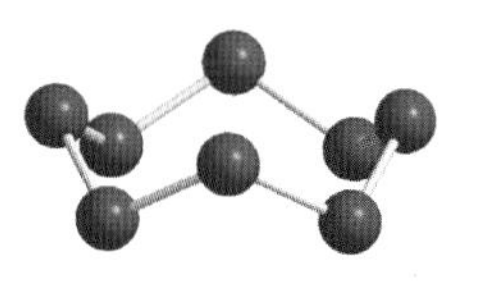
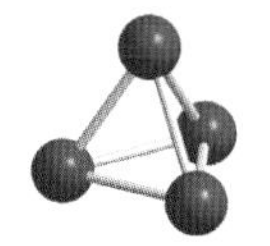
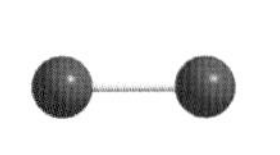

between molecules can be used to explain why the lighter halogens (F_2 and Cl_2) are gases with relatively low melting and boiling points, while the heavier halogens (Br_2 and I_2) are liquids and solids.

The magnitude of the van der Waals attraction between molecules can also be used to explain the trend in the melting points of the elements P to Ar shown in Figure 2. Phosphorus (P_4), sulfur (S_8) and chlorine (Cl_2) have molecular covalent structures. Argon (Ar) is a monatomic gas. The van der Waals attractions between the particles in these elements are much weaker than the covalent bonds between the atoms in silicon, and the metallic bonding in the metals earlier in the period. As a result, the melting points of phosphorus, sulfur, chlorine and argon are all very much lower than the melting points of the elements earlier in the period.

The melting point of sulfur is greater than the melting point of phosphorus as the S_8 molecules in sulfur have more electrons than the smaller P_4 molecules in phosphorus and therefore experience greater van der Waals attraction. Similarly, phosphorus has a higher melting point than chlorine as the P_4 molecules in phosphorus have more electrons than the Cl_2 molecules in chlorine and therefore experience greater van der Waals attraction. The melting point of chlorine is, in turn, greater than the melting point of argon (Ar) as the atoms in argon have fewer electrons than the Cl_2 molecules in chlorine and therefore experience less van der Waals attraction.

The magnitude of the van der Waals attraction between molecules can also be expressed in terms of RMM as shown in Figure 3.

The size of the van der Waals attraction between molecules is also affected by the shape of the molecules. *Increasing the surface area of a molecule increases the strength of the van der Waals attraction between molecules by increasing the amount of contact between the electrons in neighbouring molecules.* For example, pentane (C_5H_{12}) and 2,2-dimethylpropane (C_5H_{12}) have the same formula and contain similar types of bonds. The boiling point of pentane is however higher

Figure 4: Instantaneous and induced dipoles in (a) pentane, C_5H_{12} and (b) 2,2-dimethylpropane, C_5H_{12}.

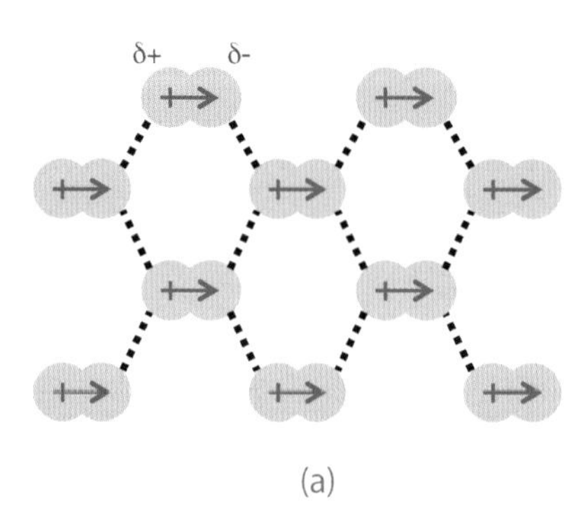

(a)

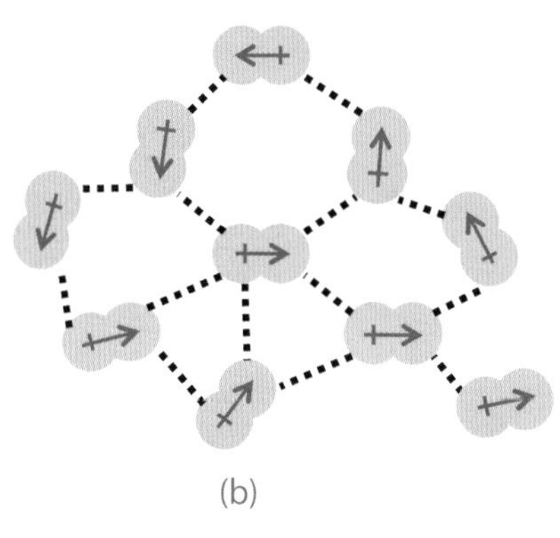
(b)

Figure 5: The interaction of permanent dipoles in (a) a solid and (b) a liquid.

as it has a larger surface area that allows for more attraction between instantaneous and induced dipoles, and as a result, stronger van der Waals attraction between molecules. The instantaneous and induced dipoles in pentane and 2,2-dimethylpropane are illustrated in Figure 4.

Exercise 1.5A

1. In which of the following liquids are the van der Waals forces greatest?

   A Argon B Krypton C Neon D Xenon

   (CCEA June 2013)

2. The melting points of silicon, phosphorus and sulfur are given in the table. Explain, with reference to the structures of silicon and sulfur, why each has a higher melting point that phosphorus.

Element	Si	P_4	S_8
Melting point (°C)	1410	44	113

(Adapted from CCEA June 2014)

Before moving to the next section, check that you are able to:

- Explain the existence of van der Waals attractions between molecules.
- Recall the effect of adding electrons and increasing surface area on the strength of van der Waals attractions.
- Explain the effect of increasing van der Waals attraction on the melting and boiling point of a substance.

Dipole Forces

In this section we are learning to:

- Explain the existence of dipole forces between molecules.
- Identify the effect of dipole forces on the melting and boiling points of a substance.

In addition to van der Waals attraction, molecules with a permanent dipole also experience **dipole forces** or 'dipole-dipole attraction' that results from the *attraction between the positive end of the permanent dipole (δ+) on one molecule and the negative end of the permanent dipole (δ-) on neighbouring molecules*. The interactions between permanent dipoles in solids and liquids are illustrated in Figure 5.

The extent to which van der Waals forces and dipole forces influence the physical properties of materials can be demonstrated by comparing the boiling points of the molecules in Figure 6. Butane (C_4H_{10}) and 2-methylpropane (C_4H_{10}) have the same formula and contain the same types of bonds. The molecules in both compounds experience only van der Waals attraction as neither molecule has a significant permanent dipole. There is, however, more contact between the electrons on neighbouring butane molecules. This results in stronger van der Waals attraction between molecules, and a higher boiling point as the more energy is needed to overcome the greater van der Waals attraction between molecules.

Molecules of acetone (C_3H_6O) and 2-methylpropane (C_4H_{10}) have a similar shape. They also have similar numbers of electrons. As a result, the van der Waals attraction between molecules of acetone is expected to be similar to the van der Waals attraction between molecules of 2-methylpropane. However, unlike butane and 2-methylpropane, a molecule of acetone

butane, C_4H_{10}
boiling point 0 °C

2-methylpropane, C_4H_{10}
boiling point −11 °C

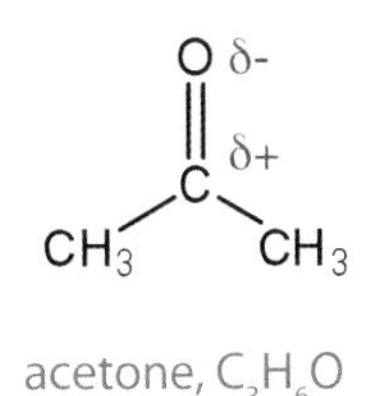

acetone, C_3H_6O
boiling point 56 °C

Figure 6: The effect of surface area and dipole forces on the boiling point of liquids.

has a permanent dipole. The extent to which the dipole forces between acetone molecules affect the boiling point of acetone can be seen by comparing the boiling points of acetone and 2-methylpropane in Figure 6.

Exercise 1.5B

1. Sulfur tetrafluoride has a boiling point of −38 °C whereas sulfur hexafluoride has a boiling point of −64 °C. (a) Which compound has the higher boiling point? (b) Explain, in terms of mass, which compound has the greater van der Waals forces. (c) Explain, in terms of intermolecular forces, the difference in boiling points.

   (CCEA January 2014)

2. A narrow stream of water flowing from a tap is attracted when a positively charged plastic rod is brought closer to the water. (a) Explain, in terms of intermolecular forces, why the water molecules are attracted to the rod. (b) Using a positive charge to represent the rod, draw a diagram to illustrate the attractive force experienced by a water molecule.

3. The diagram shows a liquid escaping from a burette and passing a charged glass rod. Which one of the following liquids will be attracted to the glass rod?

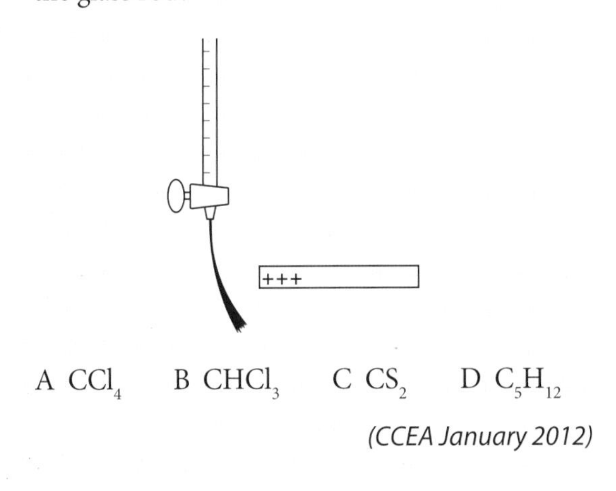

A CCl_4 B $CHCl_3$ C CS_2 D C_5H_{12}

(CCEA January 2012)

Before moving to the next section, check that you are able to:

- Describe how the interaction between the permanent dipoles on neighbouring molecules gives rise to attractive forces between molecules.
- Identify the effect of dipole forces on the melting and boiling points of substances.

Hydrogen Bonding

In this section we are learning to:

- Explain the formation of hydrogen bonds between molecules.
- Identify the effect of hydrogen bonding on the melting and boiling points of a substance.

The boiling points of the hydrides formed by the elements in Groups IV, V, VI and VII in Figure 7 reveal the presence of very strong intermolecular forces in the hydrides of nitrogen (NH_3), oxygen (H_2O) and fluorine (HF).

The Group IV hydrides (CH_4, SiH_4, GeH_4, SnH_4) have a tetrahedral shape. As a result, the bond dipoles cancel and the hydrides do not have a permanent dipole. In the absence of dipole forces the increase in the boiling

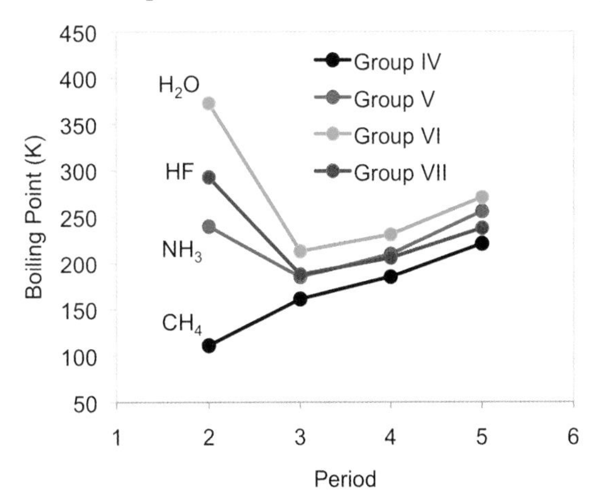

Figure 7: Boiling points for the hydrides of the Group IV, V, VI and VII elements.

points of the Group IV hydrides results from an increase in van der Waals attraction as the number of electrons in the Group IV element increases down the group.

With the exception of NH_3, H_2O and HF, the boiling points of the Group V, VI and VII hydrides also increase steadily down the group. This again reflects an increase in van der Waals attraction between molecules as the number of electrons in the hydride increases down the group.

Unlike the Group IV hydrides, the Group V, VI, and VII hydrides have a permanent dipole. The additional dipole force between the molecules increases the boiling point of the hydrides with the result that the boiling points of the Group V, VI and VII hydrides in Periods 3, 4 and 5 are higher than the boiling points of the corresponding Group IV hydrides.

Dipole forces cannot, however, account for the unexpectedly high boiling points of NH_3, H_2O, and HF. The unexpectedly high boiling points of NH_3, H_2O and HF are, instead, due to the formation of **hydrogen bonds** between molecules. *A hydrogen bond is a strong dipole-like force of attraction between a lone pair on a very electronegative atom (N, O or F) and a hydrogen atom that is covalently bonded to a very electronegative atom (N, O or F) in a neighbouring molecule*. The formation of hydrogen bonds in NH_3, H_2O, and HF is illustrated in Figure 8.

The increase in the boiling points of NH_3, H_2O and HF that results from hydrogen bonding (Figure 7) is much greater than the increase resulting from dipole forces and indicates that *hydrogen bonds are significantly stronger than dipole forces.*

The formation of hydrogen bonds between neighbouring molecules can be used to explain a number of physical properties. In liquid water the hydrogen bonds between neighbouring molecules are constantly breaking and forming. When water freezes the hydrogen bonds hold the molecules in fixed positions. This generates a highly ordered structure in which the molecules are more widely spaced, and results in the density of ice being lower than the density of liquid water. In this way the presence of hydrogen bonding between water molecules can be used to explain why ice floats on water.

Intermolecular forces arising from the interaction between hydrogen and other electronegative elements such as chlorine and bromine on a neighbouring molecule do not involve hydrogen bonding. The bonding between molecules in compounds such as hydrogen chloride (HCl) and hydrogen bromide (HBr) instead results from a combination of van der Waals attraction and dipole forces. In this way *the forces between molecules can either be a combination of van der Waals attraction and hydrogen bonding as in water, or a combination of van der Waals attraction and dipole forces as in HCl and HBr.*

Exercise 1.5C

1. Which one of the following would not form hydrogen bonds?

   A $CH_3CH_2CH_2Cl$ B $CH_3CH_2CH_2OH$

   C $CH_3CH_2CH_2NH_2$ D $CH_3CH(OH)CH_3$

   (CCEA June 2004)

2. There are three accepted types of intermolecular force: van der Waals forces, permanent dipole attractions and hydrogen bonding. Complete the following table where ✓ = present and ✗ = not present.

liquid	van der Waals	permanent dipole	hydrogen bonding
water	✓	✗	✓
ammonia			
xenon			
hydrogen chloride			

(CCEA June 2006)

Figure 8: Formation of a hydrogen bond between (a) NH_3 molecules, (b) H_2O molecules and (c) HF molecules.

3. What is the strongest intermolecular force in (a) ammonia, $NH_{3(l)}$ (b) hydrogen chloride, $HCl_{(l)}$ and (c) methane, $CH_{4(l)}$?

(Adapted from CCEA June 2009)

4. Which one of the following, in the liquid state, has van der Waals forces and permanent dipole attractions but **not** hydrogen bonds between the molecules?

A CH_4 B CO C H_2O D O_2

(CCEA January 2013)

5. The variation in the boiling point of the Group V hydrides is shown below. (a) Explain why ammonia, NH_3 has a much higher boiling point than phosphine, PH_3. (b) Explain why antimony hydride, SbH_3 has a higher boiling point than arsenic hydride, AsH_3.

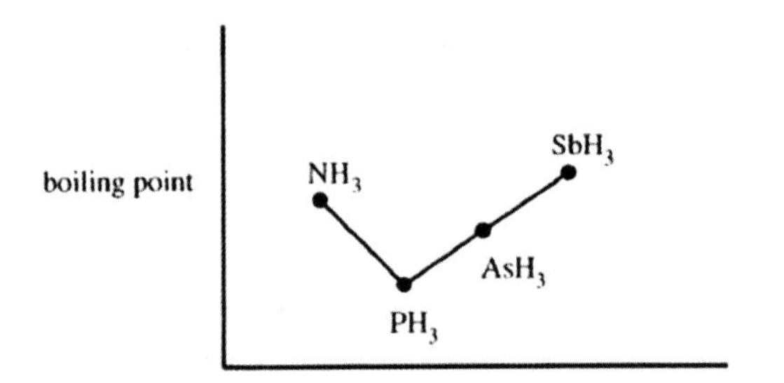

(CCEA June 2005)

6. Why does water have a higher boiling point than hydrogen sulfide?

(CCEA June 2012)

7. The boiling points of the hydrides formed by the Group VII elements (HX) are shown below. (a) Explain the trend in boiling point from hydrogen chloride to hydrogen iodide. (b) Explain why hydrogen fluoride does not follow this trend.

Element (X)	Boiling point of HX (°C)
fluorine	20
chlorine	-85
bromine	-67
iodine	-35

(Adapted from CCEA June 2015)

8. Which of the following is the reason why water boils at 100 °C while the hydrides of the other Group VI elements boil below 0 °C?

A Hydrogen bonding between water molecules

B Ionic bonding in water molecules

C The lower molar mass of water molecules

D The stability of the bonding in water molecules

(CCEA January 2014)

9. The structure of ice is shown below. The water molecules are held together by hydrogen bonds which are a type of intermolecular force.

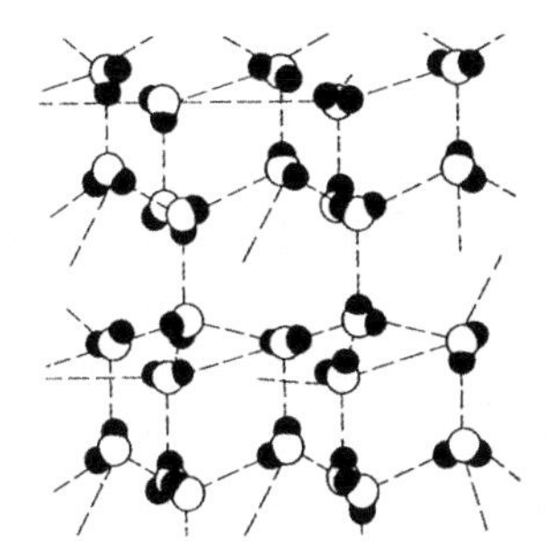

(a) Name two other types of intermolecular force. (b) Explain how hydrogen bonding takes place between water molecules in ice. (c) Explain, using the structure, why ice is less dense than water.

(CCEA January 2011)

10. Water forms hydrogen bonds between neighbouring molecules but is not capable of forming long chains of 'polywater' at room temperature. In contrast, in the liquid state, molecules such as hydrogen fluoride do form short chains. Suggest why water does not form chains and liquid hydrogen fluoride does.

(CCEA January 2011)

11. Explain why water changes to a gas at 100 °C.

(CCEA June 2015)

12. Ammonia has a pyramidal structure and can form hydrogen bonds. (a) Draw two molecules of ammonia and show the hydrogen bond between the molecules. (b) Explain why ammonia loses the ability to form hydrogen bonds when it reacts with a hydrogen ion.

(CCEA January 2011)

13. Which one of the following can **not** form hydrogen bonds?

A H_2O B H_3O^+ C NH_3 D NH_4^+

(CCEA January 2012)

14. Explain why boron trifluoride is a gas despite having a molecular mass which is much greater than that of water.

(CCEA January 2004)

Before moving to the next section, check that you are able to:

- Identify molecules capable of forming hydrogen bonds.
- Use structural formulas to illustrate the formation of hydrogen bonds between molecules.
- Identify the effect of hydrogen bonding on the melting and boiling point of a substance.

Properties of Liquids

In this section we are learning to:

- Identify the types of intermolecular force between the molecules in a liquid.
- Explain the ability of liquids to mix in terms of the types of intermolecular forces between the molecules in each liquid.

Many properties of liquids can be explained in terms of the type of intermolecular forces experienced by the molecules in the liquid. For example, the formation of hydrogen bonds in ethanol, CH_3CH_2OH (Figure 9a) can be used to account for the high boiling point of ethanol. Similarly the ability of ethanol and water to hydrogen bond (Figure 9b) can be used to explain why ethanol and water mix. The mixing of ethanol and water illustrates *the principle of 'like dissolves like' which asserts that liquids will mix if the intermolecular forces in each liquid are similar.*

Figure 9: The formation of hydrogen bonds (a) in ethanol and (b) between ethanol and water.

Liquids that mix are described as **miscible**. Alcohols such as ethanol, CH_3CH_2OH are able to form hydrogen bonds and are therefore miscible in water and other liquids capable of forming hydrogen bonds. In contrast, liquids such as dichloromethane, CH_2Cl_2 are unable to form hydrogen bonds and do not mix with water and other liquids that form hydrogen bonds. Liquids that do not mix are described as **immiscible**.

Dichloromethane, CH_2Cl_2 is however miscible with acetone, CH_3COCH_3 as the combination of van der Waals attraction and dipole attraction between molecules in acetone, and in dichloromethane, is similar in nature to the combination of van der Waals attraction and dipole attraction in a mixture of acetone and dichloromethane.

The principle of 'like dissolves like' can also be extended to molecules that do not have permanent dipoles. For example, the halogens (F_2, Cl_2, ...) do not

Nonpolar liquids	Polar liquids	Hydrogen-bonded liquids
• van der Waals forces	• van der Waals forces • dipole forces	• van der Waals forces • hydrogen bonding
Examples: Bromine, Br_2 Hexane, C_6H_{14}	Examples: Dichloromethane, CH_2Cl_2 Chloroform, $CHCl_3$ Acetone, C_3H_6O	Examples: Ammonia, NH_3 Water, H_2O Ethanol, C_2H_5OH

Polarity increasing →

Figure 10: The classification of liquids used to operate the principle of 'like-dissolves-like'. Liquids in the same category are miscible and will likely be miscible with liquids in a neighbouring category.

have a permanent dipole and cannot form hydrogen bonds. The intermolecular forces between halogen molecules consist solely of van der Waals attraction. As a result, the halogens are misible with liquids containing nonpolar molecules such as hexane, C_6H_{14}.

Thus, in order to operate the principle of like-dissolves-like, we must first classify liquids based on the types of intermolecular force operating between molecules in the liquid. The classification used to operate the principle of like-dissolves-like is illustrated in Figure 10.

Exercise 1.5D

1. Ammonia has a pyramidal structure and can form hydrogen bonds. Explain why ammonia is extremely soluble in water.

   (CCEA January 2011)

2. Carbon dioxide does not have a dipole but is soluble in water. Explain, in terms of intermolecular forces, why carbon dioxide dissolves in water.

   (Adapted from CCEA June 2011)

Before moving to the next section, check that you are able to:

- Use molecular features to identify the types of intermolecular forces operating between the molecules in a substance.
- Recall that liquids containing molecules that experience similar types of intermolecular force are more likely to mix.
- Recall that liquids containing molecules that are held together by different types of intermolecular forces are unlikely to mix.

1.6 Oxidation and Reduction

CONNECTIONS

- Batteries and fuel cells use a combination of reduction and oxidation to produce a voltage.
- Green plants use a combination of reduction and oxidation to produce sugars from carbon dioxide, water and sunlight by photosynthesis.
- Antioxidants such as Vitamin C prevent disease by acting as reducing agents. .

Redox Reactions

In this section we are learning to:

- Describe the oxidation and reduction processes that occur during a redox reaction in terms of the loss or gain of electrons.
- Write half-equations describing oxidation and reduction.
- Identify the oxidising and reducing agents in a redox reaction.
- Describe oxidation and reduction in terms of a change in oxidation state.

A **redox reaction** is *a type of chemical reaction in which oxidation and reduction occur at the same time*. The displacement reaction that occurs when an iron nail is placed in a solution of copper(II) sulfate is an example of a redox reaction.

The displacement of copper by iron:
$Fe_{(s)} + CuSO_{4(aq)} \rightarrow FeSO_{4(aq)} + Cu_{(s)}$

As the reaction proceeds iron atoms on the surface of the nail (Fe) lose electrons to form iron(II) ions (Fe^{2+}) in solution. The electrons from the iron atoms are gained by copper(II) ions (Cu^{2+}) in the solution which then form a layer of copper metal (Cu) on the surface of the nail as shown in Figure 1. The sulfate ions (SO_4^{2-}) do not get involved in the reaction and are omitted when writing the ionic equation for the reaction.

Ionic equation for the displacement of copper:
$Fe_{(s)} + Cu^{2+}_{(aq)} \rightarrow Fe^{2+}_{(aq)} + Cu_{(s)}$

On defining **oxidation** to be *a chemical change in which electrons are lost*, and **reduction** to be *a chemical change in which electrons are gained*, the

Figure 1: The deposition of copper on an iron nail placed in aqueous copper(II) sulfate.

displacement of copper by iron can be described as a redox reaction in which iron is oxidised and copper(II) ions are reduced.

The role of electrons in the redox process can be described by writing **half-equations** for the oxidation and reduction processes.

Half-equation for oxidation:
$Fe_{(s)} \rightarrow Fe^{2+}_{(aq)} + 2e^-$

Half-equation for reduction:
$Cu^{2+}_{(aq)} + 2e^- \rightarrow Cu_{(s)}$

The half-equations reveal that iron acts as a **reducing agent** by *providing electrons* for the reduction of copper(II) ions, and copper(II) ion acts as an **oxidising agent** by *accepting electrons* from iron.

$Fe_{(s)}$	+	$Cu^{2+}_{(aq)}$	$\rightarrow$	$Fe^{2+}_{(aq)}$	+	$Cu_{(s)}$
reducing agent (oxidised)		oxidising agent (reduced)				

If we consider the **oxidation state** of an element to be *the extent to which the element has been oxidised*, the oxidation of iron corresponds to an increase in the oxidation state of iron, and the reduction of copper corresponds to a decrease in the oxidation state of copper.

Exercise 1.6A

1. Define (a) *oxidation* in terms of electron transfer and (b) *reduction* in terms of changes in oxidation state.

   (CCEA June 2010)

2. Copper is displaced from solution when zinc granules are added to aqueous copper(II) sulfate. (a) Write an ionic equation for the reaction. Include state symbols. (b) Explain why the reaction is considered a redox reaction. (c) Write the half-equation for oxidation. (d) Identify the oxidising agent.

   $Zn_{(s)} + CuSO_{4(aq)} \rightarrow ZnSO_{4(aq)} + Cu_{(s)}$

3. A copper wire becomes coated in silver when dipped in aqueous silver nitrate. (a) Write an ionic equation for the reaction. Include state symbols. (b) Explain why the reaction is considered a redox reaction. (c) Write the half-equation for reduction. (d) Identify the reducing agent.

   $Cu_{(s)} + 2AgNO_{3(aq)} \rightarrow Cu(NO_3)_{2(aq)} + 2Ag_{(s)}$

Before moving to the next section, check that you are able to:

- Recall that oxidation and reduction occur at the same time in a redox reaction.
- Describe oxidation as the loss of electrons and reduction as the gain of electrons.
- Identify the oxidising agent in a redox reaction as being reduced, and the reducing agent in a redox reaction as being oxidised.
- Use the chemical equation for a redox reaction to write half-equations describing oxidation and reduction.
- Associate oxidation with an increase in the oxidation state of an element and reduction with a decrease in the oxidation state of an element.

Oxidation Numbers

In this section we are learning to:

- Use oxidation numbers to describe the oxidation states of elements.
- Associate changes in oxidation number with oxidation and reduction

Ionic Compounds

The reaction between sodium metal and chlorine gas is a redox reaction in which sodium is oxidised and chlorine is reduced to form sodium chloride.

The reaction between sodium and chlorine:
$2Na_{(s)} + Cl_{2(g)} \rightarrow 2NaCl_{(s)}$

The Na^+ ions in NaCl are assigned an **oxidation number** of +1 to indicate that a Na^+ ion is formed when a sodium atom loses one electron. Conversely, the Cl^- ions in NaCl are assigned an oxidation number of -1 to indicate that a Cl^- ion is formed when a chlorine atom gains one electron. In this way *a positive oxidation number indicates the number of electrons lost by an element and a negative oxidation number indicates the number of electrons gained by an element. The atoms in an element have neither lost or gained electrons and therefore have an oxidation number of zero.* As a result the atoms in sodium (Na) and chlorine (Cl_2) are assigned an oxidation number of zero, and the reaction between sodium and chlorine is described as a redox reaction in which the oxidation number of sodium increases from 0 to +1, and the oxidation number of chlorine decreases from 0 to -1.

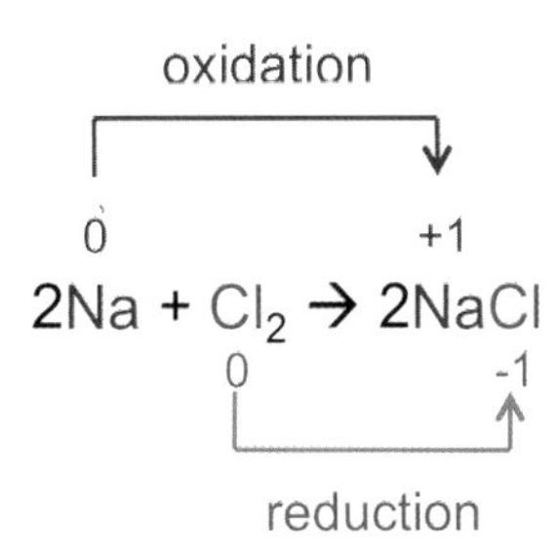

In this way *oxidation increases in the oxidation number of an element and reduction decreases the oxidation number of an element.*

When the elements towards the left and right sides of the Periodic Table combine to form simple ionic compounds such as $MgCl_2$ and CaO they form ions with full outer shells. As a result, the oxidation number of the element in each ion can be directly determined from the Periodic Table as shown in Figure 2.

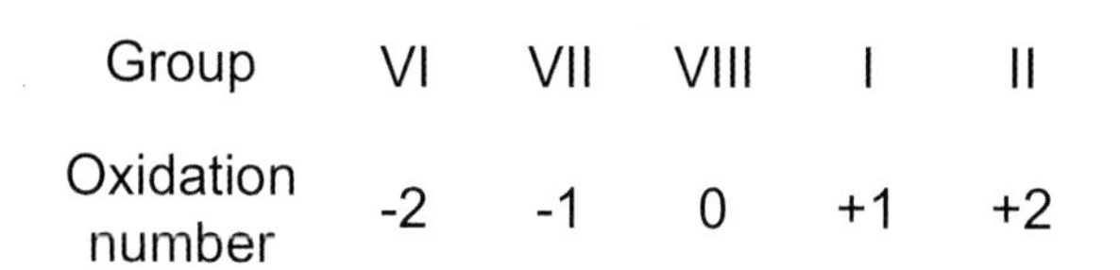

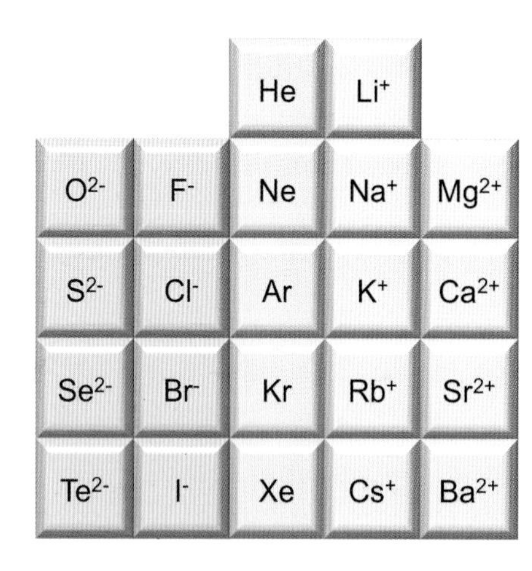

Figure 2: Using the Periodic Table to determine the oxidation states of the elements in simple ionic compounds.

Worked Example 1.6i

Determine the oxidation number (ON) of each element in (a) MgO, (b) Na_2O and (c) $AlBr_3$.

Strategy

- Use Figure 2 to determine the oxidation numbers of Mg, Na, Al, O and Br from their location in the Periodic Table.

Solution

(a) Magnesium (Group II) has an ON of +2.
Oxygen (Group VI) has an ON of –2.

(b) Sodium (Group I) has an ON of +1.
Oxygen (Group VI) has an ON of –2.

(c) Aluminium (Group III) has an ON of +3.
Bromine (Group VII) has an ON of –1.

Elements towards the middle of the Periodic Table are less likely to achieve a full outer shell of electrons when they form compounds. For example, iron routinely forms compounds containing iron(II) ions, Fe^{2+} or iron(III) ions, Fe^{3+} neither of which has a full outer shell.

When determining the oxidation number of an element in a compound it is helpful to remember that *the oxidation numbers of the atoms in one formula of the compound add to zero*. This is always true because the electrons lost by elements trying to achieve full outer shell are gained by other elements as they attempt to achieve a full outer shell.

Rules for assigning oxidation numbers in ionic compounds:

- Use the Periodic Table to assign oxidation numbers to the elements in simple ions such as K^+, Ba^{2+}, Al^{3+}, O^{2-} and Cl^-.
- Determine any remaining oxidation numbers by ensuring that the oxidation numbers of the atoms in the formula add to zero.

Worked Example 1.6ii

Determine the oxidation number of each element in the following compounds.

(a) FeO (b) Ag_2O (c) $FeBr_3$ (d) $CuCl_2$ (e) Cu_2O

Strategy

- The oxidation numbers (ONs) of the atoms in one formula add to zero.

Solution

(a) Oxygen (Group VI) has an ON of –2.
The ON of iron must be +2 if
ON(Fe) + ON(O) = 0.

(b) Oxygen (Group VI) has an ON of –2.
The ON of silver must be +1 if
2 × ON(Ag) + ON(O) = 0.

(c) Bromine (Group VII) has an ON of –1.
The ON of iron must be +3 if
ON(Fe) + 3 × ON(Br) = 0.

(d) Chlorine (Group VII) has an ON of –1.
The ON of copper must be +2 if
ON(Cu) + 2 × ON(Cl) = 0.

(e) Oxygen (Group VI) has an ON of –2.
The ON of copper must be +1 if
2 × ON(Cu) + ON(O) = 0.

Exercise 1.6B

1. Potassium hydride, KH a white crystalline solid, is formed when hydrogen gas is passed over heated potassium. (a) Write an equation for the reaction. Include state symbols. (b) Deduce the oxidation number of potassium and hydrogen in each reactant and product, and use these values to explain the redox change that occurs. (c) Explain the role of hydrogen in the reaction.
2. Solid lithium nitride, Li_3N is formed when lithium reacts with nitrogen. (a) Write an equation for the reaction. Include state symbols. (b) Describe the chemical structure of lithium nitride. (c) Deduce the oxidation number of lithium and nitrogen in each reactant and product, and use these values to explain the redox change that occurs. (d) Explain the role of lithium in the reaction.
3. Nuclear power plants use chlorine(III) fluoride, ClF_3 to produce uranium hexafluoride, UF_6 from uranium tetrafluoride, UF_4. Deduce the oxidation number of uranium in UF_4 and UF_6 and determine the role of ClF_3 in the reaction.

(CCEA June 2003)

Before moving to the next section, check that you are able to:

- Assign oxidation numbers to the atoms in elements and ionic compounds.
- Associate oxidation with an increase in oxidation number and reduction with a decrease in oxidation number.

Nonmetal Compounds

The reaction between nitrogen and hydrogen to form ammonia, NH_3 is also a redox reaction in which hydrogen is oxidised and nitrogen is reduced.

Reaction between nitrogen and hydrogen:

$N_{2(g)} + 3H_{2(g)} \rightarrow 2NH_{3(g)}$

On forming ammonia (NH_3) the nitrogen atom acquires a small negative charge (δ–) by using three electrons to form covalent bonds with the hydrogen atoms. This is indicated by assigning the nitrogen atom in ammonia an oxidation number of –3. Similarly each hydrogen atom in ammonia is assigned an oxidation number of +1 to indicate that it has acquired a small positive charge (δ+) by sharing one electron with nitrogen to form a covalent bond.

In this way the reaction can be described as a redox reaction in which the oxidation number of hydrogen increases from 0 to +1 as it is oxidised, and the oxidation number of nitrogen decreases from 0 to –3 as it is reduced.

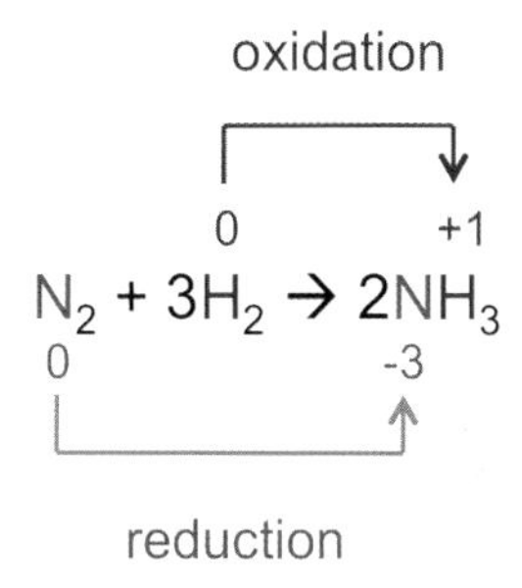

A similar procedure can be used to assign oxidation numbers in other nonmetal compounds. The oxidation numbers assigned to the elements in water, H_2O and hydrogen peroxide, H_2O_2 are shown in Figure 3.

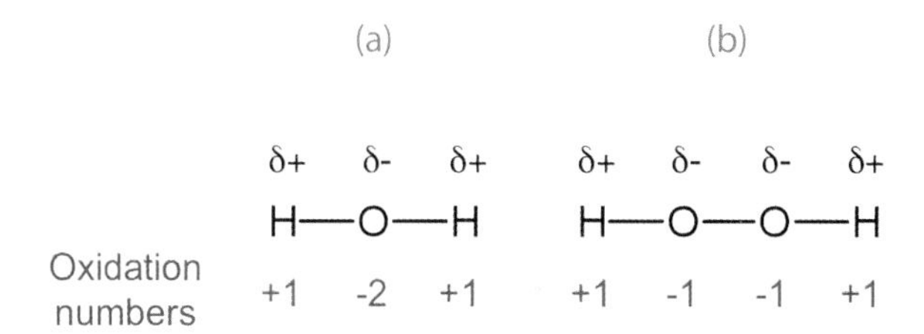

Figure 3: Oxidation numbers assigned to the atoms in (a) water, H_2O and (b) hydrogen peroxide, H_2O_2.

In water the oxygen atom is assigned an oxidation number of –2 as it has acquired a small negative charge by sharing two electrons with the hydrogen atoms. Similarly, each hydrogen atom is assigned an oxidation number of +1 to indicate that it has acquired a small positive charge by sharing one electron.

The bonding in hydrogen peroxide, H_2O_2 is similar to the bonding in water. Each hydrogen atom is assigned an oxidation number of +1 to reflect the small positive charge acquired by sharing one electron with an oxygen atom. Conversely, the oxygen atoms are each assigned an oxidation number of –1 to reflect the small negative charge acquired by sharing one electron with a hydrogen atom. The electrons in the O–O bond are shared equally between the oxygen atoms and do not affect the oxidation state of the oxygen atoms.

Rules for assigning oxidation numbers in nonmetal compounds:

- Hydrogen always has an oxidation number of +1.
- Fluorine always has an oxidation number of –1.
- Oxygen has an oxidation number of –2 except in peroxides where it has an oxidation number of –1.
- Chlorine, bromine and iodine have an oxidation number of –1 except when bonding with more electronegative elements such as oxygen and fluorine.

The rules for assigning oxidation numbers are particularly helpful when attempting to assign oxidation numbers in compounds containing elements that can exist in different oxidation states. It is also helpful to recall that, as in ionic compounds, *the oxidation numbers of the atoms in one formula of a nonmetal compound add to zero.*

For example, in dinitrogen tetroxide, N_2O_4 each oxygen atom is assigned an oxidation number of –2. It then follows that the nitrogen atoms must each have an oxidation number of +4 if their combined oxidation number (+8) balances the combined oxidation number for the four oxygen atoms (–8).

N_2O_4

Oxidation numbers: $2\times(+4) + 4\times(-2) = 0$

Similarly in quartz, SiO_2 which has a giant covalent structure, each oxygen atom is assigned an oxidation number of –2. The oxidation number of each silicon atom in the giant structure must then be +4 if the oxidation numbers of the atoms in the formula add to zero.

SiO_2

Oxidation numbers: $(+4) + 2\times(-2) = 0$

Exercise 1.6C

1. Use oxidation numbers to explain the redox change taking place in the following reaction.

   $H_{2(g)} + Cl_{2(g)} \rightarrow 2HCl_{(g)}$

2. Determine the oxidation number of each underlined element and use the oxidation numbers to explain the redox change taking place.

   $\underline{Cu}O_{(s)} + \underline{H}_{2(g)} \rightarrow \underline{Cu}_{(s)} + \underline{H}_2O_{(l)}$

3. Ammonia can act as a reducing agent. When passed over heated copper(II) oxide the following reaction occurs. Deduce the oxidation numbers for nitrogen and copper in the reactants and products, and use them to explain the redox change.

   $2NH_3 + 3CuO \rightarrow 3Cu + N_2 + 3H_2O$

   (CCEA June 2006)

Before moving to the next section, check that you are able to:

- Assign oxidation numbers to the atoms in nonmetal compounds.

Molecular Ions

The atoms in molecular ions such as sulfate (SO_4^{2-}), nitrate (NO_3^-) and bromate (BrO_3^-) are held together by covalent bonds. As a result the rules for assigning oxidation numbers in nonmetal compounds can be used to assign oxidation numbers to the atoms in a molecular ion.

If we consider the charge on a molecular ion to be the number of electrons lost or gained when the ion was formed from its elements, *the oxidation numbers of the atoms in a molecular ion must add to the charge on the ion.*

For example, in sulfate ion (SO_4^{2-}) we begin by assigning each oxygen atom an oxidation number of –2. If the oxidation numbers of all five atoms in the ion combine to give a charge of –2, sulfur must be assigned an oxidation number of +6 to counter a combined oxidation number of –8 for the four oxygen atoms.

SO_4^{2-}

$(+6) + 4\times(-2)$ = charge on ion

Similarly in bromate ion (BrO_3^-) we can begin by assigning each oxygen atom an oxidation number of −2. If the oxidation numbers of all four atoms in the ion combine to give a charge of −1, bromine must be assigned an oxidation number of +5 to counter a combined oxidation number of −6 for the three oxygen atoms.

$$BrO_3^-$$

(+5) + 3×(−2) = charge on ion

Rules for assigning oxidation numbers in molecular ions:

- Use the rules for nonmetal compounds to assign oxidation numbers to individual atoms in the molecular ion.
- Determine any remaining oxidation numbers by requiring the oxidation numbers of the atoms in the ion to add to the charge on the ion.

Worked Example 1.6iii

Determine the oxidation number of chlorine in the following compounds.

(a) NaCl (b) Cl_2O_7 (c) NaOCl (d) $NaClO_3$

(Adapted from CCEA January 2003)

Stratrgy

- Apply the rules to determine the oxidation numbers (ONs) of the elements in simple ions, molecular ions, and nonmetal compounds as needed.

Solution

(a) NaCl contains Na^+ ions and Cl^- ions.
The ON of sodium is +1.
The ON of chlorine is −1.

(b) Cl_2O_7 is a nonmetal compound.
Oxygen has an ON of −2.
The ON for chlorine must be +7 if
$2 \times ON(Cl) + 7 \times ON(O) = 0$.

(c) NaOCl contains Na^+ and OCl^- ions.
The ON of oxygen is −2.
The ON for chlorine must be +1 if
$ON(Cl) + ON(O) = -1$.

(d) $NaClO_3$ contains Na^+ and ClO_3^- ions.
The ON of oxygen is −2.
The ON for chlorine must be +5 if
$ON(Cl) + 3 \times ON(O) = -1$.

Exercise 1.6D

1. Chlorine reacts with cold dilute sodium hydroxide according to the following equation. Determine the oxidation number of chlorine in each species and explain, using the oxidation numbers of chlorine, why this is a redox reaction.

   $Cl_2 + 2NaOH \rightarrow NaCl + NaOCl + H_2O$

   (CCEA January 2008)

2. Deduce the oxidation number of nitrogen in (a) HNO_3 and (b) NO.

   (CCEA June 2010)

3. The equation for the reaction of silver with nitric acid is given below. Determine the oxidation numbers of the underlined elements and use them to explain the redox reaction taking place.

   $3\underline{Ag} + 4H\underline{N}O_3 \rightarrow 3\underline{Ag}NO_3 + 2H_2O + \underline{N}O$

   (CCEA January 2007)

4. Deduce the oxidation number of hafnium in HfF_7^{3-}.

   (Adapted from CCEA January 2014)

5. Determine the oxidation numbers of potassium and manganese in potassium permanganate, $KMnO_4$.

   (Adapted from CCEA June 2002)

6. Determine the oxidation numbers of potassium and chromium in potassium dichromate, $K_2Cr_2O_7$.

   (Adapted from CCEA June 2015)

Before moving to the next section, check that you are able to:

- Assign oxidation numbers to the atoms in molecular ions.

Additional Problems

Exercise 1.6E

1. The modern match is shown below. The head is a mixture of potassium chlorate, sulfur and phosphorus trisulfide held together by glue. The wood is soaked in ammonium phosphate which acts as a fire retardant.

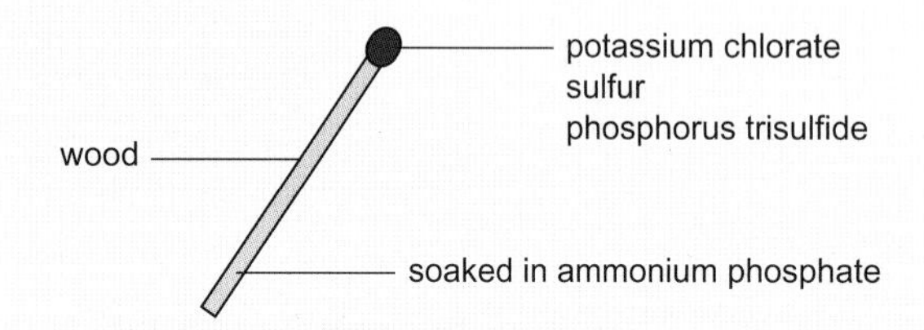

Potassium chlorate reacts with sulfur to form potassium chloride and sulfur dioxide as shown by the following equation.

$2KClO_3 + 3S \rightarrow 2KCl + 3SO_2$

(a) Deduce the oxidation number for each element in the reactants. (b) Deduce the oxidation number for each element in the products. (c) Explain, using these oxidation numbers, why this is a redox reaction.

Phosphorus trisulfide is easily ignited and provides the heat to initiate the reaction between potassium chlorate and sulfur. Phosphorus has an oxidation number of +3 in phosphorus trisulfide. (d) State the formula of phosphorus trisulfide. (e) Suggest whether phosphorus trisulfide is ionic or covalent. Explain your reasons. (f) Name the two products formed when phosphorus trisulfide is completely burnt.

(Adapted from CCEA January 2014)

2. The reaction of oxidising agents with potassium iodide can be used to prepare iodine in the laboratory. Potassium iodide liberates iodine when heated with manganese dioxide and concentrated sulfuric acid. Use oxidation numbers to explain why this is a redox reaction.

$2KI + MnO_2 + 3H_2SO_4$
$\rightarrow 2KHSO_4 + MnSO_4 + 2H_2O + I_2$

(Adapted from CCEA January 2012)

3. Which one of the following reactions is a redox reaction?

A $2NH_3 \rightarrow N_2 + 3H_2$
B $CuO + H_2SO_4 \rightarrow CuSO_4 + H_2O$
C $H_2O + H^+ \rightarrow H_3O^+$
D $AgNO_3 + KI \rightarrow KNO_3 + AgI$

(CCEA June 2008)

4. Which one of the following reactions is a redox reaction?

A $2HBr + H_2SO_4 \rightarrow Br_2 + SO_2 + 2H_2O$
B $NaOH + HCl \rightarrow NaCl + H_2O$
C $NH_3 + HCl \rightarrow NH_4Cl$
D $Ag^+ + Cl^- \rightarrow AgCl$

(CCEA June 2004)

5. Which one of the following is **not** a redox reaction?

A $2Ca(NO_3)_2 \rightarrow 2CaO + 4NO_2 + O_2$
B $Cl_2 + 2I^- \rightarrow I_2 + 2Cl^-$
C $Fe + Cu^{2+} \rightarrow Fe^{2+} + Cu$
D $H_2SO_4 + 2NaOH \rightarrow Na_2SO_4 + 2H_2O$

(CCEA June 2014)

6. Which one of the following reactions shows hydrogen peroxide, H_2O_2 behaving as a reducing agent?

A $H_2O_2 + Ag_2O \rightarrow 2Ag + O_2 + H_2O$
B $H_2O_2 + 2FeSO_4 + H_2SO_4 \rightarrow Fe_2(SO_4)_3 + 2H_2O$
C $H_2O_2 + Na_2SO_3 \rightarrow Na_2SO_4 + H_2O$
D $H_2O_2 + 2KI + 2HCl \rightarrow I_2 + 2KCl + 2H_2O$

(CCEA June 2003)

7. Which one of the following equations shows hydrogen peroxide, H_2O_2 behaving as a reducing agent?

A $2Fe^{2+} + H_2O_2 + 2H^+ \rightarrow 2Fe^{3+} + 2H_2O$
B $2I^- + H_2O_2 + 2H^+ \rightarrow I_2 + 2H_2O$
C $MnO_2 + 2H^+ + H_2O_2 \rightarrow Mn^{2+} + 2H_2O + O_2$
D $PbS + 4H_2O_2 \rightarrow PbSO_4 + 4H_2O$

(CCEA January 2014)

Balancing Redox Reactions

In this section we are learning to:

- Use half-equations to write chemical equations for redox reactions.
- Write balanced half-equations for oxidation and reduction.

The chemical equation for a redox reaction is obtained by *adding the half-equations describing oxidation and reduction in a way that makes the number of electrons generated by oxidation equal to the number of electrons used for reduction.*

For example, the reaction between aluminium powder and bromine to form aluminium bromide, $AlBr_3$ is a redox reaction in which aluminium is oxidised and bromine is reduced.

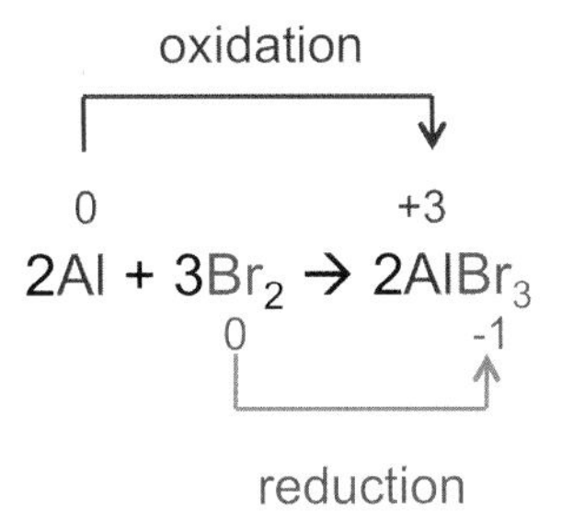

The oxidation number of aluminium increases from 0 to +3 as aluminium atoms are oxidised to form Al^{3+} ions in $AlBr_3$. Conversely, the oxidation number of bromine decreases from 0 to −1 as bromine atoms are reduced to form Br^- ions. The numbers of electrons involved in oxidation and reduction are given in the half-equations describing oxidation and reduction.

Half-equation for oxidation:
$Al \rightarrow Al^{3+} + 3e^-$

Half-equation for reduction:
$Br_2 + 2e^- \rightarrow 2Br^-$

The next step in writing a balanced equation for the reaction is to write **balanced half-equations** in which the number of electrons produced by oxidation equals the number of electrons required for reduction.

Balanced half-equations:

$2Al \rightarrow 2Al^{3+} + 6e^-$ (oxidation)

$3Br_2 + 6e^- \rightarrow 6Br^-$ (reduction)

The chemical equation for the reaction is then obtained by adding the balanced half-equations for oxidation and reduction, and combining the aluminium ions (Al^{3+}) and bromide ions (Br^-) formed by the redox process to make $AlBr_3$.

Chemical equation for the reaction:
$2Al + 3Br_2 \rightarrow 2AlBr_3$

When writing the chemical equation for the reaction, the aluminium ions (Al^{3+}) and bromide ions (Br^-) formed by the redox process are combined to form $AlBr_3$, the product of the reaction.

If a redox reaction occurs in solution, adding the balanced half-equations produces the ionic equation for the reaction. For example, when copper is placed in an aqueous solution of silver nitrate, silver is displaced from the solution as shown in Figure 4.

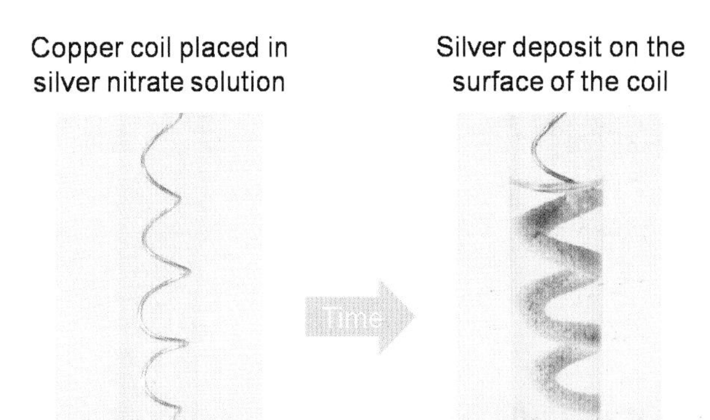

Figure 4: Using copper to displace silver from aqueous silver nitrate.

Displacing silver from aqueous silver nitrate:

$Cu_{(s)} + 2AgNO_{3(aq)} \rightarrow Cu(NO_3)_{2(aq)} + 2Ag_{(s)}$

Half-equations:

$Cu_{(s)} \rightarrow Cu^{2+}_{(aq)} + 2e^-$ (oxidation)
$Ag^+_{(aq)} + e^- \rightarrow Ag_{(s)}$ (reduction)

The half-equations are balanced when the electrons produced by the oxidation of one copper atom are used to reduce two silver ions. Adding the balanced half-equations gives the ionic equation for the reaction.

Balanced half-equations:

$Cu_{(s)} \rightarrow Cu^{2+}_{(aq)} + 2e^-$ (oxidation)

$2Ag^+_{(aq)} + 2e^- \rightarrow 2Ag_{(s)}$ (reduction)

Ionic equation:

$Cu_{(s)} + 2Ag^+_{(aq)} \rightarrow Cu^{2+}_{(aq)} + 2Ag_{(s)}$

In this reaction the nitrate ions (NO_3^-) are spectator ions. The chemical equation for the reaction is obtained by adding the spectator ions to the ionic equation.

Worked Example 1.6iv

(a) Use the following half-equations to write the ionic equation for the reaction of chlorine with a solution of potassium iodide. (b) Explain the role of iodide ions in this reaction.

$I_{2(aq)} + 2e^- \rightarrow 2I^-_{(aq)}$

$Cl_{2(aq)} + 2e^- \rightarrow 2Cl^-_{(aq)}$

(Adapted from CCEA June 2004)

Solution

(a) The half-equations for the reaction of chlorine (Cl_2) with iodide (I^-) are:

$Cl_{2(aq)} + 2e^- \rightarrow 2Cl^-_{(aq)}$

$2I^-_{(aq)} \rightarrow I_{2(aq)} + 2e^-$

The half-equations are balanced. Adding the half-equations gives the ionic equation for the reaction.

$Cl_{2(aq)} + 2I^-_{(aq)} \rightarrow I_{2(aq)} + 2Cl^-_{(aq)}$

(b) Iodide (I^-) reduces chlorine (Cl_2) to chloride ion (Cl^-). Iodide is acting as a reducing agent.

Exercise 1.6F

1. The half-equation for the reduction of concentrated nitric acid is shown below. Write a half-equation for the oxidation of iodide ions to form an iodine molecule, then combine the half-equations to give the ionic equation for the reaction.

   $HNO_3 + 3H^+ + 3e^- \rightarrow NO + 2H_2O$

   (CCEA June 2010)

2. Acidified chloric(I) acid reacts with aqueous iron(II) to form iron(III) ions. Use the following half-equations to write the equation for the reaction.

   $2HOCl_{(aq)} + 2H^+_{(aq)} + 2e^- \rightarrow Cl_{2(aq)} + 2H_2O_{(l)}$

   $Fe^{2+}_{(aq)} \rightarrow Fe^{3+}_{(aq)} + e^-$

   (CCEA January 2003)

3. Oxygen from the air will oxidise iodide ions in an acidic solution. (a) Use the following half-equations to write the equation for the reaction of iodide ions with oxygen in acidic solution. (b) Explain, using electron transfer, which equation represents an oxidation reaction. (c) Explain, using electron transfer, which equation represents a reduction reaction.

   Equation 1 $2I^- \rightarrow I_2 + 2e^-$

   Equation 2 $O_2 + 4H^+ + 4e^- \rightarrow 2H_2O$

   (Adapted from CCEA January 2012)

4. When chlorine gas is bubbled through a solution of Fe^{2+} ions, oxidation to Fe^{3+} ions occurs. Write an equation for this reaction.

   (CCEA June 2015)

5. Francium is one of the least electronegative elements in the Periodic Table and loses electrons when it reacts with chlorine. (a) Write an equation for the loss of an electron from a francium atom. (b) Write the equation for the formation of chloride ions from a chlorine molecule. (c) Write the equation for the reaction of francium with chlorine.

   (Adapted from CCEA January 2014)

6. Use the following half-equations to write an ionic equation for the reaction between acidified manganate(VII) ions and ethanedioate ions.

$MnO_4^- + 8H^+ + 5e^- \rightarrow Mn^{2+} + 4H_2O$

$C_2O_4^{2-} \rightarrow 2CO_2 + 2e^-$

(Adapted from CCEA June 2009)

7. (a) Use the following half-equations to write the equation for the reaction between hydrogen peroxide and hydrazine. (b) Use oxidation numbers to explain why this is a redox reaction.

$N_2H_4 \rightarrow N_2 + 4H^+ + 4e^-$

$H_2O_2 + 2H^+ + 2e^- \rightarrow 2H_2O$

(CCEA January 2005)

If an ionic equation has been constructed correctly the total charge on the products should be the same as the total charge on the reactants. This is also true for any half-equation and can be used to help construct half-equations.

Worked Example 1.6v

When concentrated sulfuric acid reacts with solid sodium bromide the acid reacts with bromide ion to form sulfur dioxide and bromine. (a) State how the oxidation number for sulfur changes during the reaction. (b) Write the half-equation for the formation of bromine from bromide ion. (c) Complete the following half-equation for the formation of sulfur dioxide.

$H_2SO_4 + H^+ \qquad\qquad \rightarrow SO_2 + H_2O$

(d) Write the ionic equation for the reaction of bromide ions with sulfuric acid. (e) Describe the role of bromide ions in the reaction.

(CCEA January 2010)

Solution

(a) The oxidation number of sulfur decreases from +6 in sulfuric acid (H_2SO_4) to +4 in sulfur dioxide (SO_2).

(b) The half-equation for the formation of bromine (Br_2) is: $2Br^- \rightarrow Br_2 + 2e^-$

(c) Balance the number of hydrogen atoms then add the 2 electrons needed to reduce the oxidation number of sulfur from +6 to +4.

$H_2SO_4 + 2H^+ + 2e^- \rightarrow SO_2 + 2H_2O$

(d) The half-equations in (b) and (c) are balanced. Adding gives the ionic equation for the reaction.

$2Br^- + H_2SO_4 + 2H^+ \rightarrow Br_2 + SO_2 + 2H_2O$

(e) Bromide ion acts as a reducing agent by reducing the sulfur in the sulfuric acid.

Exercise 1.6G

1. Bromate(V) ion, BrO_3^- can be reduced to bromine in acidic solution. Determine the values of x, y and z if the half-equation for the reduction has the form:

$2BrO_3^-{}_{(aq)} + xH^+{}_{(aq)} + ye^- \rightarrow Br_{2(aq)} + zH_2O_{(l)}$

(Adapted from CCEA January 2006)

2. An acidic solution of manganate(VII) ion, MnO_4^- can be used to oxidise aqueous iron(II) ions to iron(III) ions. (a) Determine the values x, y and z in the half-equation for the reduction of manganate(VII) ion. (b) Use the following half-equations to write the equation for the reaction between acidified manganate(VII) ions and iron(II) ions.

$MnO_4^- + xH^+ + ye^- \rightarrow Mn^{2+} + zH_2O$

$Fe^{2+} \rightarrow Fe^{3+} + e^-$

3. Caesium is so reactive that it will react with gold to form caesium auride, CsAu. The gold can be obtained from caesium auride by reacting it with water. Caesium hydroxide and hydrogen are the other products. (a) Write an equation for the reaction of caesium auride with water. (b) Use oxidation numbers to explain the redox change. (c) Write half-equations to explain the redox change.

(Adapted from CCEA January 2013)

Before moving to the next section, check that you are able to:

- Use half-equations to write the equation for a redox reaction.
- Write half-equations for oxidation and reduction.

Disproportionation Reactions

In this section we are learning to:

- Refer to the simultaneous oxidation and reduction of an element as disproportionation.
- Identify and explain examples of disproportionation by assigning oxidation numbers.

The term **disproportionation** refers to *a redox reaction in which the same element is oxidised and reduced*. The reaction that occurs when chlorine gas dissolves in water is an example of a disproportionation reaction. Chlorine dissolves in water to give a mixture of hydrochloric acid, $HCl_{(aq)}$ and chloric(I) acid, $HOCl_{(aq)}$.

$$Cl_{2(g)} + H_2O_{(l)} \rightarrow HCl_{(aq)} + HOCl_{(aq)}$$

The resulting solution contains chloride ions, Cl^- in which chlorine has an oxidation number of –1, and chlorate(I) ions, OCl^- in which chlorine has an oxidation number of –1. As a result, when chlorine dissolves in water, the oxidation number of the chlorine atoms forming chlorate(I) increases from 0 to +1, and the oxidation number of the chlorine atoms forming chloride decreases from 0 to –1. In this way chlorine is simultaneously oxidised to form chlorate(I) ions, and reduced to form chloride ions.

Basic oxides such as copper(II) oxide react with acids to produce a salt and water. The copper(II) ion, Cu^{2+} is stable in aqueous solution and gives the resulting solution its characteristic blue colour.

$$CuO_{(s)} + H_2SO_{4(aq)} \rightarrow CuSO_{4(aq)} + H_2O_{(l)}$$

In contrast, copper(I) ion, Cu^+ is not stable in aqueous solution. When copper(I) oxide reacts with acid the copper(I) ions disproportionate to form copper(II) ions and copper metal.

$$Cu_2O_{(s)} + H_2SO_{4(aq)} \rightarrow Cu_{(s)} + CuSO_{4(aq)} + H_2O_{(l)}$$

Half-equation for oxidation:

$$Cu_2O + 2H^+ \rightarrow 2Cu^{2+} + H_2O + 2e^-$$

Half-equation for reduction:

$$Cu_2O + 2H^+ + 2e^- \rightarrow 2Cu + H_2O$$

Worked Example 1.6vi

Which one of the following involves disproportionation?

A $Cl_2 + 2Br^- \rightarrow 2Cl^- + Br_2$

B $Cl_2 + 2OH^- \rightarrow OCl^- + Cl^- + H_2O$

C $Cl_2 + 2Fe^{2+} \rightarrow 2Cl^- + 2Fe^{3+}$

D $Cl_2 + H_2 \rightarrow 2HCl$

(CCEA June 2002)

Solution

Reactions A, B and D are simple redox reactions.

Reaction B is an example of disproportionation in which chlorine disproportionates to form chlorate(I), ClO^- and chloride, Cl^-.

Exercise 1.6H

1. The reaction between hydrogenxenate ions and hydroxide ions is given below. (a) Deduce the oxidation number of xenon in each species. (b) Explain why this is considered an example of disproportionation.

   $$2OH^- + 2HXeO_4^- \rightarrow Xe + 2H_2O + XeO_6^{4-} + O_2$$

   (CCEA January 2009)

2. The equation for the reaction of chlorine with hot concentrated sodium hydroxide is given below. (a) Deduce the oxidation number of chlorine in $NaClO_3$ and NaCl. (b) Explain why this is a disproportionation reaction.

   $$6NaOH_{(aq)} + 3Cl_{2(g)} \rightarrow NaClO_{3(aq)} + 5NaCl_{(aq)} + 3H_2O_{(l)}$$

   (CCEA June 2004)

3. Aqueous iodine reacts readily with sodium hydroxide solution to form sodium iodate, $NaIO_3$. The reaction occurs via the following two-step process.

   $$I_{2(aq)} + 2NaOH_{(aq)} \rightarrow NaI_{(aq)} + NaIO_{(aq)} + H_2O_{(l)}$$

   $$3NaIO_{(aq)} \rightarrow 2NaI_{(aq)} + NaIO_{3(aq)}$$

   (a) Write the overall equation for the reaction of iodine with sodium hydroxide to form sodium iodate. (b) Deduce the oxidation number of iodine in NaIO, NaI and $NaIO_3$ and use these oxidation numbers to explain why this process is an example of disproportionation.

   (CCEA January 2006)

4. Dinitrogen tetroxide, N_2O_4 reacts with water to form nitric acid, HNO_3 and nitrogen(II) oxide, NO. (a) Deduce the oxidation number of nitrogen in each reactant and product, and use these values to explain why the reaction is an example of disproportionation. (b) Write an equation for the reaction.

(Adapted from CCEA June 2012)

Before moving to the next section, check that you are able to:

- Identify examples of disproportionation by assigning oxidation numbers.
- Use oxidation numbers to explain disproportionation.

1.7 Group VII: The Halogens

CONNECTIONS

- The halogens are used to make a wide variety of commercial materials including plastics, pharmaceuticals, pesticides and flame retardants.
- The presence of chlorofluorocarbons (CFCs) and bromofluorocarbons (BFCs) in the atmosphere has been linked to ozone depletion.
- Hydrofluoric acid (HF) is a weak acid but is much more hazardous to humans than strong acids such as hydrochloric acid (HCl)

Properties

In this section we are learning to:

- Recall the colour and physical state of the halogens under normal conditions.
- Explain trends in melting and boiling point within the group in terms of intermolecular forces.
- Account for the solubility of the halogens in water and nonpolar solvents such as hexane in terms of intermolecular forces.

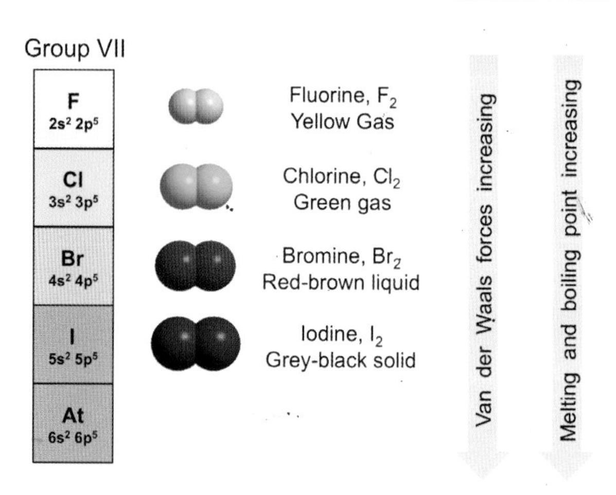

Figure 1: The effect of atomic number on the melting and boiling points of the halogens.

The halogens (Group VII) are a group of reactive nonmetals. Under normal laboratory conditions fluorine (formula: F_2) and chlorine (formula: Cl_2) are diatomic gases, bromine is a diatomic liquid (formula: Br_2), and iodine is a diatomic solid (formula: I_2). The colours and physical states of the halogens under normal laboratory conditions are summarised in Figure 1. The element astatine, At is also a halogen. It is a solid under normal laboratory conditions and has properties characteristic of a semimetal. Astatine is radioactive and has only ever been made in very small quantities. As a result, very little is known about the chemistry of astatine.

As noted in Figure 1, the van der Waals attraction between halogen molecules (X_2) increases as the number of electrons in the molecule increases down

(a)

(b)

Figure 2: (a) Bromine vapour, $Br_{2(g)}$ from the vaporisation of liquid bromine. (b) Iodine vapour, $I_{2(g)}$ formed on heating solid iodine.

the group. This increase in van der Waal attraction explains why the melting and boiling points of the halogens increase down the group, and can be used to explain why bromine vaporises readily at room temperature (Figure 2a), while solid iodine only sublimes when heated (Figure 2b).

The halogens dissolve readily in nonpolar solvents such as hexane (C_6H_{14}) and are much less soluble in polar solvents such as water. The solubility of the halogens can be explained by using the principle that 'like-dissolves-like'.

The molecules in nonpolar solvents such as hexane are held together by van der Waals attractions. In contrast, the molecules in more polar solvents such as water are held together by a combination of van der Waals attraction and dipole forces or hydrogen bonds. As a result, when halogen molecules mix with the molecules in a nonpolar solvent, much less energy is needed to disrupt the weaker van der Waals forces in the solvent, making it easier for the halogens to form a solution.

Chlorine dissolves in water to form a colourless solution known as 'chlorine water', $Cl_{2(aq)}$. In contrast, when chlorine dissolves in nonpolar solvents such as hexane, the solution has a pale-green colour. The colour of chlorine in hexane can be seen by shaking a small sample of chlorine water with an equal volume of hexane as shown in Figure 3. The results of shaking samples of bromine water, $Br_{2(aq)}$ and aqueous iodine, $I_{2(aq)}$ with hexane are also shown in Figure 3 and reveal that while the characteristic yellow/orange colour of aqueous bromine persists in hexane, the purple colour of iodine in hexane contrasts the brown colour of aqueous iodine.

Exercise 1.7A

1. Complete the table to show the colours and physical states of chlorine, bromine and iodine at room temperature and pressure.

Halogen	Colour	Physical State
Chlorine		
Bromine		
Iodine		

(CCEA June 2012)

2. Which one of the following statements is not correct?

   A Iodine has a molecular covalent structure.

   B Iodine contains nonpolar molecules.

   C Iodine exists as a grey-black shiny solid.

   D Iodine is more soluble in water than in hexane.

(CCEA January 2010)

3. Iodine is a grey-black shiny solid. Describe the bonding in solid iodine and explain the structure of iodine crystals. Also explain the relative solubility of iodine in water and hexane.

(CCEA June 2010)

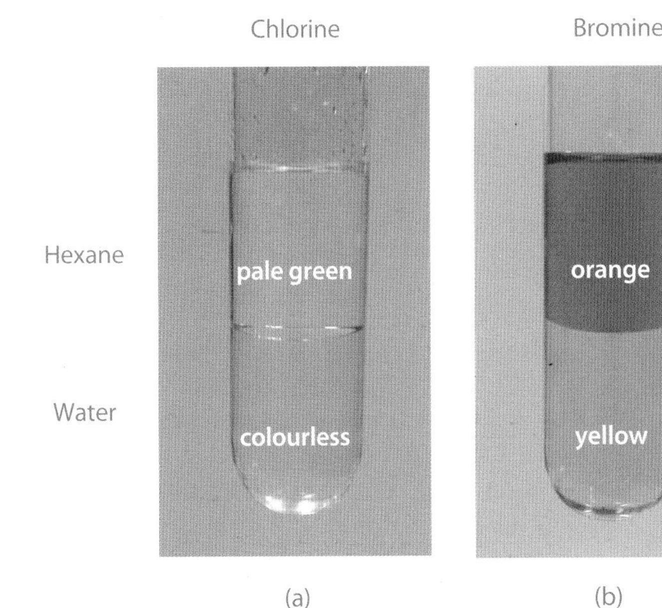

Figure 3: The colour of (a) chlorine, (b) bromine and (c) iodine in hexane (top layer) and water.

4. The boiling points of the halogens are: fluorine (–188 °C), chlorine (–35 °C), bromine (59 °C) and iodine (183 °C). Explain the trend in boiling point.

(Adapted from CCEA January 2009)

5. Astatine was predicted to exist by Mendeleev in his original Periodic Table and was given the name eka-iodine. Complete the following table by predicting some of the properties of astatine.

Property	Result for Astatine
molecular formula	
physical state at room temperature	
colour at room temperature	
colour of vapour	
solubility in water (answer yes or no)	
solubility in hexane (answer yes or no)	

(CCEA June 2011)

Before moving to the next section, check that you are able to:

- Recall the colour and physical state of the halogens.
- Account for the melting and boiling points of the halogens in terms of intermolecular forces.
- Explain the solubility of the halogens in aqueous solution and nonpolar solvents such as hexane in terms of intermolecular forces.
- Recall the colour of the halogens in aqueous solution and nonpolar solvents such as hexane.

Reactivity

In this section we are learning to:

- Recall that the halogens become less reactive as their oxidising ability decreases down the group.
- Describe the reactions of the halogens in terms of their oxidising ability.

Oxidising Ability

Reaction with Water

Fluorine reacts violently with water to produce hydrofluoric acid (HF) and oxygen gas as per Equation 1. Chlorine is much less reactive than fluorine and dissolves in water to form a mixture of hydrochloric acid, $HCl_{(aq)}$ and hypochlorous acid, $HOCl_{(aq)}$ as per Equation 2.

$$2F_{2(g)} + 2H_2O_{(l)} \rightarrow 4HF_{(aq)} + O_{2(g)} \qquad \text{Equation 1}$$

$$Cl_{2(g)} + H_2O_{(l)} \rightarrow HCl_{(aq)} + HOCl_{(aq)} \qquad \text{Equation 2}$$

Bromine reacts in a similar way to form a mixture of hydrobromic acid, $HBr_{(aq)}$ and hypobromous acid, $HOBr_{(aq)}$. Iodine, the least reactive of the halogens, does not react with water and is only sparingly soluble in cold water.

Reaction with Metals

The halogens react readily with metals in Groups I, II and III to form metal halides. For example, chlorine gas reacts vigorously with sodium metal to form solid sodium chloride (Equation 3). Similarly, aluminium turnings react vigorously with liquid bromine to produce white fumes of solid aluminium bromide, $AlBr_3$ (Equation 4). The reaction between aluminium and iodine is less vigorous, again demonstrating that iodine is lower in the group and is therefore less reactive than bromine.

$$2Na_{(s)} + Cl_{2(g)} \rightarrow 2NaCl_{(s)} \qquad \text{Equation 3}$$

$$2Al_{(s)} + 3Br_{2(l)} \rightarrow 2AlBr_{3(s)} \qquad \text{Equation 4}$$

Reaction with Hydrogen

The halogens react with hydrogen gas to form hydrogen halides, HX where X = F, Cl, Br, I (Equation 5). Fluorine gas reacts explosively when it comes into contact with hydrogen. The reaction between chlorine and hydrogen is also explosive when a mixture of hydrogen gas and chlorine gas is exposed to a spark or sunlight. Bromine and iodine react more slowly as they are lower in the group and therefore less reactive than chlorine.

$$H_2 + X_2 \rightarrow 2HX \text{ where } X = F, Cl, Br, I \qquad \text{Equation 5}$$

The reactions of the halogens with water, metals and hydrogen are redox reactions.

On introducing the term **oxidising ability** to describe the ability of a substance to act as an oxidising agent, *the reactions of the halogens are seen to become less vigorous as the oxidising ability of the halogens* *decreases down the group.*

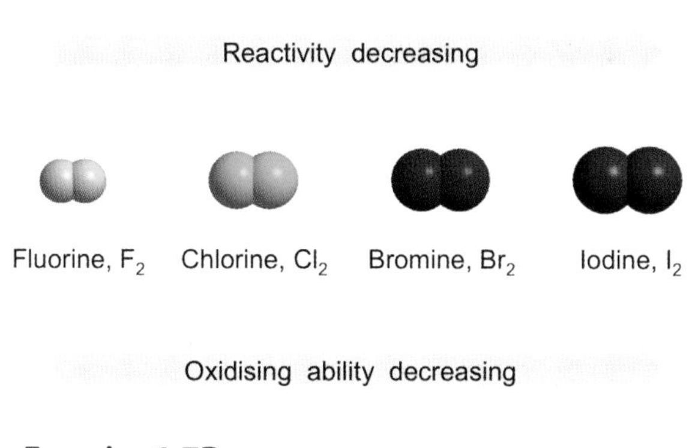

Exercise 1.7B

1. When heated, iron wool reacts with chlorine to form iron(III) chloride, and with bromine vapour to form iron(III) bromide. The reaction between iron wool and iodine vapour is less vigorous and produces iron(II) iodide. (a) Write an equation for the reaction of iron wool with iodine vapour. (b) Use oxidation numbers for iron and iodine to explain the redox change when iron reacts with iodine. (c) Explain why iron(III) iodide is **not** formed when iron reacts with iodine.

2. White phosphorus (P_4) reacts with chlorine to form phosphorus(V) chloride, PCl_5, and with bromine to produce a mixture of phosphorus(III) bromide, PBr_3 and phosphorus(V) bromide, PBr_5, an unstable solid that decomposes above 100 °C. The reaction between phosphorus and iodine produces only phosphorus(III) iodide, PI_3 which is unstable. (a) Write an equation for the reaction between chlorine and phosphorus. (b) Use oxidation numbers for phosphorus and chlorine to explain the redox change when phosphorus reacts with chlorine. (c) Explain, with reference to the oxidation state of phosphorus, how the oxidising ability of the halogens determines the product formed when phosphorus reacts with chlorine, bromine and iodine.

Halogen Displacement

A halogen displacement reaction occurs when a more reactive halogen oxidises a less reactive halogen. If the less reactive halogen is in the form of halide ions (X^-) in solution, the halide ions are oxidised to form the corresponding halogen (X_2), and the more reactive halogen is seen to displace the less reactive halogen from the solution.

For example, bromine will displace iodine from solution (Equation 6), but will not displace chlorine from the solution (Equation 7) as the chlorine would immediately reform the reactants by oxidising the bromide ions formed in the reaction.

$$Br_{2(aq)} + 2I^-_{(aq)} \rightarrow 2Br^-_{(aq)} + I_{2(aq)} \qquad \text{Equation 6}$$
$$Br_{2(aq)} + 2Cl^-_{(aq)} \rightarrow \text{no reaction} \qquad \text{Equation 7}$$

Halogen displacement can also be interpreted in terms of the ability of halide ions to act as a reducing agent. For example, bromine will displace iodine (Equation 6) as iodide is a better reducing agent than bromide, and will not react with chlorine (Equation 7) as bromide, which would be formed if the reaction occurred, is a better reducing agent than chloride.

The relationship between the oxidising ability of the halogens and the reducing ability of the corresponding halide ions is shown in Figure 4.

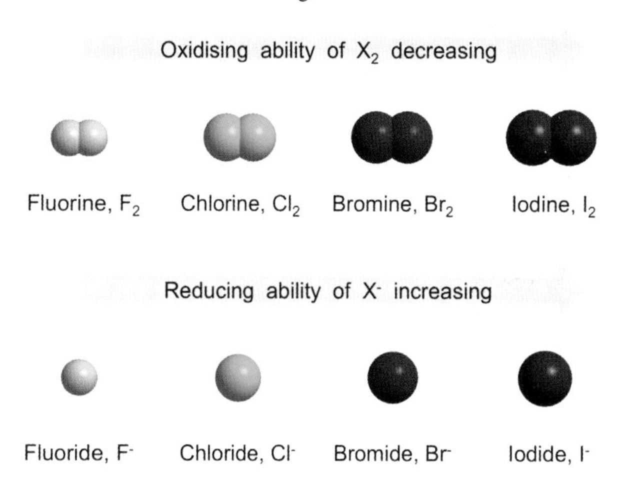

Figure 4: The relationship between the oxidising ability ofthe halogens and the reducing ability of the corresponding halide ions.

Worked Example 1.7i

A compound produces a lilac colour in a flame test. When chlorine is bubbled into an aqueous solution of the compound the solution turns from colourless to yellow/orange. Identify the compound.

(Adapted from CCEA January 2010)

Strategy

- The flame colour can be used to identify the metal in the compound. *(for more details see the chapter on 'Qualitative Analysis').*
- The colour produced by the halogen displacement

reaction can be used to identify the halide ions in the solution.

Solution

- The flame colour indicates that the compound contains potassium, and the yellow/orange colour indicates that bromide ions are present in the solution.
- The compound is potassium bromide.

Exercise 1.7C

1. (a) Write the ionic equation for the reaction between bromine solution and aqueous sodium iodide. (b) Describe the observations that would indicate a reaction has taken place when aqueous sodium iodide is added to a solution of bromine.

   (CCEA January 2011)

2. (a) Write the equation for the reaction between chlorine solution and aqueous sodium bromide. (b) Describe what is observed when chlorine solution is added to aqueous sodium bromide.

   (CCEA January 2011)

3. State what is observed, and write an equation for the reaction that occurs, when chlorine gas reacts with aqueous iodide ions.

   (CCEA June 2013)

4. Astatine, the last element in the halogen group, was synthesised in 1940. Write the equation for the reaction of iodine with sodium astatide.

   (CCEA June 2011)

5. Which one of the following properties is a characteristic of astatine?

   A It has an electronegativity value greater than that of iodine.

   B It is a solid at room temperature and pressure.

   C It oxidises bromide ions to bromine.

   D Its hydride exhibits more hydrogen bonding than hydrogen iodide.

   (CCEA January 2014)

Before moving to the next section, check that you are able to:

- Recall that the halogens become less reactive as their oxidising ability decreases down the group.
- Explain the reactions of the halogens in terms of their oxidising ability.
- Account for the colour changes that occur when halogens displace halide ions from aqueous solution.

Redox Chemistry

In this section we are learning to:

- Recall the reaction of chlorine with water and the reactions of the halogens with sodium hydroxide solution.
- Explain the redox changes that occur when the halogens react with water and sodium hydroxide solution in terms of disproportionation.
- Recall advantages and disadvantages of using chlorine and ozone to purify drinking water.
- Account for the products formed when preparing a hydrogen halide by reacting the corresponding halide salt with concentrated acids.

Reactions of Halogens with Aqueous Alkali

As the oxidising ability of the halogens decreases down the group they increasingly form compounds in which the halogen has an oxidation number of +1, +5 or even +7. This trend can be demonstrated by reacting halogens with water and aqueous solutions such as sodium hydroxide solution. For instance, chlorine gas reacts with water to form a mixture of hydrochloric acid, HCl and hypochlorous acid, HClO (Equation 8).

$$Cl_{2(g)} + H_2O_{(l)} \rightarrow HCl_{(aq)} + HClO_{(aq)} \qquad \text{Equation 8}$$

The resulting solution turns blue litmus paper red, and then white, as the hypochlorite (ClO^-) ions in the solution 'bleach' the indicator. Hypochlorite ion is the active ingredient in bleach.

Hydrochloric acid, HCl is an aqueous solution of hydrogen (H^+) ions and chloride (Cl^-) ions, and is an example of a **strong acid** as *all of the molecules dissociate to form hydrogen ions*. In contrast, hypochlorous acid, HClO is a **weak acid** as *only a fraction of the molecules dissociate to form hydrogen ions*. As a result, a solution of hypochlorous acid contains molecules of hypochlorous acid (HClO), hydrogen ions, and hypochlorite (ClO^-) ions. In this way the reaction between chlorine and water produces chloride ions that contain chlorine with an oxidation number of −1, and hypochlorite ions that contain chlorine with an oxidation number of +1, and is an example of disproportionation in which chlorine is simultaneously oxidised and reduced.

The systematic name for the hypochlorite ion, ClO^- is chlorate(I) ion, where the roman numeral is used to indicate the oxidation number of chlorine.

Chlorine also disproportionates when dissolved in sodium hydroxide solution. In cold dilute sodium hydroxide chlorine disproportionates to form a solution containing sodium chloride, NaCl and sodium chlorate(I), NaClO. The ionic equation for the reaction is given in Equation 9.

$$Cl_{2(g)} + 2OH^-_{(aq)} \rightarrow Cl^-_{(aq)} + ClO^-_{(aq)} + H_2O_{(l)}$$ Equation 9

Thus, in cold sodium hydroxide solution chlorine is simultaneously reduced to chloride ion (Cl^-) and oxidised to form chlorate(I) ions (ClO^-). If the solution is heated to 70 °C the chlorate(I) ions disproportionate to give a mixture of chloride, Cl^- and chlorate(V), ClO_3^- ions (Equation 10). The chlorate(V) ion contains chlorine with an oxidation number of +5. Chlorate(V) salts such as sodium chlorate(V), $NaClO_3$ are powerful oxidising agents.

$$3ClO^-_{(aq)} \rightarrow 2Cl^-_{(aq)} + ClO_3^-{}_{(aq)}$$ Equation 10

The disproportionation of chlorate(I) ion (Equation 10) involves the reduction of chlorate(I), ClO^- to form chloride, Cl^- (the oxidation number of Cl decreases from +1 to −1) and the oxidation of chlorate(I) to form chlorate(V), ClO_3^- (the oxidation number of Cl increases from +1 to +5).

When chlorine gas dissolves in hot concentrated sodium hydroxide, chlorate(I) ions from the disproportionation of chlorine (Equation 9) disproportionate to form a mixture of chloride ions and chlorate(V) ions (Equation 10). The overall reaction that occurs is summarised by Equation 11.

$$3Cl_{2(g)} + 6OH^-_{(aq)} \rightarrow 5Cl^-_{(aq)} + ClO_3^-{}_{(aq)} + 3H_2O_{(l)}$$ Equation 11

Bromine and iodine react in a similar way to chlorine. The corresponding bromate(I), BrO^- and iodate(I), IO^- ions are less stable than chlorate(I) with the result that bromate(I) disproportionates to bromide and bromate(V) at 15 °C, and iodate(I) rapidly disproportionates to iodide and iodate(V) at 0 °C. The relative ease with which the halate(I) ions: ClO^-, BrO^- and IO^- disproportionate to give the corresponding halate(V) ions: ClO_3^-, BrO_3^- and IO_3^- *reflects the greater stability of the +5 oxidation state as the oxidising ability of the halogens decreases down the group.*

Talking Point

In many countries chlorine is added to drinking water as part of the purification process. Ozone, O_3 is sometimes used in place of chlorine as it is more effective against microorganisms such as bacteria, viruses and algae. It is also a better oxidising agent, and will oxidise many of the compounds that produce a bad taste or bad smell when dissolved in water, making the water taste and smell better.

Unlike chlorine, ozone does not remain in the water to prevent the re-growth of microorganisms. Ozone also oxidises any bromide ions present in the water to bromate(V) ions, which may cause cancer. The oxidation of bromide ion also produces reactive forms of bromine that combine with any organic compounds present in the water to form potentially harmful bromine-containing compounds. Chlorine is therefore considered safer than ozone as bromate(V) ions and other bromine-containing compounds do not form when chlorine is added to water containing bromide ion.

Exercise 1.7D

1. Chlorine is used to sterilise water. (a) Write an equation for the reaction of chlorine with water. (b) Using changes in oxidation number explain why this is considered to be a disproportionation reaction. (c) Ultraviolet light does not react with water and is as effective as chlorine at sterilising water. Suggest the advantages and disadvantages of storing and using chlorine to sterilise water.

(CCEA June 2013)

2. Bromine reacts with water in a similar way to chlorine. (a) Suggest the equation for the reaction of bromine with water. (b) Using oxidation numbers explain why this reaction is an example of disproportionation.

(CCEA June 2014)

3. The reaction between chlorine and cold dilute sodium hydroxide is used in the manufacture of bleach. (a) Write the equation for this reaction. (b) This reaction is described as disproportionation. Explain the meaning of this term.

(CCEA June 2010)

4. Bromine water reacts with cold, dilute alkali as shown below. (a) State the colour change observed during this reaction. (b) State the oxidation states of bromine in the reaction and use them to explain why this reaction is an example of disproportionation.

$Br_{2(aq)} + 2OH^-_{(aq)} \rightarrow Br^-_{(aq)} + BrO^-_{(aq)} + H_2O_{(l)}$

(CCEA June 2015)

5. (a) Write the equation for the reaction of chlorine with hot concentrated sodium hydroxide solution. (b) Name the type of redox reaction taking place.

(CCEA June 2012)

6. Potassium chlorate, $KClO_3$ is manufactured using the reaction between chlorine and potassium hydroxide. (a) Write the equation for the reaction. (b) State the conditions under which the reaction is carried out.

(CCEA January 2014)

7. Write the ionic equation for the reaction of bromine with hydroxide ions to produce bromate(V), BrO_3^- ions.

(CCEA June 2015)

8. Chlorine does not undergo disproportionation when reacted with

A cold dilute sodium hydroxide solution

B hot concentrated sodium hydroxide solution

C potassium bromide solution

D water

(CCEA January 2013)

9. Acidified iodate(V) ions react with iodide ions according to the following equation. Which description (A-D) best describes the redox reaction and colour change that occurs?

$IO_3^-{}_{(aq)} + 5I^-_{(aq)} + 6H^+_{(aq)} \rightarrow 3I_{2(aq)} + 3H_2O_{(l)}$

	Redox	Colour Change
A	iodide ions are oxidised	brown to colourless
B	iodide ions are reduced	colourless to brown
C	iodate(V) ions are oxidised	brown to colourless
D	iodate(V) ions are reduced	colourless to brown

(CCEA June 2004)

Before moving to the next section, check that you are able to:

- Write chemical equations for the reaction of chlorine with water, cold dilute sodium hydroxide and hot concentrated sodium hydroxide.
- Explain the redox changes that occur when chlorine reacts with water and sodium hydroxide solution in terms of disproportionation.
- Recall advantages and disadvantages of using chlorine and ozone to purify drinking water.

Reaction of Halide Ions with Concentrated Acid

The hydrogen halides, HX (X = F, Cl, Br, I) are colourless gases that react with water to form acidic solutions. Aqueous solutions of HCl, HBr, and HI are strong acids. An aqueous solution of HF is a weak acid.

The hydrogen halides can be prepared by reacting the corresponding sodium halide, NaX with concentrated sulfuric acid, H_2SO_4 (Equation 12) or concentrated phosphoric acid, H_3PO_4 (Equation 13).

$$NaX + H_2SO_4 \rightarrow NaHSO_4 + HX \ (X = F, Cl, Br, I)$$

Equation 12

$$NaX + H_3PO_4 \rightarrow NaH_2PO_4 + HX \ (X = F, Cl, Br, I)$$

Equation 13

The reactions are very exothermic and the hydrogen halides form 'steamy' fumes as they come into contact with water vapour in the air.

Concentrated sulfuric acid is a better oxidising agent than concentrated phosphoric acid and oxidises a portion of the HBr and HI formed in the reaction. The oxidation of HBr produces bromine and pungent

Reducing ability of X^- increasing

	Fluoride, F^-	Chloride, Cl^-	Bromide, Br^-	Iodide, I^-
Product of H_2SO_4 + NaX	HF	HCl	HBr and some Br_2	I_2 and some HI

Figure 5: The relationship between the products formed when concentrated sulfuric acid reacts with a sodium halide salt and the reducing ability of the halide ion formed in the reaction.

fumes of sulfur dioxide (Equation 14). The bromine is observed as a red-brown vapour. Similarly, the oxidation of HI produces a mixture of iodine and sulfur dioxide (Equation 15). The iodine formed in the reaction is observed as a purple vapour and may also accumulate in the form of a grey-black solid.

$$2HBr + H_2SO_4 \rightarrow Br_2 + SO_2 + 2H_2O \quad \text{Equation 14}$$

$$2HI + H_2SO_4 \rightarrow I_2 + SO_2 + 2H_2O \quad \text{Equation 15}$$

Iodide ions from the dissociation of HI further reduce sulfur dioxide to a mixture of hydrogen sulfide (H_2S) and sulfur (S). The production of hydrogen sulfide results in the smell of rotten-eggs and the production of sulfur results in the formation of a yellow solid. The relationship between the products of the reaction and the reducing ability of the halide formed in the reaction is summarised in Figure 5.

Worked Example 1.7ii

The diagram below shows a common method of preparing hydrogen chloride gas in the laboratory.

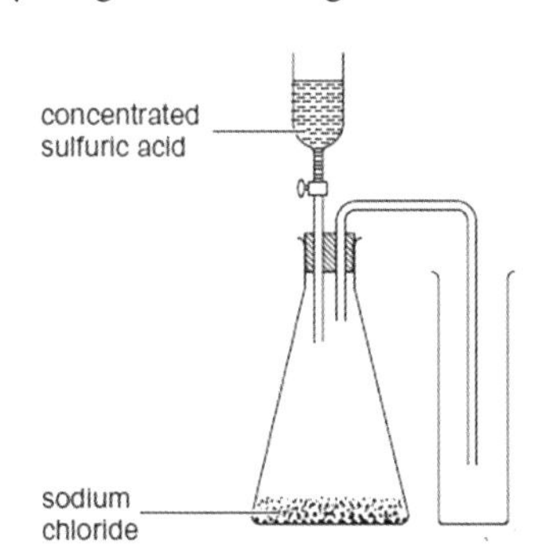

(a) Write the equation for the reaction of sodium chloride with concentrated sulfuric acid. (b) Explain why sodium hydroxide pellets would not be suitable to dry the gas produced in the reaction. (c) Explain whether this method could be used to prepare hydrogen bromide by reacting concentrated sulfuric acid with sodium bromide.

(Adapted from CCEA June 2011)

Solution

(a) $NaCl + H_2SO_4 \rightarrow NaHSO_4 + HCl$

(b) Sodium hydroxide is not a suitable drying agent as it would react with the hydrogen chloride gas produced in the reaction.

(c) The method is not suitable as hydrogen bromide would be oxidised to bromine by the sulfuric acid.

Exercise 1.7E

1. Which one of the following is the most powerful reducing agent?

   A Bromine atom

   B Chlorine atom

   C Fluoride ion

   D Iodide ion

   (CCEA June 2015)

2. Which one of the following is the strongest reducing agent?

   A F^-

   B F_2

   C I^-

   D I_2

   (CCEA June 2014)

3. Which one of the following is **not** produced when concentrated sulfuric acid reacts with sodium bromide?

   A Bromine

   B Hydrogen sulfide

   C Hydrogen bromide

   D Sulfur dioxide

   (CCEA January 2005)

4. Bromine is produced when sodium bromide reacts with concentrated sulfuric acid. Name *four other products* formed during the reaction.

   (Adapted from CCEA June 2014)

5. Which one of the following describes the reaction between solid sodium chloride and concentrated sulfuric acid?

   A Disproportionation

   B Exothermic

   C Neutralisation

   D Redox

   (CCEA June 2014)

6. Which one of the following solids will react with concentrated sulfuric acid to give hydrogen sulfide?

   A Calcium bromide

   B Magnesium iodide

   C Potassium chloride

   D Sodium fluoride

   (CCEA January 2013)

7. Which one of the following is **not** produced in the reaction between concentrated sulfuric acid and sodium iodide at room temperature?

   A I_2 B HI C $NaHSO_4$ D SO_3

   (CCEA June 2008)

8. Solid samples of sodium chloride, sodium bromide and sodium iodide can be distinguished using concentrated sulfuric acid. (a) Balance the following half-equation for the reduction of concentrated sulfuric acid to form hydrogen sulfide.

   $H_2SO_4 + H^+ + \quad\quad \rightarrow H_2S + H_2O$

   (b) Combine the half-equation with the following half-equation to produce a balanced equation for the reaction between concentrated sulfuric acid and sodium iodide.

   $2I^- \rightarrow I_2 + 2e^-$

   (c) State one observation that indicates the formation of hydrogen sulfide. (d) Name two other products formed during the reaction. (e) Suggest why iodide ions are stronger reducing agents than chloride ions.

   (Adapted from CCEA June 2012)

9. Give two observations when concentrated sulfuric acid is added to sodium iodide.

   (CCEA June 2009)

10. When concentrated sulfuric acid is added to a solid potassium halide O, a red-brown gas P, and two colourless gases Q and R are formed. Identify O, P, Q and R.

    (CCEA June 2015)

Before moving to the next section, check that you are able to:

- Write an equation for the preparation of a hydrogen halide by the reaction of concentrated sulfuric acid or concentrated phosphoric acid with the corresponding halide salt.
- Account for the products formed when preparing a hydrogen halide in terms of the reducing ability of the halide.

1.8 Volumetric Analysis

CONNECTIONS

Volumetric analysis techniques can be used to:

- Analyse the biological activity of aquatic plants by determining the amount of dissolved oxygen in water.
- Determine the amount of vitamins and minerals in food.
- Test for diabetes in humans by determining the amount of glucose in urine.

Working with Solutions

In this section we are learning to:

- Determine the level of accuracy that can be achieved when using a scale to estimate the volume of a solution.
- Recall how to use volumetric glassware to measure, prepare and transfer volumes of solution accurately.
- Identify procedures to maximise the accuracy and reliability of volumes measured, prepared and transferred.

Volumetric analysis refers to the procedures and techniques used to analyse the composition of solutions. If a solution is to be analysed accurately it is necessary to develop special techniques for handling solutions. This section of the course introduces the techniques used to handle solutions in the laboratory and accurately analyse their composition.

Volumetric Glassware

The items of glassware shown in Figure 1 are used to measure and transfer volumes of solution in the laboratory. A **volumetric pipette** (A) is used when it is necessary to accurately measure out and transfer a volume of solution. Measuring cylinders (B) are less accurate and should only be used when the volume to be measured does not affect the outcome of the analysis.

A **burette** (C) is used in experiments where it is necessary to keep an accurate record of the volume of solution used. Beakers (D) and conical flasks (E) are used to hold volumes of solution. They are not used to measure volumes of solution or prepare solutions. A **volumetric flask** (F) can be used to accurately prepare solutions of a specified concentration by dissolving a solid or diluting a concentrated solution.

Measuring cylinders and burettes come in different sizes and have a scale. The accuracy of the volumes

Figure 1: Common items of volumetric glassware include a volumetric pipette (A), measuring cylinders (B), a burette (C), beakers (D), conical flasks (E) and volumetric flasks (F).

measured using the scale depends on the size of the interval between divisions on the scale. The scales on a 50 cm^3 measuring cylinder, a 10 cm^3 measuring cylinder and a 50 cm^3 burette are shown in Figure 2. The scale on the burette (Figure 2c) is distinct from the scales on other pieces of glassware as it increases downwards, allowing the amount of liquid dispensed from the burette to be measured.

At this level of introduction, *volumes read from a scale should be considered accurate to one-half the size of the smallest division on the scale*. For example, the scale on the 50 cm^3 measuring cylinder in Figure 2a is marked at 1 cm^3 intervals and is therefore accurate to 0.5 cm^3. As a result, the volume of solution shown in Figure 2a is judged to be closer to 32.5 cm^3 than 32.0 cm^3 and is recorded as 32.5 cm^3.

Similarly, the scale on the 10 cm^3 measuring cylinder in Figure 2b is marked at intervals of 0.2 cm^3 and is accurate to $0.2 \div 2 = 0.1$ cm^3. As a result, the volume of solution shown in Figure 2b is judged to be closer to 6.3 cm^3 than 6.2 cm^3 and is recorded as 6.3 cm^3.

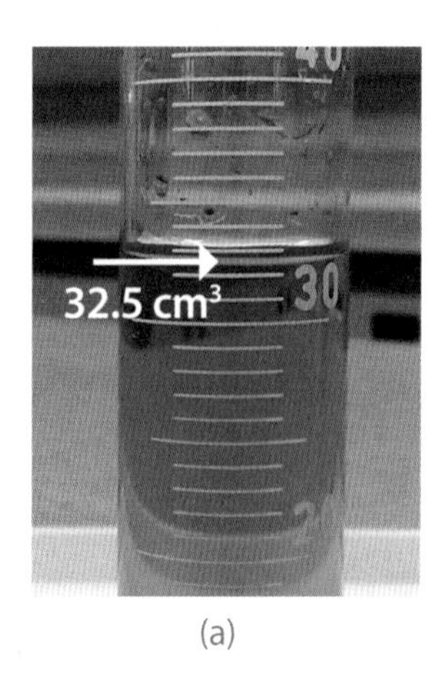

(a)

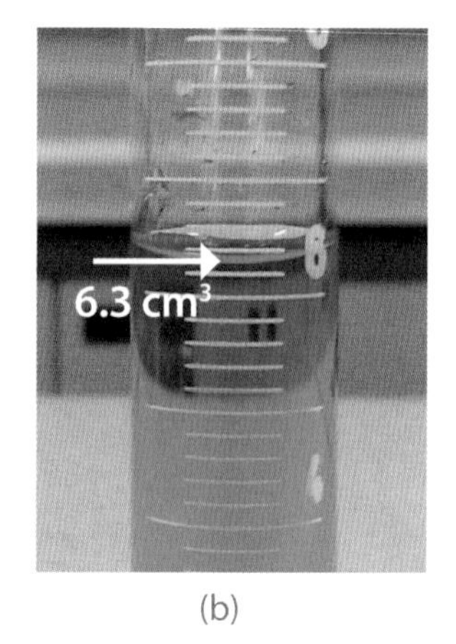

(b)

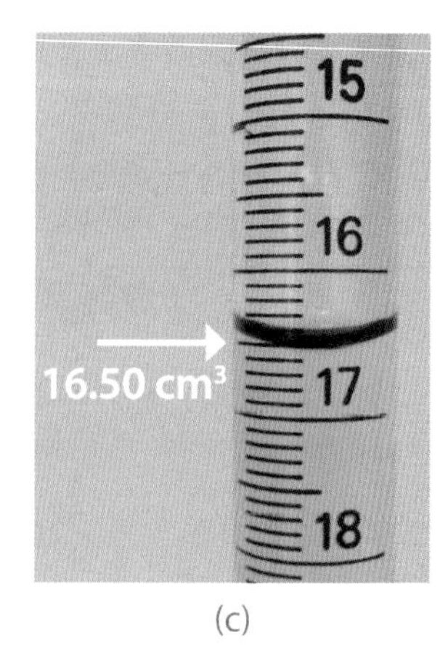

(c)

Figure 2: Scales on a (a) 50 cm³ measuring cylinder, (b) 10 cm³ measuring cylinder and (c) 50 cm³ burette.

The scale on the 50 cm³ burette in Figure 2c is even more accurate than the scale on the small measuring cylinder. The scale on the burette is marked at intervals of 0.1 cm³ and is therefore accurate to $0.1 \div 2 = 0.05$ cm³. As a result, the volume of solution shown in Figure 2c is judged to be nearer to 16.50 cm³ than 16.55 cm³ and is recorded as 16.50 cm³.

In contrast, volumetric flasks and many types of volumetric pipette do not have a scale. The volume of liquid held by the flask or pipette is instead measured by requiring the lowest point on the **meniscus**, the curved surface of the liquid, to reach the fill line or 'mark' on the glassware as shown in Figure 3.

Making Solutions

A fixed volume of solution with a specified concentration is made by transferring a known amount of substance to a volumetric flask that holds the required volume of solution and adding deionised water to make-up the required volume of solution.

Method

1. Dissolve a known (weighed) amount of solid in a minimum of deionised water and transfer the solution to a volumetric flask.
2. Use a wash bottle filled with deionised water to wash any drops of solution remaining on the glassware into the volumetric flask.
3. Add deionised water to the contents of the volumetric flask until the bottom of the meniscus lies on the fill line.
4. Stopper the flask and invert several times to mix.

The procedure used to transfer the solution to the volumetric flask is shown in Figure 4. The following steps can be taken to ensure that the concentration of the resulting solution is accurate.

Ensuring Accuracy

1. Ensure that the solid has completely dissolved.
2. Ensure that every drop of solution has been washed into the volumetric flask.

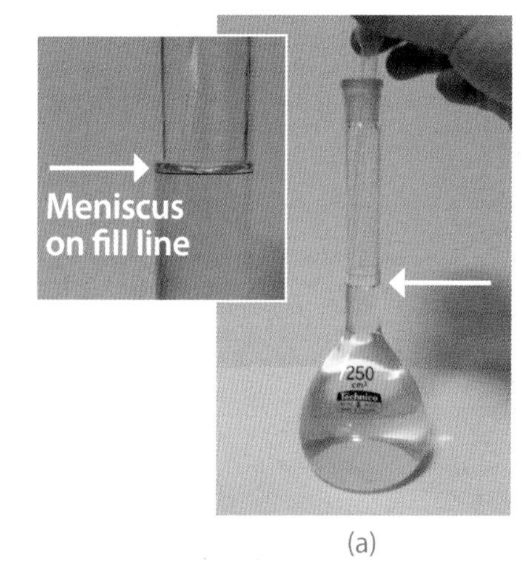

(a)

(b)

Figure 3: Use of the fill line to add the correct amount of solution to (a) a volumetric flask and (b) a volumetric pipette.

3. Wash the inside surface of the volumetric flask with deionised water when making-up the solution to the fill line.
4. Ensure that the bottom of the meniscus lies on the fill line.

Diluting Solutions

A volumetric flask can also be used to make a solution with a specified concentration by diluting a more concentrated solution whose concentration is known.

Method

1. Wash the inside surfaces of a volumetric pipette with 2–3 cm^3 of the concentrated solution.
2. Use the volumetric pipette and a pipette filler to transfer a fixed volume of the concentrated solution to the volumetric flask.
3. Dilute the solution by adding deionised water until the bottom of the meniscus lies on the fill line.
4. Stopper the flask and invert several times to mix.

The procedure used to dilute a solution is illustrated in Figure 5. The following steps can be taken to ensure that the concentration of the resulting solution is accurate.

Ensuring Accuracy

1. Wash the entire inside surface of the pipette when rinsing.
2. Use a wash bottle to wash the inside surface of the volumetric flask with deionised water when making-up the solution to the fill line.
3. Ensure that the bottom of the meniscus lies on the fill line.

Exercise 1.8A

1. Some liquid oven cleaners contain sodium hydroxide. You have been provided with a solution containing 25.0 cm^3 of oven cleaner diluted to 500 cm^3 with distilled water. Give an account of how you would prepare the diluted solution of oven cleaner and then how you would safely transfer 25.0 cm^3 of the diluted solution to a conical flask.

(CCEA June 2010)

Using a Volumetric Pipette

A volumetric pipette is used to accurately measure out a fixed volume of solution. The only markings on a volumetric pipette are the volume held by the pipette and the fill line.

Method

1. Use a pipette filler to draw 2–3 cm^3 of solution into the pipette.
2. Wash the entire inner surface of the pipette with the solution.
3. Allow the wash to drain from the pipette.
4. Use a pipette filler to draw solution into the pipette until the bottom of the meniscus lies on the fill line.

(a) (b)

Figure 4: Making a solution using a volumetric flask. (a) The solid is dissolved in a minimum of deionised water. (b) Deionised water is used to wash the solution into the volumetric flask.

(a) (b)

Figure 5: Diluting a solution using a volumetric flask. (a) A volumetric pipette is used to transfer a fixed volume of concentrated solution into the volumetric flask. (b) Deionised water is used to make-up the solution to the fill line.

The solution can then be transferred to a new container by allowing it to flow freely from the pipette as illustrated in Figure 6. The following steps can be taken to ensure that the volume of solution measured and transferred using the pipette is accurate.

Ensuring Accuracy

1. Be sure to wash the entire inside surface of the pipette when rinsing.
2. Remove any air bubbles that remain in the pipette after filling.
3. Ensure that the bottom of the meniscus lies on the fill line.
4. Do not force the final drop from the pipette when transferring the solution.

Figure 6: Using a volumetric pipette to transfer a fixed volume of solution. A drop of solution remains in the pipette when the solution flows freely from the pipette.

Using a Burette

A burette is used to dispense measured amounts of solution accurately. A burette is often used when the amount of reactant added to a reaction mixture must be carefully controlled during an experiment.

Method

1. Use a funnel to add 2–3 cm^3 of solution to the burette.
2. Wash the entire inner surface of the burette with the solution.
3. Open the tap and allow the solution to drain.
4. Use a funnel to overfill the burette by 2–3 cm^3 then remove the funnel.
5. Use the tap to fill the volume below the tap with solution.
6. Use the tap to further reduce the volume until the bottom of the meniscus is on the scale and close to the zero mark.

The procedure for filling a burette is illustrated in Figure 7. The following steps can be taken to ensure that the burette is capable of dispensing measured amounts of solution accurately.

Ensuring Accuracy

1. Be sure to wash the entire inside surface of the burette when rinsing.
2. Remove any air bubbles that have formed above and below the tap after filling.
3. Ensure that the bottom of the meniscus is on the scale after filling.

Before moving to the next section, check that you are able to:

- Use a volumetric pipette to measure and transfer a fixed volume of solution.
- Use a volumetric flask to prepare a fixed volume of solution by dissolving a solid.
- Use a volumetric flask and pipette to prepare a solution by diluting a more concentrated solution.
- Identify procedures to ensure that the volumes of solution measured, prepared and transferred using volumetric glassware are accurate and reliable.

(a)

(b)

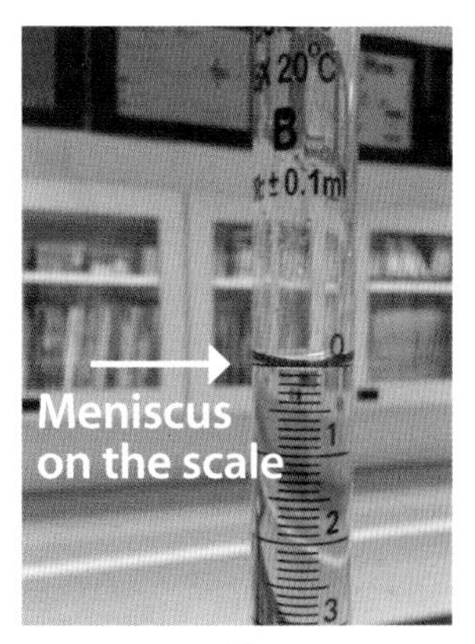

(c)

Figure 7: The procedure for filling a burette. (a) Solution is added using a funnel. (b) The volume below the tap is filled with solution. (c) The bottom of the meniscus is on the scale and close to the zero mark.

Calculations with Solutions

In this section we are learning to:

- Calculate the concentration of a solution in mol dm^{-3}, g dm^{-3} and related units as required given appropriate conversion factors.
- Use the term molarity (M) when referring to the concentration of a solution in units of mol dm^{-3}.
- Calculate amounts of solution involved in chemical reactions.
- Calculate the volume of solution required to make a specified volume of a more dilute solution.

Concentration

The **concentration** or 'strength' of a solution refers to *the amount of solute, in units of mass or moles, per unit volume of solution*. Solutions containing a lot of solute in a small amount of solvent are very concentrated. Solutions containing less solute in the same volume, or the same amount of solute in a greater volume of solvent, are less concentrated. In chemistry, the concentration of a solution is typically reported in units of moles per cubic decimetre (mol dm^{-3}) or grams per cubic decimetre (g dm^{-3}) where 1 cubic decimetre (1 dm^3) is equivalent to 1 litre (1 l = 1000 cm^3) of solution. The term **molarity** (Unit: M) is used when referring to *the concentration of a solution in units of moles per cubic decimetre (mol dm^{-3})*, and is calculated using Equation 1.

$$\text{Molarity (M)} = \frac{\text{Moles of solute (mol)}}{\text{Volume of solvent (dm}^3\text{)}} \quad \text{Equation 1}$$

Thus, according to Equation 1, a salt solution containing 2 moles of salt dissolved in 1 dm^3 of water has a concentration of 2 mol dm^{-3} = 2 M, and is referred to as a 'two-molar' solution.

Worked Example 1.8i

A solution of magnesium sulfate is made by dissolving 0.20 mol of magnesium sulfate in 500 cm^3 of water. Calculate the concentration of the solution in units of (a) mol dm^{-3} and (b) g dm^{-3}.

Strategy

- Use Equation 1 to calculate molarity.
- Use molar mass to convert moles per dm^3 to grams per dm^3.

Solution

(a) $\text{Molarity} = \frac{\text{Moles}}{\text{Volume}} = \frac{0.20\text{ mol}}{0.500\text{ dm}^3} = 0.40\text{ mol dm}^{-3}$

(b) Mass in 1 dm^3 = Molarity × Molar Mass

Mass of $MgSO_4$ in 1 dm^3

= 0.40 mol dm^{-3} × 120 g mol^{-1} = 48 g dm^{-3}

Exercise 1.8B

1. Iron(II) chloride is extremely soluble in water. 69 g of the anhydrous solid dissolve in 100 cm^3 of water at 20 °C. Assuming there is no volume change calculate the molarity of the resulting solution.

   (CCEA January 2014)

2. 2.65 g of anhydrous sodium carbonate, Na_2CO_3 was dissolved in water and the solution made up to 250 cm^3 in a volumetric flask. Calculate the molarity of the resulting solution.

   (CCEA January 2010)

3. 5.30 g of anhydrous sodium carbonate was dissolved in water and made up to 250 cm^3 in a volumetric flask. Which one of the following is the concentration of *sodium ions* in mol dm^{-3}?

   A 0.05 B 0.10

   C 0.20 D 0.40

   (CCEA June 2012)

4. 8.70 g of potassium sulfate, K_2SO_4 is dissolved in water and made up to 250 cm^3. What is the concentration of sulfate ion in mol dm^{-3}?

   (CCEA January 2003)

5. 4.35 g of potassium sulfate is dissolved in water and made up to 50.0 cm^3. Which one of the following is the concentration of potassium ions in this solution?

   A 0.025 mol dm^{-3} B 0.500 mol dm^{-3}

   C 0.644 mol dm^{-3} D 1.000 mol dm^{-3}

   (CCEA June 2014)

6. The bromine concentration in swimming pools is maintained at 4.0 mg per litre. Calculate the molarity of bromine, Br_2 in the water at this concentration.

   (CCEA June 2014)

7. Boron trichloride reacts with water to form a strongly acidic solution as shown below.

   $BCl_3 + 3H_2O \rightarrow H_3BO_3 + 3HCl$

   When 21.6 g of BCl_3 is dissolved in 250 cm^3 of water the concentration of the hydrochloric acid formed is

   A 0.55 mol dm^{-3} B 0.74 mol dm^{-3}

   C 2.21 mol dm^{-3} D 2.94 mol dm^{-3}

   (CCEA June 2015)

Concentrations may also be expressed in units of milligrams (mg), where 1g = 1000 mg, or in units of parts-per-million (ppm), where 1 ppm refers to a concentration of 1 gram in every million (10^6) grams.

Worked Example 1.8ii

A solution of sodium chloride has a concentration of 0.100 M. Calculate the concentration of the solution in units of (a) g dm^{-3} and (b) mg cm^{-3} given that there are 1000 milligrams in 1 gram (1 g = 1000 mg).

Solution

(a) Molar mass of NaCl = 58.5 g mol^{-1}

Mass of NaCl in 1 dm^3

$= 0.100 \text{ mol dm}^{-3} \times 58.5 \text{ g mol}^{-1} = 5.85 \text{ g dm}^{-3}$

(b) 1 dm^3 = 1000 cm^3 therefore:

Mass of NaCl in 1 cm^3

$$= 5.85 \text{ g dm}^{-3} \times \frac{1 \text{ dm}^3}{1000 \text{ cm}^3} = 5.85 \times 10^{-3} \text{ g cm}^{-3}$$

1 g = 1000 mg therefore:

Mass of NaCl in 1 cm^3

$$= 5.85 \times 10^{-3} \text{ g cm}^{-3} \times \frac{1000 \text{ mg}}{1 \text{ g}} = 5.85 \text{ mg cm}^{-3}$$

Exercise 1.8C

1. Sodium fluoride is added to toothpaste to strengthen tooth enamel. The data on a 50 g tube of toothpaste states that it contains '1450 ppm fluoride'; ppm means 'parts per million' i.e. there would be 1450 g of fluoride ions in 10^6 (1,000,000) g of toothpaste.

   Use the following headings to work out the concentration of sodium fluoride in the toothpaste. The density of the toothpaste is 1.6 g cm^{-3}.

   - Mass of fluoride ions in the 50 g tube.
   - Number of moles of fluoride ions in the 50 g tube.
   - Number of moles of sodium fluoride in the 50 g tube.
   - Volume of toothpaste in the 50 g tube.
   - Concentration of sodium fluoride in the toothpaste, with units.

   (CCEA January 2013)

Diluting Solutions

A solution is diluted by adding solvent. The particle picture in Figure 8 shows how the solute particles become more spread out and the solution becomes less concentrated as solvent is added. The amount of solute in the solution does not change as a result of adding solvent. As a result we are able to relate the concentration and volume of the solution before dilution (M_C and V_C) to the concentration and volume after dilution (M_D and V_D) as follows.

$$\text{Molarity } (M_C) \times \text{Volume } (V_C) = \text{Molarity } (M_D) \times \text{Volume } (V_D) \qquad \text{Equation 2}$$

Figure 8: Particle picture for diluting a solution. The concentrated solution contains more solute particles per unit of volume.

Worked Example 1.8iii

Calculate the concentration of a copper(II) sulfate solution prepared by diluting 10 cm^3 of 1.0 M copper(II) sulfate in a 100 cm^3 volumetric flask.

Solution

Molarity of diluted solution, $M_D = \frac{M_C \times V_C}{V_D}$

$= \frac{(1.0)(10)}{(100)} = 0.10 \text{ M} = 0.10 \text{ mol dm}^{-3}$

Exercise 1.8D

1. 50 cm^3 of 3.0 M hydrochloric acid was diluted with 150 cm^3 of water. What is the concentration of hydrogen ions in the resultant solution?

(CCEA January 2006)

Reacting Volumes

The relationship between molarity and moles in Equation 1 can be used to calculate the amounts of substance involved in chemical reactions that involve solutions.

Worked Example 1.8iv

Calculate the mass of calcium carbonate needed to neutralise 500 cm^3 of 0.10 M hydrochloric acid.

(CCEA June 2011)

Strategy

Write the chemical equation for the reaction then:

1. Calculate the moles of acid used.
2. Calculate the moles of $CaCO_3$ needed.
3. Calculate the mass of $CaCO_3$ needed.

Solution

$CaCO_{3(s)} + 2HCl_{(aq)} \rightarrow CaCl_{2(aq)} + CO_{2(g)} + H_2O_{(l)}$

1. Moles of HCl = Molarity × Volume
   $= 0.10 \text{ mol dm}^{-3} \times 0.500 \text{ dm}^3 = 0.050 \text{ mol}$
2. Moles of $CaCO_3 = \frac{\text{Moles of HCl}}{2}$
   $= \frac{0.050 \text{ mol}}{2} = 0.025 \text{ mol}$
3. Mass of $CaCO_3$ = Moles × Molar Mass
   $= 0.025 \text{ mol} \times 100 \text{ g mol}^{-1} = 2.5 \text{ g}$

Exercise 1.8E

1. Dilute nitric acid reacts with magnesium. (a) Calculate the volume, in cm^3, of 2.0 mol dm^{-3} nitric acid required to react with 6.0 g of magnesium. (b) Calculate the mass of magnesium nitrate produced.

   $Mg + 2HNO_3 \rightarrow Mg(NO_3)_2 + H_2$

   (Adapted from CCEA June 2012)

2. Calculate the mass of zinc chloride produced when 8.1 g of zinc oxide, ZnO is added to 150.0 cm^3 of 0.10 mol dm^{-3} hydrochloric acid.

   (Adapted from CCEA January 2013)

3. Calculate the volume of 0.20 mol dm^{-3} potassium hydroxide solution needed to neutralise 50 cm^3 of 0.20 mol dm^{-3} sulfuric acid.

   (Adapted from CCEA January 2011)

4. Sodium carbonate can form a number of hydrates, $Na_2CO_3.xH_2O$. A 6.0 g sample of hydrated sodium carbonate was dissolved in water and the solution made up to 250 cm^3. A 25.0 cm^3 portion of this solution required 24.3 cm^3 of 0.20 mol dm^{-3} sulfuric acid for complete reaction.

   $Na_2CO_3 + H_2SO_4 \rightarrow Na_2SO_4 + H_2O + CO_2$

   (a) Calculate the number of moles of sodium carbonate in 25.0 cm^3 of solution. (b) Deduce the number of moles of sodium carbonate in 250 cm^3 of solution. (c) Calculate the relative formula mass of the hydrated sodium carbonate. (d) Calculate the value of x.

   (Adapted from CCEA June 2015)

5. 0.84 g of a Group II carbonate, MCO_3 reacts with 20.0 cm^3 of a 1.0 mol dm^{-3} solution of hydrochloric acid. Identify the metal, M.

   (Adapted from CCEA June 2007)

6. When 0.28 g of a basic oxide, MO is reacted with 250 cm^3 of 0.050 mol dm^{-3} hydrochloric acid, the acid remaining after reaction required 50 cm^3 of 0.050 mol dm^{-3} sodium hydroxide solution for neutralisation. Calculate the RAM of the metal, M.

   (Adapted from CCEA June 2014)

7. 3.12 g of MCl_2 were dissolved in water and made up to one litre of solution. 25.0 cm^3 of this solution reacts with 7.5 cm^3 of 0.100 M silver nitrate solution. Identify the Group II metal, M.

$$MCl_{2(s)} \rightarrow M^{2+}{}_{(aq)} + 2Cl^{-}{}_{(aq)}$$

$$Ag^{+}{}_{(aq)} + Cl^{-}{}_{(aq)} \rightarrow AgCl_{(s)}$$

(Adapted from CCEA January 2014)

Before moving to the next section, check that you are able to:

- Use volumes and amounts of substance to calculate the concentration of a solution in mol dm^{-3}, g dm^{-3} and related units.
- Use the term molarity (M) when referring to the concentration of a solution in units of mol dm^{-3}.
- Use chemical equations to calculate the volumes of solutions involved in chemical reactions such as neutralisation.
- Calculate the volume of solution required to make a specified volume of a more dilute solution.

Titration Experiments

In this section we are learning to:

- Recall the procedures used to accurately and reliably determine the concentration of a solution by titration.
- Record and analyse the results of titration experiments using procedures designed to produce accurate and reliable outcomes.

A **titration** is *an experiment to accurately determine the concentration of a substance in solution by measuring the volume of a standard solution needed to react with a known volume of the solution*. In this context, the term **standard** refers to *a stable solution whose concentration has been accurately determined*. For example, the concentration of an acid could be determined by measuring the volume of a *standard alkali* required to neutralise a known volume of the acid. Titration experiments that make use of neutralisation reactions are referred to as **acid-base titrations**.

Acid-Base Titrations

The concentration of an acid can be determined by titrating a sample of the acid against a standard solution of a strong alkali such as sodium hydroxide. A sample of the acid is placed in a conical flask, and the alkali added in measured amounts using a burette as shown in Figure 9.

Figure 9: Using phenolphthalein to detect the end point when an acid (in the conical flask) is titrated against an alkali (in the burette).

The amount of acid in the solution decreases after each addition as the alkali reacts with the acid to form a salt and water: acid + alkali → salt + water.

The point at which the acid in the sample has been completely neutralised is known as the **equivalence point**, and the volume of alkali needed to reach the equivalence point is referred to as the **titre**.

Adding a drop of alkali at the equivalence point produces a rise in pH that can be detected by adding a few drops of a suitable **indicator** to the solution in the conical flask at the start of the titration. The point in the titration at which the indicator changes colour is known as the **end point** of the titration. An indicator can only be used to identify the equivalence point in an acid-base titration if the end point occurs at the equivalence point.

Titrating a Strong Acid

If the acid is a strong acid the pH of the solution remains low as the alkali is added from the burette and *adding a drop of alkali at the equivalence point produces a large increase in pH that is sufficient to cause most indicators to change colour*. Phenolphthalein (changes colour between pH 8 and 10) and methyl orange (changes colour between pH 3 and 5) are both suitable indicators for the titration of a strong alkali against a strong acid.

Titrating a Weak Acid

When titrating a weakly acidic solution such as vinegar against a standard solution of a strong alkali (in the burette) the pH of the solution rises steadily as the acid reacts. As a result, the change in pH at the end point begins at a high pH, and it becomes necessary to use an indicator that changes colour at even higher pH values. *The indicator phenolphthalein changes colour between pH 8 and 10, and is therefore suitable for the titration of a weak acid.*

Vinegar is the common name for an aqueous solution of ethanoic acid, CH_3COOH. The chemical equation for the neutralisation of ethanoic acid by sodium hydroxide is:

$$CH_3COOH_{(aq)} + NaOH_{(aq)} \rightarrow CH_3COONa_{(aq)} + H_2O_{(l)}$$

ethanoic acid *sodium ethanoate*

Titrating a Strong Alkali

Similarly, the concentration of an alkali can be determined by titrating the alkali against a standard solution of a strong acid such as hydrochloric acid or sulfuric acid. If the alkali is a strong alkali the pH of the solution remains high as the acid is added. *Adding a drop of acid at the equivalence point produces a large drop in pH that is sufficient to cause most indicators to change colour.* Phenolphthalein (changes colour between pH 8 and 10) and methyl orange (changes colour between pH 3 and 5) are both suitable indicators for the titration of a strong alkali against a strong acid.

Titrating a Weak Alkali

When a weak base such as sodium carbonate solution is titrated against a strong acid the pH decreases steadily as the acid is added. As a result, the decrease in pH at the equivalence point begins at a lower pH, and it is necessary to use an indicator that changes colour at even lower pH values. *The indicator methyl orange changes colour between pH 3 and 5, and is suitable for the titration of a weak base such as sodium carbonate.* Phenolphthalein would be unsuitable for the titration of a weak base as it changes colour before the equivalence point is reached.

The suitability of phenolphthalein and methyl orange as indicators for acid-base titrations is summarised in Table 1. *Indicators for the titration of a weak acid against a weak alkali are not included as the change in pH at the equivalence point is not sufficient to produce a colour change with most indicators.*

Worked Example 1.8v

Which one of the following would not be a suitable combination?

Titration	Acid	Alkali	Indicator
A	sulfuric acid	sodium carbonate	phenolphthalein
B	hydrochloric acid	sodium carbonate	methyl orange
C	hydrochloric acid	sodium hydroxide	methyl orange
D	ethanoic acid	sodium hydroxide	phenolphthalein

(CCEA June 2007)

Table 1: The suitability of phenolphthalein and methyl orange indicators for use in acid-base titrations.

Titration of a ... (in conical flask)	Against a ... (in burette)	Colour change
strong acid eg: $HCl_{(aq)}$	strong base eg: $NaOH_{(aq)}$	Methyl orange: *red* → *yellow* Phenolphthalein: *colourless* → *pink*
weak acid eg: $CH_3COOH_{(aq)}$	strong base eg: $NaOH_{(aq)}$	Methyl orange: *NOT suitable* Phenolphthalein: *colourless* → *pink*
strong base eg: $KOH_{(aq)}$	strong acid eg: $H_2SO_{4(aq)}$	Methyl orange: *yellow* → *red* Phenolphthalein: *pink* → *colourless*
weak base eg: $Na_2CO_{3(aq)}$	strong acid eg: $HCl_{(aq)}$	Methyl orange: *yellow* → *red* Phenolphthalein: *NOT suitable*

Strategy

- Classify the acids and alkalis as strong or weak.
- Recall that any indicator can be used to detect the end point of a titration that involves a strong acid and a strong alkali.
- Recall that methyl orange cannot be used in titrations involving a weak acid and that phenolphthalein cannot be used in titrations involving a weak alkali.

Solution

Titration A involves a strong acid and a weak alkali. Phenolphthalein indicator is not suitable for this titration.

Exercise 1.8F

1. (a) Explain what is meant by a *standard solution.* (b) Name a suitable indicator for the titration of hydrochloric acid with standard sodium hydroxide solution. (c) Describe the colour change observed at the end point.

   (CCEA June 2009)

2. Wood vinegar, which contains ethanoic acid, is formed when wood is heated. The percentage by mass of ethanoic acid in wood vinegar can be found by titration with standard sodium hydroxide solution. (a) What is meant by the term *standard solution*? (b) Write the equation for the reaction between ethanoic acid and sodium hydroxide. (c) Suggest a suitable indicator for the titration and state the colour change at the end point.

   (CCEA June 2013)

3. (a) Write the chemical equation for the titration of sodium carbonate solution with a standard solution of hydrochloric acid. (b) Name a suitable indicator for the titration and describe the colour change observed at the end point.

   (CCEA June 2010)

4. For which one of the following titrations would phenolphthalein be a suitable indicator?

   A Ethanoic acid and sodium carbonate
   B Ethanoic acid and sodium hydroxide
   C Hydrochloric acid and aqueous ammonia
   D Hydrochloric acid and sodium carbonate

   (CCEA January 2013)

Practical Details

The basic procedure for conducting an acid-base titration, and the steps that can be taken to improve the accuracy and reliability of the results, are detailed below.

Method

1. Rinse the volumetric pipette using 2-3 cm^3 of solution before using the pipette and a pipette filler to transfer 25 cm^3 of the solution to a conical flask.
2. Add several drops of indicator to the solution in the conical flask.
3. Rinse the burette with 2-3 cm^3 of solution before filling the burette with the solution and recording the initial reading on the burette scale.
4. Perform a rough titration by adding solution from the burette in 1 cm^3 amounts, swirling the contents of the conical flask after each addition, until the end point is reached.
5. Record the final reading on the burette and calculate the total volume of solution added (the titre).
6. Perform an accurate titration by adding solution dropwise near the end point.
7. Repeat to obtain accurate titres that are in close agreement.
8. Use the accurate titres to calculate an average titre.

Ensuring Accuracy

1. Ensure that the entire inside surface of the pipette and burette are washed.
2. Ensure that there are no bubbles in the pipette and burette after filling.
3. Gently swirl the contents of the conical flask after each addition.
4. Add solution dropwise from the burette when approaching the end point.
5. Read the burette scale at the bottom of the meniscus.

Ensuring Reliability

1. Repeat the titration to obtain accurate titres that are in close agreement.
2. Average the accurate titres to improve the reliability of the results.
3. Use only titres that are in close agreement when calculating the average titre.

Calculating Titres

A titre is obtained by subtracting an initial burette reading from the final burette reading at the end point, each with an associated uncertainty of 0.05 cm^3 that results from reading the scale on the burette. The following calculations demonstrate how *the uncertainties associated with burette readings of 21.60 cm^3 and 0.25 cm^3 combine to give a titre with an associated uncertainty of 0.10 cm^3.*

Measured Titre = Final burette reading – Initial burette reading
= 21.60 – 0.25 = 21.35 cm^3

Minimum possible titre from these readings
= 21.55 – 0.30 = 21.25 cm^3
(0.1 cm^3 lower)

Maximum possible titre from these readings
= 21.65 – 0.20 = 21.45 cm^3
(0.1 cm^3 higher)

Exercise 1.8G

1. (a) Describe how you would use a 0.10 mol dm^{-3} solution of sodium hydroxide and phenolphthalein indicator to determine the concentration of ethanoic acid in a sample of vinegar. (b) State the colour change at the end point and write the chemical equation for the reaction of sodium hydroxide with the ethanoic acid in vinegar. Include state symbols.

(CCEA June 2009)

Analysis of Results

The results of a titration experiment are recorded in a standard format. The following examples illustrate how to record and analyse the results of a titration experiment to the required level of accuracy.

Worked Example 1.8vi

The following results were obtained by titrating 25 cm^3 of sulfuric acid, H_2SO_4 against a 0.10 mol dm^{-3} solution of sodium hydroxide. Calculate the concentration of the sulfuric acid in units of g dm^{-3}.

	Initial Burette Reading (cm^3)	Final Burette Reading (cm^3)	Titre (cm^3)
Rough	0.00	26.00	26.00
First Accurate	0.25	25.60	25.35
Second Accurate	0.10	25.70	25.60
Third Accurate	0.05	25.70	25.65

Strategy

1. Use only the second and third accurate titres to calculate the average titre.
2. Calculate the moles of sodium hydroxide used in the titration.
3. Calculate the moles of acid neutralised.
4. Calculate the concentration of the acid in units of mol dm^{-3}.
5. Calculate the concentration of the acid in units of g dm^{-3}.

Solution

1. Use the accurate titres to calculate the average titre.

$$\text{Average titre} = \frac{25.60 + 25.65}{2} = 25.63\ cm^3$$

2. Calculate the moles of sodium hydroxide used in the titration.

Moles = Volume (in dm^3) × Molarity (in mol dm^{-3})

$$\text{Moles} = \frac{\text{Average Titre}}{1000} \times 0.10$$
$$= 2.563 \times 10^{-2} \times 0.10 = 2.6 \times 10^{-3}\ \text{mol}$$

3. Calculate the moles of acid neutralised.

Neutralisation reaction:

$H_2SO_4 + 2NaOH \rightarrow Na_2SO_4 + 2H_2O$

$$\text{Moles of acid} = \frac{\text{Moles of hydroxide}}{2} = \frac{2.6 \times 10^{-3}}{2}$$
$$= 1.3 \times 10^{-3}\ \text{mol}$$

4. Calculate the concentration of the acid in units of mol dm^{-3}.

$$25\ cm^3 \text{ of acid} = 25\ cm^3 \times \frac{1\ dm^3}{1000\ cm^3} = 0.025\ dm^3$$

$$\text{Molarity of acid} = \frac{\text{Moles of acid}}{\text{Volume of acid}} = \frac{1.3 \times 10^{-3}}{0.025}$$

$$= 0.052\ mol\ dm^{-3}$$

5. Calculate the concentration of the acid in units of g dm^{-3}.

Molar mass of H_2SO_4 = 98 g dm^{-1}

$$\text{Concentration in g } dm^{-3} = \text{Molarity} \times \text{Molar Mass}$$
$$= 0.052 \times 98 = 5.1\ g\ dm^{-3}$$

Exercise 1.8H

1. Analysis of a vinegar solution was carried out using the following procedure:

Transfer 25.0 cm^3 of undiluted vinegar into a 250 cm^3 volumetric flask and make the solution up to the mark using de-ionised water. Transfer 25.0 cm^3 portions of the diluted vinegar into three separate conical flasks and add a few drops of indicator to each flask. Titrate each solution with 0.10 mol dm^{-3} sodium hydroxide until an end point is reached.

A student obtained the following results:

	Initial Burette Reading (cm^3)	Final Burette Reading (cm^3)	Titre (cm^3)
Rough	0.0	21.7	21.7
First Accurate	21.7	43.1	
Second Accurate	0.0	21.3	

(a) Name a suitable indicator for this titration and state the colour change at the end point. (b) Write the equation for the reaction between vinegar and sodium hydroxide. (c) Complete the results table and calculate the average titre. (d) Use the average titre to calculate the number of moles of sodium hydroxide used in the titration. (e) Calculate the concentration of ethanoic acid in the diluted vinegar. (f) Calculate the concentration of ethanoic acid in the undiluted vinegar.

(Adapted from CCEA January 2010)

2. Wood vinegar, which contains ethanoic acid, is formed when wood is heated. The percentage by mass of ethanoic acid in wood vinegar can be found by titration with standard sodium hydroxide solution. 25.0 cm^3 of wood vinegar were diluted to 250 cm^3 in a volumetric flask. 25.0 cm^3 of the diluted wood vinegar required 30.3 cm^3 of 0.10 mol dm^{-3} sodium hydroxide solution for neutralisation.

(a) Calculate the moles of ethanoic acid in 25.0 cm^3 of diluted wood vinegar. (b) Calculate the moles of ethanoic acid in 25.0 cm^3 of undiluted wood vinegar. (c) Calculate the mass of ethanoic acid in 25.0 cm^3 of undiluted wood vinegar. (d) Calculate the percentage of ethanoic acid by mass in the wood vinegar. The density of wood vinegar is 1.02 g cm^{-3}.

(Adapted from CCEA June 2013)

Worked Example 1.8vii

The following results were obtained by titrating 25 cm^3 of a solution containing 1.20 g of hydrated sodium carbonate, $Na_2CO_3.xH_2O$ in 100 cm^3 of water against 0.100 M hydrochloric acid. Calculate the amount of water of crystallisation (x) in the hydrate.

	Initial Burette Reading (cm^3)	Final Burette Reading (cm^3)	Titre (cm^3)
Rough	0.00	22.00	22.00
First Accurate	0.10	21.40	21.30
Second Accurate	0.20	21.60	21.40

Strategy

1. Use the accurate titres to calculate the average titre.
2. Calculate the moles of acid used in the titration.
3. Calculate the moles of carbonate neutralised.
4. Calculate the mass of water in the hydrate.
5. Calculate the moles of water in the hydrate.

Solution

1. Use the accurate titres to calculate the average titre.

$$\text{Average titre} = \frac{21.30 + 21.40}{2} = 21.35\ cm^3$$

2. Calculate the moles of acid used in the titration.

   Moles = Volume (in dm^3) × Molarity (in mol dm^{-3})

   $$\text{Moles} = \frac{\text{Average Titre}}{1000} \times 0.100$$

   $$= 2.135 \times 10^{-2} \times 0.100 = 2.14 \times 10^{-3} \text{ mol}$$

3. Calculate the moles of carbonate neutralised.

   Neutralisation reaction:

   $Na_2CO_3 + 2HCl \rightarrow 2NaCl + H_2O + CO_2$

   $$\text{Moles of carbonate} = \frac{\text{Moles of acid}}{2} = \frac{2.14 \times 10^{-3}}{2}$$

   $$= 1.07 \times 10^{-3} \text{ mol}$$

4. Calculate the mass of water in the hydrate.

   Molar mass of Na_2CO_3 = 106 g mol^{-1}

   Mass of Na_2CO_3 = Moles × Molar Mass

   $$= 1.07 \times 10^{-3} \times 106 = 0.113 \text{ g}$$

   Mass of hydrate in 25 cm^3 of solution

   $$= \frac{1.20 \text{ g}}{4} = 0.300 \text{ g}$$

   Mass of water in 0.300 g of hydrate

   $$= 0.300 - 0.113 = 0.187 \text{ g}$$

5. Calculate the moles of water in the hydrate.

   Molar mass of H_2O = 18 g mol^{-1}

   Moles of water in 0.3 g of hydrate

   $$= \frac{\text{Mass}}{\text{Molar mass}} = \frac{0.187}{18} = 0.0104 \text{ mol}$$

   $$x = \frac{\text{Moles of water}}{\text{Moles of carbonate}} = \frac{0.0104}{1.07 \times 10^{-3}} = 10.2$$

   Rounding to the nearest whole number gives

   x = 10

   The formula of the hydrate is $Na_2CO_3.10H_2O$

Exercise 1.8I

1. 25.0 cm^3 of a solution made by dissolving 4.64 g of hydrated sodium carbonate, $Na_2CO_3.xH_2O$ in 1.00 dm^3 of water was neutralised by 20.0 cm^3 of 0.0500 mol dm^{-3} hydrochloric acid. Calculate x.

   (Adapted from CCEA June 2009)

2. (a) 20.0 cm^3 of a solution made by dissolving 3.05 g of hydrated barium chloride, $BaCl_2.xH_2O$ in 250 cm^3 of water was titrated with 0.100 M silver nitrate solution. The equivalence point was reached when 20.0 cm^3 of silver nitrate had been added. Write (a) the chemical equation and (b) the ionic equation for the titration reaction. (c) Calculate x.

   (Adapted from CCEA June 2011)

Before moving to the next section, check that you are able to:

- Describe how to obtain titration results that are both accurate and reliable.
- Accurately record and analyse the results of titration experiments.
- Determine the suitability of methyl orange and phenolphthalein indicators for use in acid-base titrations.

1.9 Qualitative Analysis

CONNECTIONS

- Food scientists use qualitative analysis techniques to test for the presence of impurities in food products.
- An environmental chemist uses qualitative analysis techniques to test for the presence of metal ions in polluted river water.
- Forensic scientists use qualitative analysis techniques to identify the compounds present in samples taken from a crime scene.

Qualitative analysis is the science of detecting the presence of chemical elements and compounds within mixtures. In this section the procedures used to identify the elements, compounds and ions in Table 1 are described, and key observations noted. A number of scientific terms used to record the observations are defined in Table 2.

Testing for Elements and Compounds

In this section we are learning to:

- Test for the elements hydrogen, oxygen, chlorine and iodine.
- Test for the compounds carbon dioxide, hydrogen chloride and ammonia.

The presence of a particular element or compound in a mixture can be confirmed by performing chemical tests to demonstrate that the mixture contains a substance with the same chemical properties as the element or compound.

Elements	Cations	Anions
Hydrogen, H_2 Oxygen, O_2 Chlorine, Cl_2 Iodine, I_2	*Group I and II:* Lithium, Li^+ Sodium, Na^+ Potassium, K^+ Calcium, Ca^{2+} Barium, Ba^{2+}	*Halides:* Chloride, Cl^- Bromide, Br^- Iodide, I^- *Molecular ions:* Sulfate, SO_4^{2-} Carbonate, CO_3^{2-}
Compounds		
Carbon dioxide, CO_2 Hydrogen chloride, HCl Ammonia, NH_3	*Other metals:* Copper(II), Cu^{2+} *Molecular ions:* Ammonium, NH_4^+	

Table 1: The elements, compounds and ions to be detected by qualitative analysis.

Testing for Hydrogen (H_2)

Properties	Chemical Test	Observations
Colourless gas. Odourless gas. Less dense than air.	Place a burning splint in an inverted test tube filled with the gas.	The gas burns with a 'pop' sound.

Testing for Oxygen (O_2)

Properties	Chemical Test	Observations
Colourless gas. Odourless gas.	Place a glowing splint in an inverted test tube filled with the gas.	The gas relights a glowing splint.

- The term **effervescence** describes the rapid release of gas from a reaction mixture.
- The term **evolved** describes the release of energy or matter from a reaction mixture.
- The term **miscible** describes liquids that mix in all proportions.
- A **precipitate** is an insoluble solid formed in solution as the result of a chemical reaction in the solution.
- The term **pungent** describes sharp 'acidic' odours.
- A **reagent** is a substance added to a reaction mixture to bring about a chemical reaction.
- A **suspension** is a cloudy mixture containing small particles of an insoluble solid distributed throughout a liquid.

Table 2: Glossary of terms for qualitative analysis.

Figure 1: Characteristic flame colours produced by compounds containing metal ions. The flame produced by potassium ions appears pink if viewed through blue glass.

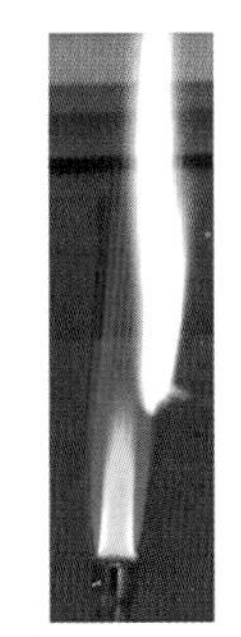
Sodium ion Na^+ *yellow/orange*

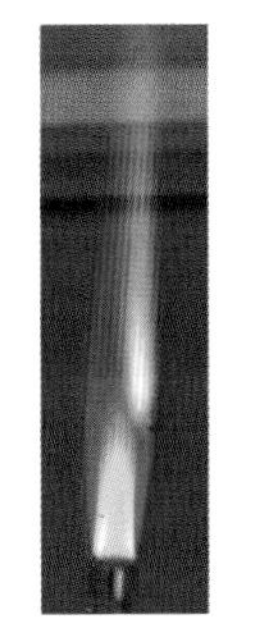
Lithium ion Li^+ *crimson*

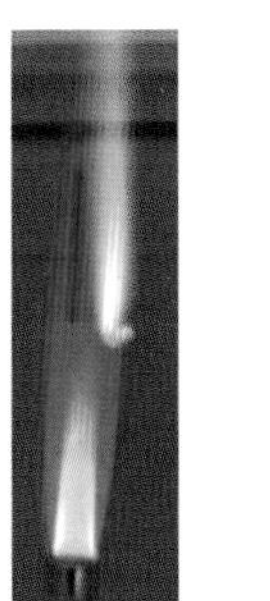
Calcium ion Ca^{2+} *brick red*

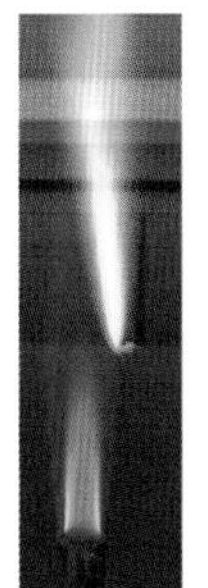
Barium ion Ba^{2+} *green*

Copper ion Cu^{2+} *blue-green*

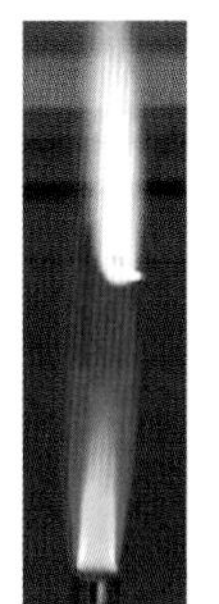
Potassium ion K^+ *lilac*

Testing for Chlorine (Cl_2)

Properties	Chemical Test	Observations
Green gas. Pungent odour of 'bleach'. Denser than air.	Bring the gas into contact with damp blue litmus paper or damp Universal Indicator paper.	Turns the damp indicator paper red then bleaches the paper.

Testing for Aqueous Iodine (I_2)

Properties	Chemical Test	Observations
Yellow solution (dilute).	Add several drops of starch solution.	A blue-black coloured solution is formed.

Testing for Carbon Dioxide (CO_2)

Properties	Chemical Test	Observations
Colourless gas. Odourless gas. Denser than air. Soluble in water.	Bubble the gas through limewater.	A cloudy white suspension turns the limewater milky.

Testing for Hydrogen Chloride (HCl)

Properties	Chemical Test	Observations
Colourless gas. Pungent odour. Denser than air. Soluble in water.	Bring the gas into contact with a glass rod dipped in concentrated ammonia solution.	White fumes of solid ammonium chloride are formed.

Testing for Ammonia (NH_3)

Properties	Chemical Test	Observations
Colourless gas. Pungent odour. Less dense than air. Soluble in water.	Bring the gas into contact with a glass rod dipped in concentrated hydrochloric acid.	White fumes of solid ammonium chloride are formed.

Exercise 1.9A

1. Describe a test for oxygen gas.

   (CCEA June 2008)

2. Describe a chemical test for ammonia. State the reagent used and the observation for a positive result.

   (CCEA January 2009)

3. Complete the table below by describing an appropriate test and the observation expected to identify each gas.

Gas	Test	Observation
Hydrogen		
Hydrogen chloride		

(Adapted from CCEA June 2008)

Before moving to the next section, check that you are able to:

- Describe the procedure used to detect the presence of each element and compound in Table 1.
- Recall the observations required to confirm the presence of each element and compound in Table 1.

Testing for Cations

In this section we are learning to:

- Test for the presence of Li^+, Na^+, K^+, Ca^{2+}, Ba^{2+}, Cu^{2+} and NH_4^+ ions in a substance.

Flame Tests

The presence of metal ions in a solid compound can be detected by performing a flame test on the compound. A flame test is conducted by suspending a small amount of the solid compound in a blue Bunsen flame. The flame colour produced by the metal ions can then be used to confirm the presence of the metal ions in the solid. The characteristic flame colours produced by compounds containing metal ions are summarised in Figure 1.

Worked Example 1.9i

Describe how you would carry out a flame test to show the presence of sodium ions in a white solid. State the flame colour expected.

(CCEA June 2010)

Solution

Wash the end of a nichrome wire in concentrated hydrochloric acid. Transfer a small amount of solid onto the wire by dipping the end of the wire in the solid. Place the solid sample in a blue Bunsen flame and record the flame colour. If the sample contains sodium ions the flame will turn a yellow or orange colour.

Exercise 1.9B

1. The presence of copper in copper(II) chloride can be shown using a flame test. The diagram shows the equipment needed for the test. (a) Identify the acid W, the metal wire X, the colour Y of the flame before the test, and the colour Z during the test. (b) State two reasons for using W.

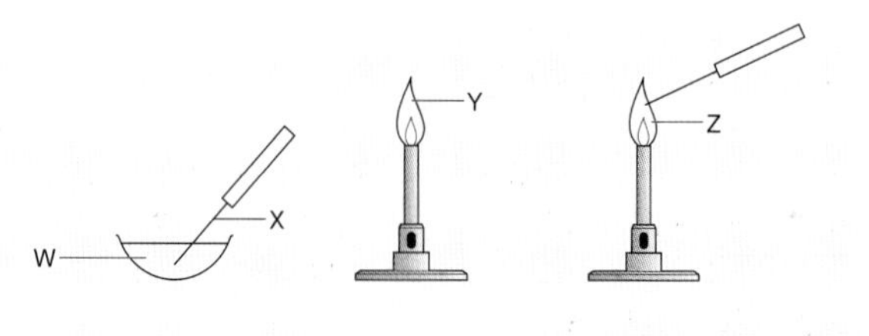

(CCEA January 2012)

2. Lithium sulfate can be used in a flame test. Explain how a flame test could be carried out and state the expected colour of the flame.

   (CCEA January 2011)

3. Describe how you would carry out a flame test to distinguish between barium carbonate and calcium carbonate, giving the result for each compound.

   (CCEA June 2012)

4. Which one of the following metal compounds will produce a lilac flame colour?

   A barium nitrate B calcium chloride
   C lithium chloride D potassium sulfate

   (CCEA June 2012)

5. Which one of the following is the colour of the flame produced when a barium compound is placed in a blue Bunsen burner flame?

   A Crimson B Green C Lilac D Orange

   (CCEA June 2014)

Testing for Ammonium Ion (NH_4^+)

Ammonium compounds such as ammonium chloride, NH_4Cl and ammonium carbonate, $(NH_4)_2CO_3$ are white solids, and dissolve to give colourless solutions. The presence of ammonium ion in solution can be detected by adding the ammonium compound to dilute sodium hydroxide solution and heating the mixture. On heating, ammonia gas is formed, and can be detected as an alkaline gas with the characteristic smell of ammonia.

$$NH_4^+(aq) + OH^-(aq) \rightarrow NH_3(g) + H_2O(l)$$

pungent smell

Confirming the presence of ammonium ion in solution:

Method	Observations
1. Add a spatula of the solid to a test tube containing 4-5 cm^3 of dilute sodium hydroxide solution.	Solid dissolves to give a colourless solution.
2. Warm the mixture gently and test any gas evolved using a glass rod dipped in concentrated hydrochloric acid.	Pungent smell. White fumes of ammonium chloride formed.

Exercise 1.9C

1. Ammonium chloride dissolves in water to form aqueous ammonium ions and chloride ions. Describe how you would test for aqueous ammonium ions.

(CCEA June 2011)

Before moving to the next section, check that you are able to:

- Describe the procedure to conduct a flame test on a solid compound.
- Recall the characteristic flame colours produced by the metal ions: Li^+, Na^+, K^+, Ca^{2+}, Ba^{2+} and Cu^{2+}.
- Describe the procedure to test for ammonium ion in a solid compound, and the observations required to confirm the presence of ammonium ion.

Testing for Anions

In this section we are learning to:

- Test for the presence of SO_4^{2-}, CO_3^{2-}, Cl^-, Br^-, and I^- ions in a substance.

Test for Sulfate

The presence of sulfate ions, SO_4^{2-} in a solid can be detected by adding a few drops of barium chloride solution (or barium nitrate solution) to a solution of the solid. Barium ions from the barium chloride solution combine with sulfate ions from the solid to form a white precipitate of barium sulfate, $BaSO_4$.

$$Ba^{2+}{}_{(aq)} + SO_4^{2-}{}_{(aq)} \rightarrow BaSO_{4(s)}$$
white solid

Confirming the presence of sulfate ions:

Method	Observations
1. Add a spatula of the solid to a test tube containing 2-3 cm^3 of water and shake to dissolve.	Solid dissolves.
2. Add 3-4 drops of barium chloride solution and shake to mix.	White precipitate formed.

Exercise 1.9D

1. (a) Name a reagent which can be used to distinguish between solutions of magnesium chloride and magnesium sulfate. (b) What is observed when this reagent is added to each solution? (c) Write the ionic equation, including state symbols, for any reaction which occurs when the reagent is added to each solution.

(CCEA June 2013)

2. The table shows the results of analysing aqueous solutions of compounds A, B and C. Identify the compounds A, B and C.

Compound	Aqueous solution	On adding aqueous barium chloride	Other information
A	blue	white precipitate	blue-green flame test
B	colourless	white precipitate	pH = 1
C	colourless	white precipitate	lilac flame test

(Adapted from CCEA June 2009)

Test for Carbonate Ion (CO_3^{2-})

Solid carbonates such as $BaCO_3$ are bases and will react with dilute acid to form a salt, water and carbon dioxide. The formation of carbon dioxide when a solid reacts with dilute acid is used to detect the presence of carbonate ion in the solid.

Confirming the presence of carbonate ion:

Method	Observations
1. Add a spatula of the solid to a test tube containing 2-3 cm^3 of water.	Solid dissolves.
2. Add 1 cm^3 of dilute nitric acid to the solution and collect any gas evolved with a pipette.	Effervescence. Colourless gas produced.
3. Bubble the contents of the pipette through 2-3 cm^3 of limewater in a separate test tube.	A cloudy white suspension turns the limewater milky.

(a)

silver chloride (white)
silver bromide (cream)
silver iodide (yellow)

(b)

	Soluble in dilute ammonia	Soluble in concentrated ammonia
Silver chloride	✓	✓
Silver bromide	✗	✓
Silver iodide	✗	✗

Figure 2: (a) Precipitates of silver chloride, silver bromide and silver iodide formed by the addition of acidified silver nitrate solution to a solution containing the corresponding halide ion. (b) The solubility of silver halides in dilute and concentrated ammonia solution.

Exercise 1.9E

1. Describe how you would use dilute hydrochloric acid to confirm the presence of carbonate ions in a solid compound and confirm the identity of the gas produced. State any observations expected.

   (Adapted from CCEA June 2010)

2. Limestone is mainly calcium carbonate, $CaCO_3$. Describe how you would confirm the presence of (a) calcium ions and (b) carbonate ions in a sample of limestone.

   (CCEA June 2008)

Testing for Halide Ions

The presence of chloride, Cl^- bromide, Br^- or iodide, I^- ions in a solid can be detected by adding dilute nitric acid, followed by a few drops of silver nitrate solution, to an aqueous solution of the solid. If chloride ion is present in the solution a white precipitate of silver chloride, AgCl forms as shown in Figure 2.

$$Ag^+_{(aq)} + Cl^-_{(aq)} \rightarrow AgCl_{(s)}$$
white solid

Similarly, the addition of acidified silver nitrate solution produces a cream precipitate of silver bromide, AgBr if bromide ions are present, and a yellow precipitate of silver iodide, AgI if iodide ions are present as shown in Figure 2.

$$Ag^+_{(aq)} + Br^-_{(aq)} \rightarrow AgBr_{(s)}$$
cream solid

$$Ag^+_{(aq)} + I^-_{(aq)} \rightarrow AgI_{(s)}$$
yellow solid

The presence of chloride, bromide or iodide in solution can then be confirmed by examining the solubility of the silver halide precipitate in dilute and concentrated ammonia solutions.

The solubility of the silver halide precipitates in dilute and concentrated ammonia solution is summarised in Figure 2.

Confirming Chloride

A precipitate of silver chloride, AgCl will dissolve when shaken with an excess of dilute ammonia solution or an excess of concentrated ammonia solution.

Method	Observations
1. Add a spatula of the solid to a test tube containing 2-3 cm^3 of water and shake to mix.	Solid dissolves.
2. Add 1 cm^3 of dilute nitric acid followed by 3-4 drops of silver nitrate solution.	White precipitate formed.
3. Add 4-5 cm^3 of dilute ammonia solution and shake to mix.	Precipitate dissolves.

Confirming Bromide

A precipitate of silver bromide, AgBr will not dissolve in dilute ammonia solution, and will only dissolve when shaken with an excess of concentrated ammonia solution.

Method	Observations
1. Add a spatula of the solid to a test tube containing 2-3 cm^3 of water and shake to mix.	Solid dissolves.
2. Add 1 cm^3 of dilute nitric acid followed by 3-4 drops of silver nitrate solution.	Cream precipitate formed.
3. Add 4-5 cm^3 of dilute ammonia solution and shake to mix.	Precipitate does not dissolve.
4. Repeat step 2 then add 4-5 cm^3 of concentrated ammonia solution and shake to mix.	Precipitate dissolves.

Confirming Iodide

A precipitate of silver iodide, AgI will not dissolve when shaken with an excess of dilute ammonia solution or an excess of concentrated ammonia solution.

Method	Observations
1. Add a spatula of the solid to a test tube containing 2-3 cm^3 of water and shake to mix.	Solid dissolves.
2. Add 1 cm^3 of dilute nitric acid followed by 3-4 drops of silver nitrate solution.	Yellow precipitate formed.
3. Add 4-5 cm^3 of dilute ammonia solution and shake to mix.	Precipitate does not dissolve.
4. Repeat step 2 then add 4-5 cm^3 of concentrated ammonia solution and shake to mix.	Precipitate does not dissolve.

Exercise 1.9F

1. Copy and complete the following table about the silver halides.

Silver halide	Formula of halide	Colour of halide	Type of bonding	Soluble in dilute ammonia	Soluble in concentrated ammonia
silver fluoride	AgF	white	ionic	yes	yes
silver chloride					
silver bromide					
silver iodide					

(Adapted from CCEA January 2014)

2. Explain how you would confirm the presence of chloride ions in an aqueous solution of copper(II) chloride.

(CCEA January 2012)

3. Describe how you could show that a solution contains bromide ions.

(CCEA June 2014)

4. Describe how you would confirm the presence of chloride or iodide ions in a solid using dilute nitric acid and silver nitrate solution. State any observations expected.

(CCEA June 2010)

5. Describe how silver chloride and silver iodide react with dilute and concentrated ammonia solutions. State any observations expected.

(CCEA June 2010)

6. If you had poured solutions of sodium iodide, bromide and chloride into beakers A, B and C and forgotten to label them, describe how, using aqueous silver nitrate and both dilute and concentrated ammonia solutions, you would determine which sodium salt was in which beaker. Each beaker must be tested.

(CCEA January 2011)

Before moving to the next section, check that you are able to:

- Describe the use of barium chloride solution to detect SO_4^{2-} ions in solution.
- Recall the use of dilute acid to detect CO_3^{2-} ions in solution.
- Describe the use of silver nitrate and ammonia solution to detect and distinguish between Cl^-, Br^- and I^- ions in solution.

Unit AS 2:

Further Physical and Inorganic Chemistry and an Introduction to Organic Chemistry

2.1 Further Calculations

CONNECTIONS

- The processes used to manufacture chemicals on an industrial scale must be carried out under conditions that produce the maximum yield of product for every £1 spent operating the process.
- The negative impact of chemical processes on the economy and the environment can be reduced by making the process 'greener'.

Percentage Yield

In this section we are learning to:

- Calculate the percentage yield for a reaction.
- Use the percentage yield for a reaction to relate the amounts of reactants and products in a chemical reaction.

The amount of product formed in a chemical reaction may be less than expected if the reactants do not completely react, or if some of the product is lost when attempting to recover it from the reaction mixture. The amount of product lost as a result of incomplete reaction and loss during recovery can be determined by calculating the **percentage yield** for the reaction.

$$\text{Percentage yield} = \frac{\text{Actual yield}}{\text{Theoretical yield}} \times 100\ \%$$

The **theoretical yield** refers to the amount of product that would be formed if the reactants completely reacted and the product was completely recovered from the reaction mixture.

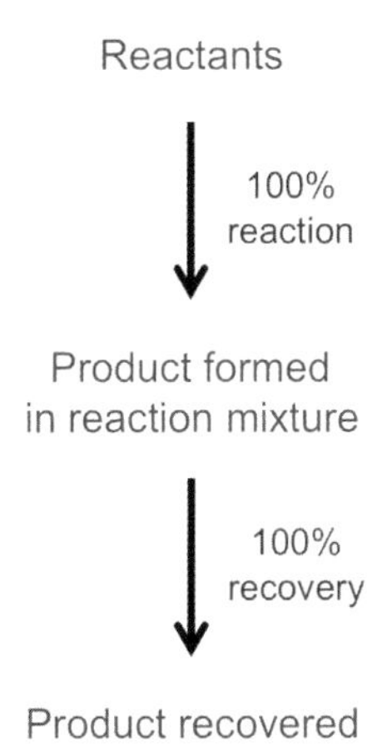

The **actual yield** refers to the amount of product recovered from the reaction mixture. The actual yield is lower than the theoretical yield as a result of incomplete reaction and loss of product during its recovery from the reaction mixture.

Reactants

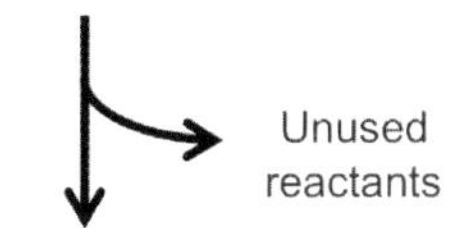

Product formed
in reaction mixture

Product recovered
from reaction mixture
(actual yield)

Worked Example 2.1i

In industry ethanol, C_2H_6O is made by reacting ethene, C_2H_4 with steam. Calculate the percentage yield of the reaction if 4.6 tonnes of ethene are needed to produce 4.7 tonnes of ethanol (1 tonne = 1000 kg).

$$C_2H_4 + H_2O \rightarrow C_2H_6O$$

Strategy

- Use the mass of ethene reacted to calculate the theoretical yield.
- Use the moles of ethanol formed to calculate the actual yield.

Solution

Mass of ethene = 4.6 tonnes = 4.6×10^6 g

$$\text{Moles of ethene reacted} = \frac{4.6 \times 10^6\ \text{g}}{26\ \text{g mol}^{-1}} = 1.8 \times 10^5\ \text{mol}$$

Theoretical yield of ethanol = 1.8×10^5 mol

$$\text{Actual yield of ethanol} = \frac{4.7 \times 10^6 \text{ g}}{46 \text{ g mol}^{-1}} = 1.0 \times 10^5 \text{ mol}$$

$$\text{Percentage yield} = \frac{1.0 \times 10^5 \text{ mol}}{1.8 \times 10^5 \text{ mol}} \times 100 = 56\%$$

Exercise 2.1A

1. 12.3 g of 1-bromobutane, C_4H_9Br was obtained from 11.1 g of butan-1-ol, C_4H_9OH. Calculate the percentage yield for the reaction.

   $C_4H_9OH + HBr \rightarrow C_4H_9Br + H_2O$

   (CCEA June 2010)

2. Calculate the percentage yield of t-butyl chloride, $(CH_3)_3CCl$ if 28 g of t-butyl chloride are obtained by reacting 25 g of t-butyl alcohol, $(CH_3)_3COH$ with hydrochloric acid.

   $(CH_3)_3COH + HCl \rightarrow (CH_3)_3CCl + H_2O$

   (Adapted from CCEA June 2011)

3. The compound cisplatin, $Pt(NH_3)_2Cl_2$ is used to treat several different forms of cancer. It is prepared by reacting the salt K_2PtCl_4 with ammonia. Calculate the percentage yield for the reaction if 2.08 g of cisplatin is formed when 3.42 g of K_2PtCl_4 react with 1.61 g of ammonia.

   $K_2PtCl_4 + 2NH_3 \rightarrow Pt(NH_3)_2Cl_2 + 2KCl$

4. Aspirin, $C_9H_8O_4$ can be prepared in the laboratory by reacting salicylic acid, $C_7H_6O_3$ with ethanoic anhydride, $C_4H_6O_3$ according to the following equation. Calculate the percentage yield if 3.08 g of aspirin were obtained when 3.00 g of salicylic acid were reacted with 6.0 cm^3 of ethanoic anhydride. The density of ethanoic anhydride is 1.08 g cm^{-3}.

   $C_7H_6O_3 + C_4H_6O_3 \rightarrow C_9H_8O_4 + C_2H_4O_2$

   (Adapted from CCEA June 2012)

The percentage yield can also be used to calculate the amount of product formed in a reaction, or the amount of reactants needed to form a given amount of product.

Worked Example 2.1ii

The solvent dichloromethane, CH_2Cl_2 is used to make decaffeinated coffee by extracting the compound caffeine from coffee beans. Dichloromethane is formed by reacting methane, CH_4 with chlorine.

$$CH_4 + 2Cl_2 \rightarrow CH_2Cl_2 + 2HCl$$

Calculate the mass of dichloromethane formed when 1.00 tonne of methane reacts with an excess of chlorine. The percentage yield for the reaction is 43.1%.

Strategy

- Use the mass of methane to calculate the theoretical yield of dichloromethane.
- Use the percentage yield to calculate the actual yield of dichloromethane.

Solution

$$\text{Moles of } CH_4 \text{ reacted} = \frac{1.00 \times 10^6 \text{ g}}{16 \text{ g mol}^{-1}} = 6.25 \times 10^4 \text{ mol}$$

Theoretical yield of $CH_2Cl_2 = 6.25 \times 10^4$ mol

$$\text{Actual Yield} = \text{Theoretical yield} \times \frac{\text{Percentage yield}}{100}$$

$= 2.69 \times 10^4$ mol

$$\text{Mass of } CH_2Cl_2 = 2.69 \times 10^4 \text{ mol} \times 85 \text{ g mol}^{-1} = 2.29 \times 10^6 \text{ g} = 2.29 \text{ tonnes}$$

Exercise 2.1B

1. Ethanoic acid, $C_2H_4O_2$ reacts with isopentyl alcohol, $C_5H_{12}O$ to form isopentyl acetate, $C_7H_{14}O_2$ an artificial flavour that smells and tastes like bananas. The percentage yield for the reaction is 45.0%. Calculate the mass of isopentyl acetate formed when 3.58 g of ethanoic acid reacts with 4.75 g of isopentyl alcohol.

   $C_2H_4O_2 + C_5H_{12}O \rightarrow C_7H_{14}O_2 + H_2O$

2. Butan-1-ol, C_4H_9OH is used to prepare 1-bromobutane, C_4H_9Br according to the equation:

   $C_4H_9OH + HBr \rightarrow C_4H_9Br + H_2O$

   (a) Write an equation to explain the term *percentage yield.* (b) Assuming a 40 % yield, calculate the mass of butan-1-ol needed to produce 5.48 g of 1-bromobutane.

   (Adapted from CCEA June 2014)

3. Decane, $C_{10}H_{22}$ can be converted to octane, C_8H_{18} by heating at 500 °C in the presence of a

catalyst. Ethene, C_2H_4 is a useful by-product of the reaction. Calculate the mass of decane needed to produce 1 tonne of ethene by this method if the percentage yield for the reaction is 94 %.

$C_{10}H_{22} \rightarrow C_8H_{18} + C_2H_4$

Before moving to the next section, check that you are able to:

- Calculate the percentage yield for a reaction given the amounts of reactants used and the amounts of products formed.
- Use the percentage yield for a reaction to relate the amounts of reactants used and the amounts of products formed in the reaction.

Atom Economy

In this section we are learning to:

- Use atom economy to measure the amount of product produced by a chemical reaction.

Percentage yield measures the amount of product produced by a chemical reaction, and does not take into account the amount of waste generated by the reaction. For example, calcium oxide or 'lime' is produced by heating limestone (mostly $CaCO_3$), and is used by the farming industry to reduce the acidity of soil.

Heating limestone: $CaCO_{3(s)} \rightarrow CaO_{(s)} + CO_{2(g)}$

The percentage yield for the reaction is 75%. As a result, every 1 kg of $CaCO_3$ produces 420 g of CaO and 330 g of carbon dioxide.

The amount of useful product generated by the reaction can also be measured by calculating the **atom economy** for the reaction.

$$\text{Atom economy} = \frac{\text{Mass of useful products}}{\text{Mass of all products}} \times 100\%$$

If the carbon dioxide is not useful the reaction produces 420 g of useful product (CaO) and 330 g of waste (carbon dioxide) for every 1 kg of limestone that reacts, and the atom economy for the reaction is calculated to be:

$$\frac{420 \text{ g}}{420 \text{ g} + 330 \text{ g}} \times 100\% = 56\%$$

In this example the actual yield of each product has been used to calculate the atom economy for the reaction. The atom economy can also be calculated using the theoretical yield of each product. For example, when 1 kg of limestone reacts it is expected to form 560 g of CaO and 440 g of CO_2, and the atom economy is again calculated to be:

$$\frac{560 \text{ g}}{560 \text{ g} + 440 \text{ g}} \times 100\% = 56\%$$

An atom economy of 56% indicates that 100% – 56% = 44 % of the products are waste. In this way a low atom economy (56%) reveals that reactions with a fairly high percentage yield (75%) can still generate a lot of waste.

(a) A chemical process with a low atom economy

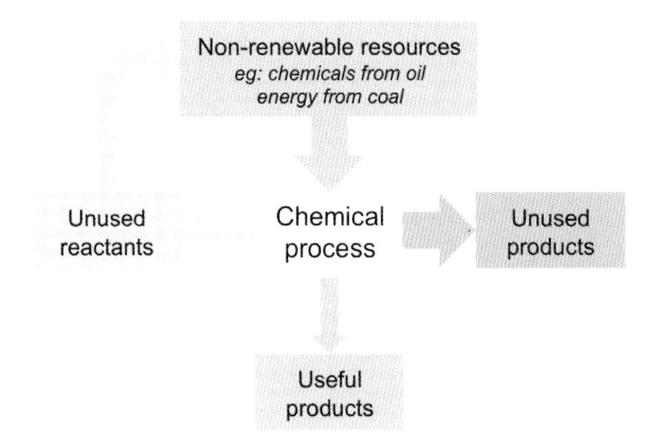

(b) A chemical process with a high atom economy

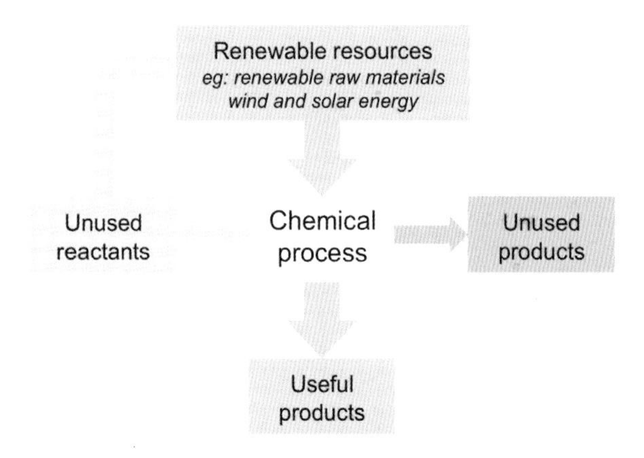

Figure 1: (a) A chemical process with a low atom economy. (b) The same process with a higher atom economy achieved by: reducing the amount of unused products and using renewable resources.

The flow schemes in Figure 1 demonstrate how the concept of atom economy can be used to help develop more environmentally friendly chemical processes.

Worked Example 2.1iii

In industry, iron is extracted from iron(III) oxide by reducing the oxide with carbon monoxide. Calculate the atom economy for the reaction if the carbon dioxide formed in the reaction is (a) released into the atmosphere and (b) used to produce more carbon monoxide.

$$Fe_2O_3 + 3CO \rightarrow 2Fe + 3CO_2$$

Strategy

- Assume that one mole of Fe_2O_3 reacts.
- Use the theoretical yield to calculate the atom economy.
- The carbon dioxide formed is considered waste.

Solution

(a) One mole of Fe_2O_3 produces 2 moles of Fe and 3 moles of CO_2:

Mass of 2 moles of Fe = 2 mol × 56 g mol^{-1} = 112 g

Mass of 3 moles of CO_2 = 3 mol × 44 g mol^{-1} = 132 g

$$\text{Atom economy} = \frac{\text{Mass of iron formed}}{\text{Total mass of products}}$$

$$= \frac{112\text{ g}}{112\text{ g} + 132\text{ g}} \times 100\% = 45.9\%$$

(b) The reaction produces no waste.

Atom economy = 100%

Exercise 2.1C

1. Barium carbonate is roasted with carbon to produce barium oxide. Calculate the atom economy of the reaction.

   $$BaCO_3 \rightarrow BaO + CO_2$$

   (CCEA June 2011)

2. In industry decane, $C_{10}H_{22}$ is 'cracked' to produce octane, C_8H_{18} and ethene, C_2H_4. The octane produced by the process is used to produce gasoline (petrol). Calculate the atom economy of the reaction if ethene is discarded.

   $$C_{10}H_{22} \rightarrow C_8H_{18} + C_2H_4$$

3. Calculate the atom economy for the production of iron in the Thermite Process.

   $$2Al + Fe_2O_3 \rightarrow 2Fe + Al_2O_3$$

   (Adapted from CCEA January 2014)

4. Methane reacts with excess chlorine to form tetrachloromethane, CCl_4. Calculate the atom economy for the formation of CCl_4.

   $$CH_4 + 4Cl_2 \rightarrow CCl_4 + 4HCl$$

   (Adapted from CCEA June 2013)

5. Titanium can be extracted from its ore (mostly TiO_2) by reacting the ore with a more reactive metal such as magnesium. Titanium can also be extracted from its ore by electrolysis. Calculate the atom economy when titanium is produced from its ore by (a) reaction with magnesium and (b) electrolysis.

   Reaction with magnesium:

   $$TiO_2 + 2Mg \rightarrow Ti + 2MgO$$

   Electrolysis: $TiO_2 \rightarrow Ti + O_2$

6. Butan-1-ol, C_4H_9OH is used to prepare 1-bromobutane, C_4H_9Br according to the equation:

   $$C_4H_9OH + HBr \rightarrow C_4H_9Br + H_2O$$

   (a) Write an equation to explain the term *atom economy*. (b) Calculate the atom economy for the formation of 1-bromobutane from butan-1-ol.

   (Adapted from CCEA June 2014)

7. Aspirin, $C_9H_8O_4$ can be prepared in the laboratory by reacting salicylic acid, $C_7H_6O_3$ with ethanoic anhydride, $C_4H_6O_3$ according to the following equation. (a) Explain what is meant by the term *atom economy* of a reaction. (b) Calculate the atom economy of the reaction to prepare aspirin.

   $$C_7H_6O_3 + C_4H_6O_3 \rightarrow C_9H_8O_4 + C_2H_4O_2$$

   (Adapted from CCEA June 2012)

Before moving to the next section, check that you are able to:

- Calculate the atom economy for a chemical reaction.
- Recall that reactions with a high atom economy generate little waste and have a more positive impact on the environment.

Calculating Gas Volumes

In this section we are learning to:

- Use Avogadro's law to determine the volumes of gases involved in chemical reactions.

The volume of gas produced or used by a chemical reaction can be calculated using **Avogadro's law** which states that *the volume of one mole of any gas is exactly 24 dm^3 if measured at 20 °C and 1 atmosphere pressure*. This relationship between the amount of gas and its volume is summarised by the equation:

$$\text{Moles of gas} = \frac{\text{Volume of gas (in } dm^3)}{24\ dm^3}$$

The volume of one mole of gas at a specified temperature and pressure is referred to as the **molar volume** of the gas. Thus, according to Avogadro's law, *the molar volume of any gas is exactly 24 dm^3 when measured at 20 °C and 1 atmosphere.*

Worked Example 2.1iv

Mercury(II) oxide decomposes on heating. Calculate the mass of mercury(II) oxide needed to produce 10.0 dm^3 of oxygen at 20 °C and 1 atmosphere pressure.

$$2HgO_{(s)} \rightarrow 2Hg_{(l)} + O_{2(g)}$$

Strategy

- Calculate the moles of oxygen in 10.0 dm^3.
- Use the chemical equation to calculate the moles of mercury(II) oxide needed.

Solution

Moles of oxygen in 10.0 dm^3

$$= \frac{10.0\ dm^3}{24\ dm^3\ mol^{-1}} = 0.417\ mol$$

The molar volume (24 dm^3 mol^{-1}) is considered to be an exact number, and does not limit the number of significant figures in the answer.

Moles of mercury(II) oxide needed

= 0.417 mol × 2 = 0.834 mol

Mass of mercury(II) oxide needed

= 0.834 mol × 217 g mol^{-1} = 181 g

Exercise 2.1D

1. Sodium nitrate decomposes when heated to produce sodium nitrite and oxygen. Calculate the volume of oxygen produced at 20 °C and 1 atmosphere pressure when 4.25 g of sodium nitrate decomposes.

   $$2NaNO_3 \rightarrow 2NaNO_2 + O_2$$

   (Adapted from CCEA January 2013)

2. Calcium nitrate decomposes when heated. Calculate the total volume of gas formed when 8.20 g of calcium nitrate completely decomposes at room temperature and pressure.

   $$2Ca(NO_3)_{2(s)} \rightarrow 2CaO_{(s)} + 4NO_{2(g)} + O_{2(g)}$$

   (Adapted from CCEA June 2012)

3. Ammonium chloride, NH_4Cl is quite soluble in water. 37.2 g dissolve in 100 cm^3 of water at 20 °C. Calculate the maximum volume of ammonia gas that could be obtained from this solution at 20 °C and 1 atmosphere pressure.

   (CCEA June 2011)

4. Calculate the volume of hydrogen produced at 20 °C and 1 atmosphere pressure when 8.00 g of magnesium reacts with an excess of steam.

   $$Mg_{(s)} + H_2O_{(g)} \rightarrow MgO_{(s)} + H_{2(g)}$$

   (CCEA January 2006)

5. Airbags in cars contain sodium azide, NaN_3. The airbag inflates when sodium azide decomposes to produce nitrogen gas. When inflated, an airbag holds 50 dm^3 of nitrogen at 20 °C and 1 atmosphere pressure. What mass of sodium azide is needed to inflate the airbag?

   $$2NaN_{3(s)} \rightarrow 2Na_{(s)} + 3N_{2(g)}$$

   (CCEA January 2011)

6. Calculate the volume of hydrogen produced at 20 °C and 1 atmosphere pressure when an excess of magnesium is added to 50.0 cm³ of 2.0 M hydrochloric acid.

$$Mg_{(s)} + 2HCl_{(aq)} \rightarrow MgCl_{2(aq)} + H_{2(g)}$$

(Adapted from CCEA January 2008)

7. Manganese(IV) oxide catalyses the decomposition of hydrogen peroxide. Calculate the volume of oxygen produced at 20 °C and 1 atmosphere pressure when 50.0 cm³ of 2.0 mol dm⁻³ hydrogen peroxide solution decomposes.

$$2H_2O_2 \rightarrow 2H_2O + O_2$$

(CCEA June 2010)

8. Nitrogen dioxide reacts with sodium hydroxide to form a mixture of sodium nitrate and sodium nitrite in solution. Calculate the volume of 2.0 M NaOH needed to react with 720 dm³ of nitrogen dioxide at 20 °C and 1 atmosphere pressure.

$$2NO_2 + 2NaOH \rightarrow NaNO_3 + NaNO_2 + H_2O$$

(CCEA January 2007)

Calculating Gas Density

By using Avogadro's law to calculate the number of moles in 1 cm^3 of a gas, the number of moles can be converted into grams to give the density of the gas in units of g cm^{-3}.

Worked Example 2.1v

Calculate the density of nitrogen gas at 20 °C and 1 atmosphere to two significant figures.

Solution

Moles in 1 cm^3 of any gas $\dfrac{0.001\ dm^3}{24\ dm^3} = 4.2 \times 10^{-5}$ mol

Mass of N_2 in 1 cm^3 = 4.2×10^{-5} mol × 28 g mol^{-1}
= 1.2×10^{-3} g

Density of nitrogen gas (N_2) = 1.2×10^{-3} g cm^{-3}

Exercise 2.1E

1. Calculate the density of (a) hydrogen, (b) oxygen and (c) carbon dioxide at 20 °C and 1 atmosphere to two significant figures.

2. You have two identical flasks: one contains nitrogen and the other contains hydrogen. How could you tell which is nitrogen without opening the flasks? Explain your answer.

Before moving to the next section, check that you are able to:

- Recall Avogadro's law and use it to relate the amount of a gas to its volume.
- Use Avogadro's law to calculate volumes of gases involved in chemical reactions.

Percentage Composition Calculations

In this section we are learning to:

- Calculate the mass percentage of an element in a compound.
- Describe the amount of each element in a compound by writing the percentage composition for the compound.

Mass Percentage

The **mass percentage** of an element in a substance is *the mass of the element in 100 g of the substance expressed as a percentage*. The mass percentages of the elements in aspirin, $C_9H_8O_4$ are: 4.4% hydrogen, 60.0% carbon and 35.6% oxygen. This is equivalent to stating that 100g of aspirin contains 4.4 g of hydrogen, 60.0 g of carbon and 35.6 g of oxygen.

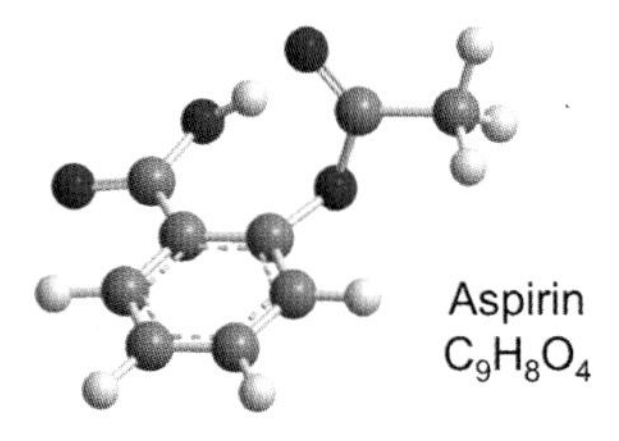

The mass percentage of an element in a compound is calculated using the formula:

$$\text{Mass percentage of an element} = \frac{\text{Mass of the element in 1 mole}}{\text{Molar mass of the compound}} \times 100\%$$

Worked Example 2.1vi

Calculate the mass percentage of carbon in aspirin, $C_9H_8O_4$.

Solution

One mole of aspirin contains 9 moles of carbon.

Mass of carbon in 1 mol of aspirin

$$= 9 \text{ mol} \times 12 \text{ g mol}^{-1} = 108 \text{ g of C}$$

Molar mass of aspirin = 180 g mol^{-1}

$$\text{Mass percentage carbon} = \frac{108 \text{ g}}{180 \text{ g mol}^{-1}} \times 100\ \% = 60.0\ \%$$

A list detailing the mass percentage of each element in a substance is known as the **percentage composition** of the substance. For example, the percentage composition of aspirin is 60.0% C, 4.4% H, and 35.6% O.

Exercise 2.1F

1. Calculate the percentage composition of (a) propane, C_3H_8 and (b) hexane, C_6H_{14}.
2. Calculate the percentage composition of glucose, $C_6H_{12}O_6$.

Before moving to the next section, check that you are able to:

- Use the formula of a compound to calculate the mass percentage of an element in the compound.
- Describe the amount of each element in a compound by writing the percentage composition for the compound.

Empirical Formula

The chemical formula of a newly discovered substance is often determined with the help of percentage composition data. The formula obtained from percentage composition data is known as the **empirical formula** for the compound and is *the simplest whole number ratio of atoms of each element in the compound.*

The empirical formula for an ionic compound is same as the chemical formula for the compound as it describes the ratio of ions in the lattice. For example, the empirical formula for magnesium chloride is $MgCl_2$ as the ionic lattice contains two chloride ions for every magnesium ion.

In contrast, the chemical formula for a molecular compound is often a multiple of the empirical formula, and is referred to as the **molecular formula** as it *represents the number of atoms of each element in one molecule of the compound.*

Worked Example 2.1vii

The percent composition of ascorbic acid is: 40.9% C, 4.6% H, 54.5% O. Calculate the empirical formula for ascorbic acid.

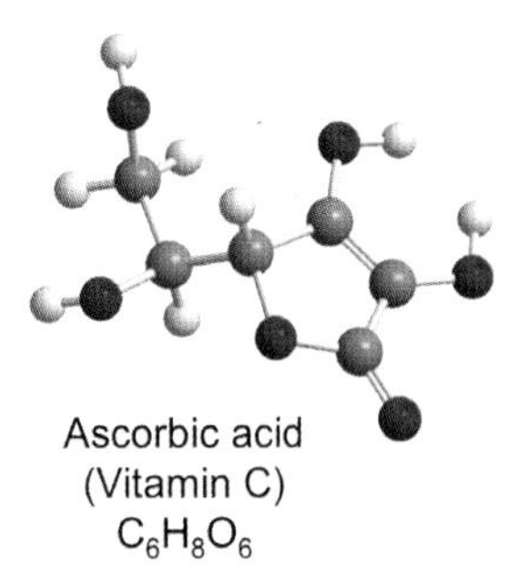

Ascorbic acid (Vitamin C) $C_6H_8O_6$

Strategy

- Calculate the moles of each element in 100 g of the compound.
- Use the mole ratio for the elements to write a formula for the compound.
- Convert the mole ratio to the simplest whole number ratio.

Solution

100 g of ascorbic acid contains:

$$40.9 \text{ g of C} = \frac{40.9 \text{ g}}{12 \text{ g mol}^{-1}} = 3.41 \text{ mol of C}$$

$$4.6 \text{ g of H} = \frac{4.6 \text{ g}}{1 \text{ g mol}^{-1}} = 4.6 \text{ mol of H}$$

$$54.5 \text{ g of O} = \frac{54.5 \text{ g}}{16 \text{ g mol}^{-1}} = 3.41 \text{ mol of O}$$

The mole ratio of the elements corresponds to the formula $C_{3.41}H_{4.6}O_{3.41}$

Converting to whole numbers:

$$C_{\frac{3.41}{3.41}}H_{\frac{4.6}{3.41}}O_{\frac{3.41}{3.41}} = C_1H_{\frac{4}{3}}O_1 = C_3H_4O_3$$

The empirical formula for ascorbic acid is $C_3H_4O_3$

Table 1: The molecular and empirical formulas for several compounds.

Substance	Molecular Formula	Empirical Formula
Methane (natural gas)	CH_4	CH_4
Octane (petrol)	C_8H_{18}	C_4H_9
Ethylene glycol (anti-freeze)	$C_2H_6O_2$	CH_3O
Glucose (sugar)	$C_6H_{12}O_6$	CH_2O
Ascorbic acid (Vitamin C)	$C_6H_8O_6$	$C_3H_4O_3$

The molecular ion peak in the mass spectrum for ascorbic acid corresponds to a RMM of 176. This value is exactly twice the RMM for the empirical formula, $C_3H_4O_3$ and indicates that the molecular formula for ascorbic acid is $C_6H_8O_6$ ($= 2 \times C_3H_4O_3$). In this way *the RMM obtained from the mass spectrum of a new compound can be combined with percentage composition data to determine the molecular formula of the compound.* The empirical formulas and molecular formulas for several compounds are given in Table 1.

Exercise 2.1G

1. An oxide of phosphorus contains 43.7% by mass of phosphorus. The relative molecular mass of the oxide is 284. This oxide reacts with sodium hydroxide to form sodium phosphate (Na_3PO_4) and water as the only products. (a) Calculate the empirical formula of the oxide. (b) Calculate the molecular formula of the oxide. (c) Write an equation for the reaction of the oxide with sodium hydroxide.

   (Adapted from CCEA January 2013)

2. (a) Explain the term empirical formula as applied to a molecular compound. (b) Explain the term molecular formula as applied to a molecular compound.

   (Adapted from CCEA January 2013)

3. An organic compound consists of 40.7% carbon, 5.1% hydrogen and 54.2% oxygen, and has a relative molecular mass of 118. (a) Calculate the empirical formula of the compound. (b) Determine the molecular formula of the compound.

   (Adapted from CCEA June 2012)

4. The refrigerant CFC-114 is 14.0% carbon and 44.4% fluorine by mass. The remainder is chlorine. Calculate the empirical formula of CFC-114.

   (Adapted from CCEA January 2011)

5. Analysis of a compound containing carbon, hydrogen and bromine showed that the compound is 22.2% carbon and 3.7% hydrogen by mass. The relative molecular mass of the compound is 216. Calculate the molecular formula of the compound.

   (CCEA June 2010)

6. The compound tetraethyllead (TEL) contains carbon, hydrogen and lead and is the chemical form of lead in petrol. TEL is 29.7% carbon and 6.23% hydrogen by mass. Calculate the empirical formula of tetraethyllead.

7. 0.8 moles of a liquid X was found to contain 19.2 g of carbon, 3.2 g of hydrogen and 56.8 g of chlorine. Calculate the empirical formula of X.

   (Adapted from CCEA June 2007)

Before moving to the next section, check that you are able to:

- Define what is meant by the term *empirical formula.*
- Use percentage composition data to determine the empirical formula for a compound.
- Explain why the empirical formula for an ionic compound is the same as the chemical formula for the compound.
- Define what is meant by the term *molecular formula.*
- Use the RMM and empirical formula for a molecular compound to determine the molecular formula of the compound.

2.2 Organic Chemistry

CONNECTIONS

- Organic chemistry is the branch of chemistry concerned with the preparation and study of organic compounds.
- Many plastics and fibres such as polythene and Nylon are made from the organic compounds found in crude oil.
- Many medicines contain organic compounds with structures based on Natural Products obtained directly from plants.

Organic Compounds

In this section we are learning to:

- Recognise organic compounds and recall how they can exist in different structural forms known as isomers.
- Recall that organic compounds have systematic names and common 'everyday' names.
- Explain how organic compounds can be grouped together to form a homologous series of compounds.

An **organic compound** is a substance whose structure is based on the element carbon. Many of the compounds found in nature are organic compounds. For instance, the compound carvone is an organic compound, and gives spearmint leaves their distinctive smell. Much of the food and medicine we consume is also made-up of organic compounds. Common pain-killers such as aspirin, dietary supplements such as Vitamin C, and sugars such as glucose are all organic compounds. The structures and molecular formulas for carvone, Vitamin C and Aspirin are shown in Figure 1.

The mirror image of the spearmint form of carvone produces the odour of caraway seeds. The relationship between the spearmint and caraway forms of carvone is shown in Figure 2. The spearmint and caraway forms of carvone are referred to as **isomers** as they are *compounds with the same molecular formula that have a different arrangement of atoms*. They are distinguished by including the prefix *D*- or *L*- when naming the isomers.

(a) Carvone (Spearmint) Formula: $C_{10}H_{14}O$

(b) Ascorbic acid (Vitamin C) Formula: $C_6H_8O_6$

(c) Acetylsalicylic acid (Aspirin) Formula: $C_9H_8O_4$

Figure 1: Structures and molecular formulas for (a) carvone (spearmint), $C_{10}H_{14}O$ (b) ascorbic acid (Vitamin C), $C_6H_8O_6$ and (c) acetylsalicylic acid (Aspirin), $C_9H_8O_4$.

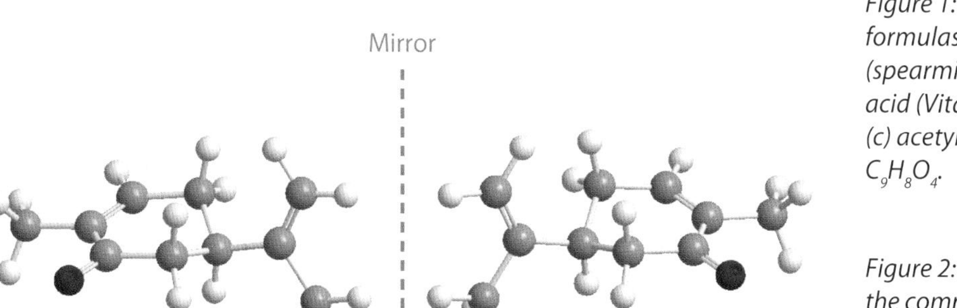

Figure 2: Mirror image isomers of the compound carvone. The isomers have the same molecular formula ($C_{10}H_{14}O$) but have a different arrangement of atoms in space and cannot be superimposed.

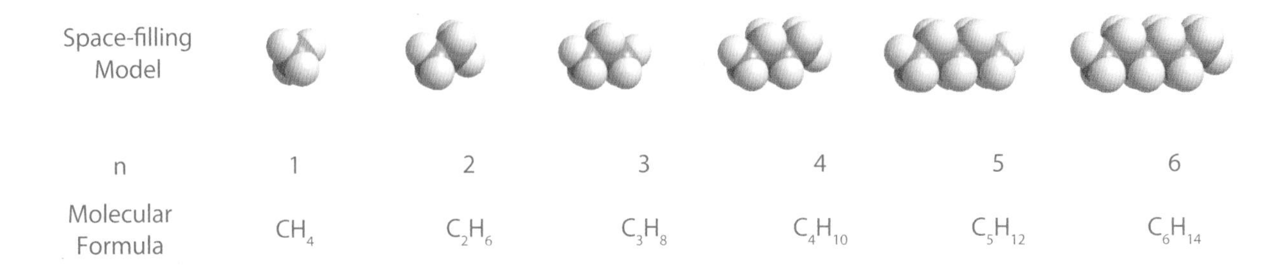

Figure 3: Space-filling models for the first six alkanes. Carbon atoms are shown in grey and hydrogen atoms are shown in white.

Talking Point

Every organic compound has a systematic name based on the rules developed and maintained by the International Union of Pure and Applied Chemistry (IUPAC). A large number of organic compounds such as lactic acid and glucose are still referred to by their common 'everyday' names.

Lactic acid, $C_3H_6O_3$

2-hydroxypropanoic acid

Glucose, $C_6H_{12}O_6$

2, 3, 4, 5, 6-pentahydroxyhexanal

Organic compounds can be very large. The carbohydrates and proteins in our food are constructed from much smaller organic molecules such as simple sugars and amino acids. Plastics such as polythene and fibres such as Nylon are also organic compounds, and are constructed from many smaller organic molecules.

Many of the plastics and other materials we use in our everyday lives are made from the organic compounds in crude oil. Crude oil is a complex mixture of organic compounds known as hydrocarbons where the term **hydrocarbon** refers to *a compound that contains only carbon and hydrogen.*

A large number of hydrocarbons belong to a family of compounds known as the *alkanes*. The molecular formulas for the alkanes can be obtained by setting n = 1, 2, 3 ... in the **general formula** C_nH_{2n+2}.

The effect of increasing n on the size of the alkanes can be seen by constructing **space-filling models** in which *each atom is represented by a sphere centred on the nucleus of the atom*. The space-filling models in Figure 3 clearly show that the alkane molecules become longer as the number of carbon atoms (n) increases.

The alkanes form a **homologous series** as they are *a family of compounds with similar chemical properties, whose formulas derive from the same general formula and differ by CH_2 when arranged by mass, and whose physical properties vary smoothly as the molecules get bigger.*

Ethanol, C_2H_6O is used to make alcoholic drinks and belongs to a different homologous series known as the *alcohols*. The molecular formulas for the alcohols are obtained by setting n = 1, 2 ... in the general formula $C_nH_{2n+2}O$. The space-filling models in Figure 4 clearly show that, as with the alkanes, the alcohols become longer as the number of carbon atoms in the molecule (n) increases.

Exercise 2.2A

1. In many countries ethanol, C_2H_6O is added to petrol to produce a blend called Gasohol. (a) Suggest how you would compare the size of an ethanol molecule and the size of a molecule of dimethylether, an isomer of ethanol. (b) Explain what is meant by the term *isomer*. (c) Explain what is meant by the term *hydrocarbon*. (d) Explain why ethanol is not a hydrocarbon.

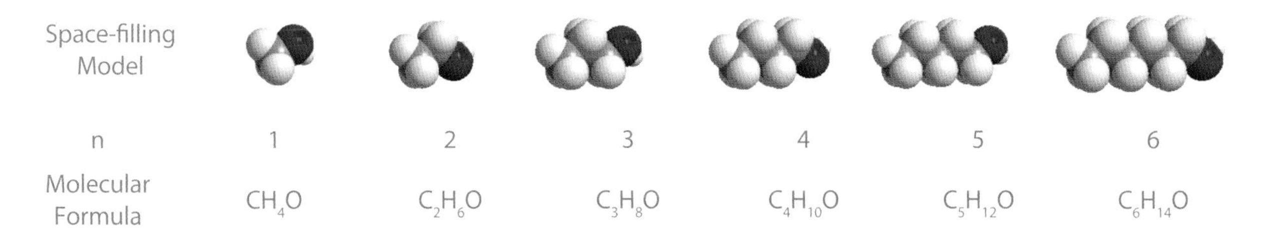

Figure 4: Space-filling models for the first six alcohols. The oxygen atom in each molecule is shown in red.

2. (a) Explain what is meant by the term *homologous series*. (b) Write the general formula for an alkane. (c) Write the molecular formula for an alkane with six carbon atoms.

3. Which one of the following is an alkane?

   A $C_{10}H_{18}$ B $C_{10}H_{20}$ C $C_{10}H_{22}$ D $C_{10}H_{24}$

   (CCEA January 2006)

4. Which one of the following is the *empirical* formula of an alkane?

   A C_4H_8 B C_4H_8O C C_3H_8 D C_4H_{10}

5. Which one of the following is the *molecular* formula of an alcohol?

   A C_2H_5O B C_2H_6 C C_2H_6O D C_2H_4O

6. Use the general formula for an alcohol in the form $C_nH_{2n+1}OH$ to write formulas for alcohols with up to four carbon atoms.

Before moving to the next section, check that you are able to:

- Recognise carbon-based compounds as organic compounds and explain why a compound can exist as isomers.
- Explain what is meant by the term *hydrocarbon* and recall that most hydrocarbons are obtained from crude oil.
- Recall the use of space-filling models to describe the size of molecules.
- Explain what is meant by a *homologous series* and use a general formula to generate molecular formulas for the compounds in a homologous series.

Structural Formulas

In this section we are learning to:

- Recall the use of molecular models to describe the size and shape of molecules.
- Describe the use of structural formulas and condensed formulas to represent the bonding within an organic compound.

Space-filling models are useful when comparing the size of molecules but cannot be used to describe the bonding within a compound. The bonding in a molecule or ion is instead described by constructing a **structural formula** in which *a shared pair of electrons is represented by a line between the atoms sharing the electrons.*

Comparing the space-filling model, the molecular shape, and the structural formula for ethane, C_2H_6 reminds us that *the structural formula for a molecule describes how the atoms are bonded and does not describe how the atoms are arranged in space.*

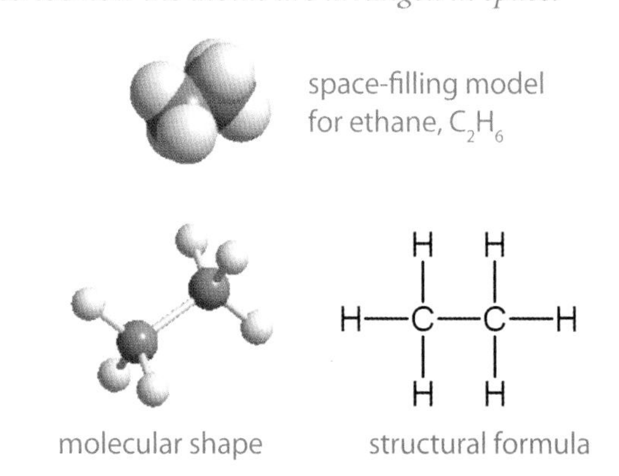

In practice, a structural formula is only used when it is necessary to detail the bonds formed between individual atoms in the structure. In many molecules the nature of the bonding between atoms can be inferred by writing the **condensed formula** for the compound in which *formulas are used to represent groups of atoms within the structure.*

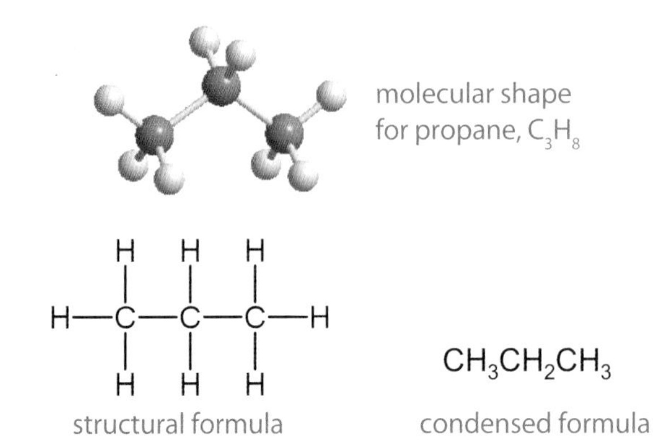

A condensed formula can also be used to quickly convey the arrangement of the atoms in a compound and the size of the molecule, making it a convenient tool to describe the structure of larger molecules such as hexane, C_6H_{14}.

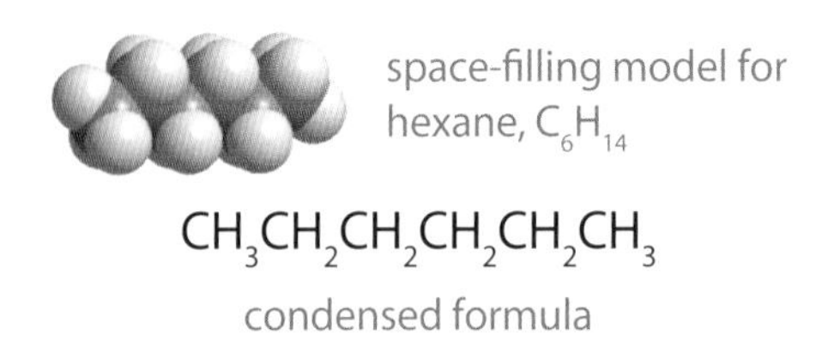

Exercise 2.2B

1. Propanone, C_3H_6O is the principal ingredient in nail-polish remover. (a) With reference to the condensed formula for propanone, explain what is meant by the *condensed formula* of a compound. (b) Explain why the structural formula for propanone cannot be used to describe the shape of a propanone molecule.

```
   H  O  H
   |  ‖  |
H—C—C—C—H
   |     |
   H     H
```

CH_3COCH_3

structural formula for propanone — condensed formula for propanone

> **Before moving to the next section, check that you are able to:**
>
> - Distinguish the use of molecular models to describe the shapes of molecules and the use of structural formulas to describe the bonding in molecules.
> - Explain how a condensed formula can be used to describe the structure of an organic compound.

Functional Groups

In this section we are learning to:

- Recall that the reactions of an organic compound are determined by the functional groups present in the compound.
- Describe the use of skeletal formulas to highlight the functional groups present in a compound.

The reactions of an organic compound are determined by the functional groups present in the molecule. The term **functional group** is used when referring to *a group of atoms in a molecule that together determine the reactions of the compound.*

For example, when wine 'goes bad' some of the alcohol (ethanol) in the wine is oxidised to form ethanoic acid.

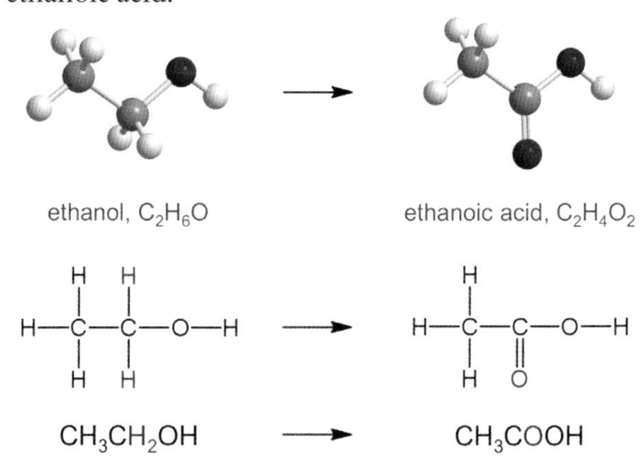

The reaction transforms the $-CH_2-$ group adjacent to the hydroxyl (-OH) group into part of the larger carboxyl (-COOH) group in the acid.

The hydroxyl (-OH) group is a functional group and is responsible for ethanol behaving as an alcohol. Similarly the carboxyl (-COOH) group makes it possible for ethanoic acid to react with metals, bases and other compounds in the same way as acids such as sulfuric acid and nitric acid.

Oxidation to form a carboxyl (-COOH) group is one of several reactions that can occur when a hydroxyl (-OH) group is present as $-CH_2OH$, and demonstrates how functional groups determine the reactions of a compound.

The nature and location of the functional groups in a molecule can be efficiently described by drawing a **skeletal formula** in which *the hydrogen atoms are omitted, and each shared pair of electrons in the remaining 'skeleton' is represented by a line or 'bond' between two atoms.* For example, the skeletal formula for carvone (spearmint) quickly reveals the presence of three functional groups: a C=O bond and two C=C bonds.

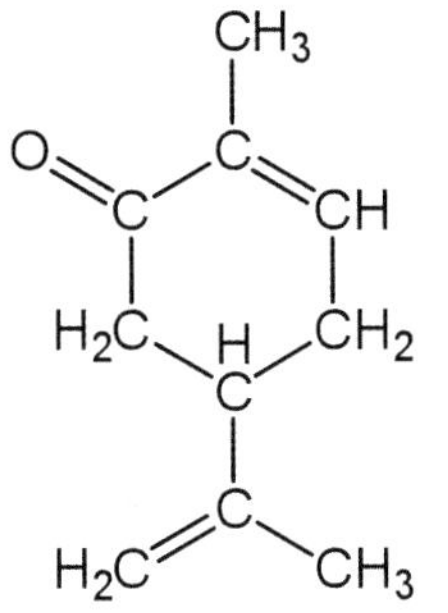

structure of carvone

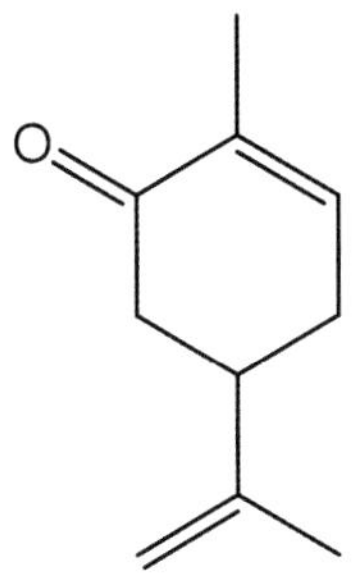

skeletal formula for carvone

Similarly, on using the condensed formula -COOH to represent a carboxyl group, the skeletal formula for traumatic acid, a wound healing agent found in plants, effectively reveals three functional groups: two carboxyl (-COOH) groups and a C=C bond.

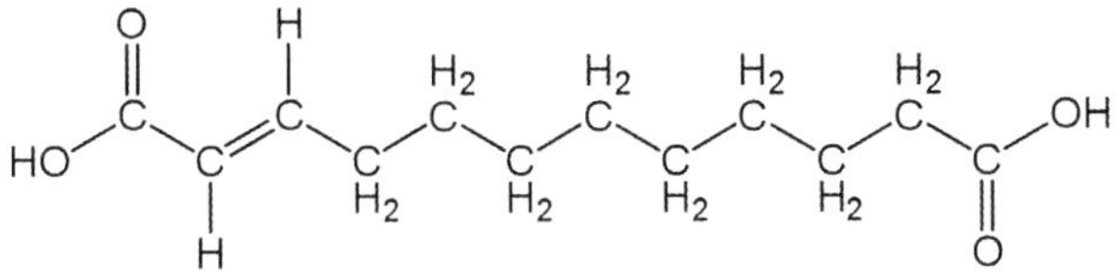

structure of traumatic acid

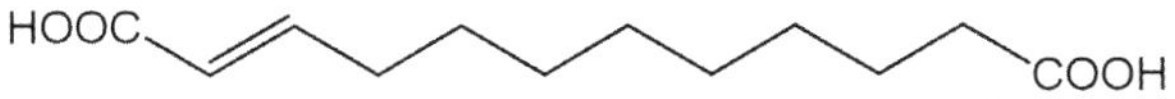

HOOC ... COOH

skeletal formula for traumatic acid

Exercise 2.2C

1. The compound cinnamyl alcohol contains several functional groups. Explain the meaning of the term *functional group*.

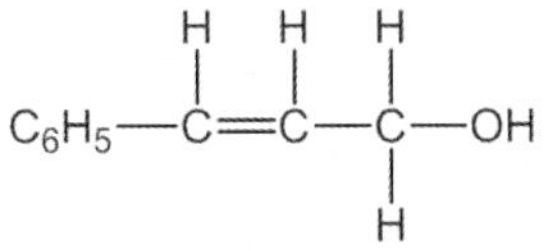

C_6H_5—CH=CH—CH_2—OH cinnamyl alcohol

(CCEA January 2003)

2. The structure of a molecule of lactic acid is shown below. (a) Write the molecular formula for lactic acid. (b) Draw the skeletal formula for a molecule of lactic acid using the condensed formulas -OH and -COOH to represent the functional groups present in the molecule.

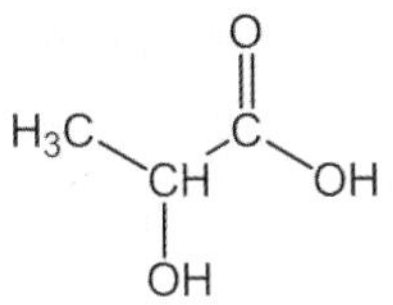

lactic acid

3. The structure of a molecule of glucose is shown below. (a) Write the molecular formula for glucose. (b) Write the empirical formula for glucose. (c) Draw the skeletal formula for a molecule of glucose using the condensed formulas -OH and -CHO to represent functional groups present in the molecule.

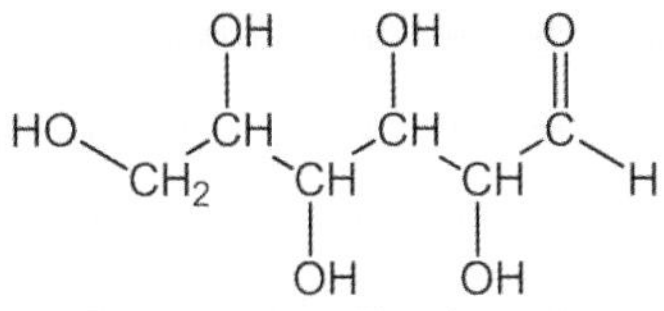

glucose

4. The structure of a molecule of aspirin is shown below. (a) Write the molecular formula for aspirin. (b) Write the empirical formula for aspirin. (c) Draw the skeletal formula for a molecule of aspirin using the condensed formula -OH to represent part of the structure.

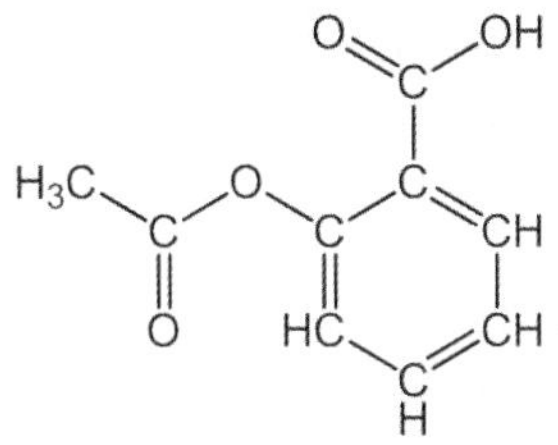

aspirin

Before moving to the next section, check that you are able to:

- Explain what is meant by the term *functional group*.
- Describe how to construct the skeletal formula for a compound.

2.3 Hydrocarbons: Alkanes

CONNECTIONS

- Crude oil is a major source of the hydrocarbons used to make plastics and fibres such as Nylon. It is also a nonrenewable resource and cannot be replaced at a rate that would sustain its use.
- Crude oil is also a major source of hydrocarbon-based fuels such as gasoline (petrol) and kerosene (jet fuel). Many scientists believe that the burning of hydrocarbon-based fuels is contributing to global warming; a global problem that cannot be effectively addressed without international cooperation.

Structure and Properties

In this section we are learning to:

- Recognise alkanes as a homologous series of hydrocarbons.
- Account for the properties of alkanes in terms of intermolecular forces.
- Recognise and draw structural isomers of alkanes.
- Use systematic (IUPAC) rules to name alkanes.

Physical Properties

The alkanes are a homologous series of hydrocarbons with the general formula C_nH_{2n+2}. The formulas for the first six alkanes (n = 1–6) in Table 1 reveal that alkanes are **saturated hydrocarbons** as they *contain only carbon and hydrogen, and do not contain any C=C or C≡C bonds*.

The C-C and C-H bonds in alkanes are strong and difficult to break. As a result, alkanes are unreactive and will only react when in contact with very reactive substances.

The bond dipoles associated with C-H bonds are very small. As a result, alkanes do not have a significant permanent dipole, and experience only van der Waals attraction between neighbouring molecules. The nonpolar nature of the alkanes makes liquids such as hexane, C_6H_{14} good solvents for other nonpolar molecules such as bromine, Br_2.

The van der Waals attraction between alkane molecules increases as the number of electrons in the molecule increases. This explains why the lighter alkanes (small n) are gases while the heavier alkanes (large n) are solids. The effect of increasing van der Waals attraction on boiling point is illustrated in Figure 1.

The 'straight-chain' alkanes in Table 1 are based on a simple chain of carbon atoms and are referred to as the **parent alkanes** on which more branched structures such as neopentane in Figure 2 are based.

The alkanes in Figure 2 are isomers; both have the formula C_5H_{12}. Pentane has a higher boiling point as the more branched structure of neopentane reduces the amount of contact between the electrons on neighbouring molecules. This reduces the van der Waals attraction between neopentane molecules and ensures that neopentane has a lower boiling point than pentane.

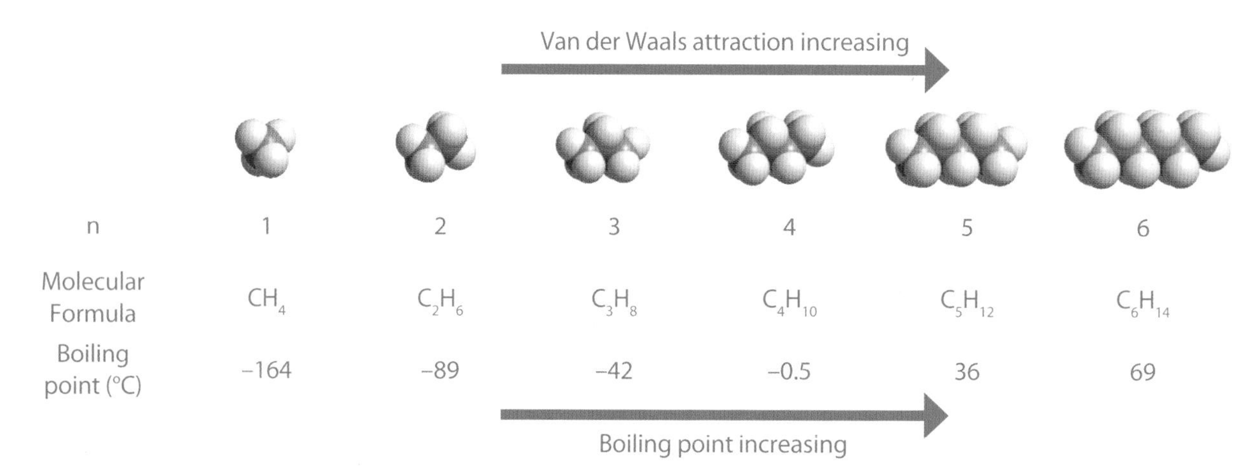

Figure 1: The effect of van der Waals attraction on boiling point for the first six alkanes.

n	Structural formula, name and molecular formula	Condensed formula	Skeletal formula
1	Methane CH_4	CH_4	
2	Ethane C_2H_6	CH_3CH_3	
3	Propane C_3H_8	$CH_3CH_2CH_3$	
4	Butane, C_4H_{10}	$CH_3CH_2CH_2CH_3$	
5	Pentane C_5H_{12}	$CH_3CH_2CH_2CH_2CH_3$	
6	Hexane C_6H_{14}	$CH_3CH_2CH_2CH_2CH_2CH_3$	

Table 1: Structures and formulas for the first six alkanes.

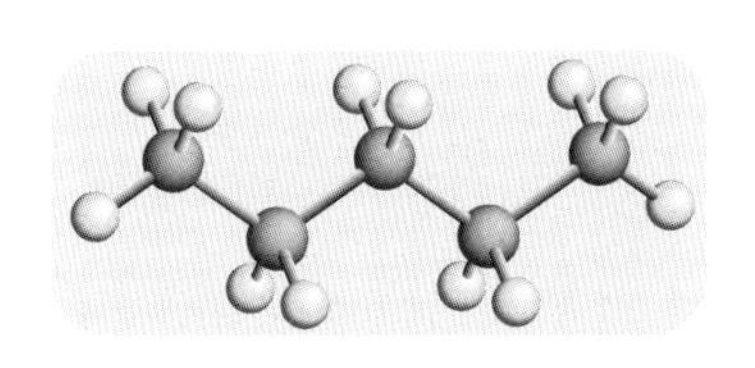

Pentane, C_5H_{12}
Boiling point 36 °C

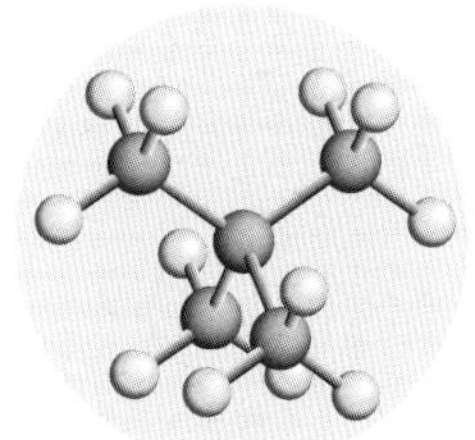

Neopentane, C_5H_{12}
Boiling point 10 °C

Figure 2: The effect of molecular shape on the boiling point of pentane isomers.

Exercise 2.3A

1. The alkane with 20 carbon atoms is known as icosane. (a) Write the general formula for an alkane. (b) Write the molecular formula for icosane. (c) Alkanes are saturated. Explain what the term *saturated* means. (d) Explain why icosane is a solid at room temperature whereas ethane, C_2H_6 is a gas.

(Adapted from CCEA January 2014)

Naming Alkanes

The name of an alkane is based on the name of the parent (straight-chain) alkane on which its structure is based. The names of the parent alkanes consist of a prefix such as *meth* or *prop* that describes the number of carbon atoms in the chain, followed by the suffix *ane* to indicate that the compound is an alkane. The prefixes used to name alkanes with up to six carbon atoms in the longest carbon chain are summarised in Table 2.

Carbon atoms in longest chain	Naming prefix	Name of parent alkane
1	meth	meth**ane**
2	eth	eth**ane**
3	prop	prop**ane**
4	but	but**ane**
5	pent	pent**ane**
6	hex	hex**ane**

Table 2: Naming prefixes for alkanes based on a chain of up to six carbon atoms

The compound 2-methylbutane is an isomer of pentane, C_5H_{12} and is named as a butane as its structure is based on a chain of four carbon atoms. The fifth carbon atom, in the form of a methyl group ($-CH_3$), is attached to the second carbon atom in the chain.

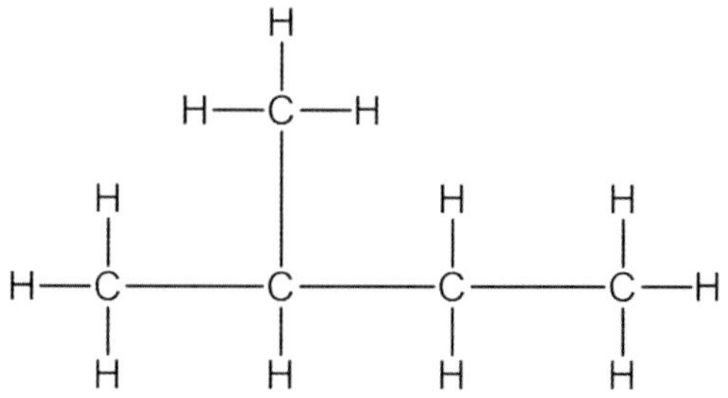

Structural formula for 2-methylbutane, C_5H_{12}

$CH_3CH(CH_3)CH_2CH_3$

Condensed formula for 2-methylbutane, C_5H_{12}

In 2-methylbutane, the prefix 2-*methyl* is used to indicate that the methyl group is attached to the second carbon atom in the longest carbon chain. Use of the prefix *2-methyl* is, however, potentially confusing. The methyl group is only attached to the second carbon atom if we choose to number the carbon atoms in the chain from left to right. If we instead choose to number the carbon atoms from right to left, the location of the methyl group is described by the prefix *3-methyl.* As a result, and to avoid confusion, *the carbon atoms in the longest carbon chain are always numbered in a way that produces the lowest numbers in prefixes.*

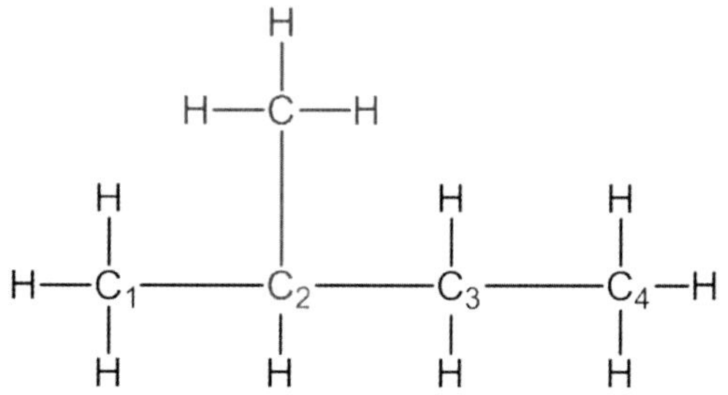

2-methylbutane ✓

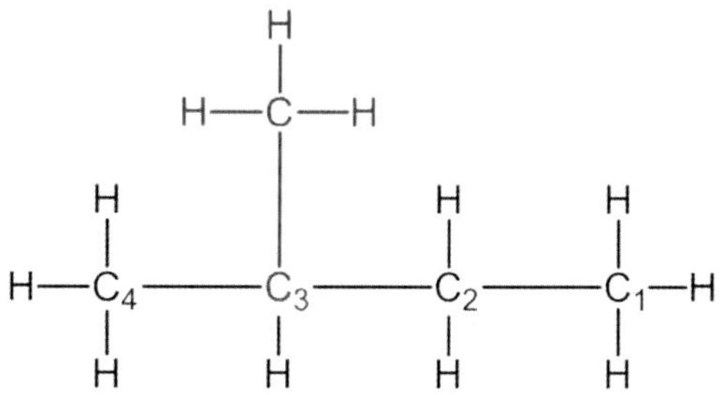

3-methylbutane ✗

The compound 2,2-dimethylpropane is also an isomer of pentane, C_5H_{12}. It is named as a propane as its structure is based on the chain of three carbon atoms in propane. The remaining two carbon atoms are in the form of methyl (-CH_3) groups attached to the second carbon atom in the chain.

```
            H
            |
         H—C—H
            |
  H         |         H
  |         |         |
H—C————————C————————C—H
  |         |         |
  H         |         H
            |
         H—C—H
            |
            H
```

Structural formula for 2,2-dimethylpropane, C_5H_{12}

$CH_3C(CH_3)_2CH_3$

Condensed formula for 2,2-dimethylpropane, C_5H_{12}

The prefix *2,2-dimethyl* is used to indicate that both methyl groups are attached to the second carbon atom in the longest carbon chain.

The presence of additional methyl groups in the molecule could be described by using the number prefixes *tri, tetra, penta, hexa, ...* to construct naming prefixes such as *trimethyl* and *tetramethyl.* An additional number prefix such as *1,2,2-* or *2,2,3,3-* could then be added to describe the location of each methyl group as in 2,2,4-trimethylpentane.

```
   H  CH3 H  CH3 H
   |   |  |   |  |
H—C———C——C———C——C—H
   |   |  |   |  |
   H  CH3 H   H  H
```

2,2,4-trimethylpentane

skeletal formula

Worked Example 2.3i

Deduce the systematic name and skeletal formula for the compound:

```
  CH3 H  H  CH3 H
   |  |  |   |  |
H—C——C——C———C——C—H
   |  |  |   |  |
  CH3 H  H   H  H
```

(Adapted from CCEA June 2001)

Strategy

- Determine the longest carbon chain in the molecule.
- Locate the methyl groups on the longest chain using the lowest number prefixes.

Solution

```
  CH3 H  H  CH3 H
   |  |  |   |  |
H—C——C——C———C——C—H
   |  |  |   |  |
  CH3 H  H   H  H
```

The molecule is a hexane with methyl groups attached to the second and fifth carbon atoms in the chain.

The compound is 2,5-dimethylhexane.

Skeletal formula:

Exercise 2.3B

1. Which one of the following is the IUPAC name for the hydrocarbon shown below?

```
       H  H  CH3
       |  |  |
H3C——C——C——C—H
       |  |  |
       H  H  CH3
```

A 1,1-dimethylbutane

B 4,4-dimethylbutane

C 2-methylpentane

D 1,1,3-trimethylpropane

(CCEA January 2013)

2. What is the IUPAC name for the hydrocarbon shown below?

```
      CH3
      |
CH3CHCHCH3
        |
        CH3
```

A 1,4-dimethylbutane

B 2,3-dimethylbutane

C 1,4-dimethylhexane

D 2,3-dimethylhexane

(CCEA June 2011)

3. The correct name for the following compound is

```
      CH3 H   H   CH3
       |  |   |   |
   H — C — C — C — C — H
       |  |   |   |
      CH3 H   H   H
```

A 1,1,4-trimethylbutane
B 1,4-dimethylpentane
C 1,4,4-trimethylbutane
D 2-methylhexane

(CCEA June 2010)

4. Which one of the following compounds has the highest boiling point?

A 2,3-dimethylbutane
B Hexane
C 2-methylpentane
D 3-methylpentane

(CCEA June 2013)

Structural Isomers

The term **structural isomer** is used when referring to *molecules with the same molecular formula that have different structural formulas*. The compounds pentane and neopentane are structural isomers with the molecular formula C_5H_{12}.

```
                          CH3
                           |
                    H3C — C — CH3
CH3CH2CH2CH2CH3            |
                          CH3

    pentane            neopentane
```

Worked Example 2.3ii

Draw a structural formula for each butane with the molecular formula C_6H_{14}. Label each isomer with its condensed formula and systematic (IUPAC) name.

Strategy

- Begin by drawing a chain of four carbon atoms (butane).
- Add the remaining atoms to generate a saturated hydrocarbon in which the longest carbon chain contains four carbon atoms.
- Repeat until all possible structures have been found.

Solution

There are two butanes with the molecular formula C_6H_{14}

```
       H   CH3  H   H
       |   |    |   |
   H — C — C — C — C — H
       |   |    |   |
       H   CH3  H   H
```

$CH_3C(CH_3)_2CH_2CH_3$ 2,2-dimethylbutane

```
       H   CH3  H   H
       |   |    |   |
   H — C — C — C — C — H
       |   |    |   |
       H   H   CH3  H
```

$CH_3CH(CH_3)CH(CH_3)CH_3$ 2,3-dimethylbutane

Exercise 2.3C

1. (a) Write the general formula for an alkane. (b) The pentanes in the table are structural isomers. Explain the term *structural isomer*. (c) Write IUPAC names for isopentane and neopentane. (d) Explain why the pentanes have different boiling points.

structure	name	boiling point
$CH_3CH_2CH_2CH_2CH_3$	normal pentane	36 °C
$(H_3C)_2CHCH_2CH_3$	isopentane	28 °C
$H_3C-C(CH_3)_2-CH_3$	neopentane	9 °C

(CCEA January 2010)

2. How many structural isomers have the formula C_4H_{10}?

(CCEA June 2009)

3. How many structural isomers have the formula C_5H_{12}?

A Two B Three C Four D Five

(CCEA January 2013)

Alkyl Groups

An **alkyl group** is *a functional group with the general formula C_nH_{2n+1} where n = 1, 2, 3 ...* The structures and condensed formulas for simple (straight-chain) alkyl groups with up to three carbon atoms (n=1-3) are summarised in Table 3.

n	Structural formula for alkyl group	Condensed formula	Name of alkyl group
1	H —C—H H	$-CH_3$	Methyl
2	H H —C—C—H H H	$-C_2H_5$	Ethyl
3	H H H —C—C—C—H H H H	$-C_3H_7$	Propyl

Table 3: Structures and formulas for simple (straight-chain) alkyl groups.

The presence of an alkyl group in a molecule is indicated by a prefix that describes the type of alkyl group and its location in the molecule. For example, the prefix *3,3-diethyl* could be used to locate two ethyl groups on the third carbon atom as in 3,3-diethylhexane.

```
          C2H5
           |
CH3CH2—C—CH2CH2CH3
           |
          C2H5
```

3,3-diethylhexane

skeletal formula

Similarly, the prefix *4-ethyl-2-methyl* could be used to locate an ethyl group on the fourth carbon atom and a methyl group on the second carbon atom as in 4-ethyl-2-methylhexane.

```
      CH3  H    C2H5
       |   |    |
H3C—C—C—C—CH2CH3
       |   |    |
       H   H    H
```

4-ethyl-2-methylhexane

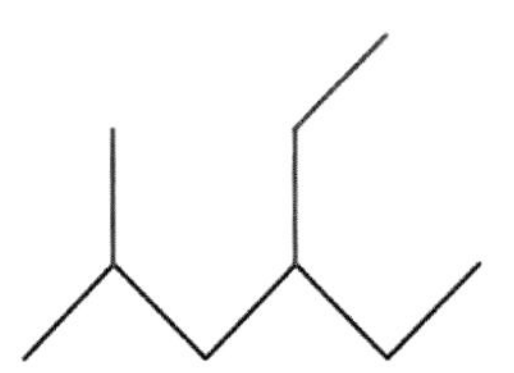

skeletal formula

The same compound could also have been named using the prefix *2-methyl-4-ethyl*. As a result, and to avoid confusion, *the prefixes used to locate different functional groups in a molecule are written in alphabetical order when naming the compound.*

Rules for naming alkanes:

- The name of an alkane is based on the name of the parent (straight-chain) alkane on which its structure is based.
- A prefix is added to describe the type and location of each alkyl group. The prefixes are written in alphabetical order and are based on a numbering of the longest carbon chain that produces the lowest numbers in the prefixes.

Exercise 2.3D

1. What is the systematic name for the molecule shown below?

```
H3C               CH2CH3
    \            /
     CHCH2CH
    /            \
H3C               CH3
```

A 2,4-dimethylhexane
B 2-ethyl-4-methylpentane
C 2-methyl-4-ethylpentane
D 3,5-dimethylhexane

(CCEA June 2008)

2. (a) Write the systematic name for the following compound. (b) Draw the skeletal formula for a molecule of the compound.

```
       H   CH2CH3
       |   |
H3C—C—C—CH2CH3
       |   |
      CH3  H
```

3. (a) Write the systematic name for the following compound. (b) Draw the skeletal formula for a molecule of the compound.

```
      CH3 H   CH2CH3
       |  |   |
H3C—C—C—C—CH2CH3
       |  |   |
      CH3 H   H
```

4. (a) Write the molecular formula for compound A. (b) Deduce the systematic name for compounds A and B. (c) Explain why compounds A and B are structural isomers.

```
          CH2CH2CH3
          |
CH3CH2—C—CH2CH3
          |
          CH3
```

Compound A

```
       H  H   CH2CH3
       |  |   |
H3C—C—C—C—CH2CH3
       |  |   |
      CH3 H   H
```

Compound B

5. Which one is the general formula of an alkyl group?

A C_nH_{2n} B C_nH_{2n-1} C C_nH_{2n+1} D C_nH_{2n+2}

(CCEA June 2015)

Before moving to the next section, check that you are able to:

- Recall the general formula for an alkane and use the general formula to generate molecular formulas for alkanes.
- Explain how molecular size and shape affects the boiling point of alkanes.
- Write structural formulas and condensed formulas for alkanes.
- Recognise and draw structural isomers of alkanes.
- Deduce systematic names for alkanes.

Combustion of Hydrocarbons

In this section we are learning to:

- Describe the complete and incomplete combustion of hydrocarbons.
- Recall the environmental problems associated with the combustion of hydrocarbons.
- Explain the use of a catalytic converter to reduce environmental pollution resulting from the burning of hydrocarbon-based fuels in motor vehicles.

Complete Combustion

The term combustion refers to the reaction between a compound and oxygen. If a compound is burnt in an excess of air (oxygen), the compound completely reacts with oxygen, and the reaction is referred to as **complete combustion**. The complete combustion of hydrocarbons produces carbon dioxide (CO_2), water (H_2O) and energy in the form of heat.

The **COMPLETE** combustion of butane, C_4H_{10}:

$$C_4H_{10(g)} + \frac{13}{2}O_{2(g)} \rightarrow 4CO_{2(g)} + 5H_2O_{(l)}$$

or

$$2C_4H_{10(g)} + 13O_{2(g)} \rightarrow 8CO_{2(g)} + 10H_2O_{(l)}$$

Worked Example 2.3iii

10 cm^3 of a hydrocarbon was reacted with 70 cm^3 of oxygen. After reaction the mixture contained 30 cm^3 of carbon dioxide and 20 cm^3 of oxygen. Calculate the formula of the hydrocarbon.

(Adapted from CCEA January 2010)

Strategy

- Write a balanced equation for the combustion of a hydrocarbon C_xH_y.
- Use Avogadro's law (the volume of a gas is proportional to the moles of gas present) to relate the amounts of gas present.

Solution

$$C_xH_y + zO_2 \rightarrow xCO_2 + \frac{y}{2}H_2O$$

Moles of CO_2 in 30 cm³ = 3 × Moles of C_xH_y in 10 cm³ (Avogadro's law). Therefore x = 3 and the formula of the hydrocarbon is C_3H_y.

The hydrocarbon reacts with 50 cm³ of oxygen.
Moles of O_2 in 50 cm³ = 5 × Moles of C_xH_y in 10 cm³ (Avogadro's law).
Therefore z = 5 and the equation for the combustion becomes:

$$C_3H_y + 5O_2 \rightarrow 3CO_2 + \frac{y}{2}H_2O$$

Balancing oxygen in the equation gives y = 8.
The formula of the hydrocarbon is C_3H_8.

Exercise 2.3E

1. Branched hydrocarbons such as 2,2,4-trimethylpentane, C_8H_{18} burn smoothly in car engines. (a) Draw the structure of 2,2,4-trimethylpentane. (b) Write an equation for the complete combustion of 2,2,4-trimethylpentane.

   (Adapted from CCEA January 2002)

2. Paraffin wax is a mixture of alkanes containing 20-40 carbon atoms. The alkane with 20 carbon atoms is known as icosane. Write an equation for the complete combustion of icosane.

   (CCEA January 2014)

3. Which one of the following compounds will produce equal volumes of carbon dioxide and water vapour when burnt completely in oxygen?

   A CH_4 B C_2H_6 C C_2H_4 D C_3H_8

   (Adapted from CCEA January 2014)

4. Which one of the following is the volume of oxygen required for the complete combustion of 100 cm³ of butane at room temperature and pressure?

   A 400 cm³ B 500 cm³ C 650 cm³ D 1300 cm³

   (CCEA January 2012)

5. 20 cm³ of a gaseous hydrocarbon reacts with exactly 90 cm³ of oxygen to produce 60 cm³ of carbon dioxide. Calculate the formula of the hydrocarbon.

   (Adapted from CCEA January 2008)

Before moving to the next section, check that you are able to:

- Recall the conditions for the complete combustion of hydrocarbons.
- Write equations for the complete combustion of hydrocarbons.

Incomplete Combustion

If a hydrocarbon burns in a limited supply of oxygen, the carbon within the hydrocarbon is converted to carbon monoxide (CO) and the process referred to as **incomplete combustion**. The incomplete combustion of alkanes and other hydrocarbons may also produce soot (carbon) as there is not enough oxygen available to react with all of the carbon in the compound.

The **INCOMPLETE** combustion of butane, C_4H_{10}:

$$C_4H_{10(g)} + \frac{9}{2}O_{2(g)} \rightarrow 4CO_{(g)} + 5H_2O_{(l)}$$

or

$$2C_4H_{10(g)} + 9O_{2(g)} \rightarrow 8CO_{(g)} + 10H_2O_{(l)}$$

Exercise 2.3F

1. Write an equation for (a) the *complete* combustion of pentane and (b) the *incomplete* combustion of pentane to form carbon monoxide.

   (CCEA January 2010)

2. Write an equation for the *incomplete* combustion of heptadecane, $C_{17}H_{36}$ to form carbon monoxide.

   (CCEA June 2010)

3. Carbon (soot) can be formed by the incomplete combustion of hydrocarbons. (a) Name two other products formed by the incomplete combustion of hydrocarbons. (b) What causes incomplete combustion to occur?

   (CCEA January 2009)

4. Branched-chain alkanes are less likely to cause 'knocking' when a mixture of the alkane and air is ignited in car engines. (a) Write the systematic name for the alkane shown below. (b) The branched-chain alkane is a structural isomer of a straight-chain alkane. Explain the term *structural isomer*. (c) Write the equation for the *incomplete* combustion of this alkane. (d) Explain why incomplete combustion produces smoke.

```
     H3C         CH3
      |           |
   CH3CHCH2CH2CHCH3
```

   (Adapted from CCEA June 2005)

5. The incomplete combustion of propane according to the following equation produced gases with a total volume of 9000 dm^3. The molar gas volume, under these conditions, is 30 dm^3.

   $C_3H_8(g) + 4O_2(g) \rightarrow CO_2(g) + 2CO(g) + 4H_2O(l)$

   (a) Define the term *molar gas volume*. (b) Calculate the number of moles of carbon monoxide produced. (b) Calculate the number of moles of oxygen used. (c) Calculate the mass of propane burned in kg. (d) Under a different set of conditions, methane undergoes incomplete combustion to produce carbon dioxide and carbon monoxide in a 2:1 ratio. Write an equation for this reaction.

   (Adapted from CCEA June 2014)

Before moving to the next section, check that you are able to:

- Recall the conditions for the incomplete combustion of hydrocarbons.
- Write equations for the incomplete combustion of hydrocarbons.

Air Pollution

Since the beginning of the modern industrial era the amount of carbon dioxide in the atmosphere has increased from around 0.03% to 0.04% of all gases in the atmosphere. The increase is primarily the result of burning fossil fuels such as coal and oil, and has been linked to an increase in global temperatures; an effect known as **Global Warming**.

The exhaust fumes produced by motor vehicles also contain significant amounts of carbon monoxide (CO), nitrogen oxides (NO and NO_2), unburnt hydrocarbons and particulate matter that is mostly carbon in the form of soot. Carbon monoxide is very toxic; even brief exposure to elevated levels results in acute breathing difficulties. In the presence of sunlight the nitrogen oxides and hydrocarbons react to produce a hazardous mixture of **volatile** organic compounds that combine with the particulates to form a think smoky fog known as photochemical smog. Photochemical smog causes short-term breathing difficulties and, with repeated exposure, longer term respiratory problems in humans.

The amount of carbon monoxide, nitrogen oxides and hydrocarbons in exhaust emissions can be reduced by passing the exhaust gases through a catalytic converter. A catalytic converter contains a 'honeycomb' made of a ceramic material as shown in Figure 3. The inside surfaces of the honeycomb are coated in small particles of metals such as palladium (Pd), rhodium (Rh) and platinum (Pt). When exhaust gases from the engine pass through the ceramic honeycomb they bond to the metal particles dispersed on the honeycomb and are transformed into less toxic products. The metal particles act as a **catalyst** by speeding up the reactions involving the exhaust gases without being consumed by the reactions.

The metal catalyst in a catalytic converter is known as a three-way catalyst as it catalyses three types of reactions: the oxidation of carbon monoxide, the reduction of nitrogen oxides and the combustion of unburnt hydrocarbons.

Oxidation of carbon monoxide

$$2CO + O_2 \rightarrow 2CO_2$$

Reduction of nitrogen oxides

$$2NO_x \rightarrow N_2 + xO_2$$

Combustion of unburnt hydrocarbons

hydrocarbon + oxygen → carbon dioxide + water

CHARLES D. WINTERS/SCIENCE PHOTO LIBRARY

ASTRID & HANNS-FRIEDER MICHLER/SCIENCE PHOTO LIBRARY

Figure 3: (a) A catalytic converter for use in the exhaust system of a motor vehicle. (b) A cross-section of the ceramic honeycomb inside the catalytic converter.

Worked Example 2.3iv

Write an equation to show how (a) unburnt heptane, C_7H_{16} is removed, and (b) carbon monoxide reacts with nitrogen(II) oxide to form nitrogen, when vehicle exhaust emissions are passed through a catalytic converter.

(Adapted from CCEA January 2011)

Solution

(a) Hydrocarbons react to form carbon dioxide and water:

$C_7H_{16} + 11O_2 \rightarrow 7CO_2 + 8H_2O$

(b) Combining the reactions $2CO + O_2 \rightarrow 2CO_2$ and $2NO \rightarrow N_2 + O_2$ gives:

$2CO + 2NO \rightarrow 2CO_2 + N_2$

The formation of a bond between a molecule and a surface is referred to as **chemisorption**, and is the first step in the mechanism for the oxidation of carbon monoxide. The mechanism is illustrated in Figure 4 and reveals that the carbon monoxide molecules chemisorb on the metal surface before reacting with oxygen atoms from the dissociation of nitrogen oxides to form carbon dioxide. When CO chemisorbs on the surface the bond between the C and O atoms in the molecule becomes weaker, making it easier for the CO molecule to react. Chemisorption also brings the reactants closer together and arranges them in the correct orientation, making it easier for the reaction to occur.

Worked Example 2.3v

The catalytic converter in a car exhaust system may contain metals such as platinum and palladium. The metals act as heterogeneous catalysts. Explain their catalytic behaviour in terms of chemisorption.

(Adapted from CCEA June 2008)

Solution

The molecules in the exhaust gases chemisorb on the surface of the metal. Chemisorption brings the reactants closer together, aligns them in the correct orientation for a reaction to occur, and weakens the bonds within the gases. In this way chemisorption makes it easier for the reaction to occur. Once formed, the products desorb from the surface.

The chemisorption of exhaust gases in a catalytic converter is also an example of **heterogeneous catalysis**; a term that describes a reaction in which the catalyst is in a different physical state than the reactants. The ceramic honeycomb greatly increases the efficiency of the metal catalyst by creating a larger surface over which the exhaust gases can come into contact with the metal catalyst. The efficiency of the metal catalyst is further improved by using **finely divided** metals consisting of small particles with a large surface area on which the reacting gases can chemisorb and react.

The efficiency of the catalytic converter can be maintained if the catalytic converter is used in

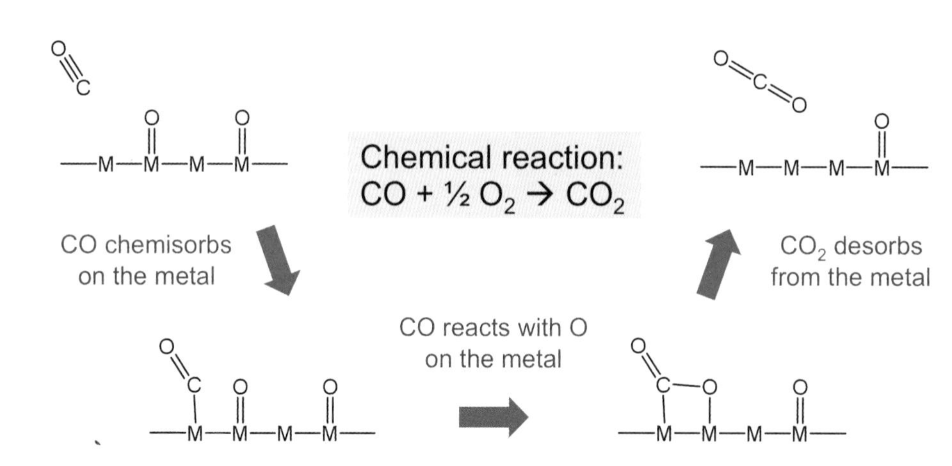

Figure 4: The chemisorption and subsequent oxidation of carbon monoxide to carbon dioxide on the surface of a metal catalyst.

conjunction with unleaded fuels. A catalytic converter cannot be used with leaded fuels as the lead in additives such as tetraethyllead, $Pb(C_2H_5)_4$ bonds to the metal particles and 'poisons' the catalyst by preventing exhaust gases from bonding to the metal particles.

Exercise 2.3G

1. Catalytic converters are used to reduce the environmental damage due to vehicle emissions. Discuss the environmental problems associated with the combustion of hydrocarbons.

   (CCEA June 2015)

2. What is formed when (a) carbon monoxide (b) unburnt hydrocarbons and (c) nitrogen oxides are passed through a catalytic converter?

   (CCEA January 2010)

3. Write equations for the conversion of carbon monoxide and nitrogen monoxide to less polluting products in a catalytic converter.

   (CCEA June 2015)

4. The reactions that occur in a catalytic converter are examples of heterogeneous catalysis involving chemisorption. (a) Explain the term *chemisorption*. (b) Explain why the ceramic support in a catalytic converter has a 'honeycomb' structure.

   (CCEA June 2005)

5. A catalytic converter contains a ceramic honeycomb structure coated in finely divided metals. (a) Suggest why the metals are finely divided. (b) Explain how chemisorption takes place on the surface of the metals and leads to a reaction. (c) Explain why lead-free petrol must be used with a catalytic converter.

   (CCEA January 2009)

6. (a) Write the equation for the reaction between CO and NO in a catalytic converter. (b) Name one metal used in catalytic converters. (c) Suggest why the metal particles in a catalytic converter are spread very thinly on a ceramic support. (d) Why are the reactions in a catalytic converter described as examples of heterogeneous catalysis?

   (Adapted from CCEA January 2008)

Before moving to the next section, check that you are able to:

- Recall the environmental problems associated with the combustion of hydrocarbon-based fuels.
- Explain how a catalytic converter can be used to reduce environmental pollution resulting from the burning of hydrocarbon- based fuels in motor vehicles.
- Describe the role of chemisorption in the reactions that occur in a catalytic converter and why the catalysis is an example of heterogeneous catalysis.
- Explain why the efficiency of a catalytic converter is increased by the use of a honeycomb coated in finely divided metals, and decreased by the use of leaded fuels.

Halogenation of Alkanes

In this section we are learning to:

- Describe the mechanism for the free radical substitution of methane and explain the role of ultraviolet light in the reaction.

Alkanes are generally unreactive and will only react when they come into contact with very reactive substances. A **free radical** is an *atom, molecule or ion with an unpaired electron.* Free radicals are very reactive and will react with alkanes.

When a halogen molecule absorbs ultraviolet (UV) light, the energy provided by the UV light breaks the covalent bond in the halogen molecule and produces two halogen atoms, each with an unpaired electron in its outer shell.

Homolytic fission of the Cl-Cl bond in chlorine:

$$Cl_{2(g)} \rightarrow 2Cl\bullet_{(g)}$$

Homolytic fission of the Br-Br bond in bromine:

$$Br_{2(g)} \rightarrow 2Br\bullet_{(g)}$$

The formulas for the halogen atoms are written Cl• and Br• to indicate that they are free radicals with an unpaired electron (•) in their outer shell. This bond-breaking process is an example of **homolytic fission** as *each atom forming the covalent bond receives one electron from the shared pair when the bond breaks.*

When a mixture of methane, CH_4 and chlorine, Cl_2 is exposed to UV light, chlorine radicals formed by the homolytic fission of the Cl-Cl bond in chlorine react with methane to form chlorinated hydrocarbons such as chloromethane, CH_3Cl.

$$CH_{4(g)} + Cl_{2(g)} \rightarrow CH_3Cl_{(g)} + HCl_{(g)}$$

Smaller amounts of more substituted hydrocarbons such as dichloromethane, CH_2Cl_2 and trichloromethane, $CHCl_3$ are also formed in the reaction.

$$CH_3Cl_{(g)} + Cl_{2(g)} \rightarrow CH_2Cl_{2(g)} + HCl_{(g)}$$

$$CH_2Cl_{2(g)} + Cl_{2(g)} \rightarrow CHCl_{3(g)} + HCl_{(g)}$$

The chlorination of methane is an example of a **photochemical reaction** as the *reaction occurs when light is absorbed by one or more of the reactants.* The light provides the energy needed for the reaction to occur and is consumed by the reaction. As a result, light cannot be considered a catalyst for the reaction as a catalyst would speed up the reaction without being consumed by the reaction.

The photochemical chlorination of methane occurs in three stages: initiation, propagation and termination. When the reactions taking place at each stage are written in the order they occur, the resulting scheme is referred to as the **reaction mechanism**.

Mechanism for the photochemical chlorination of methane:

INITIATION

Chlorine molecules absorb UV light and undergo homolytic fission to produce chlorine radicals:

$$Cl_2(g) \rightarrow 2Cl\bullet(g)$$

PROPAGATION

Chlorine radicals abstract hydrogen from methane to form hydrogen chloride (HCl).

$$CH_{4(g)} + Cl\bullet_{(g)} \rightarrow CH_3\bullet_{(g)} + HCl_{(g)}$$

A methyl radical ($CH_3\bullet$) then abstracts a chlorine atom from a chlorine molecule (Cl_2) to form chloromethane.

$$CH_3\bullet_{(g)} + Cl_{2(g)} \rightarrow CH_3Cl_{(g)} + Cl\bullet_{(g)}$$

The chlorine radical (Cl•) produced in this step 'propagates' the reaction by reacting with another molecule of methane.

TERMINATION

Termination reactions slow the reaction by removing the methyl radicals and chlorine radicals needed for propagation.

$$2CH_3\bullet_{(g)} \rightarrow C_2H_{6(g)}$$

$$2Cl\bullet_{(g)} \rightarrow Cl_{2(g)}$$

$$CH_3\bullet_{(g)} + Cl\bullet_{(g)} \rightarrow CH_3Cl_{(g)}$$

Bromine (Br_2) reacts with alkanes in the same way as chlorine to produce brominated alkanes such as bromomethane, CH_3Br and dibromomethane, CH_2Br_2.

The photochemical halogenation of an alkane is an example of a **substitution reaction** as it involves *the replacement of an atom or group of atoms with a different atom or group of atoms.*

The reaction is also an example of **free radical substitution** as the *substitution reaction is propagated by free radicals.*

Talking Point

Halogen radicals such as Cl• are produced when chlorofluorocarbons (CFCs) such as freon-11, CCl_3F and freon-12, CCl_2F_2 are exposed to UV radiation from the sun in the upper atmosphere. The halogen radicals react with ozone, O_3 reducing the levels of ozone in the upper atmosphere, and increasing levels of UV radiation at the earth's surface. The reaction is an example of a photochemical chain reaction with one halogen radical destroying thousands of ozone molecules.

$Cl\bullet + O_3 \rightarrow ClO\bullet + O_2$

$ClO\bullet + O_3 \rightarrow Cl\bullet + 2O_2$

Exercise 2.3H

1. Which one of the following can act as a free radical?

   A Cl   B Cl^- C Cl^+ D Cl_2

   (CCEA January 2014)

2. The bond between atoms I and Br consists of two electrons, I:Br. Which one of the following equations represents homolytic fission?

   A $IBr \rightarrow I^+ + Br^-$

   B $IBr \rightarrow I^- + Br^+$

   C $IBr \rightarrow I\bullet + Br\bullet$

   D $IBr \rightarrow I\!: + Br$

   (CCEA June 2015)

3. The chlorination of methane in sunlight occurs by a free radical mechanism. (a) Explain what is meant by the term *free radical*. (b) Write the equation for the formation of chloromethane from chlorine and methane. (c) Write equations for (i) the initiation step, (ii) both propagation steps and (iii) one possible termination step.

   (CCEA June 2008)

4. (a) Explain how chlorine radicals, Cl• are formed in the free radical substitution reaction that occurs between methane and chlorine. (b) Methyl radicals, $\bullet CH_3$ are formed during the propagation step. Draw a dot-cross diagram for a methyl radical using only the outer electrons on each atom.

   (Adapted from CCEA January 2002)

5. Which one of the following substances is **not** formed when methane reacts with chlorine in the presence of ultraviolet light?

   A dichloromethane   B ethane
   C hydrogen   D hydrogen chloride

   (CCEA June 2010)

6. Light is sometimes wrongly regarded as a catalyst in the chlorination of methane. Using an appropriate definition of a catalyst, explain why light cannot be considered a catalyst for this reaction.

   (Adapted from CCEA January 2007)

7. The reaction between methane and chlorine involves free radicals created by the action of light. Free radicals can also be created by the use of high temperatures which is known as pyrolysis. The following reactions occur when ethane is pyrolysed at 700 °C. (a) Write the formulae of the species, in these equations, which are free radicals. (b) Classify these reactions, using the letters P-U, under the headings initiation, propagation and termination.

   P $C_2H_6 \rightarrow 2CH_3\bullet$
   Q $CH_3\bullet + C_2H_6 \rightarrow C_2H_5\bullet + CH_4$
   R $C_2H_5\bullet \rightarrow C_2H_4 + H\bullet$
   S $H\bullet + C_2H_6 \rightarrow H_2 + C_2H_5\bullet$
   T $2C_2H_5\bullet \rightarrow C_2H_6 + C_2H_4$
   U $2C_2H_5\bullet \rightarrow C_4H_{10}$

   (CCEA June 2015)

8. Chloroethane, C_2H_5Cl can also be made by reacting chlorine with ethane in a reaction that is similar to that of chlorine with methane. (a) State the experimental condition required for the reaction to occur. (b) Write equations for the propagation steps of the reaction. (c) Name the alkane produced in a termination step.

   (Adapted from CCEA January 2014)

Before moving to the next section, check that you are able to:

- Explain how homolytic fission results in the formation of free radicals.
- Recall the mechanism for the reaction between chlorine and methane, and explain the role of UV light in the reaction.

2.4 Hydrocarbons: Alkenes

CONNECTIONS

- Hydrocarbons containing one or more C=C bonds can combine to form the long chain molecules found in plastics such as polythene and polystyrene.
- Fats and oils containing two or more C=C bonds are known as polyunsaturates and are an essential part of our diet.

Structure and Properties

In this section we are learning to:

- Recognise alkenes as a homologous series of hydrocarbons.
- Explain the properties of alkenes in terms of intermolecular forces.
- Deduce structural isomers of alkenes.
- Use systematic (IUPAC) rules to name alkenes.

Simple alkenes with the general formula C_nH_{2n} form a homologous series of compounds. The structures for ethene (C_2H_4 ; n=2), propene (C_3H_6 ; n=3), but-1-ene (C_4H_8 ; n=4) and but-2-ene (C_4H_8 ; n=4) in Table 1 reveal that simple alkenes are **unsaturated hydrocarbons** as they *are made of carbon and hydrogen, and contain one or more C=C or C≡C bonds*. The structures of the butenes (C_4H_8 ; n=4) also reveal that structural isomers such as but-1-ene and but-2-ene result from the positioning of the C=C bond in the carbon chain.

The bond dipoles associated with the C-H bonds in alkenes are very small. As a result, *alkenes do not have a permanent dipole and experience only van der Waals attraction between molecules*. The strength of the van der Waals attraction between molecules increases as the number of electrons in the molecule increases, and explains why the lighter alkenes (small n) are gases while the heavier alkenes (large n) are solids. The van der Waals attraction between molecules also explains why alkenes are miscible with nonpolar substances such as hydrocarbons, and are immiscible with water and other substances that form hydrogen bonds between molecules.

n	Structural formula, name and molecular formula	Condensed formula
2	H–C(H)=C(H)–H Ethene, C_2H_4	$CH_2{=}CH_2$
3	H–C(H)=C(H)–C(H)(H)–H Propene C_3H_6	$CH_2{=}CHCH_3$
4	H–C(H)=C(H)–C(H)(H)–C(H)(H)–H But-1-ene, C_4H_8	$CH_2{=}CHCH_2CH_3$
4	H–C(H)(H)–C(H)=C(H)–C(H)(H)–H But-2-ene, C_4H_8	$CH_3CH{=}CHCH_3$

Table 1: Structures and formulas for alkenes with n = 2, 3 and 4.

Exercise 2.4A

1. Which one of the following hydrocarbons contains a double bond?

   A CH_4 B C_2H_2 C C_2H_4 D C_2H_6

   (CCEA June 2009)

2. Which one of the following is a correct statement for the homologous series of hydrocarbons called alkenes?

   A Each member of the series has a different empirical formula and different molecular formula.

   B Each member of the series has the same empirical formula and same molecular formula.

   C Each member of the series has a different empirical formula and different structural formula.

   D Each member of the series has the same empirical formula and the same general formula.

   (CCEA June 2015)

3. The compound hex-1-ene is a colourless liquid which boils at 64 °C. (a) Explain, using the molecular formula for the compound, why hex-1-ene is an alkene. (b) With reference to intermolecular forces explain why hex-1-ene forms two layers when mixed with bromine water, $Br_{2(aq)}$.

   $CH_2{=}CHCH_2CH_2CH_2CH_3$ hex-1-ene

Naming Alkenes

Alkenes are named according to the parent (straight-chain) alkene on which their structure is based. The names of straight-chain alkenes with up to six carbon atoms are given in Table 2. Each consists of a prefix that indicates the number of carbon atoms in the chain, and the suffix *ene* to indicate that the compound is an alkene. The structural isomers resulting from the position of the C=C bond in the carbon chain are also given in Table 2.

Consider, for example, the structural isomers pent-1-ene and pent-2-ene. Both are named as pentenes as their structure is based on a chain of five carbon atoms that contains the C=C bond.

pent-1-ene
$CH_2{=}CHCH_2CH_2CH_3$

pent-2-ene
$CH_3CH{=}CHCH_2CH_3$

In pent-1-ene the suffix *1-ene* is used to locate the C=C bond between the first pair of carbon atoms ($C_1{=}C_2$) in the chain. Similarly, in pent-2-ene the suffix *2-ene* is used to locate the C=C bond between the second pair of carbon atoms ($C_2{=}C_3$). In each case *the carbon atoms in the chain are numbered from the end that produces the lowest number in the suffix.*

The alkenes 2-methylbut-1-ene and 2-methylbut-2-ene are structural isomers of pent-1-ene and pent-2-ene.

Carbon atoms in longest chain	Parent alkene	Structural isomers of parent alkene
2	eth**ene**	$CH_2{=}CH_2$
3	prop**ene**	$CH_2{=}CHCH_3$
4	but**ene**	$CH_2{=}CHCH_2CH_3$ $CH_3CH{=}CHCH_3$
5	pent**ene**	$CH_2{=}CHCH_2CH_2CH_3$ $CH_3CH{=}CHCH_2CH_3$
6	hex**ene**	$CH_2{=}CHCH_2CH_2CH_2CH_3$ $CH_3CH{=}CHCH_2CH_2CH_3$ $CH_3CH_2CH{=}CHCH_2CH_3$

Table 2: Parent alkenes with up to six carbon atoms.

Both are named as butenes as their structure is based on a chain of four carbon atoms that contains the C=C bond. As in pent-1-ene and pent-2-ene, the suffix *1-ene* or *2-ene* is used to locate the C=C bond within the carbon chain. With the location of the C=C bond established, the same numbering of the carbon chain is used to locate the methyl group on the second carbon atom.

```
     H  CH3 H   H
     |   |  |   |
  H—C=C—C—C—H
            |   |
            H   H
```

2-methylbut-1-ene
$CH_2{=}C(CH_3)CH_2CH_3$

```
     H  CH3 H   H
     |   |  |   |
  H—C—C=C—C—H
     |          |
     H          H
```

2-methylbut-2-ene
$CH_3C(CH_3){=}CHCH_3$

Alkenes containing two C=C bonds are known as **dienes** and are named using the suffix *diene*. The compound penta-1,4-diene is an example of a diene. It is considered a pentadiene as it is based on a chain of five carbon atoms that contains two C=C bonds. The suffix *1,4-diene* is used to locate the C=C bonds between the first and fourth pairs of carbon atoms in the chain.

```
     H   H   H
     |   |   |
  H—C=C—C—C=C—H
             |   |   |
             H   H   H
```

penta-1,4-diene
$CH_2{=}CHCH_2CH{=}CH_2$

```
     H  CH3
     |   |
  H—C=C—C=C—H
             |   |
             H   H
```

2-methylbuta-1,3-diene
$CH_2{=}C(CH_3)CH{=}CH_2$

Similarly, the compound 2-methylbuta-1,3-diene is considered a butadiene as it is based on a chain of four carbon atoms that contains two C=C bonds. The suffix *1,3-diene* is used to locate the C=C bonds between the first and third pairs of carbon atoms in the chain.

Rules for naming alkenes (including dienes):

- The name of an alkene is based on the name of the parent (straight-chain) alkene on which its structure is based.
- The location of each C=C bond is described by numbering the carbon atoms in a way that produces the smallest numbers.
- Additional prefixes are then added to describe the type and location of each functional group on the carbon chain. The prefixes are written in alphabetical order and are based on the numbering used to locate the C=C bonds.

Exercise 2.4B

1. The systematic name for the alkene is

```
CH3CH2CH2C=CH2
          |
          CH3
```

A 2-propylprop-1-ene
B 2-methylpent-1-ene
C 2,4-dimethylbut-1-ene
D 2-methylhex-1-ene

(CCEA June 2003)

2. The IUPAC name for the alkene is

```
CH3C=CH2
    |
    CH2CH3
```

A 1,2-dimethylpropene
B 2-ethylpropene
C 2-methylbut-1-ene
D 1-ethyl-1-methylethene

(CCEA January 2003)

3. (a) Explain what is meant by the term *structural isomer*. (b) Draw and name all possible structural isomers with the formula C_4H_8.

Before moving to the next section, check that you are able to:

- Recall the general formula for an alkene and use the general formula to generate molecular formulas for alkenes.
- Explain what is meant by the term *unsaturated.*
- Explain how molecular size and shape affects the boiling point of alkenes.
- Write structural formulas and condensed formulas for alkenes.
- Recognise and draw structural isomers of alkenes.
- Deduce systematic names for alkenes (including dienes).

The C=C Functional Group

In this section we are learning to:

- Describe the electronic structure of C-C and C=C bonds.
- Explain the properties of C-C and C=C bonds in terms of their electronic structure.
- Describe how a C=C bond may give rise to geometric isomers.
- Represent geometric isomers using structural formulas and skeletal formulas.
- Use *cis-trans* notation and *E-Z* notation to distinguish geometric isomers.

Structure and Properties

In ethene, CH_2=CH_2 the repulsion between the electrons in the C-H and C=C bonds is minimised when the atoms around each carbon atom are directed towards the corners of a triangle in a trigonal planar arrangement. The structural formulas used to represent the trigonal planar arrangement of atoms about each carbon in ethene are shown in Figure 1.

A C=C bond is formed when a pair of carbon atoms share two pairs of electrons. One bonding pair forms a **sigma (σ) bond**, a type of *covalent bond that results from the linear or 'head-on' overlap of orbitals on neighbouring atoms*. When the orbitals overlap, the electron in each orbital combines to form a shared pair of electrons that occupies the region of space between the carbon atoms where the orbitals overlap.

linear overlap of orbitals — C-C σ-bond

The second bonding pair in the C=C bond forms a **pi (π) bond**, a type of *covalent bond that results from the sideways overlap of p-orbitals on neighbouring atoms*. In a π-bond, the electrons form a shared pair that occupies the space above and below the C-C σ-bond where the p-orbitals that form the π-bond overlap.

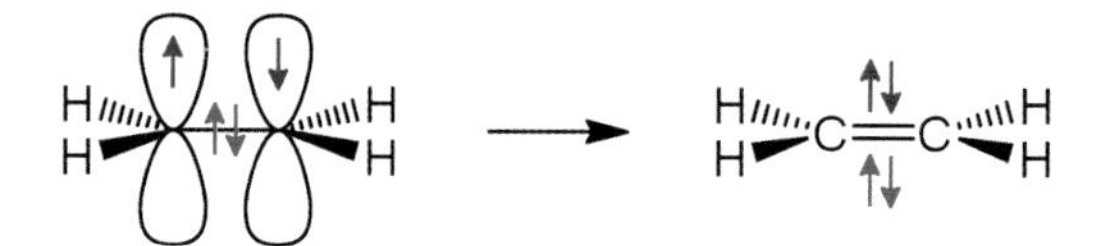

sideways overlap of p-orbitals — C-C π-bond

Worked Example 2.4i

A C=C bond contains a σ (sigma) bond and a π (pi) bond. Draw a labelled diagram to show the formation of (a) a σ-bond and (b) a π-bond from p orbitals.

(Adapted from CCEA June 2003)

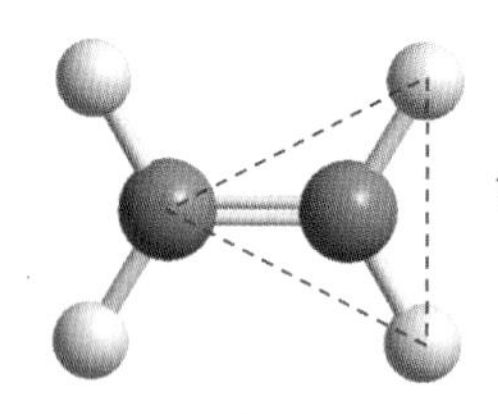

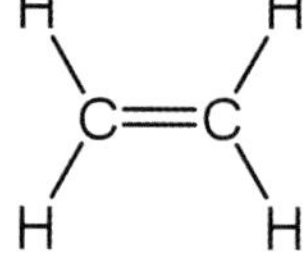

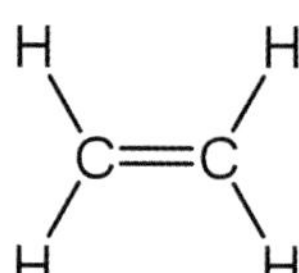

Top view

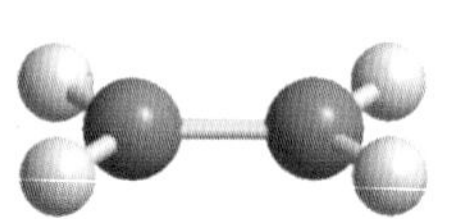

Side view

Figure 1: The structural formulas used to represent the arrangement of atoms about each carbon atom in ethene, CH_2=CH_2. Solid and dashed wedges are used to indicate bonds coming out of the page (solid wedge) and going behind the page (dashed wedge).

Solution

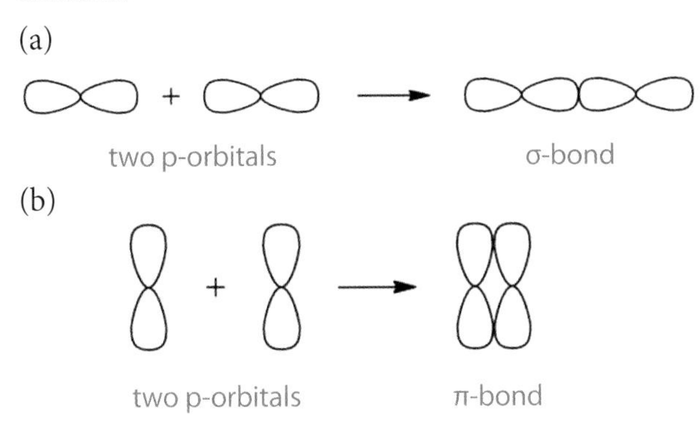

Exercise 2.4C

1. Ethene, C_2H_4 contains

   A 4 sigma (σ) and 2 pi (π) bonds

   B 5 sigma (σ) and 1 pi (π) bonds

   C 2 sigma (σ) and 4 pi (π) bonds

   D 1 sigma (σ) and 5 pi (π) bonds

   (CCEA January 2008)

2. State the number of π (pi) bonds in propene.

   (Adapted from CCEA January 2013)

A single bond between two carbon atoms (C-C) is a σ-bond. The additional bonds in a C=C bond or C≡C bond are π-bonds. The combination of σ and π-bonds in a C=C bond results in the carbon atoms being held more tightly in the bond than the carbon atoms in a C-C (σ) bond. As a result, a C=C bond has a shorter bond length than a C-C bond where the term **bond length** refers to *the distance between the nuclei of two covalently bonded atoms.*

The electrons in a σ-bond behave differently than the electrons in a π-bond. The electrons in a σ-bond are closer to the nuclei of the atoms forming the bond. As a result they are held more tightly in the bond making a σ-bond more difficult to break than a π-bond. The strength of a C-C π-bond can be estimated by subtracting the average bond energy for a C-C (σ) bond (350 kJ mol^{-1}) from the average bond energy for a C=C (σ+π) bond (590 kJ mol^{-1}) where **average bond energy** refers to *the energy needed to break one mole of bonds of a specified type averaged over many compounds.*

By this calculation the bond energy for a C-C π-bond is approximately 590 - 350 = 240 kJ mol^{-1}. In this way we see that while a C=C (σ+π) bond is significantly stronger than a C-C (σ) bond, the π-bond within the C=C bond (240 kJ mol^{-1}) is weaker than a C-C (σ) bond (350 kJ mol^{-1}). As a result, *the π-bond within the C=C bond will break under more moderate reaction conditions than a C-C (σ) bond, making alkenes considerably more reactive than alkanes.*

The exposed nature of the electrons in the C-C π-bond also makes *the π-bond a region of high electron density that is vulnerable to attack by atoms, molecules or ions seeking electrons to bond with.*

Exercise 2.4D

1. But-1-ene contains a carbon-carbon double bond, C=C as well as carbon-carbon single bonds, C-C. (a) Compare and explain the difference in bond strength and bond length of C=C and C-C bonds. (b) Explain why but-1-ene is more reactive than butane.

   (CCEA June 2012)

2. The bond length and bond energy for the carbon-carbon bonds in ethane and ethene are compared in the following table. (a) Explain the difference in bond length and bond energy in terms of sigma (σ) and pi (π) bonding. (b) Explain why ethene is more reactive despite its bond energy being greater.

	Ethane	Ethene
Bond length / nm	0.154	0.134
Bond energy / kJ mol^{-1}	346	598

(CCEA June 2008)

Before moving to the next section, check that you are able to:

- Describe with the aid of diagrams how the overlap of orbitals results in the formation of C-C σ-bonds and C-C π-bonds.
- Explain the relative strength and length of C-C and C=C bonds in terms of the formation of σ- and π-bonds.
- Explain why the presence of a C=C bond makes an alkene more reactive than an alkane.

Geometric Isomers

The electrons in a C-C (σ) bond are not disturbed when the molecule rotates about the bond. This makes it possible for molecules to rotate freely about each of the σ-bonds in the molecule. In contrast, rotating the molecule about a C=C bond involves breaking the π-bond and results in molecules being unable to rotate about a C=C bond unless the bond is being broken as part of a reaction.

As a result an alkene may exist as **geometric isomers** which have *the same structural formula but a different arrangement of atoms in space as the molecule cannot rotate about one or more C=C bonds.* The compound but-2-ene has two geometric isomers: *cis*-but-2-ene and *trans*-but-2-ene. The prefix *cis-* or *trans-* is used to indicate that the methyl groups are on the same side (*cis-*) or opposite sides (*trans-*) of the C=C bond.

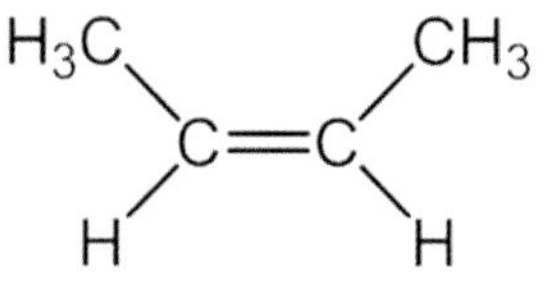

cis-but-2-ene 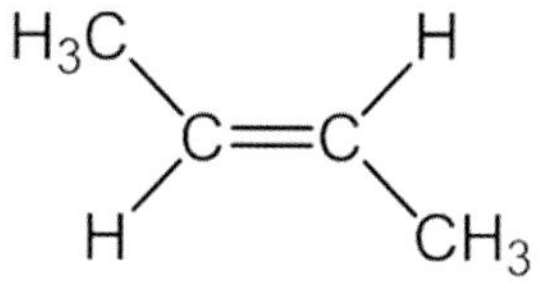 *trans*-but-2-ene

Worked Example 2.4ii

Cinnamyl alcohol exhibits *cis-trans* isomerism. (a) Draw the structure of each isomer. (b) Explain why cinnamyl alcohol exhibits *cis-trans* isomerism.

C_6H_5—CH=CH—CH_2—OH

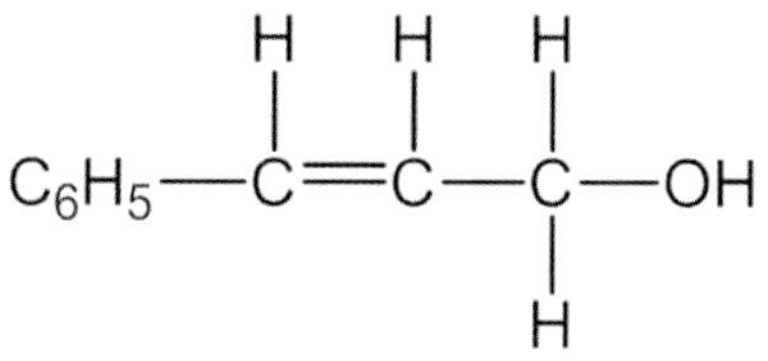

(CCEA January 2003)

Solution

(a)

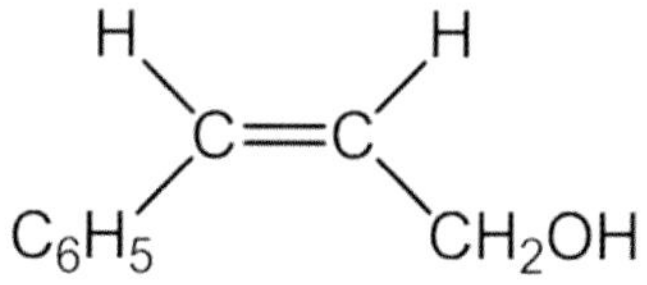

cis isomer

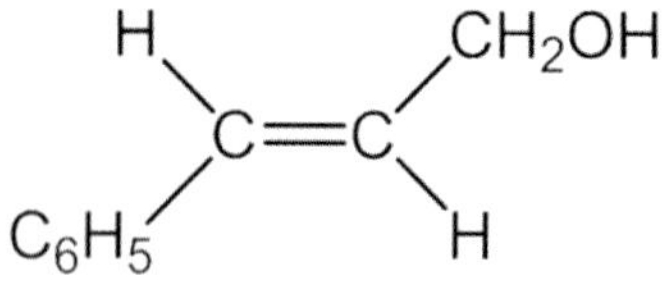

trans isomer

(b) The molecule cannot rotate about the C=C bond.

If the groups attached at one end of the C=C bond are the same the C=C bond will not be able to generate geometric (cis-trans) isomers. For example, the C=C bond in 2-methylbut-2-ene is unable to generate geometric isomers as two methyl groups are attached to the same end of the C=C bond.

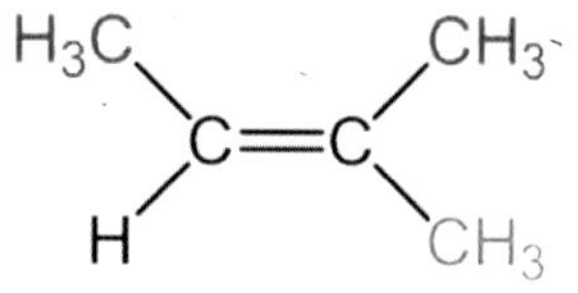

2-methylbut-2-ene

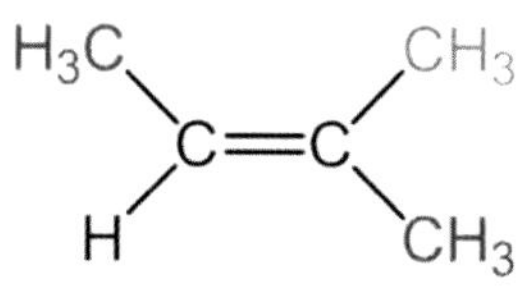

switching the groups produces the same structural isomer

Exercise 2.4E

1. Which one of the following exists as *cis-trans* isomers?

   A CH_2=$CHCH(CH_3)CH_3$
   B CH_2=$CHCH_2CH_2CH_3$
   C CH_3CH=$CHCH_2CH_3$
   D $CH_3C(CH_3)$=$CHCH_3$

   (CCEA January 2011)

2. Which one of the following molecules exists as *cis-trans* isomers?

   A $(CH_3)_2C$=$C(CH_3)_2$
   B CH_3CH_2CH=$CHCH_2CH_3$
   C CH_3CH_2CH=$C(CH_3)_2$
   D $(CH_3)_2CHCH_2CH$=CH_2

   (CCEA June 2008)

3. (a) Draw the structure of a branched alkene C_6H_{12} which exhibits geometric isomerism. (b) Name the alkene. (c) Draw the geometric isomers of the alkene.

   (CCEA June 2010)

E-Z Notation

The prefixes *cis-* and *trans-* can be used to distinguish geometric isomers when the same type of group is attached at both ends of a C=C bond. In the case of 3-methylpent-2-ene the prefixes *cis-* and *trans-* describe the relative positioning of the methyl groups about the C=C bond.

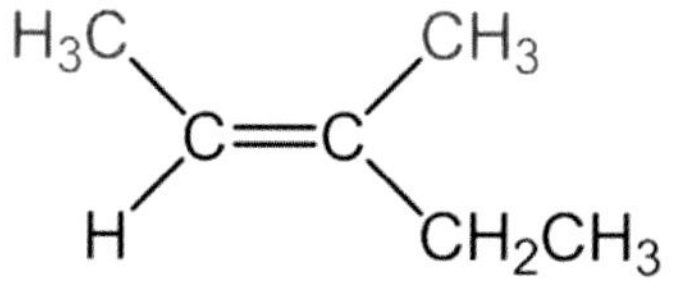

cis-3-methylpent-2-ene

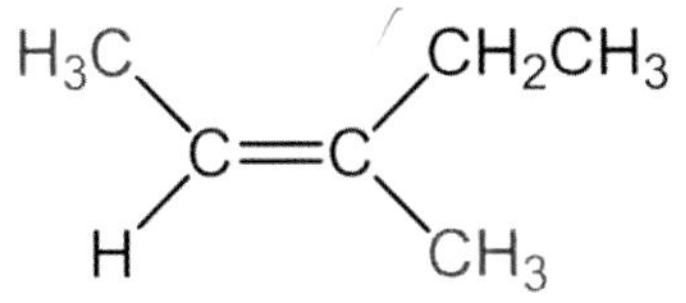

$H_3C(H)C=C(CH_2CH_3)(CH_3)$

trans-3-methylpent-2-ene

In cases where the same type of group is not attached at either end of the C=C bond, the prefixes *cis-* and *trans-* cannot be used, and we must use the Cahn-Ingold-Prelog rules to assign the prefixes *E-* and *Z-* to the isomers.

The first step in assigning the prefix *E-* or *Z-* is to identify which of the two groups attached at either end of the C=C bond will be used to describe the positioning of the groups about the C=C bond. This is accomplished by assigning one of the groups attached to each end of the C=C bond a *high priority* and the other group a *low priority*.

If the high priority groups are located on the same side of the C=C bond the isomer is assigned the prefix Z- and if the high priority groups are on opposite sides of the C=C bond the isomer is assigned the prefix E-. The relationship between the high priority groups in *E-Z* isomers is illustrated by 3-methylpent-2-ene.

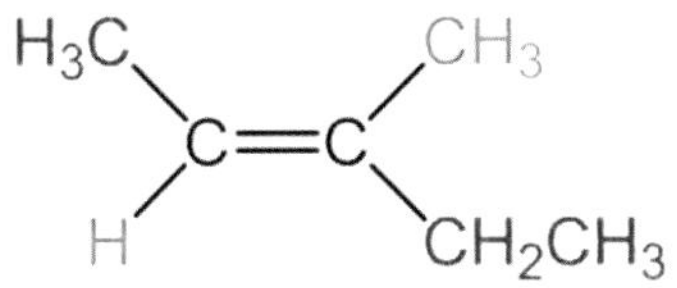

$H_3C(H)C=C(CH_3)(CH_2CH_3)$

E-3-methylpent-2-ene

$H_3C(H)C=C(CH_2CH_3)(CH_3)$

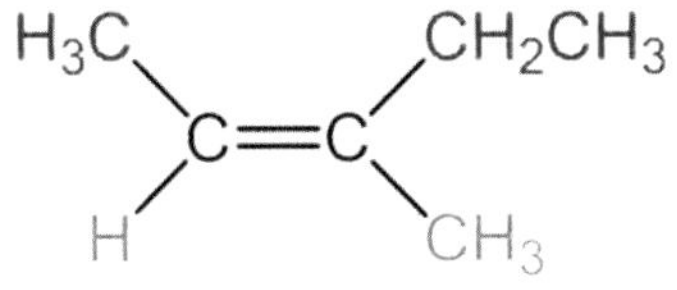

Z-3-methylpent-2-ene

In order to assign high and low priority to the groups attached at either end of the C=C bond it is necessary to determine that methyl ($-CH_3$) has a higher priority than hydrogen (-H) and that ethyl ($-CH_2CH_3$) has a higher priority than methyl ($-CH_3$).

The relative priority of the methyl and hydrogen groups can be determined by considering the atomic number of the atoms attached to the C=C bond. The methyl group is attached to the C=C bond via a carbon atom (C) and the hydrogen group is attached to the C=C bond via a hydrogen atom (H). The atomic number of carbon is greater than the atomic number of hydrogen and, as a result, the methyl group is assigned a higher priority than the hydrogen group.

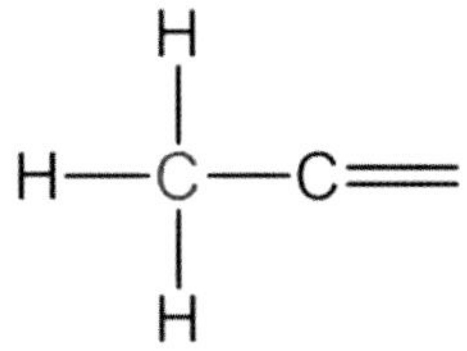

H—C(H)(H)—C=   H—C=

methyl, $-CH_3$ 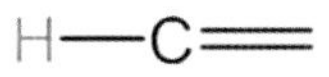   hydrogen, -H

The ethyl and methyl groups attached at the other end of the C=C bond are both connected to the C=C bond via a carbon atom and, as a result, cannot be distinguished by the type of atom connected to the C=C bond. In cases such as this it becomes necessary to compare the types of atoms further from the C=C bond.

In an ethyl group ($-CH_2CH_3$) the carbon attached to the C=C bond is bonded to two hydrogen atoms and one carbon atom. In contrast, the carbon atom in a methyl group ($-CH_3$) is bonded to three hydrogen atoms. The combination of two hydrogen atoms and a carbon atom is used to assign a higher priority to the ethyl group as the carbon atom (C) has a greater atomic number than the hydrogen atom in the corresponding position within the methyl group (H).

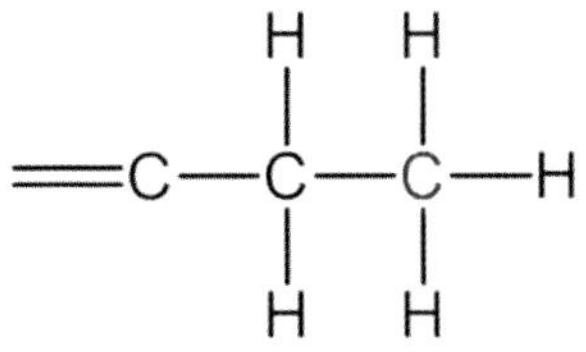

=C—C(H)(H)—C(H)(H)—H   =C—C(H)(H)—H

ethyl, $-CH_2CH_3$ 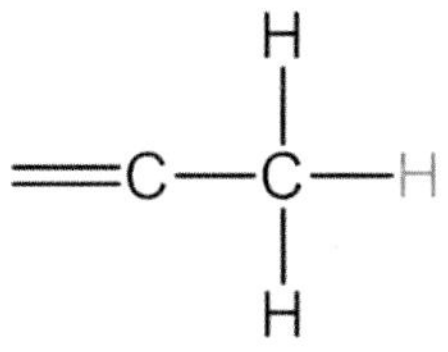   methyl, $-CH_3$

Worked Example 2.4iii

Draw and label the structural (E-Z) isomers for the compound $CH_3CH=C(OH)CH_3$.

Solution

On the left end of the C=C bond $-CH_3$ has a higher priority than -H as carbon (C) has a higher atomic number than hydrogen (H).

On the right end of the C=C bond $-CH_3$ has a lower priority than -OH as oxygen (O) has a higher atomic number than carbon (C).

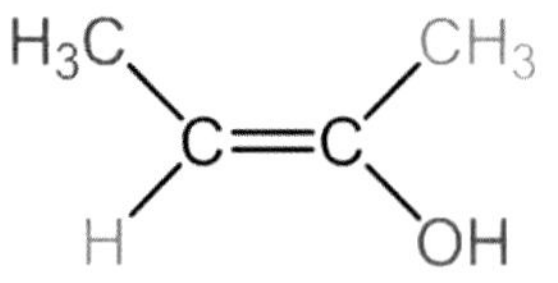

E isomer (high priority on opposite sides) | Z isomer

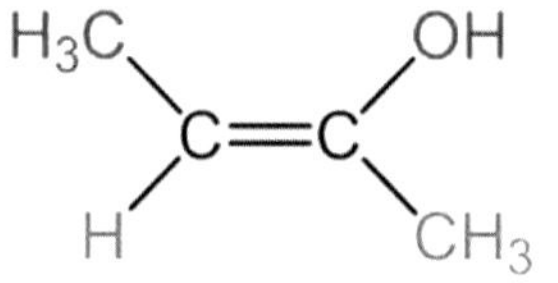

Exercise 2.4F

1. (a) Draw and label the E and Z isomers of but-2-ene. (b) Explain why but-2-ene can exist as E and Z isomers.

   (CCEA June 2012)

2. The compound 3-methylpent-2-ene, $CH_3CH{=}C(CH_3)CH_2CH_3$ exists as geometric isomers. (a) Explain what is meant by the term *geometric isomer.* (b) Give *two* reasons why 3-methylpent-2-ene exists as geometric isomers. (c) Draw the E-isomer of 3-methylpent-2-ene.

   (Adapted from CCEA January 2013)

3. The fluorohydrocarbon $CH_3CH_2CH{=}CFCH_2CH_3$ exists in two isomeric forms. (a) Draw and label the structures of the E and Z isomers of this fluorohydrocarbon. (b) Explain why one of the structures you have drawn is classified as the Z isomer.

   (Adapted from CCEA June 2014)

4. Draw and label the geometric (E-Z) isomers of the compound $CH_3CCl{=}CHCH_3$.
5. Draw and label the geometric (E-Z) isomers of the compound $CH_3CH{=}CHCH_2OH$.
6. Draw and label the geometric (E-Z) isomers of the compound $CHCl{=}C(CH_3)CH_2OH$.
7. The dibromoalkene CHBr=CH-CH=CHBr exists in E and Z forms. Draw all possible E-Z isomers of the compound.

   (Adapted from CCEA June 2015)

Skeletal Formulas for Alkenes

The E-Z isomers of hex-2-ene demonstrate how a skeletal formula can be used to quickly and clearly communicate the arrangement of the groups attached to a C=C bond.

E-hex-2-ene

Structural formula: $H_3C(H)C{=}C(H)CH_2CH_2CH_3$

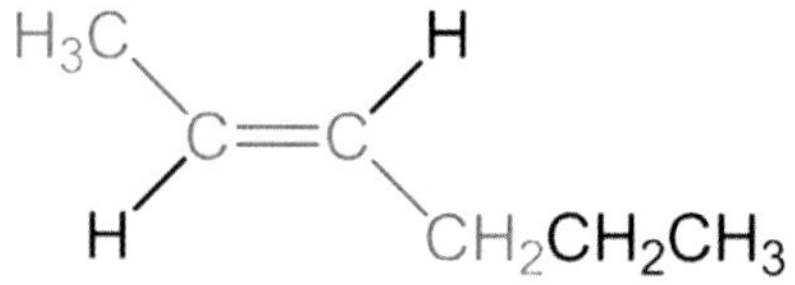

Skeletal formula

Z-hex-2-ene

Structural formula: $H(H_3C)C{=}C(H)CH_2CH_2CH_3$

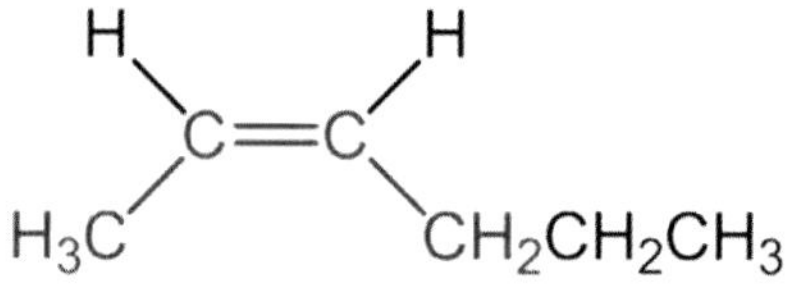

Skeletal formula

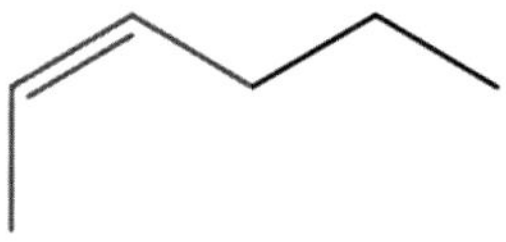

Exercise 2.4G

1. The compound myrcene is found in oils extracted from the rind of citrus fruits. (a) Draw the skeletal formula for myrcene. (b) State the number of geometric isomers formed by myrcene. Explain your reasoning.

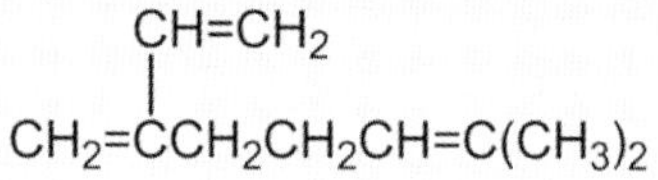

   $CH_2{=}C(CH{=}CH_2)CH_2CH_2CH{=}C(CH_3)_2$

   myrcene

2. Limonene belongs to a family of compounds known as terpenes. It is a colourless liquid found in significant quantities in the rind of citrus fruits. (a) Draw the skeletal formula for limonene. (b) Suggest why the structure of limonene does not give rise to geometric isomers.

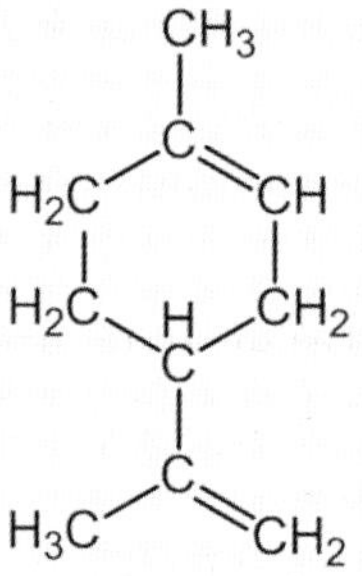

   limonene

3. Citral is a major component of the oil extracted from plants such as lemon myrtle, lemon tea-tree and lemongrass. The citral found in natural oils is a mixture of geometric isomers. The E-isomer of citral is known as geranial. The Z-isomer of citral is known as neral. (a) Draw skeletal formulas for geranial and neral using -CHO to represent the functional group attached to the C=C bond. Label each isomer as E or Z. (b) Explain why geranial and neral are geometric isomers.

$C(CH_3)_2{=}CHCH_2CH_2C(CH_3){=}C(CHO)(H)$

citral

4. The compound retinol is one form of Vitamin A and is consumed as part of a balanced diet. The different forms of retinol found within the body play important roles in processes such as bone growth, maintaining good skin health, and maintaining good vision. Draw the skeletal formula for the *all-trans* isomer of retinol in which the arrangement of groups about each of the four C=C bonds in the carbon chain corresponds to the E-isomer. Use -OH to represent the functional group at the end of the carbon chain.

H_3C, CH_3, H_2C, H_2C, H_2, CH_3 — ring C—$CH{=}CHC(CH_3){=}CHCH{=}CHC(CH_3){=}CHCH_2OH$

retinol

Before moving to the next section, check that you are able to:

- Describe how the presence of one or more C=C bonds in a molecule can result in geometric isomers.
- Draw structural formulas for geometric isomers.
- Use both *cis-trans* and *E-Z* notation to identify geometric isomers.
- Draw skeletal formulas for geometric isomers.

Reactions of Alkenes

In this section we are learning to:

- Use structural formulas to describe the addition of hydrogen, halogens and hydrogen halides across a C=C bond.
- Describe the use of hydrogenation to harden unsaturated oils and the use of bromine water to test for the presence of C=C bonds in a compound.
- Predict the structure of the product formed when a hydrogen halide adds across a C=C bond.

Hydrogenation

When a mixture of an alkene and hydrogen gas is heated in the presence of a catalyst, hydrogen molecules 'add across' the C=C bonds in the alkene to form the corresponding alkane. For example, when a mixture of propene, $CH_3CH{=}CH_2$ and hydrogen is heated, a molecule of hydrogen gas (H_2) adds across the C=C bond to form propane, $CH_3CH_2CH_3$. In the process the pair of electrons in the H-H bond and the pair of electrons in the C-C π-bond are used to form two new C-H bonds. The C-C σ-bond within the C=C bond is not involved in the reaction.

$$CH_3CH{=}CH_2 + H{-}H \rightarrow CH_3CH_2CH_3$$

propene $CH_3CH{=}CH_2$

propane $CH_3CH_2CH_3$

The reaction is referred to as **hydrogenation** as it involves *the addition of a hydrogen molecule (H_2) across a C=C or C≡C bond*. The reaction is also an example of an **addition reaction** as the product results from *the addition of a molecule across a C=C or C≡C bond*.

In the food industry hydrogenation is used to convert unsaturated vegetable oils containing several C=C bonds into the harder saturated fats and oils used in margarine. The hydrogenation is carried out by bubbling hydrogen gas through the oil in the presence of a finely divided nickel catalyst, while maintaining the oil at a temperature that is typically in the range 100-200 °C. The process is an example of heterogeneous catalysis as the catalyst (solid) is in a different physical phase than the reactants (liquid or gas).

Worked Example 2.4iv

The Smoky Mountains in the United States are enveloped by a blue haze which contains simple gaseous hydrocarbons emitted by trees and plants. One such hydrocarbon is isoprene.

(a) Write an equation for the hydrogenation of isoprene and name the product.
(b) Name the catalyst used.

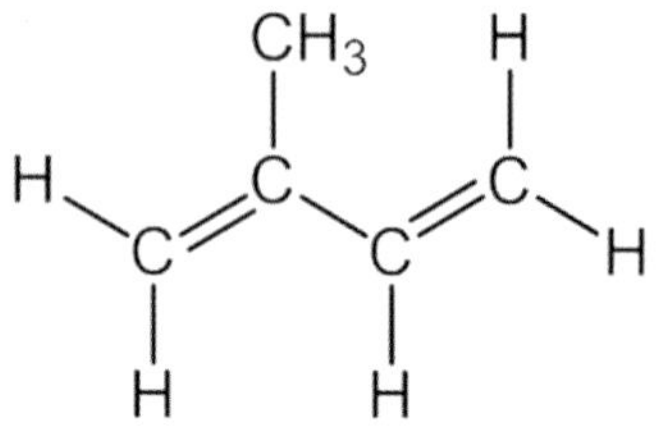

isoprene

(CCEA June 2007)

Solution

(a) $C_5H_8 + 2H_2 \rightarrow C_5H_{12}$
The product is methylbutane:
$CH_3CH(CH_3)CH_2CH_3$.
(b) Finely divided nickel.

Exercise 2.4H

1. Methylpropene is an unsaturated hydrocarbon. (a) Draw the structure of methylpropene. (b) Explain what is meant by the term *unsaturated hydrocarbon*. (c) Calculate the volume of hydrogen gas, measured at 20 °C and one atmosphere pressure, required to saturate 4.2 g of methylpropene.

(Adapted from CCEA June 2013)

2. 3-methylpent-2-ene is an unsaturated hydrocarbon. It can be converted into a saturated hydrocarbon by catalytic hydrogenation. (a) Name the catalyst used in the hydrogenation reaction and describe its state. (b) Write an equation for the reaction and use it to calculate the percentage atom economy for the reaction.

(CCEA January 2013)

3. (a) Draw the structure for penta-1,4-diene. (b) Write an equation for the conversion of penta-1,4-diene to pentane. (c) State the catalyst used.

(Adapted from CCEA June 2006)

4. Rubber is a polymer of isoprene, C_5H_8. Isoprene is fully hydrogenated when it reacts with hydrogen in the presence of a metal catalyst. (a) Write the equation for the reaction. (b) Name the metal catalyst. (c) In what form is the solid metal used? (d) Isoprene has a boiling point of 34 °C. The product, pentane, has a boiling point of 24 °C. What does this suggest about the intermolecular forces in pentane and isoprene?

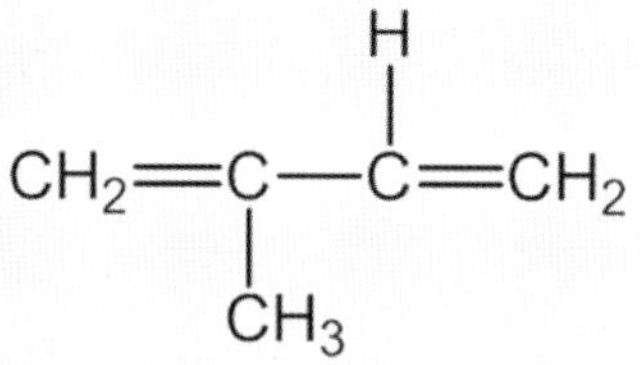

$CH_2=C(CH_3)-CH=CH_2$

Isoprene, C_5H_8

(Adapted from CCEA January 2012)

5. (a) Write an equation for the hydrogenation of cinnamyl alcohol. (b) Suggest a suitable catalyst for the reaction.

$C_6H_5CH=CHCH_2OH + H_2 \rightarrow$
cinnamyl alcohol

(Adapted from CCEA January 2003)

Before moving to the next section, check that you are able to:

- Use structural formulas and equations to describe the addition of hydrogen across a C=C bond.
- Recall the conditions used for the hydrogenation of alkenes and the use of hydrogenation to harden unsaturated oils.

Addition of Halogens

Halogens such as chlorine (Cl_2) and bromine (Br_2) will add across a C=C bond to form a halogenated alkane. When bromine adds to propene, $CH_3CH=CH_2$ the pair of electrons in the Br-Br bond and the pair of electrons in the C-C π-bond are used to form two new

C-Br bonds in the product.

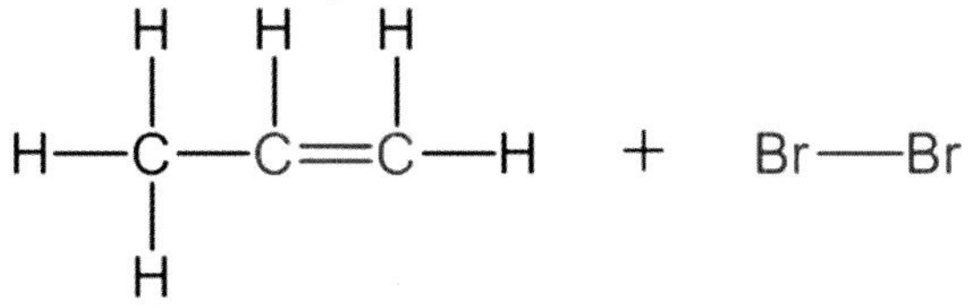

$$CH_3CH=CH_2 + Br-Br \rightarrow CH_3CHBrCH_2Br$$

propene $CH_3CH=CH_2$

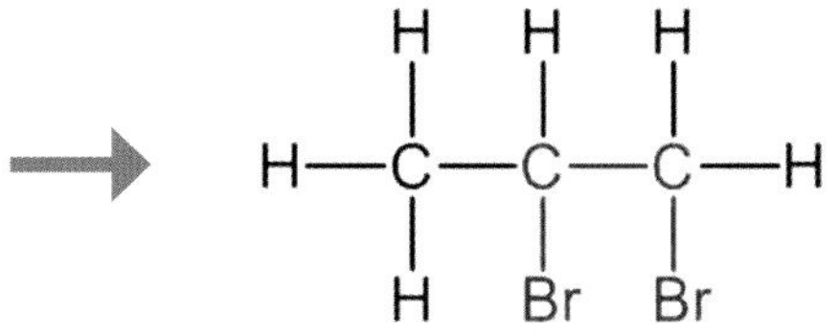

1,2-dibromopropane $CH_3CHBrCH_2Br$

The addition of bromine to a C=C bond can be used to test for the presence of one or more C=C bonds in a molecule. When a molecule containing one or more C=C bonds is shaken with bromine water, $Br_{2(aq)}$ the colour of the bromine water fades as the molecules of bromine in the solution react.

Worked Example 2.4v

Limonene belongs to a family of compounds known as terpenes. It is a colourless liquid found in significant quantities in the rind of citrus fruits.

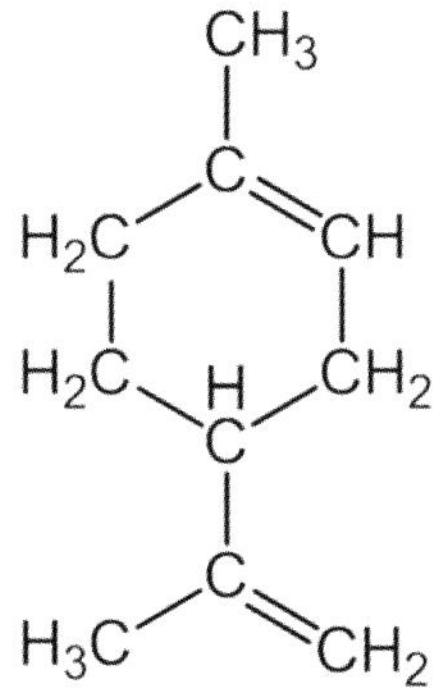

limonene

(a) State what would be observed when a sample of limonene was shaken with an excess of bromine water.

(b) Write an equation for the reaction that occurs when limonene is shaken with an excess of bromine water.

Solution

(a) The bromine water is decolourised when a mixture of bromine water and limonene is shaken.

(b) $C_{10}H_{16} + 2Br_2 \rightarrow C_{10}H_{16}Br_4$

Exercise 2.4I

1. (a) Write a chemical equation for the reaction that occurs when but-1-ene is shaken with chlorine water, $Cl_{2\,(aq)}$. (b) Draw the structure of the product showing all bonds present.

2. Describe a test, including any observations, to show the presence of C=C in but-1-ene.

   (CCEA June 2012)

3. Vinyl chloride, CH_2CHCl is used to manufacture polyvinylchloride (PVC). The first stage in the production of vinyl chloride is the reaction of ethene with chlorine. The product of the reaction then decomposes on heating to form vinyl chloride. (a) Write an equation for the reaction of ethene with chlorine. (b) Name the product formed when ethene reacts with chlorine. (c) What is the other product when the product of the reaction between ethene and chlorine decomposes to form vinyl chloride?

   (Adapted from CCEA January 2014)

4. (a) Write an equation for the reaction that occurs when penta-1,4-diene is shaken with an excess of bromine water. (b) Draw the structure of the product showing all bonds present.

5. Myrcene is found in many natural oils including lemon oil.

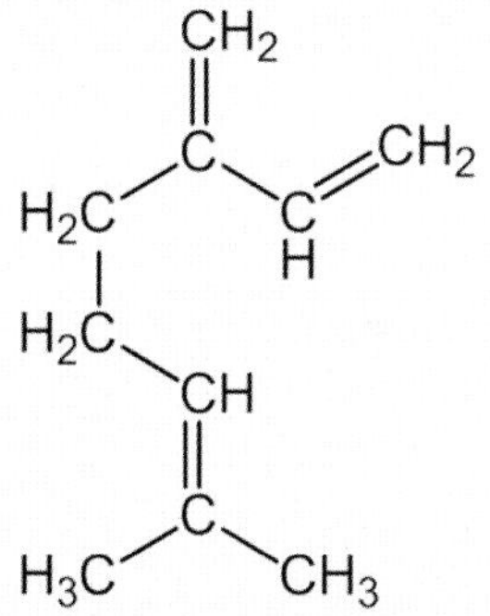

   myrcene

   (a) Write the molecular formula and empirical formula for myrcene. (b) State what would be observed when a small amount of myrcene was shaken with an excess of bromine water. (c) Draw the structure of the product when myrcene is reacted with an excess of chlorine in the absence of light.

   (Adapted from CCEA January 2006)

6. 2.1 g of an alkene reacts with 8.0 g of bromine to produce a dibromoalkane. Write the formula for the alkene.

   (Adapted from CCEA January 2007)

7. 1.4 g of an alkene reacts with chlorine to produce 3.8 g of a dichloroalkane. Write the formula for the alkene.

   (Adapted from CCEA June 2011)

8. Calculate the volume of liquid bromine (density = 3.2 g cm^{-3}) required to react with 120 cm^3 of ethene at 20 °C and 1 atmosphere pressure.

   $C_2H_4 + Br_2 \rightarrow C_2H_4Br_2$

   (Adapted from CCEA January 2008)

Before moving to the next section, check that you are able to:

- Use structural formulas and equations to describe the addition of halogens (Cl_2, Br_2 ...) across a C=C bond.
- Describe the use of bromine water to test for the presence of one or more C=C bonds in a molecule.

Step 1

Step1

Heterolytic fission of HBr as H^+ adds to the C=C bond.

Step 2

Step 2

Bromide ion adds to the carbocation intermediate.

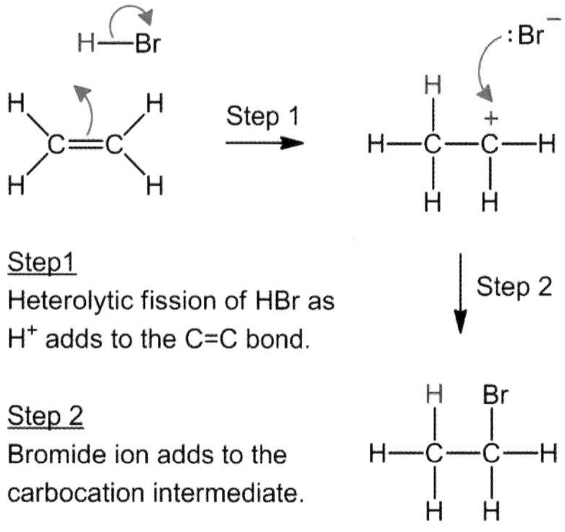

Figure 2: Flow scheme detailing the mechanism for the addition of hydrogen bromide to ethene: $C_2H_4 + HBr \rightarrow C_2H_5Br$.

Addition of Hydrogen Halides

Hydrogen halides (HCl, HBr ...) will also add across a C=C bond to produce a halogenated alkane. The flow scheme in Figure 2 reveals that the reaction occurs in two steps. In the first step, the pair of electrons in the C-C π-bond is used to bond the hydrogen in HBr to one of the carbon atoms. The C-H bond results from the attraction between the positive end of the H-Br dipole ($H^{\delta+}$) and the electrons in the C-C π-bond. In this way HBr acts as an **electrophile** by 'attacking' the region of high electron density produced by the C-C π-bond. Formation of the C-H bond results in **heterolytic fission** of the H-Br bond to form a bromide ion.

Heterolytic fission of the H-Br bond
$H\text{-}Br \rightarrow H^+ + Br^-$

Homolytic fission of the H-Br bond
$H\text{-}Br \rightarrow H\bullet + Br\bullet$

When drawing the mechanism for a reaction, the movement of a pair of electrons is described using a 'curly arrow', ↺. In the reaction between HBr and ethene, curly arrows are used to describe the electron pair in the C-C π-bond moving to form a bond between carbon and hydrogen, and the electron pair in the H-Br bond moving onto bromine during heterolytic fission of the H-Br bond.

The cation formed by the addition of H^+ to the C=C bond contains an electron deficient carbon atom with only six electrons in its outer shell (C^+). A cation containing an electron deficient carbon (C^+) is known as a **carbocation**. Carbocations are unstable and will react quickly. In the second step of the reaction, bromide (Br^-) donates a pair of electrons to the electron deficient carbon (C^+) to form a coordinate bond. The second step is much faster than the first due to the attraction between the positive and negative ions.

The addition of a hydrogen halide across a C=C bond is an example of an **electrophilic addition** reaction as an electrophile, in this case a hydrogen halide (HCl, HBr ...), adds across a C=C bond.

Exercise 2.4J

1. The double bond in ethene allows the molecule to undergo an electrophilic addition reaction with hydrogen bromide. Explain the term *electrophilic addition.*

   $C_2H_4 + HBr \rightarrow C_2H_5Br$

   (CCEA June 2003)

2. Which one of the following equations represents a step in the mechanism for the reaction between hydrogen bromide and ethene?

A $C_2H_4 + Br^+ \rightarrow C_2H_4Br^+$
B $C_2H_4 + HBr \rightarrow C_2H_5^+ + Br^-$
C $C_2H_4 + HBr \rightarrow C_2H_5\bullet + Br\bullet$
D $C_2H_4 + HBr \rightarrow C_2H_4Br^- + H^+$

(CCEA January 2009)

3. (a) Explain why alkenes are more reactive than alkanes. (b) What name is given to the reaction between propene and hydrogen bromide? (c) Use a flow diagram to suggest the mechanism for the reaction between an alkene and hydrogen bromide. Represent the alkene as:

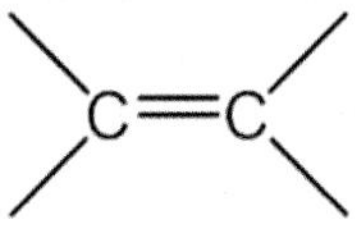

C=C

(Adapted from CCEA January 2011)

4. The Smoky Mountains in the United States are enveloped by a blue haze which contains simple gaseous hydrocarbons emitted by trees and plants. One such hydrocarbon is isoprene:

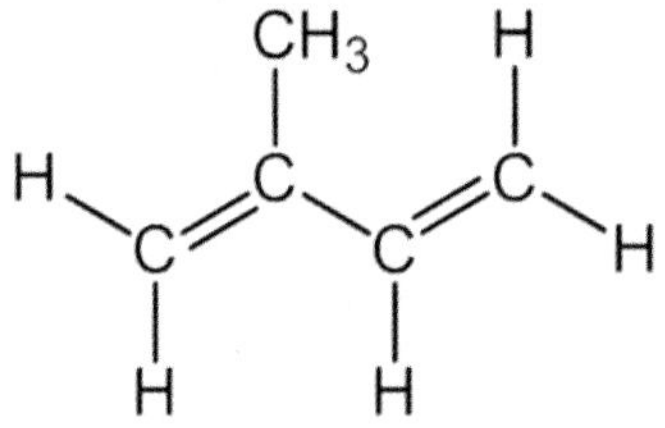

CH_3 H
H C C H
C C
H H

Isoprene reacts as a typical unsaturated hydrocarbon. Draw a flow scheme to show the mechanism for the reaction between one molecule of hydrogen bromide and isoprene to form 3-bromo-3-methylbut-1-ene, $CH_3CBr(CH_3)CH{=}CH_2$.

(Adapted from CCEA June 2007)

Before moving to the next section, check that you are able to:

- Use structural formulas and equations to describe the addition of hydrogen halides (HCl, HBr ...) across a C=C bond.
- Draw the mechanism for the addition of HX (X = Cl, Br) to a C=C bond.
- Recall the role of heterolytic fission in the addition of HBr to a C=C bond and explain why the reaction is described as electrophilic addition.

Markovnikov's Rule

The electrophilic addition of HBr to propene, $CH_3CH{=}CH_2$ might be expected to yield a mixture of 1-bromopropane and 2-bromopropane.

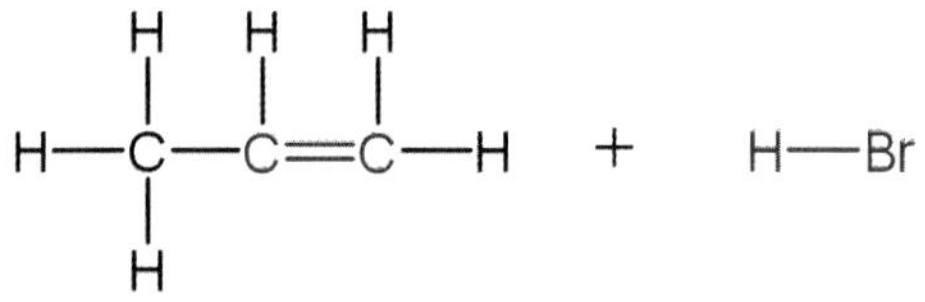

H—C—C=C—H + H—Br

propene $CH_3CH{=}CH_2$

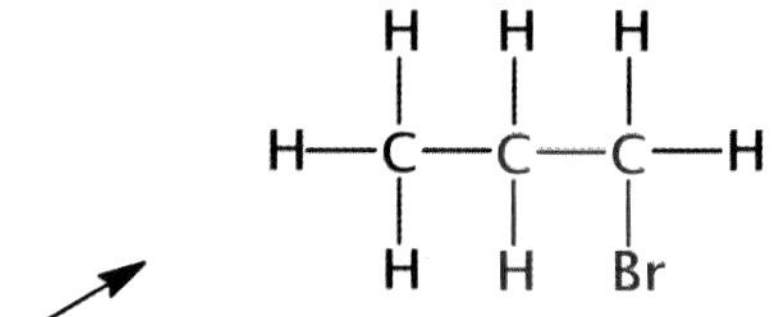

1-bromopropane $CH_3CH_2CH_2Br$

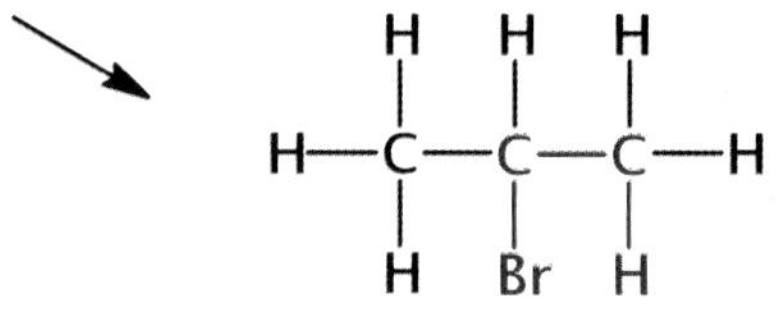

2-bromopropane $CH_3CHBrCH_3$

In reality 2-bromopropane is the major product as the secondary carbocation formed during the reaction (P) is more stable than the primary carbocation (Q) that would lead to the formation of 1-bromopropane.

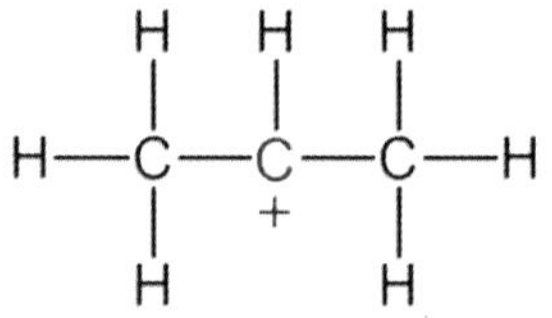

Carbocation P

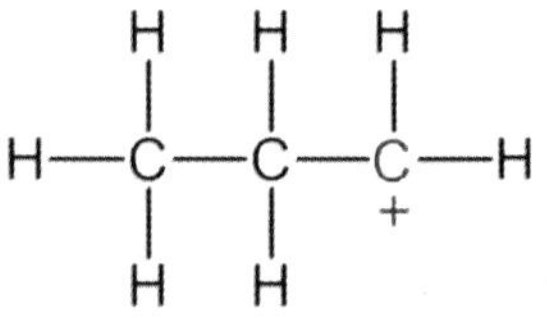

Carbocation Q

Similarly, the addition of HCl to 2-methylpropene might be expected to yield a mixture of 1-chloro-2-methylpropane and 2-chloro-2-methylpropane.

```
     H   CH3 H
     |    |  |
H ── C ── C═C ── H   +   H ── Cl
     |
     H
```

2-methylpropene
$C(CH_3)_2{=}CH_2$

```
     H   CH3 H
     |    |  |
H ── C ── C ── C ── H
     |    |  |
     H    H  Cl
```

1-chloro-2-methylpropane
$CH_3CH(CH_3)CH_2Cl$

```
     H   CH3 H
     |    |  |
H ── C ── C ── C ── H
     |    |  |
     H   Cl  H
```

2-chloro-2-methylpropane
$CH_3C(CH_3)ClCH_3$

Here again, 2-chloro-2-methylpropane is the major product as the tertiary carbocation formed during the reaction (R) is more stable than the primary carbocation (S) that would lead to the formation of 1-chloro-2-methylpropane.

```
     H   CH3 H
     |    |  |
H ── C ── C ── C ── H
     |    +  |
     H       H
```

Carbocation R

```
     H   CH3 H
     |    |  |
H ── C ── C ── C ── H
     |    |  +
     H    H
```

Carbocation S

The relationship between the structure and stability of the carbocations formed during the reaction is summarised in Figure 3, and gives rise to **Markovnikov's Rule** which asserts that *when HX (X = Cl, Br) adds across a C=C bond the halogen (X) attaches to the carbon with more alkyl groups (R) attached to it.*

According to the classification in Figure 3, a **primary carbocation** has *one carbon atom bonded to the positively charged carbon atom (C^+)*. Similarly, a **secondary carbocation** has *two carbon atoms bonded to the positively charged carbon atom*, and a **tertiary carbocation** has *three carbon atoms bonded to the positively charged carbon atom within the carbocation.*

```
     H              R'              R'
     |              |               |
R ── C ── H    R ── C ── H     R ── C ── R"
     +              +               +
```

primary carbocation | secondary carbocation | tertiary carbocation

Stability increasing →

Figure 3: The relationship between the structure and stability of the carbocations formed during the reaction.

Exercise 2.4K

1. But-1-ene, $CH_2{=}CHCH_2CH_3$ reacts with hydrogen bromide to form two isomers, the major product being 2-bromobutane. (a) Explain why hydrogen bromide is attracted to but-1-ene. (b) Suggest flow schemes for the mechanisms of the reaction of hydrogen bromide with but-1-ene to form 1-bromobutane and 2-bromobutane, showing the structure of both intermediates. (c) Suggest how you would separate a mixture of 1-bromobutane and 2-bromobutane. (d) Suggest why 2-bromobutane is the major product. (e) Name the mechanism of these reactions.

```
     H    H
     |    |
H ── C ── C ── CH2CH3
     |    |
     Br   H
```

1-bromobutane

```
     H    H
     |    |
H ── C ── C ── CH2CH3
     |    |
     H    Br
```

2-bromobutane

(Adapted from CCEA June 2014)

2. (a) Draw the structures of all possible carbocations formed when HCl adds across the C=C bond in 2-methylbut-2-ene. (b) Draw the structure of the product that is most likely to be formed. (c) Explain, with reference to the

structure of the intermediate carbocation, why this product is most likely to be formed.

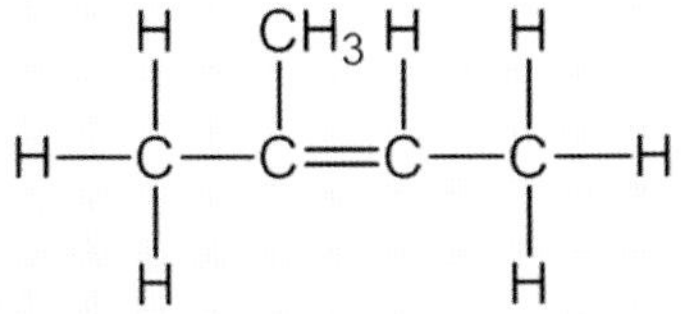

2-methylbut-2-ene

3. The addition of HBr to pent-2-ene, $CH_3CH{=}CHCH_2CH_3$ might be expected to yield a mixture of 2-bromopentane and 3-bromopentane. (a) Draw the structure of the carbocation that leads to the formation of (i) 2-bromopentane and (ii) 3-bromopentane. (b) Explain, in terms of the carbocations formed, why the reaction is expected to yield a mixture of products.

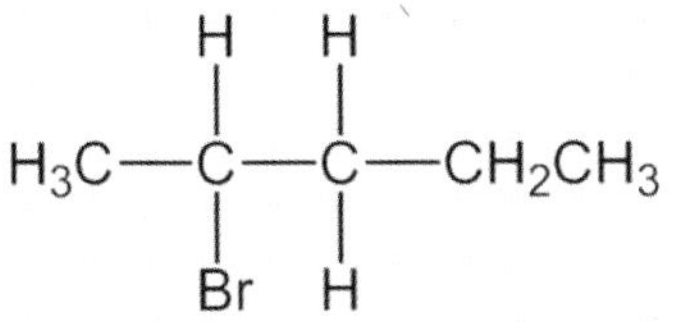

2-bromopentane

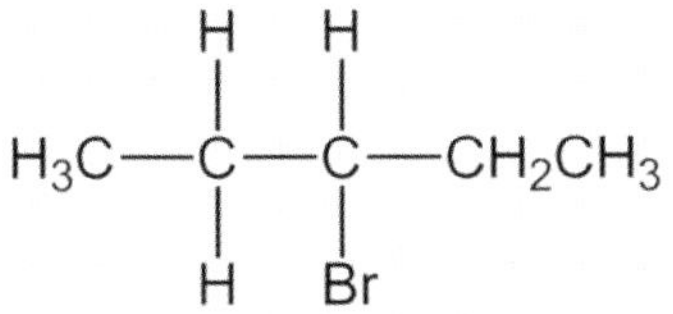

3-bromopentane

Before moving to the next section, check that you are able to:

- Use Markovnikov's Rule to predict the product(s) of electrophilic addition across a C=C bond.
- Account for the product(s) of an electrophilic addition reaction in terms of the stability of the carbocation(s) formed during the reaction.

Polymerisation

When ethene, $CH_2{=}CH_2$ is heated in the presence of a catalyst many ethene molecules combine to form a molecule of 'polythene'. A molecule of polythene is referred to as a **polymer** as it is *a large molecule formed by the joining together of many smaller monomers* where the term **monomers** refers to *the small molecules that join together to form a polymer.*

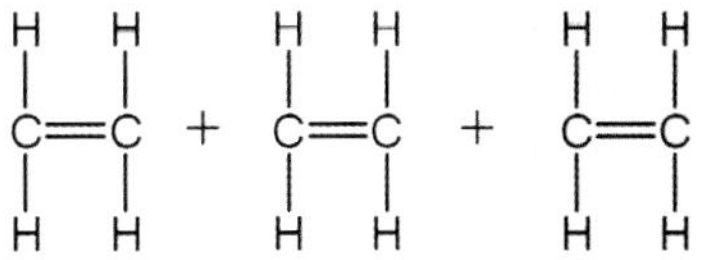

three ethene monomers

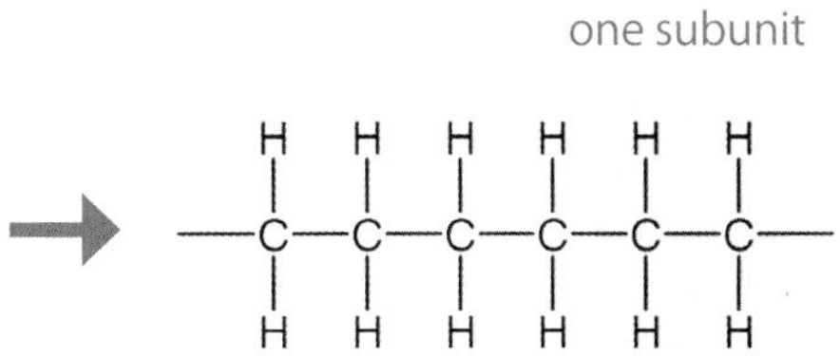

one subunit

a section of a polythene molecule containing three identical subunits

The reaction between ethene monomers to form a molecule of polythene is an example of **polymerisation** as it involves *the formation of a polymer by the joining together of many monomers to form a large molecule.* The formation of polythene from ethene is also an example of **addition polymerisation** as *each polymer molecule is formed as a result of addition reactions between monomers.* The formation of a polythene molecule containing a large number of monomers (n) can be summarised by the equation:

one subunit in a polythene molecule formed from n monomers

Similarly, the addition polymerisation of substituted ethene molecules, $CHR{=}CH_2$ can be used to form polymers such as polypropene ($R = CH_3$), polyvinylchloride ($R = Cl$) and polystyrene (R = phenyl, C_6H_5).

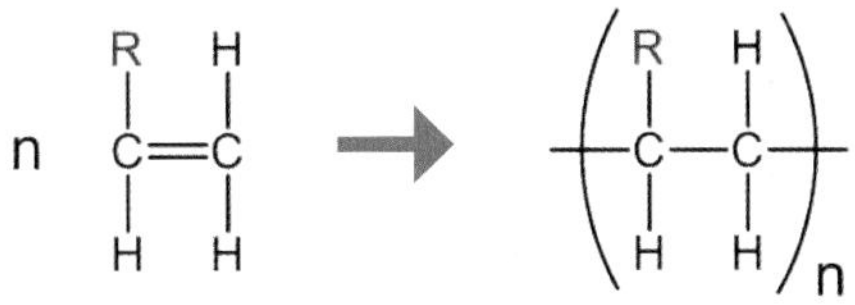

Worked Example 2.4vii

Natural rubber is a polymer of isoprene (2-methylbuta-1,3-diene). (a) Explain the meaning of the term *polymer*. (b) State the functional group which enables isoprene to polymerise. (c) Draw a section of natural rubber showing how isoprene units combine to form the polymer.

$$H_2C{=}C(H){-}C(CH_3){=}CH_2 \quad \textit{isoprene}$$

(Adapted from CCEA January 2005)

Strategy

The electrons from one of the two C-C π-bonds in isoprene are needed to add isoprene to the polymer. The electrons in the second C-C π-bond are used to complete the octet around each carbon when isoprene adds to the polymer.

Solution

(a) A polymer is a large molecule that is formed when many identical molecules known as monomers combine.

(b) The C=C functional group.

(c)

$$-CH_2-CH{=}C(CH_3)-CH_2-CH_2-CH{=}C(CH_3)-CH_2-$$

Exercise 2.4L

1. Propene undergoes addition polymerisation to form polypropene. Draw the structure of polypropene showing *three* repeating units.
(CCEA June 2014)

2. But-1-ene can be polymerised to form poly(but-1-ene). (a) What type of polymerisation does but-1-ene undergo? (b) Draw a section of poly(but-1-ene) showing *two* repeating units.
(CCEA June 2012)

3. Under suitable conditions, methylpropene can be converted to poly(methylpropene). (a) Name the type of reaction occurring. (b) Draw the structure of poly(methylpropene) showing three repeating units.
(CCEA June 2013)

4. 3-methylpent-2-ene, $CH_3CH{=}C(CH_3)CH_2CH_3$ can undergo addition polymerisation. Draw a section of the resulting polymer chain showing *two* repeating units.
(CCEA January 2013)

5. Propenonitrile, $CH_2{=}CHCN$ can be polymerised to form a fibre known as Orlon that is used for making clothes. (a) Draw the structure of propenonitrile, showing all of the bonds present. (b) Explain if propenonitrile can exist as *cis* and *trans* (E-Z) forms. (c) Draw the structure of two repeating units in Orlon. (d) State the type of polymerisation taking place.
(CCEA June 2009)

6. Polytetrafluoroethene is made by the polymerisation of tetrafluoroethene, $CF_2{=}CF_2$. (a) Write the equation for the polymerisation reaction. (b) Why is tetrafluoroethene able to polymerise? (c) State the type of reaction that occurs when tetrafluoroethene polymerises.
(Adapted from CCEA January 2010)

7. Ethene and propene can be converted into polythene and polypropene respectively. (a) Name the type of reaction involved in the conversion of propene to polypropene. (b) Suggest the equation for the conversion of styrene to polystyrene.

$$n\ HHC{=}CH(C_6H_5) \longrightarrow$$

(CCEA June 2010)

Before moving to the next section, check that you are able to:

- Explain what is meant by the term *polymerisation.*
- Recall that addition polymerisation can only occur when the monomer contains one or more C=C bonds.
- Deduce the structure of an addition polymer from the structure of the monomer used to form the polymer.
- Write equations to describe the formation of polymers containing a large number (n) of monomers.

2.5 Halogenoalkanes

CONNECTIONS

- Very few compounds in nature contain halogens. Most are toxic and are used by living organisms to fight disease or defend against predators.
- Halogenated alkanes such as halothane, $CF_3CHClBr$ and chloroform, $CHCl_3$ have been widely used as anaesthetics.
- Halogenated alkanes are also used in fire suppression systems as they terminate the free radical reactions that sustain flames.

Structure and Properties

In this section we are learning to:

- Explain the properties of halogenoalkanes in terms of intermolecular forces.
- Classify halogenoalkanes as primary, secondary or tertiary.
- Use systematic (IUPAC) rules to name halogenoalkanes.

Factors Affecting Solubility

A **halogenoalkane** is a saturated hydrocarbon containing one or more halogen atoms. Small halogenoalkanes such as chloromethane, CH_3Cl and bromomethane, CH_3Br are gases, while larger halogenoalkanes such as bromobutane, C_4H_9Br are oily liquids with a sweet smell. The carbon-halogen bonds within a halogenoalkane are the functional groups that determine the reactions of the halogenoalkane. The remainder of the molecule consists of alkyl groups and is nonpolar as demonstrated by the examples in Figure 1.

Smaller halogenoalkanes such as chloromethane, CH_3Cl and bromoethane, C_2H_5Br experience a combination of attractive dipole forces and van der Waals attraction between molecules. As a result they are soluble in nonpolar solvents such as hexane, and are only sparingly soluble in water. As the halogenoalkanes become larger the van der Waals attraction dominates and molecules as small as bromobutane, C_4H_9Br become immiscible with water.

The solubility of halogenoalkanes in polar and nonpolar solvents can be demonstrated by shaking a mixture of bromobutane and copper sulfate solution with hexane as shown in Figure 2. When the mixture is shaken the bromobutane (bottom layer) dissolves in the hexane to form a single nonpolar layer that is immiscible with the copper sulfate solution.

Bromine and iodine are considerably heavier than the atoms in most liquids. As a result, halogenoalkanes containing these elements have a greater density than water and other liquids made from lighter elements such as C, H, N and O. When a halogenoalkane containing bromine or iodine is mixed with an aqueous solution it forms a separate non-aqueous layer below the aqueous solution as shown in Figure 2.

(a) (b)

Figure 1: Polar and nonpolar regions within (a) 1-bromobutane, C_4H_9Br and (b) 1,5-dichloropentane, $C_5H_{10}Cl_2$.

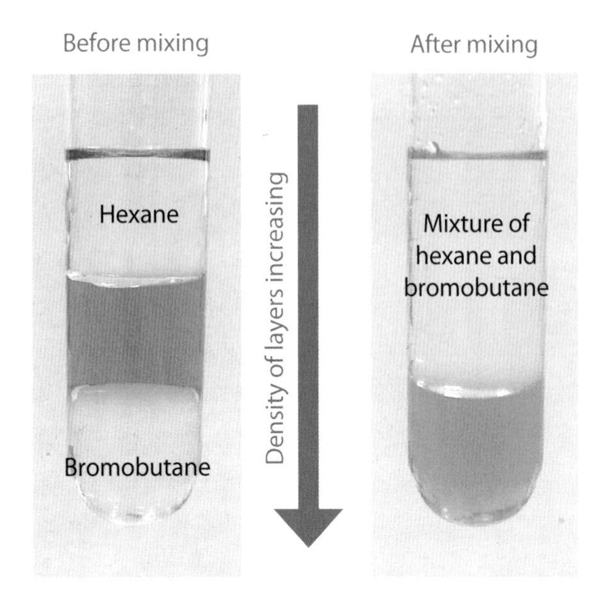

Figure 2: The effect of mixing hexane and bromobutane with aqueous copper sulfate. Bromobutane mixes with hexane to form a single nonpolar layer.

Factors Affecting Boiling Point

The chloroalkanes are a homologous series of halogenoalkanes. The formula for each member of the series is obtained by inserting n = 1, 2, 3 ... in the general formula $C_nH_{2n+1}Cl$. The boiling points for the straight-chain chloroalkanes in Figure 3 demonstrate that the boiling points of the chloroalkanes increase as the molar mass of the compound increases. As in alkanes, the increase in boiling point is due to an increase in the van der Waals attraction between neighbouring molecules as the number of electrons in the molecule increases. Other homologous series such as the fluoroalkanes ($C_nH_{2n+1}F$), bromoalkanes ($C_nH_{2n+1}Br$) and iodoalkanes ($C_nH_{2n+1}I$) behave in a similar way.

The compounds 2-bromobutane and 2-bromo-2-methylpropane in Figure 4 are structural isomers. The shapes of the isomers reveal that, as in alkanes, the more branched structure of 2-bromo-2-methylpropane lowers the boiling point by reducing the van der Waals attraction between molecules.

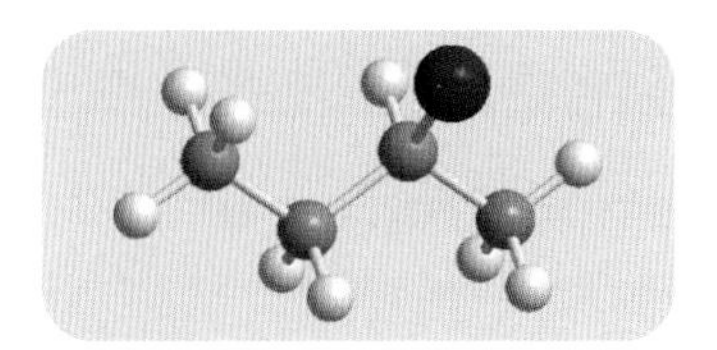

2-bromobutane, C_4H_9Br
Boiling point 91 °C

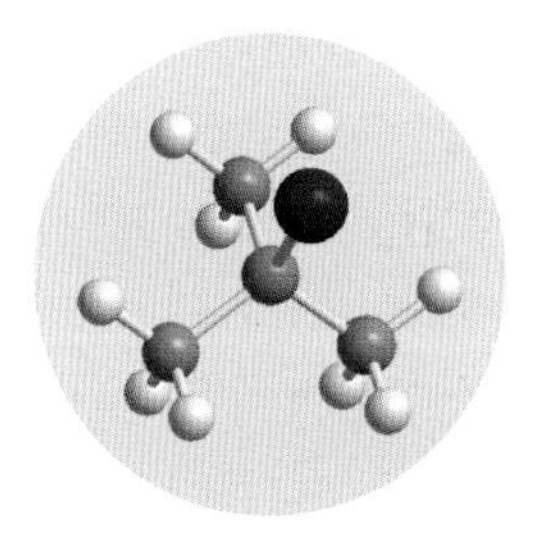

2-bromo-2-methylpropane, C_4H_9Br
Boiling point 73 °C

Figure 4: The effect of van der Waals attraction on the boiling points of bromobutane isomers.

The magnitude of the van der Waals attraction between molecules is also determined by the nature of the halogen atoms in the molecule. The boiling points for the ethyl halides, C_2H_5X (X = F, Cl, Br, I) in Figure

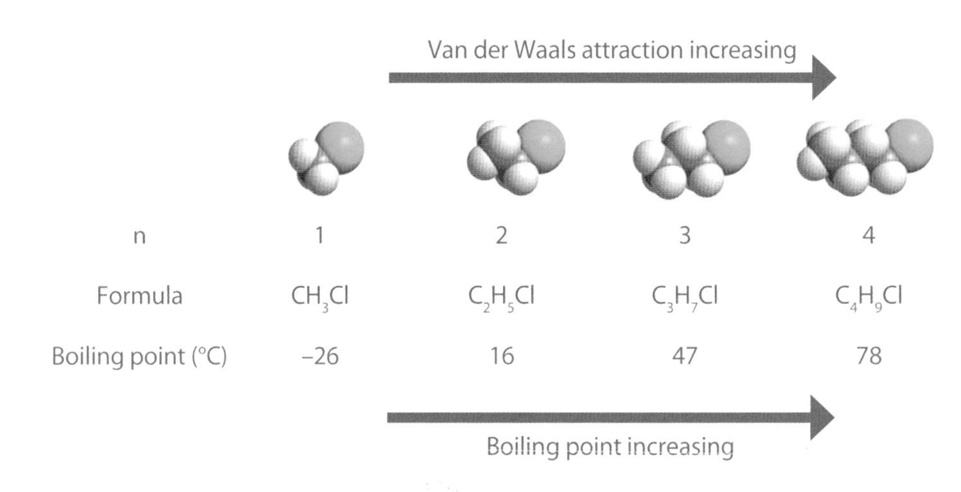

n	1	2	3	4
Formula	CH_3Cl	C_2H_5Cl	C_3H_7Cl	C_4H_9Cl
Boiling point (°C)	−26	16	47	78

Figure 3: The effect of van der Waals attraction on the boiling points of the straight-chain chloroalkanes, $C_nH_{2n+1}Cl$.

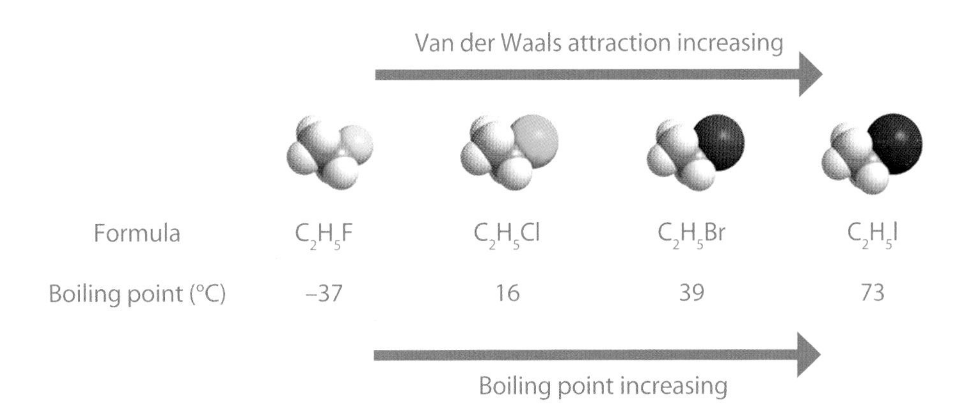

Figure 5: The effect of van der Waals attraction on the boiling points of the ethyl halides, C_2H_5X (X = F, Cl, Br, I)

5 demonstrate that the van der Waals attraction between molecules increases as the number of electrons in the molecule increases.

Adding halogens to a molecule further increases the van der Waals attraction between neighbouring molecules. The increase in boiling point that results from adding chlorine atoms is illustrated by the sequence of chloropropanes in Figure 6a. The even more substantial increases in boiling point that result from adding larger halogens are illustrated by the corresponding bromopropanes in Figure 6b.

Exercise 2.5A

1. (a) Use δ+ and δ– to indicate the polarity of the carbon-bromine bond in 2-bromobutane.
(b) Explain why the carbon-bromine bond is polar.

```
   H  H  H  H
   |  |  |  |
H—C—C—C—C—H   2-bromobutane
   |  |  |  |
   H  H  Br H
```

(CCEA June 2005)

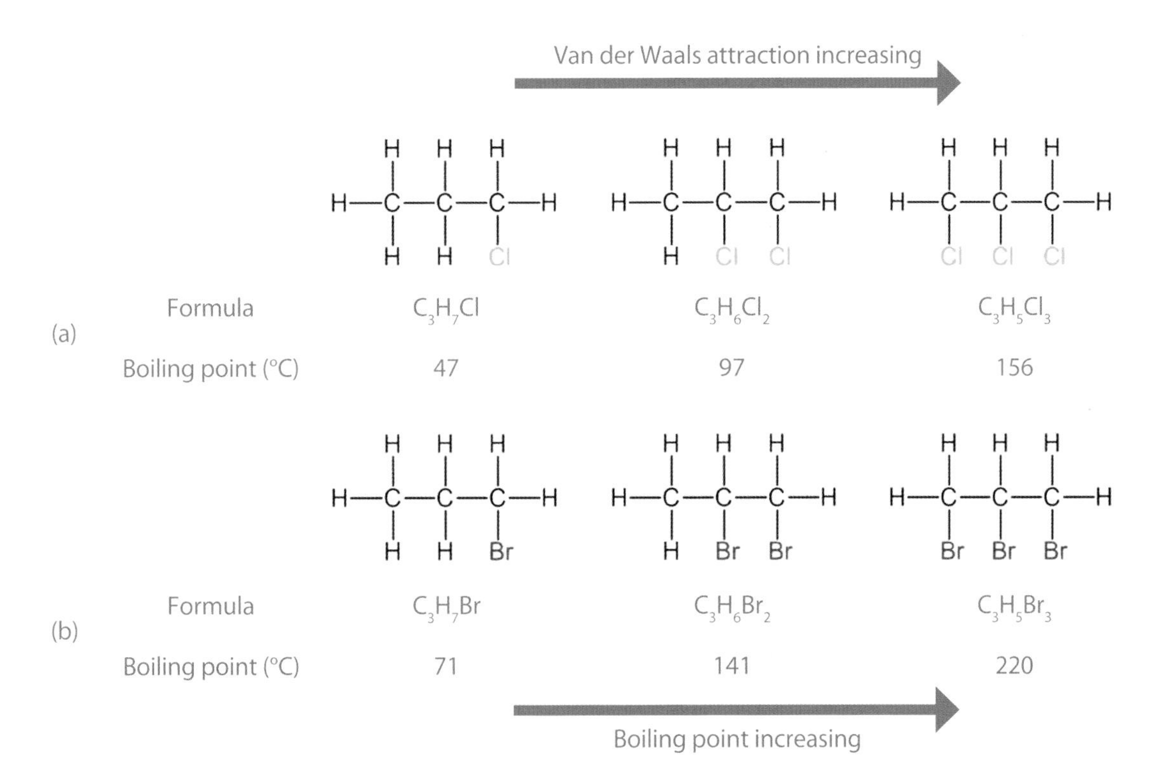

Figure 6: The effect of additional halogen atoms on the boiling point of (a) chloropropane, C_3H_7Cl and (b) bromopropane, C_3H_7Br.

2. 1-bromobutane is a structural isomer of 2-bromo-2-methylpropane. (a) Define the term *structural isomer.* (b) State, giving reasons for your choice, which of these two compounds would be expected to have the higher boiling point.

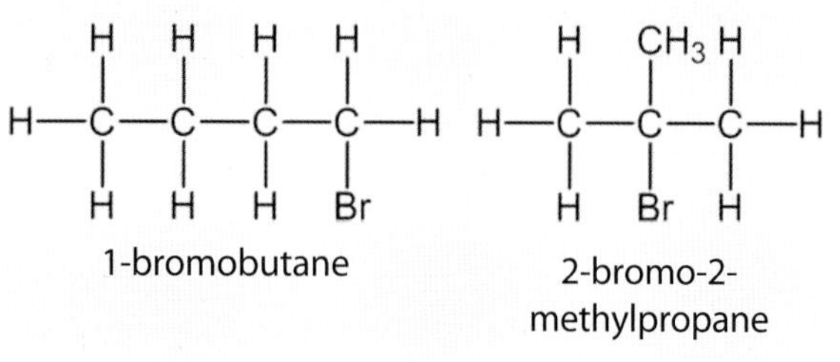

1-bromobutane 2-bromo-2-methylpropane

(CCEA January 2013)

3. Account for the boiling points of the following halogenoalkanes in terms of the intermolecular forces operating between the molecules.

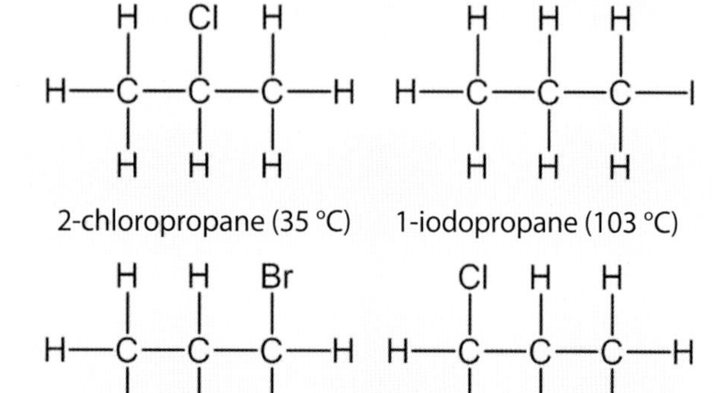

2-chloropropane (35 °C) 1-iodopropane (103 °C)

1-bromopropane (71 °C) 1,1-dichloropropane (88 °C)

4. Account for the boiling points of the following halogenoalkanes in terms of the intermolecular forces operating between the molecules.

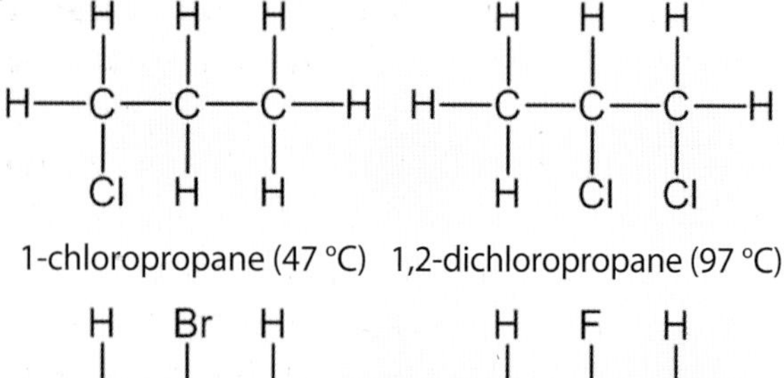

1-chloropropane (47 °C) 1,2-dichloropropane (97 °C)

2-bromo-1-chloropropane (117 °C) 1-chloro-2-fluoropropane (69 °C)

Factors Affecting Reactivity

The relationship between the structure of a halogenoalkane and the reactivity of the carbon-halogen bond can be understood if we classify halogenoalkanes as primary, secondary or tertiary according to the number of carbon atoms attached to the carbon-halogen bond.

The compound 1-bromobutane is an example of a **primary halogenoalkane** *as the halogen atom is bonded to a carbon (C) with one carbon atom (C) attached.* The isomer 1-bromo-2-methylpropane is also a primary halogenoalkane.

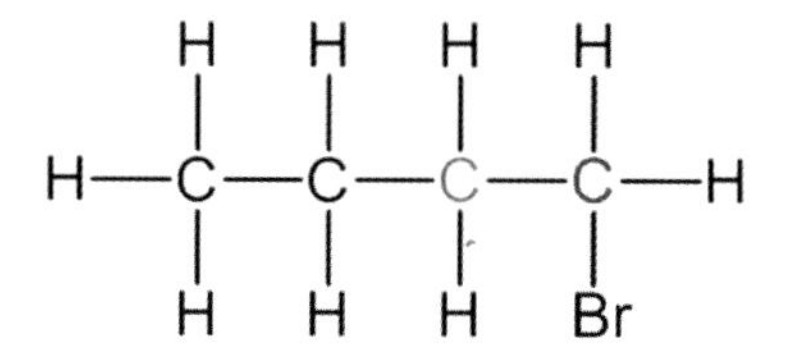

1-bromobutane (*primary*)

$CH_3CH_2CH_2CH_2Br$

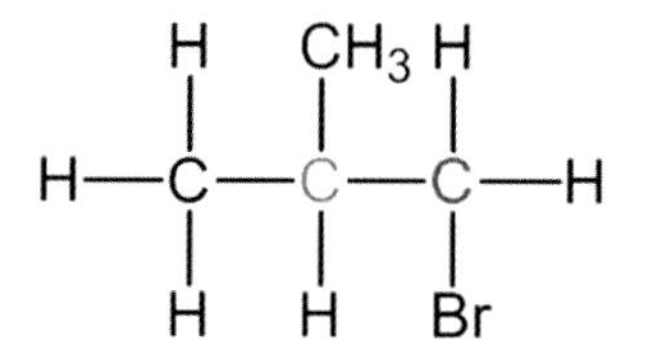

1-bromo-2-methylpropane (*primary*)

$CH_3CH(CH_3)CH_2Br$

In contrast, the isomer 2-bromobutane is an example of a **secondary halogenoalkane** as *the halogen atom is bonded to a carbon (C) with two carbon atoms (C) attached.* Similarly, the isomer 2-bromo-2-methylpropane is an example of a **tertiary halogenoalkane** as *the halogen atom is bonded to a carbon (C) with three carbon atoms (C) attached.*

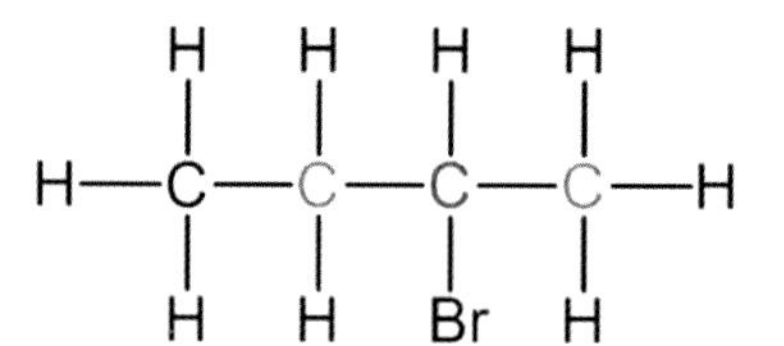

2-bromobutane (secondary)

$CH_3CH_2CHBrCH_3$

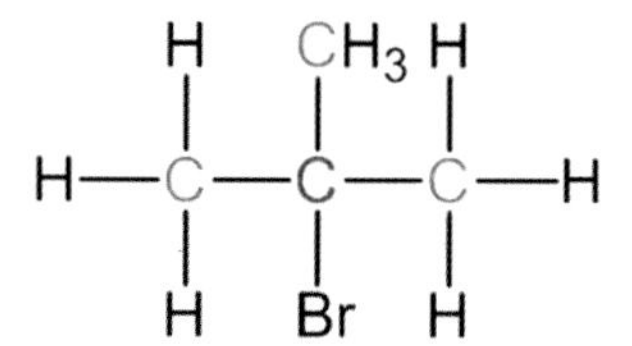

2-bromo-2-methylpropane (*tertiary*)

$CH_3C(CH_3)BrCH_3$

The conditions under which primary, secondary and tertiary halogenoalkanes react are often quite different. As a result the primary, secondary, or tertiary nature of the carbon-halogen bonds in a molecule is one of several factors that must be taken into account when describing the reactions of the compound.

Worked Example 2.5i

Draw the structure of 2-chloro-2-methylpropane, C_4H_9Cl and explain why the compound is described as a *tertiary* halogenoalkane.

(Adapted from CCEA January 2003)

Solution

```
      CH3
      |
Cl----C----CH3
      |
      CH3
```

The compound is described as a tertiary halogenoalkane as the carbon atom in the carbon-halogen bond is bonded to three other carbon atoms.

Exercise 2.5B

1. (a) Draw the structural formula for a secondary chloroalkane with a molecular formula C_3H_7Cl.
   (b) Explain what is meant by the term *secondary*.
2. (a) Draw the structure of each iodoalkane with a molecular formula C_3H_7I. Classify each isomer as a primary, secondary or tertiary iodoalkane.
   (b) Explain what is meant by the term *isomer*.
3. (a) Draw the structure of each primary fluoroalkane with a molecular formula C_4H_9F.
   (b) Explain what is meant by the term *primary*.

Naming Halogenoalkanes

The systematic (IUPAC) name for a halogenoalkane is based on the systematic name for the parent (straight-chain) alkane on which its structure is based. For example, 1-bromobutane and 2-bromobutane are considered bromobutanes as their structure is based on butane, $CH_3CH_2CH_2CH_3$.

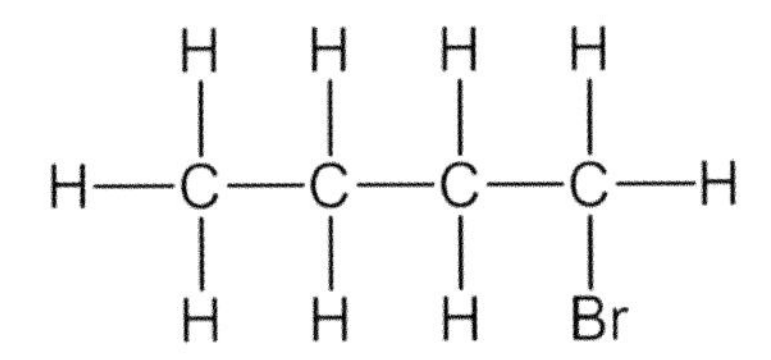

1-bromobutane $CH_3CH_2CH_2CH_2Br$

```
   H   H   H   H
   |   |   |   |
H--C---C---C---C--H
   |   |   |   |
   H   H   Br  H
```

2-bromobutane $CH_3CH_2CHBrCH_3$

The prefix *1-bromo* or *2-bromo* is added to locate bromine on the carbon chain. As in alkanes, the prefix used to locate the bromine atom is based on a numbering of the carbon atoms that generates the lowest numbers.

The compounds 1-bromo-2-methylpropane and 2-bromo-2-methylpropane are structural isomers of the bromobutanes. Both are considered propanes as their structure is based on the three carbon chain in propane, $CH_3CH_2CH_3$.

```
   H  CH3  H
   |   |   |
H--C---C---C--H
   |   |   |
   H   H   Br
```

1-bromo-2-methylpropane $CH_3CH(CH_3)CH_2Br$

```
   H  CH3  H
   |   |   |
H--C---C---C--H
   |   |   |
   H   Br  H
```

2-bromo-2-methylpropane $CH_3C(CH_3)BrCH_3$

As in alkanes, the prefixes describing the location of bromine and the methyl group are listed in alphabetical order, and are based on a numbering of the carbon atoms that produces the lowest number prefixes.

In the same way the prefixes *fluoro*, *chloro*, and *iodo* are used to locate fluorine, chlorine and iodine atoms, and can be combined to locate several halogens within a molecule. For example, the prefix *2-bromo-1-chloro* could be used to locate bromine on the second carbon and chlorine on the first carbon in the molecule. Similarly, the prefix *1-chloro-1-fluoro* could be used to locate chlorine and fluorine on the first carbon in the molecule. Once again the prefixes are listed in alphabetical order and are based on a numbering of the carbons that produces the lowest number prefixes. The correct use of multiple prefixes is further illustrated by the following examples.

```
    Cl  H   H
    |   |   |
H — C — C — C — I
    |   |   |
    H   H   H
```

1-chloro-3-iodopropane

$CH_2ClCH_2CH_2I$

```
    H   CH3 H
    |   |   |
H — C — C — C — H
    |   |   |
    H   Br  Cl
```

2-bromo-1-chloro-2-methylpropane

$CH_3C(CH_3)BrCH_2Cl$

The location of two or more halogen atoms of the same type can be described by using the number prefixes *di, tri, tetra* ... to construct naming prefixes such as *dibromo* and *trichloro*. An additional prefix such as *1,2-* or *1,1,2-* can then be added to locate each halogen atom as in 1,1,1-trichloroethane, CCl_3CH_3. The correct use of number prefixes is illustrated further by the following examples.

```
    H   H   H
    |   |   |
H — C — C — C — H
    |   |   |
    Br  H   Br
```

1,3-dibromopropane

$CH_2BrCH_2CH_2Br$

```
    Cl  H
    |   |
F — C — C — H
    |   |
    F   H
```

1-chloro-1,1-difluoroethane

$CClF_2CH_3$

Rules for naming halogenoalkanes:

- The name of a halogenoalkane is based on the name of the parent (straight-chain) alkane on which its structure is based.
- The prefixes *fluoro, chloro, bromo* and *iodo* are combined with the prefixes *di, tri, tetra* ... to describe the type and location of each halogen atom.
- Prefixes such as *methyl* and *ethyl* are combined with the prefixes *di, tri, tetra* ... to describe and locate other functional groups.
- The prefixes are listed in alphabetical order and the carbon atoms numbered in a way that produces the lowest number prefixes.

Exercise 2.5C

1. The compound trichlorofluoromethane, CCl_3F is known as CFC-11. Write systematic names for CFC-12, CCl_2F_2 and CFC-13, $CClF_3$.
2. The halogenoalkane CCl_3CH_3 is used as a solvent in correction fluid. (a) Draw the structure of the molecule and write the IUPAC name for the compound. (b) Draw the remaining structural isomers of the compound and write the systematic name for each isomer.
3. (a) Draw the structure of the halogenoalkane CCl_2BrCH_3 and write the systematic name for the compound. (b) Draw the remaining structural isomers of the compound and write the systematic name for each isomer.
4. There are four bromoalkanes with the formula, C_4H_9Br. Copy and complete the following table by drawing the missing structures, writing the name for each structure, and classifying each structure as primary, secondary or tertiary.

Structure	Name	Classification
H H H H H—C—C—C—C—H H H H Br	1-bromobutane	primary
H CH3 H H—C—C—C—H H Br H	2-bromo-2-methylpropane	
	1-bromo-2-methylpropane	

(CCEA June 2010)

5. The chlorobutanes in the table are isomers. (a) Write the molecular formula for a chlorobutane. (b) Write the general formula for a chloroalkane. (c) Deduce the systematic name for t-butyl chloride. (d) Suggest what is meant by the term *sec*. (e) Suggest why the boiling point of t-butyl chloride is so different from the other chlorobutanes.

butyl chloride	structure	boiling point /°C
t-butyl chloride	$(CH_3)_3CCl$	51–52
n-butyl chloride	$CH_3CH_2CH_2CH_2Cl$	78–79
sec-butyl chloride	$CH_3CH_2CHClCH_3$	68–70
iso-butyl chloride	$(CH_3)_2CHCH_2Cl$	68–69

(Adapted from CCEA June 2011)

Before moving to the next section, check that you are able to:

- Explain why halogenoalkanes are denser than water and become less soluble in water as the molecules get bigger.
- Explain how the boiling points of halogenoalkanes are affected by the size and shape of the molecule, the type of halogen atoms in the molecule, and the number of halogen atoms in the molecule.
- Write structural and condensed formulas for halogenoalkanes.
- Classify the structures of halogenoalkanes as primary, secondary, or tertiary.
- Deduce systematic names for halogenoalkanes.

Preparation

In this section we are learning to:

- Describe how to prepare a pure, dry sample of a halogenoalkane.

The process of making a pure organic compound from the starting materials is known as **organic synthesis**. In the first stage of a synthesis one or more chemical reactions are used to prepare a crude (impure) sample of the compound. The crude product is then extracted from the reaction mixture and purified. The synthesis of 1-bromobutane, C_4H_9Br illustrates the process and techniques used to prepare a pure, dry sample of a halogenoalkane in the laboratory.

Organic Synthesis

STAGE 1
The product is formed in the reaction mixture.

STAGE 2
The product is extracted from the reaction mixture.

STAGE 3
Purification of the crude product.

Stage 1: Forming the product

Crude 1-bromobutane, C_4H_9Br is synthesised by reacting the alcohol butan-1-ol, C_4H_9OH with hydrogen bromide.

$$CH_3CH_2CH_2CH_2OH + HBr \rightarrow CH_3CH_2CH_2CH_2Br + H_2O$$

butan-1-ol → 1-bromobutane

The hydrogen bromide is produced within the reaction mixture, a technique referred to as ***in situ***, by adding concentrated sulfuric acid to a mixture of solid sodium bromide and butan-1-ol as the reaction mixture is stirred.

$$NaBr + H_2SO_4 \rightarrow NaHSO_4 + HBr$$

The concentrated sulfuric acid is added one drop at a time using a dropping funnel as shown in Figure 7a. The concentrated sulfuric acid oxidises a portion of the hydrogen bromide to bromine, giving the reaction mixture a characteristic orange colour. The dropping funnel is then removed and several small pieces of ground glass or porcelain added to the reaction mixture. The reaction mixture is then refluxed for an extended period in a round-bottom flask fitted with a Liebig condenser in the vertical position as shown in Figure 7b.

In this context **reflux** describes *the continuous boiling and condensing of a reaction mixture using a condenser fitted in the vertical position*. Refluxing increases the amount of product formed during the reaction by making it possible for the reactants to react at an elevated temperature over an extended period of time. The pieces of ground glass or porcelain added to

(a)

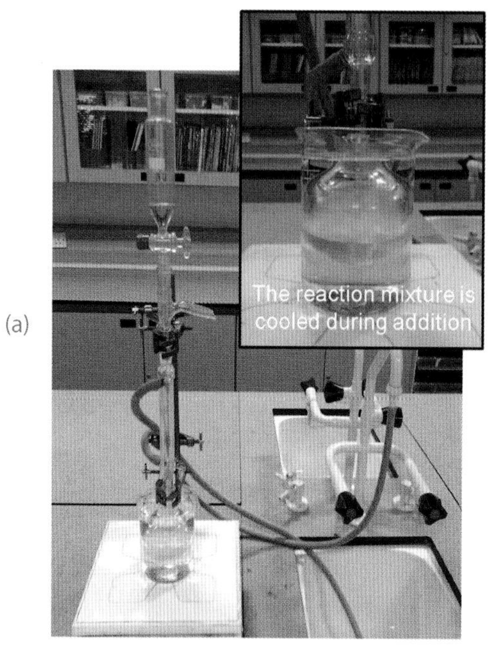

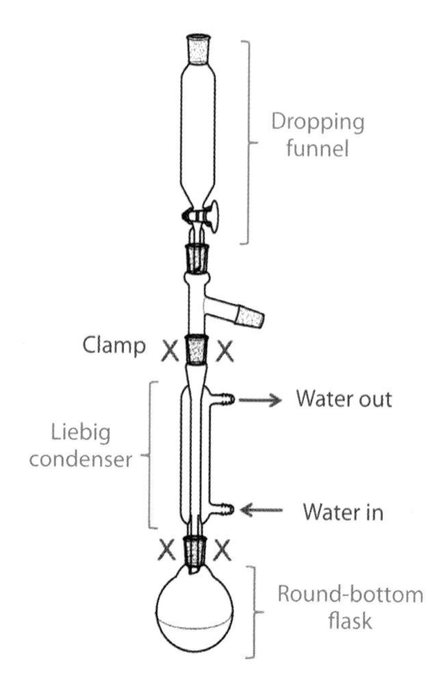

Figure 7: (a) Using a dropping funnel to control the addition of concentrated sulfuric acid to the reaction mixture. (b) Refluxing the reaction mixture.

(b)

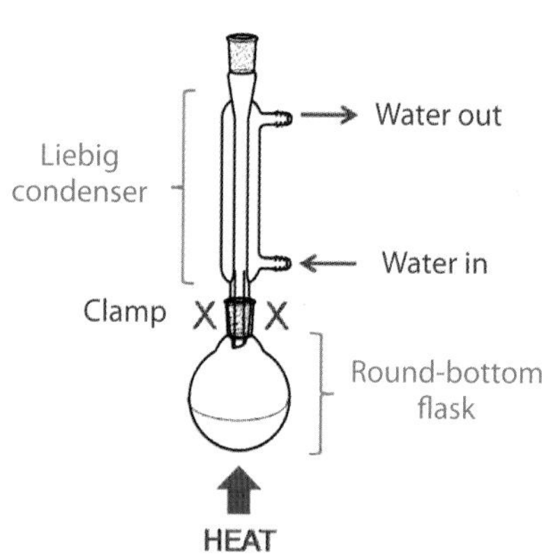

the reaction mixture are referred to as **boiling chips** or 'anti-bumping granules' as they ensure smooth boiling when the reaction mixture is refluxed.

Worked Example 2.5ii

When bromobutane is prepared by the reaction of butanol with concentrated hydrobromic acid in the presence of concentrated sulfuric acid which one of the following is not found in the reaction flask?

A bromine

B butanol

C butane

D hydrogen bromide

(CCEA June 2011)

Solution

The hydrobromic acid (HBr) is formed *in situ* and traces of the reactants (butanol and hydrobromic acid, HBr) may remain after reaction. Concentrated sulfuric acid will also oxidise some of the hydrobromic acid to bromine.

Answer C.

Stage 2: Extracting the product

With the reflux complete the reaction mixture is left to cool. The apparatus is then reorganised with the condenser in the horizontal position and 1-bromobutane extracted from the reaction mixture by **distillation** as shown in Figure 8.

Distillation involves slowly heating the reaction mixture until the 1-bromobutane in the mixture boils (boiling point 101 °C). The vapour produced by the boiling rises, enters the condenser, and condenses to form a liquid **distillate**. The distillate is referred to as

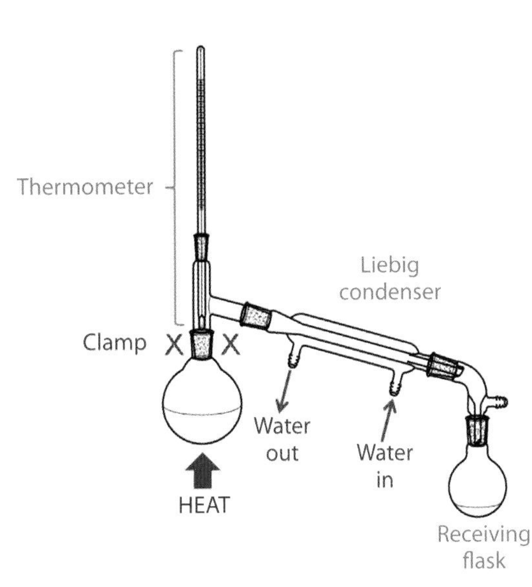

Figure 8: Separating crude 1-bromobutane from the reaction mixture by distillation.

crude 1-bromobutane as it contains 1-bromobutane and small amounts of impurities from the reaction mixture. A pure sample of 1-bromobutane would be collected over a range of 1–2 °C. The impurities in the crude distillate increase the temperature range over which the distillate is collected.

Worked Example 2.5iii

Separating a product from a mixture by distillation is an important practical technique in organic chemistry. (a) Draw a labelled diagram of the apparatus used to carry out a distillation. (b) Why are anti-bumping granules added to a mixture being distilled?

(CCEA June 2009)

Solution

(a)

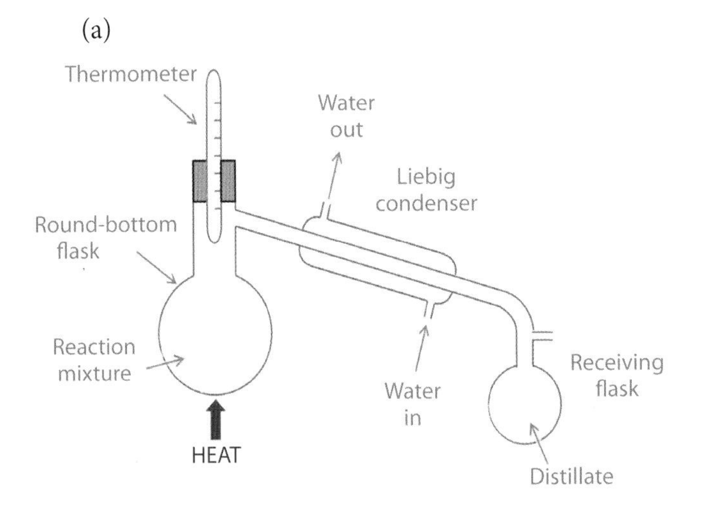

(b) To ensure that the reaction mixture boils smoothly.

Stage 3: Purifying the product

The first step in the purification process involves shaking the crude 1-bromobutane with dilute sodium hydrogencarbonate solution in a **separating funnel**. Acidic impurities in the crude 1-bromobutane react with the dilute alkali to form a salt and carbon dioxide gas which must be released to avoid the build-up of pressure in the separating funnel. This is achieved by inverting the separating funnel and using the tap to release the pressure inside the funnel at intervals as the mixture is being shaken. The technique of inverting and shaking to mix the contents of the separating funnel is illustrated in Figure 9a.

The stopper is then removed and the mixture left to settle. The 1-bromobutane is immiscible in water and forms a separate (organic) layer as shown in Figure 9b. The lower (organic) layer containing the product can then be collected from the funnel and the upper (aqueous) layer discarded.

The lower (organic) layer containing the 1-bromobutane is then shaken with distilled water in a separating funnel to remove salts and other water soluble impurities before being collected in a small dry conical flask. A solid **drying agent** such as anhydrous calcium chloride or anhydrous sodium sulfate is then added to remove traces of water in the

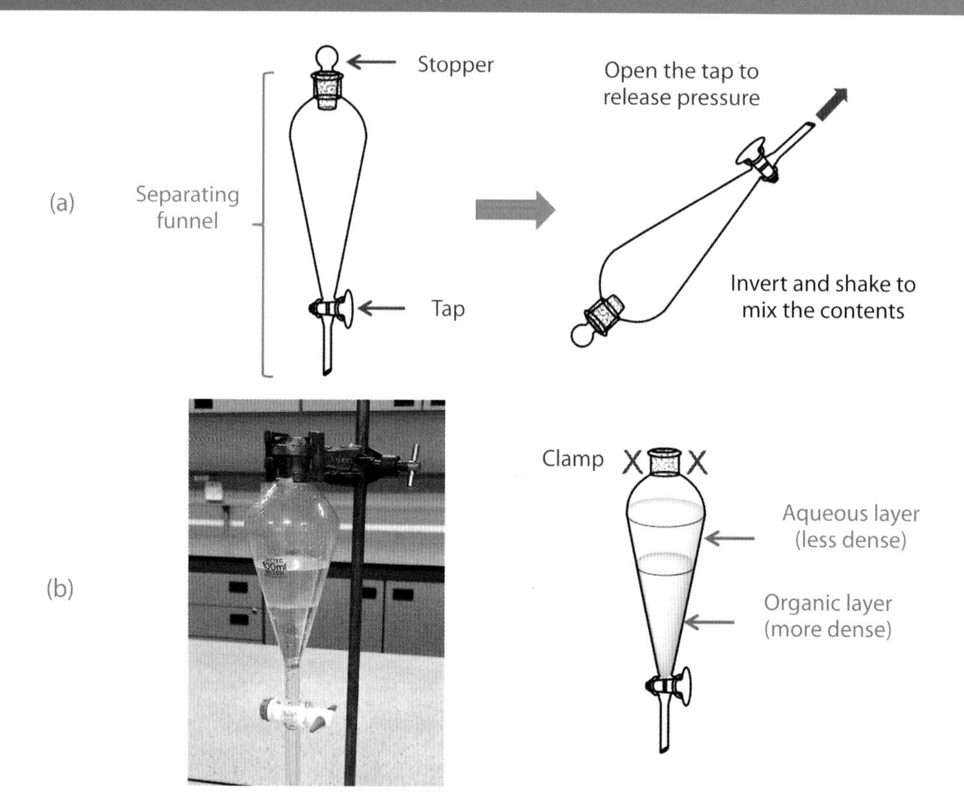

Figure 9: The use of a separating funnel to (a) mix and (b) separate the layers formed during the purification process.

product. If the product contains small amounts of water it will appear cloudy and will become increasingly clear as the water is absorbed by the drying agent. The drying agent is then removed by filtering the mixture into a clean round-bottom flask and the product distilled to obtain a pure, dry sample of bromobutane at its boiling point.

Worked Example 2.5iv

The compound 1-bromobutane is prepared by refluxing a mixture containing butan-1-ol, water, sodium bromide and concentrated sulfuric acid for 45 minutes. Describe the remaining steps required to produce a pure, dry sample of this haloalkane.

(CCEA June 2010)

Solution

Distil the reaction mixture and collect crude 1-bromobutane at its boiling point.

Add dilute sodium hydrogencarbonate solution to the crude product in a separating funnel. Stopper, invert and shake the separating funnel. Open the tap every few seconds to release any carbon dioxide formed as the mixture is shaken.

Clamp the separating funnel, remove the stopper, and allow the layers to separate. Collect the lower (organic) layer containing the product.

Wash the product with distilled water using a separating funnel and collect the lower (organic) layer. Add anhydrous sodium sulfate to the product in a small conical flask and swirl the mixture until the product becomes clear.

Filter the mixture into a clean, dry round-bottom flask and distil the product to obtain a pure, dry sample of 1-bromobutane at its boiling point.

Exercise 2.5D

1. Amphetamines can be synthesised by the following sequence. The liquid product obtained in Step 1 is impure. Explain, giving experimental details, how the product can be separated, dried and purified.

$$R\text{—}C(CH_3)(H)\text{—}OH \xrightarrow{\text{Step 1}} R\text{—}C(CH_3)(H)\text{—}Br \xrightarrow{\text{Step 2}} R\text{—}C(CH_3)(H)\text{—}NH_2$$

(CCEA January 2011)

2. The following procedure is used to make t-butyl chloride from the corresponding alcohol.

   $(CH_3)_3COH + HCl \rightarrow (CH_3)_3CCl + H_2O$

   25 g of t-butyl alcohol and 85 cm³ of concentrated hydrochloric acid (an excess) are mixed in a separating funnel. The mixture is shaken from time to time over 20 minutes. The mixture is then allowed to stand for a few minutes and the acid layer removed. The crude t-butyl chloride is washed with sodium hydrogencarbonate solution and then with water. Anhydrous calcium chloride is swirled with the t-butyl chloride in a conical flask. The liquid is decanted, 2–3 chips of porous porcelain added and distilled to collect 28 g of t-butyl chloride at 51–52 °C.

   Explain why the t-butyl chloride is (a) shaken with sodium hydrogencarbonate solution, (b) shaken with water and (c) swirled with anhydrous calcium chloride. Also explain why (d) porous porcelain is added before distillation and (e) the separating funnel is inverted and the tap opened after each shaking. (f) Calculate the percentage yield of t-butyl chloride.

   (Adapted from CCEA June 2011)

3. 17.8 g of 1-bromobutane (RMM 137) was formed when 14.8 g of butan-1-ol (RMM 74) was refluxed with sodium bromide and sulfuric acid. Calculate the percentage yield for the reaction.

   (Adapted from CCEA June 2007)

4. Which one of the following is not used in the laboratory preparation of a pure chloroalkane from the corresponding alcohol?

   A Aqueous sodium chloride

   B Aqueous hydrogencarbonate ions

   C Distillation

   D Refluxing

   (CCEA June 2015)

Before moving to the next section, check that you are able to:

- Describe the use of reflux and distillation in the preparation of a halogenoalkane, and the use of a separating funnel, drying agent and distillation in the purification of a halogenoalkane.
- Draw labelled diagrams of the apparatus used to reflux and distil the reaction mixture, and explain the role of each process in the preparation of a halogenoalkane.
- Explain the role of the separating funnel, drying agent and distillation in the purification of a halogenoalkane.
- Recall examples of drying agents.

Reactions

In this section we are learning to:

- Use structural formulas to describe the reactions of halogenoalkanes with dilute alkali, ammonia and cyanide ions.
- Explain how the mechanism for the reaction between a halogenoalkane and dilute alkali is determined by the structure of the halogenoalkane, and how the rate of the reaction is affected by the strength of the carbon-halogen bond.
- Deduce the structure of the products formed when a halogenoalkane undergoes an elimination reaction.

Substitution Reactions

When a halogenoalkane is refluxed with dilute alkali, the halogen atom is replaced by a hydroxyl (-OH) group to form the corresponding alcohol. For example, butan-1-ol is formed when a mixture of 1-bromobutane and dilute sodium hydroxide is refluxed.

```
     CH3 H   H
     |   |   |
H—C—C—C—H   +   NaOH   ⟶
     |   |   |
     H   H   Br
```

1-bromobutane

$CH_3CH_2CH_2CH_2Br$

```
     CH3 H   H
     |   |   |
H—C—C—C—H   +   NaBr
     |   |   |
     H   H   OH
```

butan-1-ol

$CH_3CH_2CH_2CH_2OH$

The reaction is an example of a substitution reaction as it involves replacement of the halogen by another group. It is also an example of **nucleophilic substitution** as *an atom or group of atoms is replaced by a nucleophile when the nucleophile uses a lone pair to form a coordinate bond with an electron-deficient atom.*

The term **nucleophile** is used to describe *a molecule or ion that attacks regions of low electron density by using a lone pair to form a coordinate bond with an electron-deficient atom.* The hydroxide ions in the alkali are nucleophiles. Water (H_2O), ammonia (NH_3), halide ions (F^-, Cl^-, Br^-, I^-), and cyanide ion, CN^- are also nucleophiles as each is capable of using a lone pair of electrons to form a coordinate bond with an electron deficient atom (E).

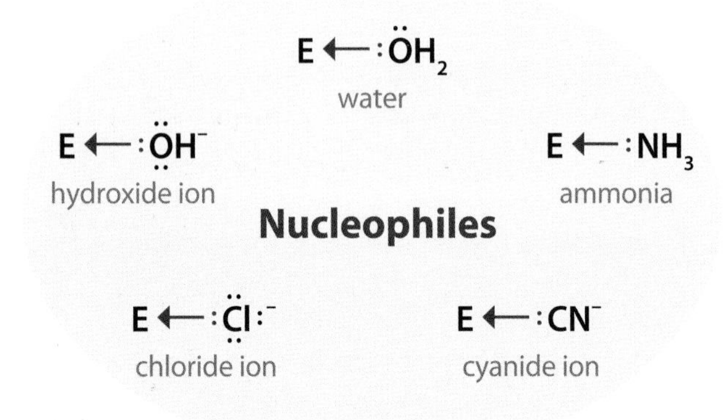

Exercise 2.5E

1. Explain why an ammonia molecule can act as a nucleophile but an ammonium ion cannot act as a nucleophile.

   (CCEA June 2003)

2. Which one of the following molecules can **not** act as a nucleophile?

   A CH_3NH_2 B CH_4 C H_2O D NH_3

   (CCEA January 2012)

If nucleophilic substitution occurs at a primary carbon, as in 1-bromobutane, $CH_3CH_2CH_2CH_2Br$ the reaction occurs in a single step. The mechanism for the reaction is described by the flow scheme in Figure 10. As hydroxide ion approaches the halogenoalkane it uses a lone pair to begin forming a coordinate bond with the electron-deficient carbon in the carbon-halogen bond ($C^{\delta+}$-$Br^{\delta-}$). In this instance the carbon is considered electron-deficient as the electrons in the carbon-halogen bond are not shared equally, leaving carbon with fewer than eight electrons in its outer shell.

As the coordinate bond continues to form (C---OH) the carbon-halogen bond begins to break (C---Br). The process continues until the coordinate bond is fully formed (C-OH) and the carbon-halogen bond has undergone heterolytic fission to produce bromide ion. The intermediate structure (HO---C---Br) is referred to as a transition state. The transition state is an unstable structure and cannot be isolated as a compound.

In contrast, if nucleophilic substitution occurs at a tertiary carbon, as in 2-bromo-2-methylpropane, $CH_3C(CH_3)BrCH_3$ the substitution reaction occurs in two steps. The flow scheme in Figure 11 details the mechanism for the reaction. The first step in the mechanism involves heterolytic fission of the carbon-halogen bond to produce a tertiary carbocation. Hydroxide ion then acts as a nucleophile by forming a coordinate bond with the electron-deficient carbon in the carbocation (C^+).

The situation at a secondary carbon atom is frequently more complex with one or both of these mechanisms operating simultaneously.

HO:⁻ → H–C(H)(C₃H₇)–Br → [HO------C(H)(H)(C₃H₇)-------Br]⁻ → HO–C(H)(H)(C₃H₇) + Br⁻

Transition State

Figure 10: The mechanism for the nucleophilic substitution reaction that occurs when 1-bromobutane (a primary halogenoalkane) reacts with dilute alkali. Dashed bonds (---) are used to indicate partially formed and partially broken bonds.

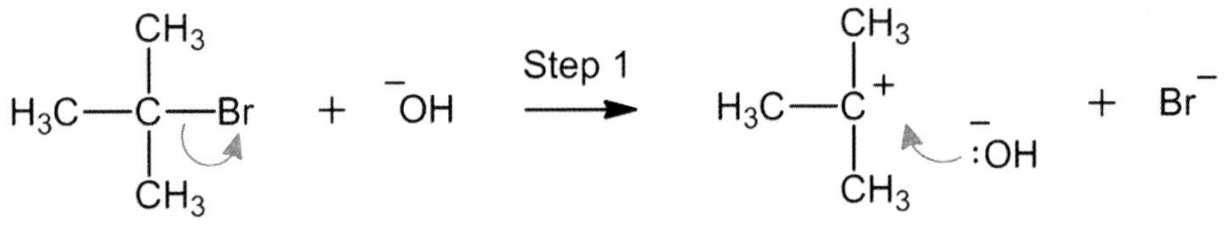

$$(CH_3)_3C\text{—}Br + {}^{-}OH \xrightarrow{\text{Step 1}} (CH_3)_3C^{+} \;\; {}^{-}\!:OH + Br^{-}$$

STEP 1

Heterolytic fission of the C-Br bond to form a tertiary carbocation

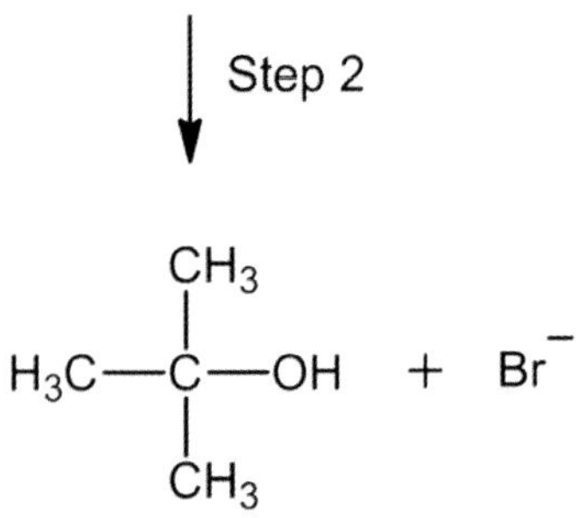

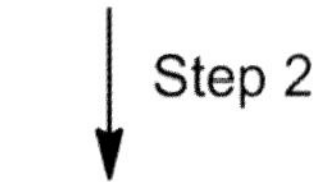

STEP 2

Hydroxide ion forms a coordinate bond with the tertiary carbon (C^+)

$$(CH_3)_3C\text{—}OH + Br^{-}$$

Figure 11: The mechanism for the nucleophilic substitution reaction that occurs when 2-bromo-2-methylpropane (a tertiary halogenoalkane) reacts with dilute alkali.

Exercise 2.5F

1. Bromoethane undergoes nucleophilic substitution when it reacts with aqueous potassium hydroxide. (a) Write an equation for the reaction. (b) Explain the meaning of the term *nucleophilic substitution*. (c) Draw a flow scheme to illustrate the mechanism for the reaction of bromoethane with hydroxide ions.

   (CCEA January 2012)

2. Draw a mechanism for the reaction between chloromethane and sodium hydroxide.

   (Adapted from CCEA June 2004)

3. When 1-bromopropane reacts with aqueous potassium hydroxide the intermediate structure formed is:

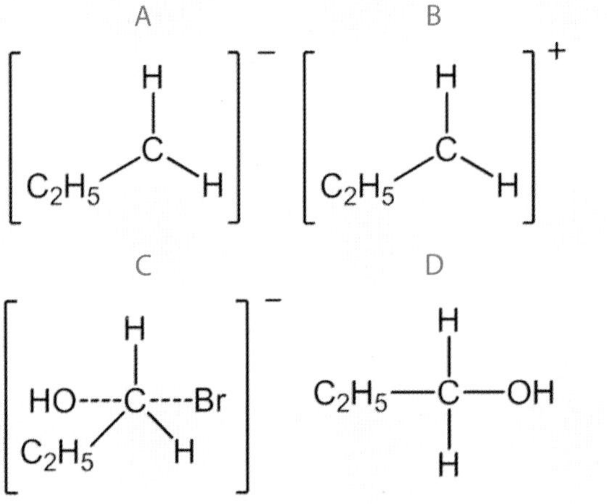

   A $[C_2H_5CH_2]^{-}$

   B $[C_2H_5CH_2]^{+}$

   C $[HO\text{----}CH(C_2H_5)\text{----}Br]^{-}$

   D $C_2H_5\text{—}CH_2\text{—}OH$

   (CCEA January 2003)

4. The compound 2-chloro-2-methylpropane is an isomer of 1-chlorobutane. (a) Explain why the compounds are *structural isomers*. (b) Write the mechanism for the reaction between 2-chloro-2-methylpropane and hydroxide ions.

   (Adapted from CCEA June 2006)

5. 2-bromo-2-methylpropane reacts with aqueous potassium hydroxide to produce a tertiary alcohol. (a) Name the type of mechanism for this reaction. (b) Write an equation for the first step in the mechanism. (c) Draw the structure of the organic product in the first step. (d) Write an equation for the second step in the mechanism.

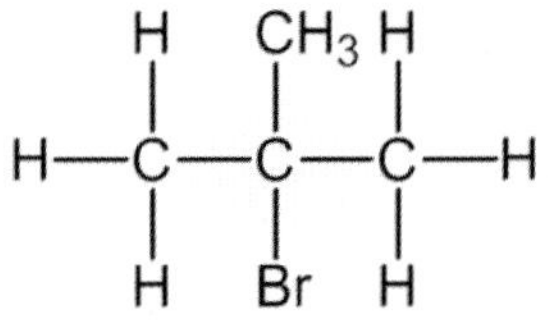

   $$CH_3\text{—}C(CH_3)(Br)\text{—}CH_3$$

   2-bromo-2-methylpropane

   (CCEA January 2013)

6. The product from the bromination of hex-1-ene is a dibromoalkane with the following structure. Draw the structure of the product when the dibromoalkane is reacted with an excess of aqueous alkali.

   $CH_3CH_2CH_2CH_2CHBrCH_2Br$

   (CCEA June 2015)

Before moving to the next section, check that you are able to:

- Recall what is meant by the term *nucleophilic substitution.*
- Write and explain the mechanism for the reaction between a primary halogenoalkane and dilute alkali.
- Write and explain the mechanism for the reaction between a tertiary halogenoalkane and dilute alkali.

Amines and Nitriles

Heating a halogenoalkane with an excess of concentrated ammonia solution in a sealed glass tube produces the corresponding **amine**. The reaction is an example of substitution as the halogen is replaced by an amino (-NH_2) group.

$CH_3CH_2CH_2Br + 2NH_3 \rightarrow CH_3CH_2CH_2NH_2 + NH_4Br$

1-bromopropane $CH_3CH_2CH_2Br$

1-aminopropane $CH_3CH_2CH_2NH_2$

The reaction is also an example of nucleophilic substitution as the ammonia molecule displaces the halogen by using its lone pair to form a coordinate bond with carbon. The substitution produces HBr which reacts with a second molecule of ammonia to form ammonium bromide, NH_4Br.

Similarly, when a halogenoalkane is refluxed with an aqueous solution of sodium cyanide, the halogen is replaced by a cyano (-CN) group to form the corresponding **nitrile**. The reaction is considered nucleophilic substitution as cyanide ion (CN^-) displaces the halogen by using a lone pair to form a coordinate bond with carbon.

$CH_3CH_2CH_2Br + NaCN \rightarrow CH_3CH_2CH_2CN + NaBr$

1-bromopropane $CH_3CH_2CH_2Br$

butanenitrile $CH_3CH_2CH_2CN$

Exercise 2.5G

1. Chloroethane will react with ammonia in solution. Write an equation for this reaction and name the organic product.

   (CCEA January 2014)

2. Write the structure for the product when 2-bromobutane reacts with (a) ammonia and (b) aqueous sodium cyanide.

3. Which one of the following mixtures will react to produce a compound with molecular formula C_4H_7N?

   A 1-bromobutane and ammonia
   B 1-bromobutane and potassium cyanide
   C 1-bromopropane and ammonia
   D 1-bromopropane and potassium cyanide

   (CCEA June 2014)

4. The product from the bromination of hex-1-ene is a dibromoalkane with the following structure. (a) Draw the structure of the product when the dibromoalkane is reacted with an *excess* of ammonia. (b) Draw the structure of the product when the dibromoalkane is reacted with an *excess* of cyanide ions.

   $CH_3CH_2CH_2CH_2CHBrCH_2Br$

   (CCEA June 2015)

5. Excess ammonia reacts with 1,5-dichloropentane, $Cl(CH_2)_5Cl$ to produce cadaverine, $H_2N(CH_2)_5NH_2$ a putrid product formed from the decay of flesh. (a) Suggest an equation for this reaction. (b) The smell of cadaverine can be eliminated by adding hydrochloric acid. What does this suggest about the chemical nature of cadaverine? *(CCEA June 2006)*

Before moving to the next section, check that you are able to:

- Deduce the structure of the product when a halogenoalkane is refluxed with ammonia or cyanide ion and write an equation for the reaction.
- Explain why the displacement of a halogen to form the corresponding amine or nitrile is an example of nucleophilic substitution.

Hydrolysis Reactions

The term **hydrolysis** describes *a reaction in which bonds are broken as the result of a compound reacting with water*. Forming an alcohol by refluxing a halogenoalkane with water is an example of hydrolysis.

$$CH_3CH_2CH_2CH_2Br + H_2O \rightarrow CH_3CH_2CH_2CH_2OH + HBr$$

The reaction is also an example of nucleophilic substitution as it involves water forming a coordinate bond with the electron-deficient carbon in the carbon-halogen bond ($C^{\delta+}$-$Br^{\delta-}$). The same reaction occurs when a halogenoalkane is refluxed with dilute alkali, but is considerably faster as hydroxide ion is a better nucleophile than water. The reaction between a halogenoalkane and dilute alkali to produce the corresponding alcohol is known as **alkaline hydrolysis**.

$$CH_3CH_2CH_2CH_2Br + OH^- \rightarrow CH_3CH_2CH_2CH_2OH + Br^-$$

Exercise 2.5H

1. Write the equation for the reaction that occurs when chloromethane is hydrolysed with 'heavy water', D_2O where deuterium, D is an isotope of hydrogen. *(Adapted from CCEA June 2011)*

Hydroxide ion is a better nucleophile than water as it is more strongly attracted to the electron-deficient carbon in the carbon-halogen bond ($C^{\delta+}$). As a result, the rate of hydrolysis is expected to increase as the halogen becomes more electronegative, and the corresponding carbon-halogen bond becomes more polar. In reality the rate of hydrolysis increases as the carbon-halogen bond becomes weaker down the group.

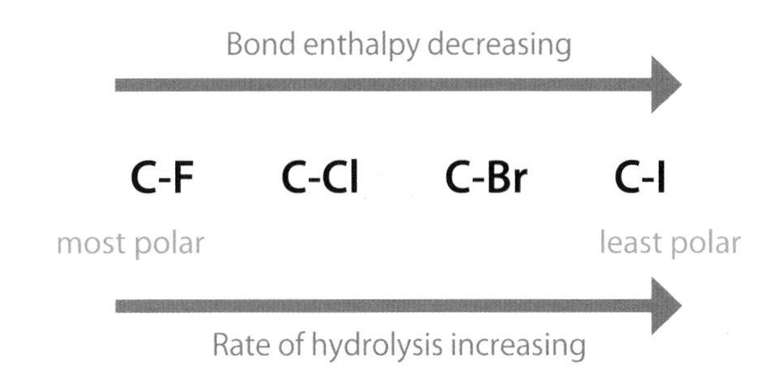

The hydrolysis of a halogenoalkane produces halide ions which can be detected by adding silver nitrate solution to the reaction mixture. As the reaction proceeds halide ions formed in the reaction combine with silver ions from the silver nitrate solution to form a precipitate of the corresponding silver halide.

Hydrolysis of the carbon-halogen bond:

$$R\text{-}X + H_2O \rightarrow R\text{-}OH + H^+ + X^-$$

where X = Cl, Br, I

Formation of a silver halide precipitate:

$$Ag^+ + X^- \rightarrow AgX$$

The rate at which a carbon-halogen bond undergoes hydrolysis can then be calculated by measuring the time it takes for the silver halide precipitate to form.

$$\text{Rate of hydrolysis} = \frac{1}{\text{Time for precipitate to form}}$$

The formation of silver halide precipitates during the hydrolysis of 1-chlorobutane, 1-bromobutane and 1-iodobutane is shown in Figure 12. The precipitates form in the order: AgI (yellow), AgBr (cream), AgCl (white) indicating that, as expected, *the rate of hydrolysis increases as the carbon-halogen bond becomes weaker down the group*.

(a)

silver
iodide

(b)

Figure 12: The formation of silver halide precipitates during the hydrolysis of (from left to right): 1-chlorobutane, 1-bromobutane and 1-iodobutane. (a) Silver iodide forms first. (b) Silver bromide forms second.

Time to form precipitate decreasing →

C-F C-Cl C-Br C-I

strongest bond — weakest bond

Rate of hydrolysis increasing →

Worked Example 2.5v

Samples of 1-chloro, 1-bromo and 1-iodobutane were added to separate test tubes and heated gently. State what is observed when silver nitrate solution is added to each test tube and explain the relative rates of reaction in terms of bond strength.

(Adapted from CCEA June 2003)

Solution

A yellow precipitate forms in the mixture containing 1-iodobutane. A cream precipitate then forms in the mixture containing 1-bromobutane and is followed by a white precipitate in the mixture containing 1-chlorobutane.

The results show that the rate of hydrolysis increases as the carbon-halogen bond becomes weaker down the group.

Exercise 2.5I

1. Explain, in terms of bond enthalpy, why the hydrolysis of chloroethane is slower than that of bromoethane.

   (Adapted from CCEA January 2012)

2. Equal volumes of 1-chlorobutane and 1-iodobutane are warmed with aqueous silver nitrate in the presence of ethanol. Which one of the following is the reason why the 1-chlorobutane reacts more slowly?

   A The C-Cl bond is more polar than the C-I bond

   B The C-Cl bond is stronger than the C-I bond

   C The C-I bond is more polar than the C-Cl bond

   D The C-I bond is stronger than the C-Cl bond

   (CCEA June 2014)

3. Samples of 1-iodobutane, 1-bromobutane and 1-chlorobutane are placed in separate test tubes labelled A, B and C respectively. Silver nitrate is added to each test tube and the mixtures warmed gently. (a) Describe the changes that occur in test tube A. (b) Explain the relative rates of hydrolysis in terms of the relative strength of the carbon-halogen bonds. (c) Write equations for the two reactions that occur in test tube B.

   (Adapted from CCEA June 2006)

4. Bromochlorodifluoromethane, $BrClF_2C$ is used to extinguish fires. Suggest which carbon-halogen bond is most readily hydrolysed when $BrClF_2C$ is heated with aqueous sodium hydroxide.

   (CCEA June 2007)

5. When added to water t-butyl chloride, $(CH_3)_3CCl$ undergoes immediate hydrolysis to produce the corresponding alcohol. Suggest why the hydrolysis of n-butyl chloride, $CH_3CH_2CH_2CH_2Cl$ is significantly slower than the hydrolysis of t-butyl chloride.

(CCEA June 2011)

Before moving to the next section, check that you are able to:

- Describe how to measure the relative rates of hydrolysis for halogenoalkanes.
- Explain how the rate of hydrolysis for a halogenoalkane is affected by the strength of the carbon-halogen bond in the molecule.

Elimination Reactions

Refluxing 2-bromopropane with a solution of potassium hydroxide dissolved in ethanol results in the elimination of hydrogen bromide (HBr) from the molecule.

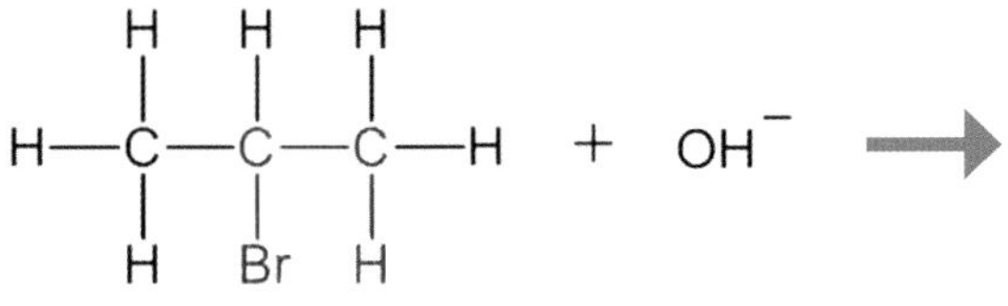

$$CH_3CHBrCH_3 + OH^- \rightarrow CH_3CH{=}CH_2 + Br^- + H_2O$$

2-bromopropane $CH_3CHBrCH_3$

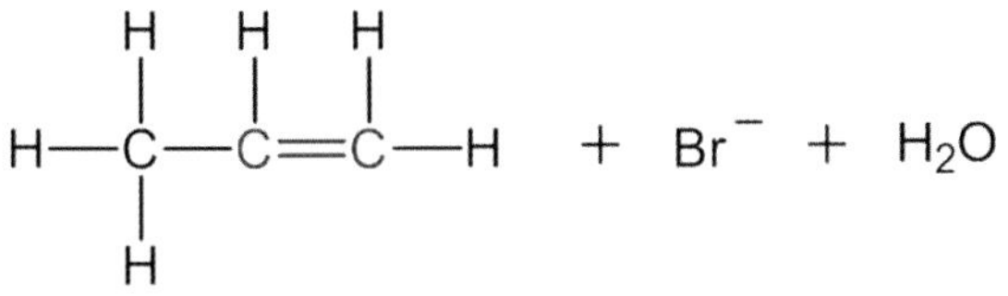

propene $CH_3CH{=}CH_2$

The solution of potassium hydroxide in ethanol is referred to as ethanolic potassium hydroxide where the term **ethanolic** describes a solution in which ethanol is the solvent.

Reactions of this type are referred to as **elimination reactions** as they involve *the removal of a small molecule from a larger molecule*. In this case a C=C bond forms in the product when a molecule of a hydrogen halide (HX) is eliminated from a halogenoalkane (RX).

OH⁻ in ethanol, Reflux

halogenoalkane (X = halogen) → alkene

Worked Example 2.5vi

Which one of the following is an elimination reaction?

A $CH_2{=}CH_2 + Br_2 \rightarrow CH_2BrCH_2Br$

B $CH_3Br + NaOH \rightarrow CH_3OH + NaBr$

C $CH_3CH_2Br + NaOH \rightarrow CH_2{=}CH_2 + NaBr + H_2O$

D $CH_3CH_2Br + NaCN \rightarrow CH_3CH_2CN + NaBr$

(CCEA June 2002)

Solution

Reaction A is an addition reaction.
Reactions B and D are substitution reactions.
Reaction C involves elimination of HBr to form a C=C bond.
Answer C.

Exercise 2.5J

1. (a) Write an equation for the reaction of 1-bromobutane with sodium hydroxide in alcohol. (b) Name the product of the reaction. (c) State the type of reaction that has occurred.

(Adapted from CCEA June 2007)

2. 2-bromo-2-methylpropane reacts with ethanolic potassium hydroxide. (a) Write an equation for this reaction. (b) Name the organic product. (c) Name the type of reaction.

(CCEA January 2013)

3. The following experiment shows the reaction of bromoethane with ethanolic potassium hydroxide. (a) Write the equation for the reaction. (b) Name the gas collected in the jar. (c) Describe a chemical test to confirm the identity of the gas collected. (d) Explain how the gas collected proves that an *elimination* reaction has taken place.

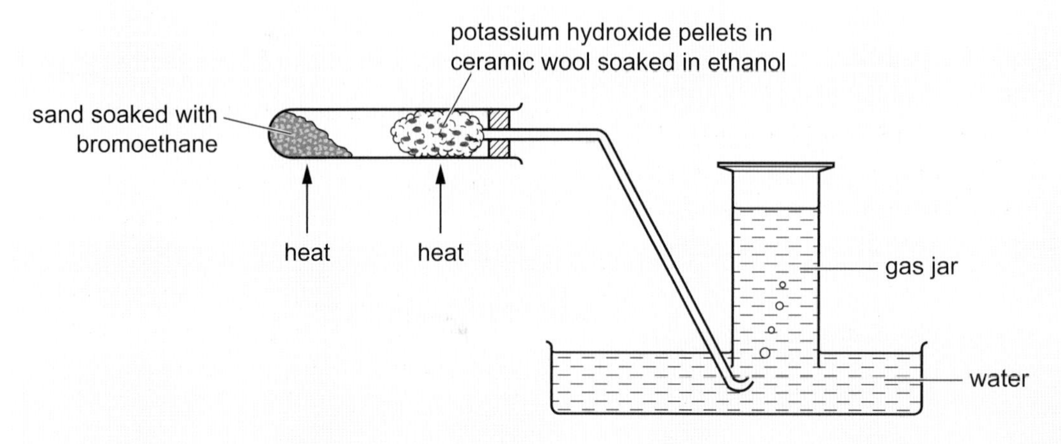

(Adapted from CCEA January 2012)

4. State the type of reaction that occurs when 1,5-dichloropentane reacts with sodium hydroxide in ethanol to produce penta-1,4-diene.

   $2NaOH + Cl(CH_2)_5Cl \rightarrow CH_2{=}CHCH_2CH{=}CH_2 + 2NaCl + 2H_2O$

(Adapted from CCEA June 2006)

An elimination reaction will often produce more than one product. For example, refluxing 2-bromobutane with ethanolic potassium hydroxide produces a mixture of but-1-ene and but-2-ene. Both isomers are produced by the elimination of hydrogen bromide.

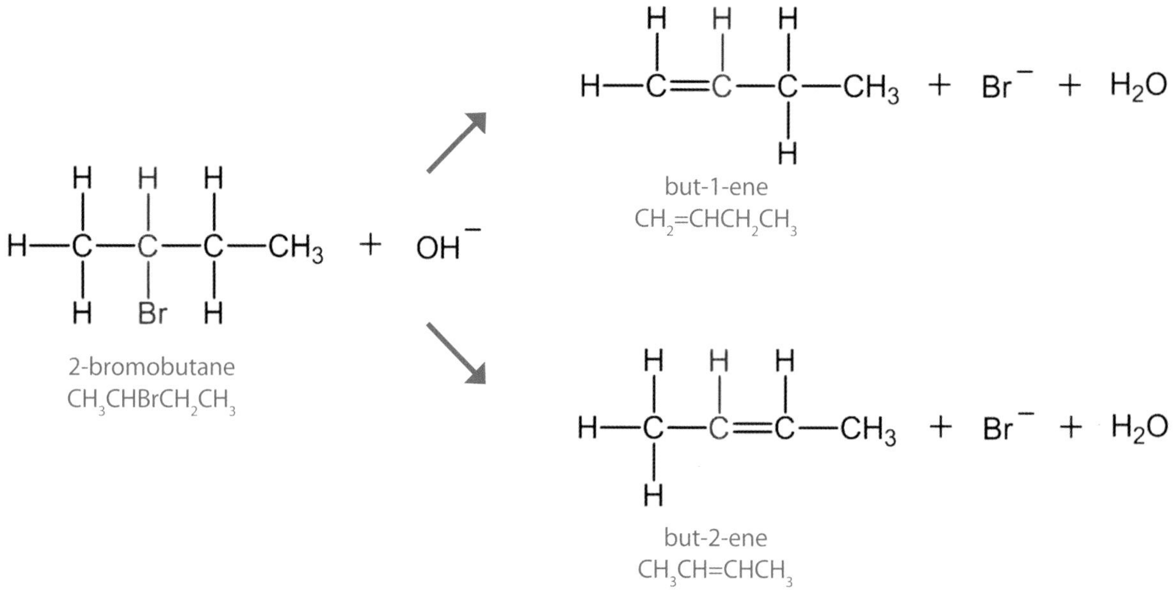

$$H{-}CH_2{-}CHBr{-}CH_2{-}CH_3 + OH^- \nearrow H_2C{=}CH{-}CH_2{-}CH_3 + Br^- + H_2O$$

2-bromobutane $CH_3CHBrCH_2CH_3$ → but-1-ene $CH_2{=}CHCH_2CH_3$

$$H{-}CH_2{-}CHBr{-}CH_2{-}CH_3 + OH^- \searrow H{-}CH_2{-}CH{=}CH{-}CH_3 + Br^- + H_2O$$

2-bromobutane $CH_3CHBrCH_2CH_3$ → but-2-ene $CH_3CH{=}CHCH_3$

Worked Example 2.5vii

(a) Write the equation for the reaction that occurs when 2-chlorobutane reacts with ethanolic potassium hydroxide. (b) Explain the term *ethanolic*. (c) Draw the structure of all possible products and write the systematic name for each product.

(Adapted from CCEA June 2005)

Solution

(a) $C_4H_9Cl + KOH \rightarrow C_4H_8 + KCl + H_2O$

(b) The term ethanolic describes a solution in which the solvent is ethanol.

(c)

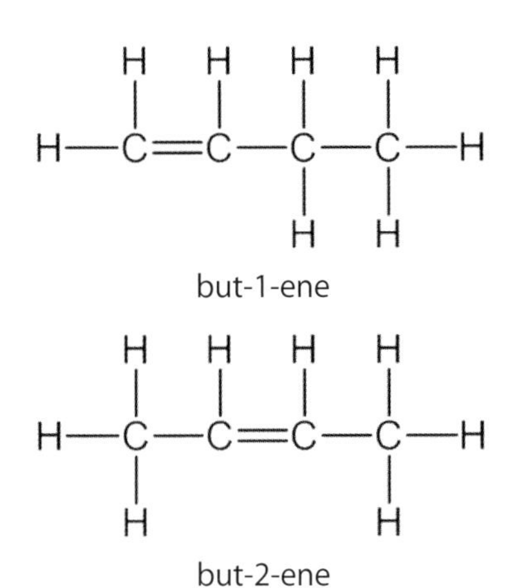

Exercise 2.5K

1. Draw the structure of the product formed when 1-bromo-2-methylbutane is heated with (a) aqueous sodium hydroxide and (b) alcoholic sodium hydroxide.

   (CCEA January 2008)

2. Which one of the following alkenes may be formed when 3-bromo-2-methylpentane reacts with ethanolic potassium hydroxide?

   A 1,1-dimethylbut-1-ene
   B 2-methylpent-3-ene
   C 4-methylpent-2-ene
   D 4-methylpent-3-ene

   (CCEA June 2013)

Before moving to the next section, check that you are able to:

- Explain what is meant by the term *elimination reaction*.
- Recall the conditions for the elimination of a hydrogen halide from the corresponding halogenoalkane.
- Recognise that elimination may result in more than one product and deduce the structure of the products formed by an elimination reaction.

2.6 Alcohols

CONNECTIONS

- The 'sugar alcohols' sorbitol, xylitol and mannitol are found in many types of plants and are used as artificial sweeteners in many foods and drinks.
- In many countries a blend of gasoline and alcohol known as E10 or 'Gasohol' is used to fuel motor vehicles.

Structure and Properties

In this section we are learning to:

- Account for the properties of alcohols in terms of the bonding within alcohols.
- Classify the structures of alcohols as primary, secondary or tertiary.
- Use systematic (IUPAC) rules to name alcohols.

An **alcohol** is *a compound that contains a hydroxyl (-OH) functional group*. The hydroxyl (-OH) group generates a polar region within the molecule as shown in Figure 1, and determines the reactions and properties of the alcohol.

Factors Affecting Solubility

Smaller alcohols such as methanol, CH_3OH and ethanol, C_2H_5OH experience a combination of van der Waals attraction and hydrogen bonding between molecules. The ability of alcohols to form hydrogen bonds results in smaller alcohols being miscible with water in all proportions. The formation of hydrogen bonds between molecules in methanol, CH_3OH and a mixture of methanol and water is illustrated in Figure 2. As the alcohols become larger the van der Waals attraction dominates to the extent that some isomers of hexanol, $C_6H_{13}OH$ are immiscible in water.

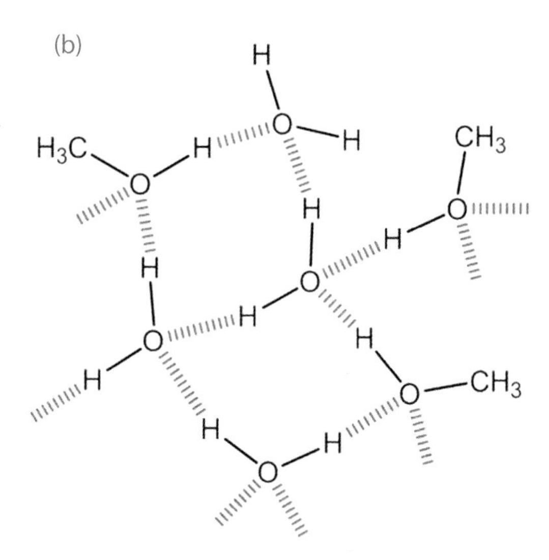

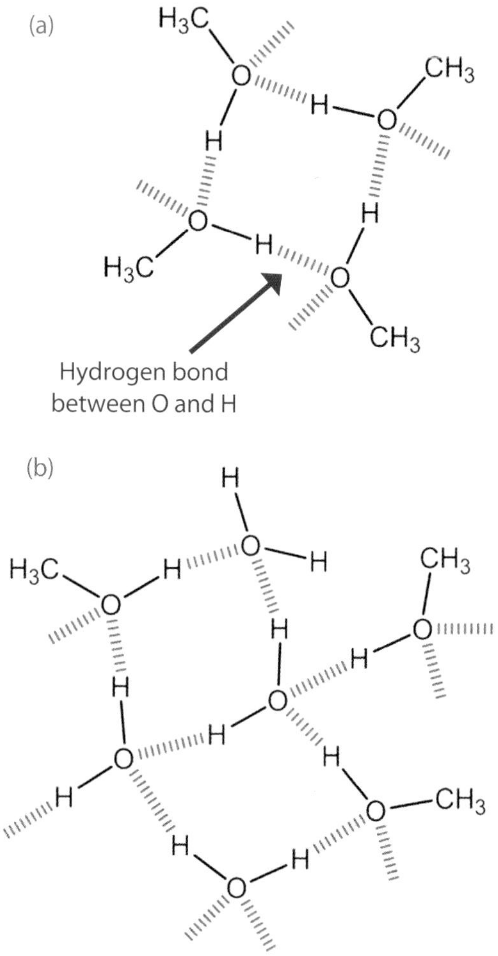

Figure 2: The formation of hydrogen bonds between neighbouring molecules in (a) methanol, CH_3OH and (b) a mixture of methanol and water.

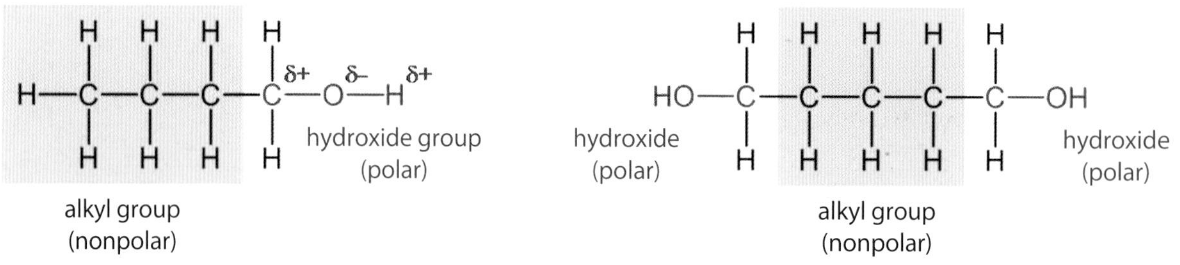

δ+ δ– δ+
hydroxide group (polar)
alkyl group (nonpolar)
(a)

hydroxide (polar)
hydroxide (polar)
alkyl group (nonpolar)
(b)

Figure 1: Polar and nonpolar regions within (a) butan-1-ol and (b) pentan-1,5-diol.

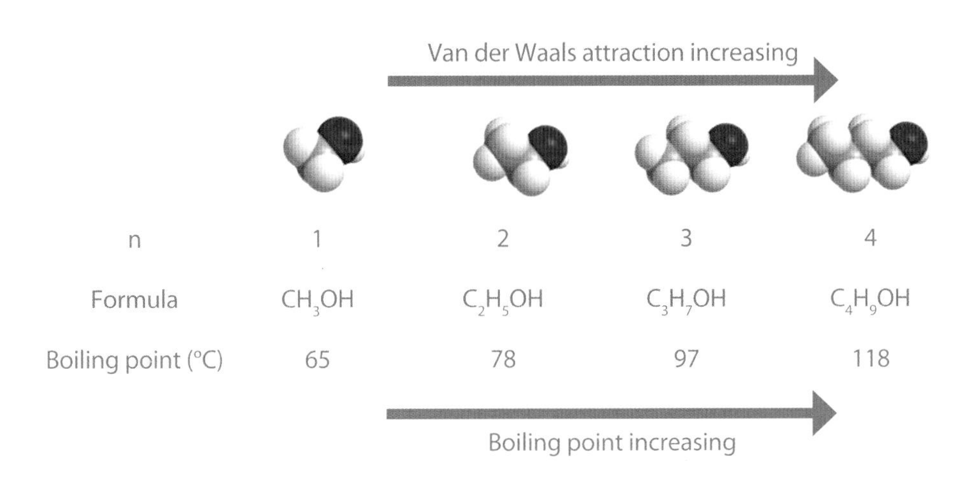

Figure 3: The effect of van der Waals attraction on the boiling points of the straight-chain alcohols, $C_nH_{2n+1}OH$.

Exercise 2.6A

1. (a) Draw the structure of ethanol, C_2H_5OH showing all the bonds present. (b) Explain why ethanol is soluble in water.

   (Adapted from CCEA June 2004)

2. The attraction between adjacent molecules in ethanol is due to

   A covalent bonds only.

   B hydrogen bonds only.

   C hydrogen bonds and van der Waals forces.

   D van der Waals forces only.

   (CCEA June 2011)

3. Explain why ethanol is soluble in water but ethene is immiscible with water.

   (Adapted from CCEA June 2003)

4. Methylated spirits contains ethanol (boiling point 78 °C), methanol (boiling point 64 °C) and water. (a) Explain why ethanol, methanol and water are miscible. (b) Suggest how the components of methylated spirits could be separated.

   (CCEA June 2007)

Before moving to the next section, check that you are able to:

- Describe the nature of the attraction between molecules in smaller alcohols.
- Explain why alcohols become less soluble in polar solvents as they get larger.

Factors Affecting Boiling Point

Alcohols containing a single hydroxyl group form a homologous series with the general formula $C_nH_{2n+1}OH$. The formulas for the members of the series are obtained by inserting n = 1, 2, ... in the general formula. The sequence of straight-chain alcohols in Figure 3 demonstrates that the boiling points of the alcohols increase as the molar mass of the compound increases. The increase in boiling point is due to an increase in the van der Waals attraction between neighbouring molecules as the number of electrons in the molecule increases.

As with alkanes, a straight-chain alcohol will have a higher boiling point than a more branched isomer. For example, the alcohols in Figure 4 are structural

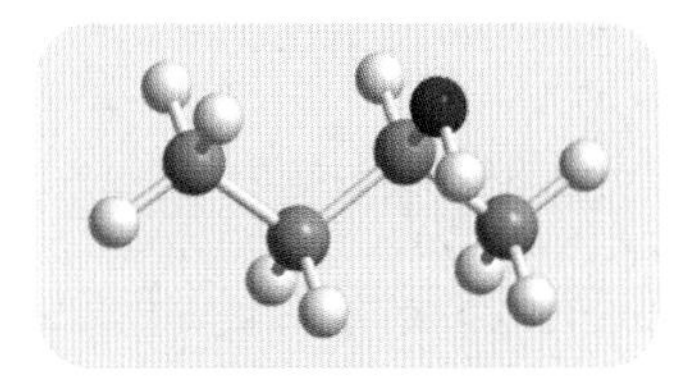

Butan-2-ol, C_4H_9OH
Boiling point 99 °C

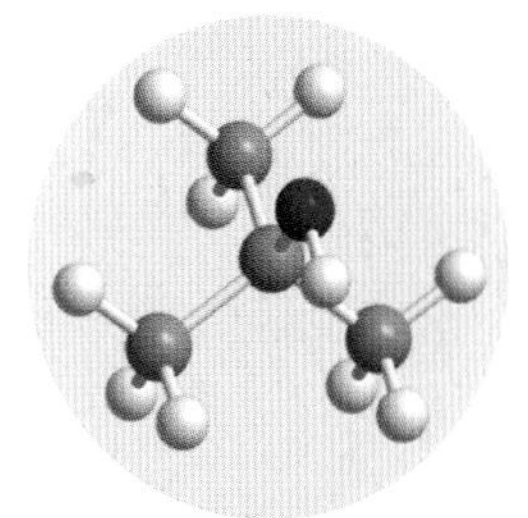

2-methylpropan-2-ol, C_4H_9OH
Boiling point 83 °C

Figure 4: The effect of van der Waals attraction on the boiling points of the isomers butan-2-ol and 2-methylpropan-2-ol.

```
     H   H   H   H
     |   |   |   |
 H — C — C — C — C — H
     |   |   |   |
     H   H   H   OH
```

Butan-1-ol, C_4H_9OH
Boiling point 118 °C

```
     H   H   H   H
     |   |   |   |
 H — C — C — C — C — H
     |   |   |   |
     H   H   OH  H
```

Butan-2-ol, C_4H_9OH
Boiling point 99 °C

Figure 5: The effect of van der Waals attraction on the boiling points of the isomers butan-1-ol and butan-2-ol.

isomers. The boiling point of 2-methylpropan-2-ol is lower as its branched structure reduces the amount of contact between electrons in neighbouring molecules, which in turn decreases the strength of the van der Waals attraction between molecules.

The butanol isomers in Figure 5 demonstrate that the boiling point of an alcohol is also affected by the location of the hydroxyl group in the molecule. The boiling point of butan-1-ol is higher as it is a primary alcohol making it easier for the hydroxyl group to form hydrogen bonds between molecules that are packed closely together.

The extent to which van der Waals attraction and hydrogen bonding influence boiling point can be assessed by the boiling point comparison in Figure 6.

Comparing the boiling points for the alkanes (n = 1-4) and chloroalkanes (n = 1-4) reveals that the boiling point of chloromethane, CH_3Cl (–24 °C) resembles the boiling point of propane, $CH_3CH_2CH_3$ (–42 °C) and the boiling point of chloroethane, CH_3CH_2Cl (13 °C) resembles the boiling point of butane, $CH_3CH_2CH_2CH_3$ (0 °C). In this way adding a chlorine atom increases the boiling point by an amount that is roughly equivalent to adding two carbon atoms to the compound.

Figure 6 also reveals that the increase in boiling point on adding a bromine atom is greater, and is roughly equivalent to the effect of adding three carbon atoms to the compound. The increase in boiling point on adding a hydroxyl group is even greater, and is equivalent to adding four or more carbon atoms to the compound.

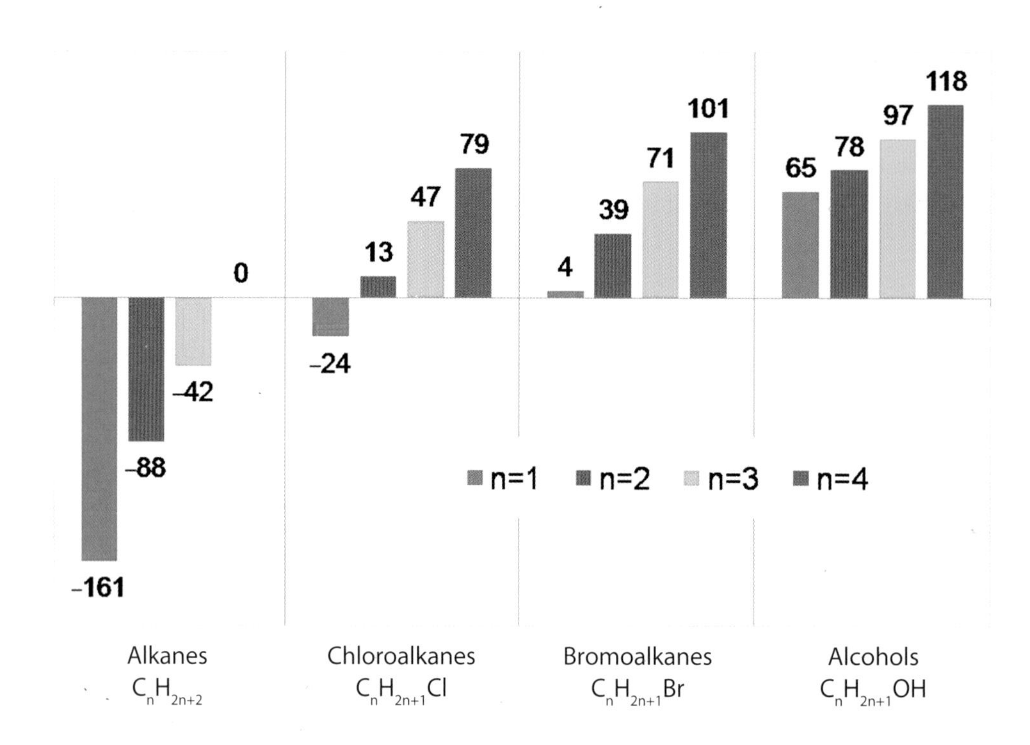

Figure 6: Comparison of the boiling points for the first four straight-chain alkanes, chloroalkanes, bromoalkanes and alcohols.

Exercise 2.6B

1. The diagram shows the formation of a bond between molecule A and molecule B. (a) Name the molecules A and B. (b) State the type of bond formed between A and B. (c) State one physical property of substance A affected by this type of bond.

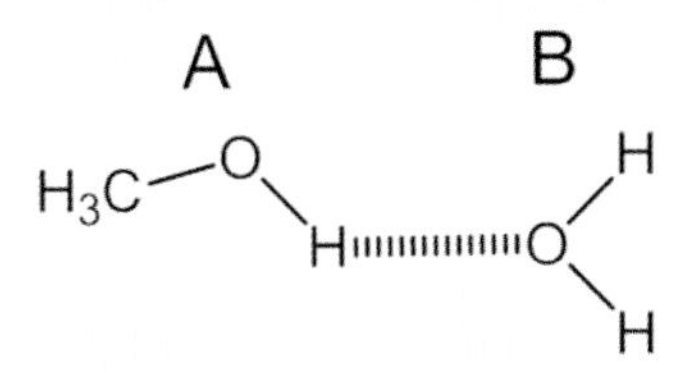

(Adapted from CCEA June 2002)

2. The forces of attraction between ethanol molecules are

   A permanent dipole-dipole attractions only.
   B permanent dipole-dipole attractions and hydrogen bonds.
   C hydrogen bonds.
   D hydrogen bonds and van der Waals forces.

(CCEA June 2012)

3. Which one of the following has the highest boiling point?

   A CH_3OH B CH_3CH_2OH
   C CH_3Cl D $CH_3CH_2CH_3$

(CCEA January 2003)

4. Which one of the following lists the compounds in order of increasing boiling point?

   A $CH_3CH_2CH_3$ CH_3CH_2F CH_3CH_2OH
   B $CH_3CH_2CH_3$ CH_3CH_2OH CH_3CH_2F
   C CH_3CH_2F CH_3CH_2OH $CH_3CH_2CH_3$
   D CH_3CH_2OH CH_3CH_2F $CH_3CH_2CH_3$

(CCEA June 2014)

5. Which one of the following has the highest boiling point?

   A CH_3CH_2Cl
   B $CH_3CH_2CH_2OH$
   C $CH_3CH_2CH_2CH_3$
   D $CH_3CH_2CH_2CH_2CH_3$

(CCEA June 2009)

6. The formula and boiling point for each isomer of butanol, C_4H_9OH is given in the table. Explain why butan-1-ol has the highest boiling point.

Compound	Formula	Boiling point (°C)
butan-1-ol	$CH_3CH_2CH_2CH_2OH$	118
butan-2-ol	$CH_3CH_2CHOHCH_3$	99
2-methylpropan-1-ol	$(CH_3)_2CHCH_2OH$	108
2-methylpropan-2-ol	$(CH_3)_3COH$	83

(CCEA June 2015)

7. The boiling point of ethylene glycol is 197 °C and that of ethanol is 78 °C. Explain this large difference in boiling points.

CH_2OH
|
CH_2OH

ethylene glycol

(CCEA June 2014)

Before moving to the next section, check that you are able to:

- Recall that simple alcohols containing one hydroxyl group form a homologous series, and use the general formula $C_nH_{2n+1}OH$ to generate the formula for each member of the series.
- Explain how the boiling point of an alcohol is determined by the size and shape of the molecule.
- Recall that the boiling point of an alcohol is affected by the location of the hydroxyl group within the molecule.
- Rank the boiling points of alcohols, halogenoalkanes and alkanes.

Factors Affecting Reactivity

The extent to which a hydroxyl group will participate in chemical reactions is determined by the structure of the alcohol. Alcohols can be classified as *primary*, *secondary* or *tertiary* alcohols on the basis of their structure.

The compound butan-1-ol is an example of a **primary alcohol** as *the hydroxyl (-OH) group is bonded to a carbon (C) with one carbon atom (C) attached*. The isomer 2-methylpropan-1-ol is also a primary alcohol.

```
    H   H   H   H
    |   |   |   |
H — C — C — C — C — H
    |   |   |   |
    H   H   H   OH
```

butan-1-ol (*primary*)

$CH_3CH_2CH_2CH_2OH$

```
    H  CH3  H
    |   |   |
H — C — C — C — H
    |   |   |
    H   H   OH
```

2-methylpropan-1-ol (*primary*)

$CH_3CH(CH_3)CH_2OH$

The compounds butan-2-ol and 2-methylpropan-2-ol are also isomers of butan-1-ol. Butan-2-ol is an example of a **secondary alcohol** as *the hydroxyl group is bonded to a carbon (C) with two carbon atoms (C) attached.* In contrast, the isomer 2-methylpropan-2-ol is an example of a **tertiary alcohol** as *the hydroxyl group is bonded to a carbon (C) with three carbon atoms (C) attached.*

```
    H   H   H   H
    |   |   |   |
H — C — C — C — C — H
    |   |   |   |
    H   OH  H   H
```

butan-2-ol (*secondary*)

$CH_3CH(OH)CH_2CH_3$

```
    H  CH3  H
    |   |   |
H — C — C — C — H
    |   |   |
    H   OH  H
```

2-methylpropan-2-ol (*tertiary*)

$CH_3C(CH_3)(OH)CH_3$

Exercise 2.6C

1. (a) Ethylene glycol, $HOCH_2CH_2OH$ contains primary alcohol groups. Explain the term *primary alcohol.* (b) Explain why ethylene glycol is very soluble in water.

(CCEA January 2009)

2. Which one of the following is a secondary alcohol?

A
```
    H
    |
H — C — OH
    |
H — C — OH
    |
    H
```

B
```
          CH3
           |
C2H5 — C — OH
           |
          CH3
```

C
```
        H
        |
H3C  —  C — OH
        |
        H
```

D
```
       CH3
        |
H   —   C — OH
        |
       C2H5
```

(CCEA January 2007)

3. Classify the following alcohols as primary, secondary or tertiary.

(a)
```
H3C — OH
```

(b)
```
       OH
        |
H3C  —  C — CH3
        |
        H
```

(c)
```
        H
        |
H3C  —  C — CH3
        |
       CH2OH
```

(d)
```
       CH3  OH
        |    |
H3C  —  C  — C — CH3
        |    |
       CH3  CH3
```

(Adapted from CCEA January 2006)

Naming Alcohols

The systematic (IUPAC) name for an alcohol is based on the name of the parent (straight-chain) alkane on which its structure is based. For example methanol, CH_3OH is named using the prefix *methan* as its structure is based on methane, CH_4. Similarly ethanol, CH_3CH_2OH is named using the prefix *ethan* as its structure is based on ethane, CH_3CH_3.

In larger alcohols such as butanol, C_4H_9OH the hydroxyl group generates structural isomers and it becomes necessary to describe the location of the hydroxyl group when naming the compound. For example, the suffix *1-ol* or *2-ol* is added to locate the hydroxyl group in butan-1-ol and butan-2-ol, and is *based on a numbering of the carbon atoms that generates the lowest numbers in the suffix.*

The systematic names for structural isomers of straight-chain alcohols with up to six carbon atoms are summarised in Table 1.

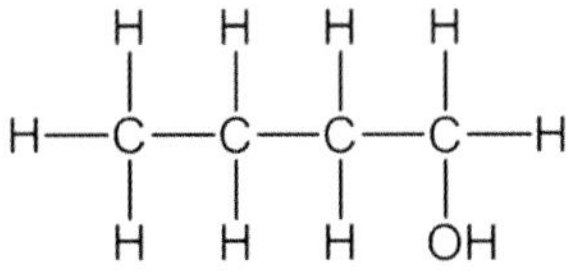

butan-1-ol
$CH_3CH_2CH_2CH_2OH$

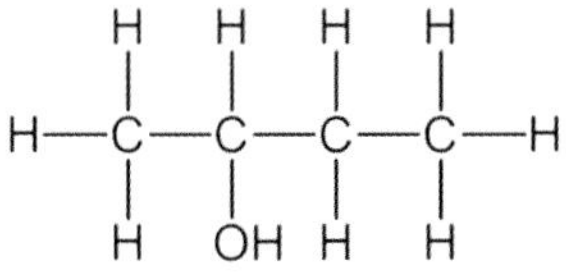

butan-2-ol
$CH_3CH(OH)CH_2CH_3$

n	Parent alcohol	Structural isomers	Skeletal formula
1	CH_3OH **methan**ol	CH_3OH	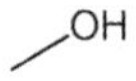 OH
2	C_2H_5OH **ethan**ol	CH_3CH_2OH	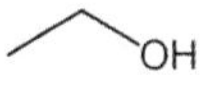 OH
3	C_3H_7OH **propan**ol	$CH_3CH_2CH_2OH$ propan-1-ol $CH_3CH(OH)CH_3$ propan-2-ol	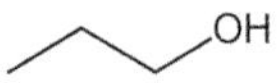 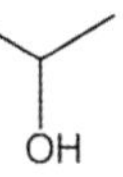 OH OH
4	C_4H_9OH **butan**ol	$CH_3CH_2CH_2CH_2OH$ butan-1-ol $CH_3CH_2CH(OH)CH_3$ butan-2-ol	 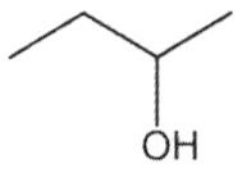 OH OH
5	$C_5H_{11}OH$ **pentan**ol	$CH_3CH_2CH_2CH_2CH_2OH$ pentan-1-ol $CH_3CH_2CH_2CH(OH)CH_3$ pentan-2-ol $CH_3CH_2CH(OH)CH_2CH_3$ pentan-3-ol	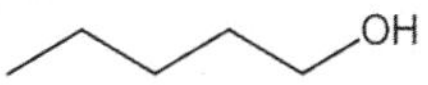 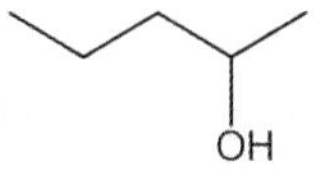 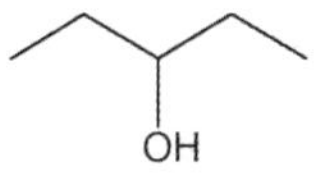 OH OH OH
6	$C_6H_{13}OH$ **hexan**ol	$CH_3CH_2CH_2CH_2CH_2CH_2OH$ hexan-1-ol $CH_3CH_2CH_2CH_2CH(OH)CH_3$ hexan-2-ol $CH_3CH_2CH_2CH(OH)CH_2CH_3$ hexan-3-ol	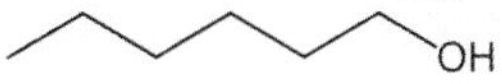 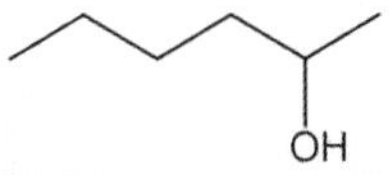 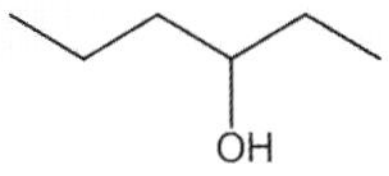 OH OH OH

Table 1: Systematic names for the structural isomers of straight-chain alcohols (ROH) with up to six carbon atoms.

The alcohols 2-methylpropan-1-ol and 2-methylpropan-2-ol are isomers of butan-1-ol and are named using the prefix *propan* as their structure is based on propane, $CH_3CH_2CH_3$. The location of the hydroxyl group is described by the suffix *1-ol* or *2-ol* and is again based on a numbering of the carbon atoms that produces the lowest numbers. The prefix *2-methyl* is then used to locate the methyl group, and *is based on the numbering of carbon atoms used to locate the hydroxyl group.*

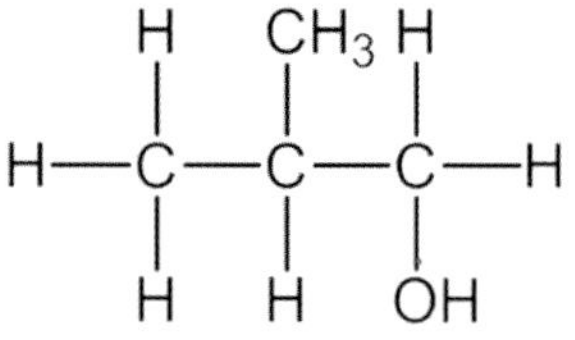

2-methylpropan-1-ol

$CH_3CH(CH_3)CH_2OH$

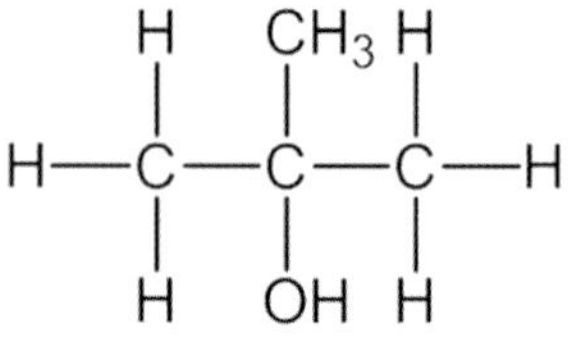

2-methylpropan-2-ol

$CH_3C(CH_3)(OH)CH_3$

The location of two or more functional groups can be described by combining prefixes as in the isomers 1-chloro-3-methylpentan-3-ol and 3-chloro-4-methylpentan-2-ol. Both are named using the prefix *pentan* as their structure is based on the chain of five carbon atoms in pentane, $CH_3CH_2CH_2CH_2CH_3$. *The prefixes used to locate the chlorine and methyl groups are listed in alphabetical order, and are based on the numbering of carbon atoms used to locate the hydroxyl group.*

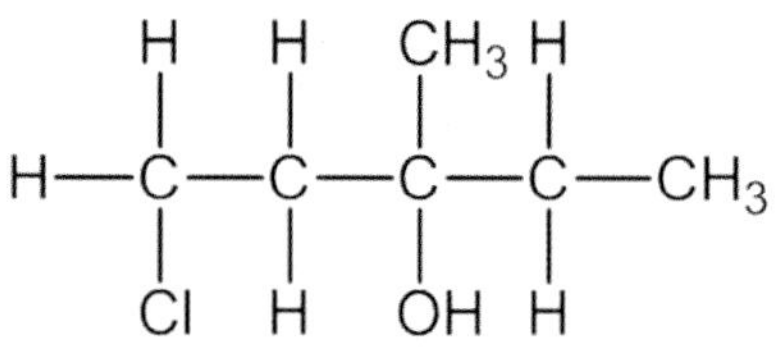

1-chloro-3-methylpentan-3-ol

$CH_2ClCH_2C(CH_3)(OH)CH_2CH_3$

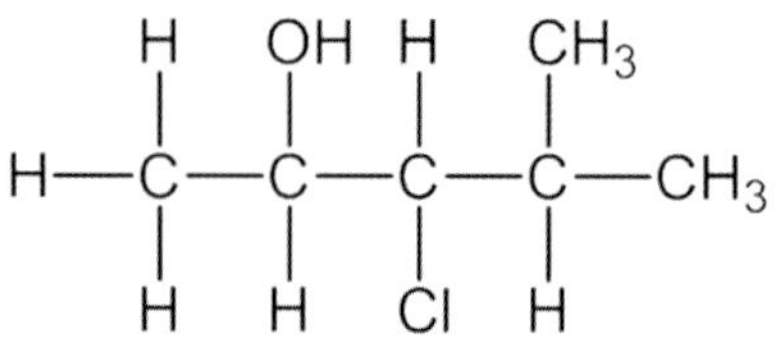

3-chloro-4-methylpentan-2-ol

$CH_3CH(OH)CHClCH(CH_3)_2$

A compound containing two hydroxyl groups is known as a **diol** and is named using the suffix *diol*. When naming a diol the name of the parent alkane is followed by a suffix such as *1,2-diol* or *1,3-diol* to locate the hydroxyl groups. Here again the carbon atoms are numbered to produce the lowest numbers in the suffix. Prefixes such as *2-methyl* and *1-bromo-3-methyl* are then added to locate additional groups on the carbon chain. As in other alcohols *the prefixes are listed in alphabetical order and are based on the numbering used to locate the hydroxyl groups.*

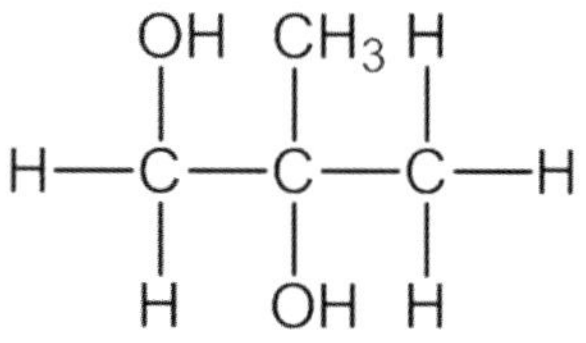

2-methylpropane-1,2-diol

$CH_2(OH)C(OH)(CH_3)_2$

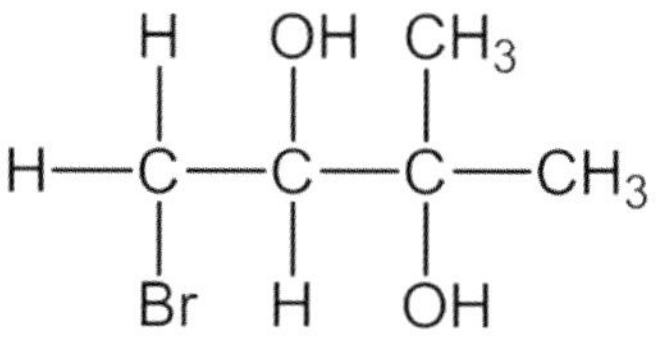

1-bromo-3-methylbutane-2,3-diol

$CH_2BrCH(OH)C(CH_3)_2(OH)$

Similarly, a compound containing three hydroxyl groups is known as a **triol** and is named using the suffix *triol*. When naming a triol the name of the parent alkane is followed by a suffix such as *1,2,3-triol* to locate the hydroxyl groups.

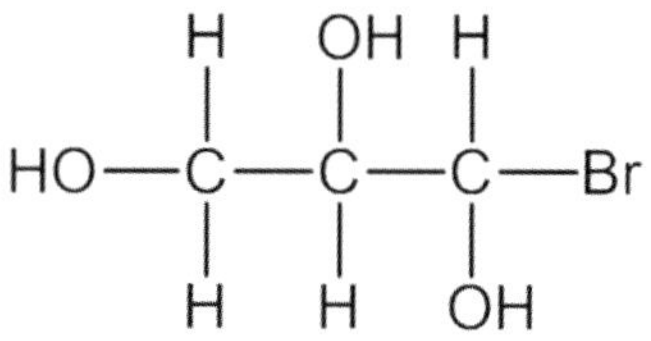

1-bromopropane-1,2,3-triol

$CH_2(OH)CH(OH)CH(OH)Br$

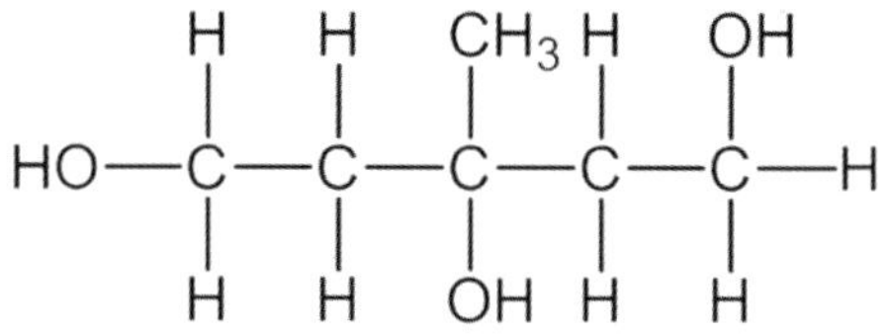

3-methylpentane-1,3,5-triol

$CH_2(OH)CH_2C(CH_3)(OH)CH_2CH_2OH$

Rules for naming alcohols (including diols and triols):

- The name of an alcohol is based on the name of the parent (straight-chain) alkane on which its structure is based.
- The location of each hydroxyl group is described by numbering the carbon atoms in a way that produces the smallest numbers.
- Additional prefixes are then added to describe the type and location of each functional group on the carbon chain. The prefixes are written in alphabetical order and are based on the numbering used to locate the hydroxyl groups.

Exercise 2.6D

1. There are four alcohols with the formula, C_4H_9OH. Copy and complete the following table by drawing the missing structures, writing the name for each structure, and classifying each structure as primary, secondary or tertiary.

Structure	Name	Classification
H H H H H—C—C—C—C—H H H H OH	butan-1-ol	primary
H CH_3 H H—C—C—C—H H OH H	2-methyl propan-2-ol	
	2-methyl propan-1-ol	

(CCEA January 2008)

2. Which one of the following is the name of the alcohol below?

```
      H    H    H    CH3
      |    |    |    |
 H—   C —  C —  C —  C —H
      |    |    |    |
      CH3  H    OH   H
```

A 1,4-dimethylbutan-2-ol
B 1,4-dimethylbutan-3-ol
C Hexan-3-ol
D Hexan-4-ol

(CCEA June 2013)

3. Which one of the following is a tertiary alcohol?

A 2-methylbutan-1-ol
B 2-methylbutan-2-ol
C 3-methylbutan-1-ol
D 3-methylbutan-2-ol

(CCEA January 2009)

4. Which one of the following is a tertiary alcohol?

A pentan-2-ol
B pentan-3-ol
C 2-methylpentan-1-ol
D 2-methylpentan-2-ol

(CCEA January 2013)

5. (a) Write the systematic name for ethylene glycol. (b) Write the empirical formula for ethylene glycol.

```
 CH2OH
 |
 CH2OH
```

ethylene glycol

(CCEA June 2014)

6. Compounds A and B are triols. (a) Write the systematic names for compounds A and B. (b) Explain what is meant by the term *triol.*

```
        H    OH
        |    |
 HO —   C —  C — H
        |    |
        OH   H
```

A

```
      H    OH   H
      |    |    |
 H —  C —  C —  C — OH
      |    |    |
      OH   H    H
```

B

7. (a) Write the systematic names for compounds A and B. (b) Explain why compound A is referred to as a *diol*.

```
     H   H   CH3
     |   |   |
  H—C———C———C—CH3
     |   |   |
     OH  H   OH
         A

     H   H   CH2OH
     |   |   |
  H—C———C———C—CH3
     |   |   |
     OH  H   OH
         B
```

Before moving to the next section, check that you are able to:

- Write structural and condensed formulas for alcohols.
- Classify the structures of alcohols as primary, secondary or tertiary.
- Deduce systematic names for alcohols, diols and triols.

Combustion of Alcohols

In this section we are learning to:

- Describe the complete and incomplete combustion of alcohols.
- Explain the use of ethanol as an alternative fuel.

In many countries ethanol is manufactured on an industrial scale by the fermentation of sugars obtained from sugar cane or corn. Much of the ethanol produced by fermentation is then added to gasoline to produce blended fuels that can be burnt in motor vehicles. A mixture of ethanol and gasoline that is 10% ethanol by volume is referred to as E10 and is commonly known as 'Gasohol'. The engines in modern cars can burn Gasohol without modification.

Ethanol contains less carbon per gram than the hydrocarbons in gasoline. As a result, burning ethanol produces less carbon dioxide for every mole of water formed.

Complete combustion of octane (gasoline):
$2C_8H_{18(l)} + 25O_{2(g)} \rightarrow 16CO_{2(g)} + 18H_2O_{(l)}$

Complete combustion of ethanol:
$2C_2H_5OH_{(l)} + 6O_{2(g)} \rightarrow 4CO_{2(g)} + 6H_2O_{(l)}$

The amount of carbon in a fuel is reflected in the colour of the flame when the fuel burns. Fuels with a relatively high carbon content such as the hydrocarbons in gasoline burn with a 'smoky' yellow flame. A smoky flame contains soot (carbon) from the incomplete combustion of the fuel. In contrast, fuels with a relatively low carbon content such as ethanol burn with a cleaner 'blue flame'. The flames produced by the combustion of hexane, C_6H_{14} and ethanol, C_2H_5OH are compared in Figure 7.

Incomplete combustion of octane (gasoline):
$2C_8H_{18(l)} + 17O_{2(g)} \rightarrow 16CO_{(g)} + 18H_2O_{(l)}$

Incomplete combustion of ethanol:
$2C_2H_5OH_{(l)} + 4O_{2(g)} \rightarrow 4CO_{(g)} + 6H_2O_{(l)}$

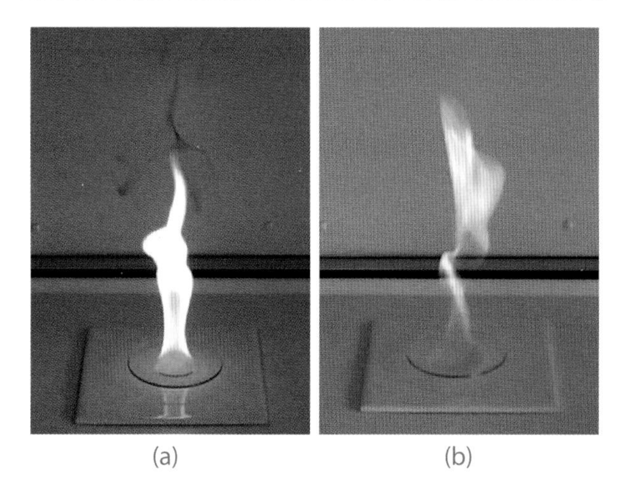

Figure 7: Flames produced by the combustion of (a) hexane, C_6H_{14} and (b) ethanol, C_2H_5OH.

The combustion of gasoline releases large quantities of carbon dioxide into the atmosphere that contribute to global warming. Gasoline also contains a significant amount of sulfur that is converted to sulfur dioxide when the gasoline is burnt. The sulfur dioxide then dissolves in rain droplets to form acid rain that damages the environment. In contrast, the ethanol used in blended fuels has a lower carbon content than gasoline and does not contain sulfur. As a result, blended fuels release less carbon dioxide and less sulfur dioxide into the atmosphere when they burn. Blended fuels are also better for the environment as the supply of ethanol can be sustained by growing more crops such as sugar cane and corn.

Worked Example 2.6i

(a) Write an equation for the complete combustion of ethanol. (b) Write an equation for the incomplete combustion of ethanol. (c) Explain the environmental advantage of using ethanol instead of petrol in cars.

(CCEA June 2006)

Solution

(a) $C_2H_5OH + 3O_2 \rightarrow 2CO_2 + 3H_2O$

(b) $C_2H_5OH + 2O_2 \rightarrow 2CO + 3H_2O$

(c) Ethanol has a lower carbon content than the hydrocarbons in petrol and does not contain sulfur. As a result, ethanol produces less carbon dioxide and less sulfur dioxide when it burns. Reducing carbon dioxide emissions reduces damage to the environment from global warming and a reduction in sulfur dioxide emissions reduces damage to the environment caused by acid rain. Ethanol is also better for the environment as its use can be sustained by growing more crops such as sugar cane and corn.

Exercise 2.6E

1. (a) Write an equation for the complete combustion of methanol. (b) Write an equation for the incomplete combustion of methanol. (c) Explain the environmental advantage of using methanol instead of petrol in cars.

(Adapted from CCEA January 2006)

Before moving to the next section, check that you are able to:

- Write equations for the complete and incomplete combustion of alcohols.
- Account for the nature of the flames produced by the combustion of hydrocarbons and alcohols.
- Discuss the advantages and disadvantages of using blended fuels.

Preparation

In this section we are learning to:

- Describe the formation of an alcohol by the alkaline hydrolysis of a halogenoalkane.

An alcohol can be prepared by the alkaline hydrolysis of the corresponding halogenoalkane. For example, butan-1-ol can be prepared by refluxing 1-bromobutane with aqueous sodium hydroxide.

$$\underset{\text{1-bromobutane}}{CH_3CH_2CH_2CH_2Br} + OH^- \rightarrow \underset{\text{butan-1-ol}}{CH_3CH_2CH_2CH_2OH} + Br^-$$

The reaction involves the substitution of bromine by hydroxide ion. If the structure of the bromoalkane is represented by the condensed formula RBr the equation can be written in a more general form that describes the hydrolysis of any bromoalkane.

Alkaline hydrolysis of a bromoalkane:

$$RBr + OH^- \rightarrow ROH + Br^-$$

Alcohols can also be prepared by the alkaline hydrolysis of fluoroalkanes (RF), chloroalkanes (RCl) and iodoalkanes (RI). For example, butan-2-ol can be prepared by refluxing 2-chlorobutane with aqueous sodium hydroxide.

$$\underset{\text{2-chlorobutane}}{CH_3CHClCH_2CH_3} + OH^- \rightarrow \underset{\text{butan-2-ol}}{CH_3CH(OH)CH_2CH_3} + Cl^-$$

Alkaline hydrolysis of a chloroalkane:

$$RCl + OH^- \rightarrow ROH + Cl^-$$

Exercise 2.6F

1. (a) Write the equation for the reaction that occurs when 2-chloropropane is refluxed with dilute potassium hydroxide. (b) Draw and name the product.
2. (a) State the type of reaction that occurs when an alcohol is formed by refluxing a halogenoalkane with aqueous alkali. (b) Explain why the mixture is refluxed.
3. (a) Write the equation for the reaction that occurs when 1,2-dichloroethane is refluxed with dilute sodium hydroxide. (b) Draw and name the product.

Before moving to the next section, check that you are able to:

- Recall that an alcohol can be prepared by the alkaline hydrolysis of the corresponding halogenoalkane.
- Deduce the structure of the alcohol formed by the alkaline hydrolysis of a halogenoalkane and write an equation for the reaction.

Reactions

In this section we are learning to:

- Describe the reactions of alcohols with sodium, hydrogen bromide and phosphorus(V) chloride.

Reaction with Sodium

The O-H bond in a hydroxyl (-OH) group behaves in the same way as the O-H bonds in water. As a result, reactive metals such as sodium will react with alcohols in the same way they react with water.

Reaction of sodium with water:

$2H_2O_{(l)} + 2Na_{(s)} \rightarrow 2NaOH_{(aq)} + H_{2(g)}$

Reaction of sodium with ethanol:

$2CH_3CH_2OH_{(l)} + 2Na_{(s)}$

$\rightarrow 2CH_3CH_2ONa_{(alc)} + H_{2(g)}$

sodium ethoxide

Observations: The solid disappears, a colourless solution is formed, bubbles of colourless gas are produced, and the reaction mixture warms up as it 'fizzes'.

The resulting 'ethanolic solution' containing sodium ethoxide dissolved in ethanol is described using the state symbol (alc) to indicate that the solvent is an alcohol.

The sodium ethoxide formed in the reaction is a strong base and reacts with water to form a colourless solution that is strongly alkaline.

$CH_3CH_2ONa_{(alc)} + H_2O_{(l)}$

$\rightarrow CH_3CH_2OH_{(aq)} + NaOH_{(aq)}$

Exercise 2.6G

1. Ethanol is a neutral liquid. When sodium is added to ethanol it sinks and reacts to form a strong base. Describe what is observed during this reaction.

   (Adapted from CCEA June 2005)

2. (a) Write an equation for the reaction between propan-1-ol and sodium. (b) Name the salt formed in the reaction.

3. Sodium *tert*-butoxide is formed when sodium reacts with the alcohol shown below. (a) Write an equation for the reaction. (b) Name the alcohol. (c) Suggest the meaning of the prefix *tert*-.

```
        CH3
        |
 H3C — C — CH3
        |
        OH
```

Halogenation of Alcohols

The term **halogenation** refers to a reaction in which one or more halogen atoms are added to a compound. A number of different reagents can be used to halogenate alcohols.

Reaction with Hydrogen Bromide

Refluxing an alcohol (ROH) with hydrogen bromide, HBr produces the corresponding bromoalkane (RBr). The reaction is an example of halogenation as the hydroxyl (-OH) group in the alcohol is replaced by a bromine atom.

$ROH + HBr \rightarrow RBr + H_2O$

The hydrogen bromide is prepared *in situ* by reacting sodium bromide with concentrated sulfuric acid.

$NaBr + H_2SO_4 \rightarrow NaHSO_4 + HBr$

For example, 2-bromobutane is produced when butan-2-ol is refluxed with an excess of sodium bromide and concentrated sulfuric acid for an extended period.

$CH_3CH_2CH(OH)CH_3 + HBr$

butan-2-ol

$\rightarrow CH_3CH_2CHBrCH_3 + H_2O$

2-bromobutane

Reaction with Phosphorus(V) Chloride

Phosphorus(V) chloride, PCl_5 is a white solid that reacts vigorously with alcohols (ROH) to produce the corresponding chloroalkane (RCl).

Chlorination of an alcohol by PCl_5:

$$ROH_{(l)} + PCl_{5(s)} \rightarrow RCl_{(l)} + HCl_{(g)} + POCl_{3(l)}$$

For example, when phosphorus(V) chloride is added to propan-1-ol it reacts to form 1-chloropropane.

$CH_3CH_2CH_2OH_{(l)} + PCl_{5(s)}$
propan-1-ol

$\rightarrow CH_3CH_2CH_2Cl_{(l)} + HCl_{(g)} + POCl_{3(l)}$
1-chloropropane

Observations: The solid disappears, the reaction mixture warms up, and steamy fumes of a pungent smelling gas are produced as the reaction mixture 'fizzes'.

The reaction is carried out in a fume cupboard as it is highly exothermic and produces a pungent smelling gas (HCl).

The addition of phosphorus(V) chloride can be used to test for the presence of an alcohol as the HCl gas turns damp blue litmus paper red, and produces white fumes when it contacts the vapour produced by a glass rod dipped in concentrated ammonia solution.

Exercise 2.6H

1. Complete the flow scheme by writing the formula for each product.

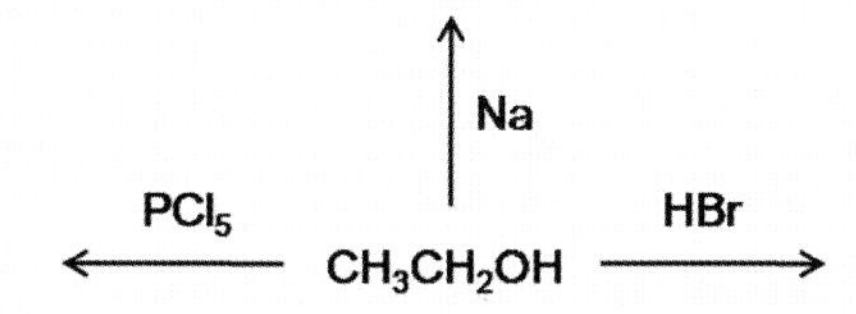

(Adapted from CCEA January 2010)

2. (a) Write the equation for the reaction of butan-2-ol with hydrogen bromide. (b) Calculate the theoretical yield of 2-bromobutane if 11.1 g of butan-2-ol is used. (c) Calculate the mass of 2-bromobutane obtained if the percentage yield for the reaction is 85 %.

(CCEA June 2005)

3. (a) Name the alcohol used to make 1-chlorobutane by reacting an alcohol with phosphorus pentachloride. (b) Write the equation for the reaction. (c) State two observations when phosphorus pentachloride is added to the alcohol.

(CCEA June 2001)

4. Write an equation for the reaction of ethylene glycol with an excess of phosphorus pentachloride.

CH_2OH
|
CH_2OH
ethylene glycol

(CCEA June 2014)

5. Which one of the following reagents could be used to detect the presence of ethanol in Gasohol – a mixture of alkanes, alkenes and ethanol.

A bromine B hydrogen chloride
C sodium D universal indicator

(CCEA June 2009)

Before moving to the next section, check that you are able to:

- Deduce the structure of the product, explain what is observed, and write an equation for the reaction when an alcohol reacts with sodium, hydrogen bromide or phosphorus(V) chloride.

Oxidation of Alcohols

In this section we are learning to:

- Describe the oxidation of alcohols and the use of acidified potassium dichromate to distinguish between primary, secondary and tertiary alcohols.

Oxidation of Primary Alcohols

A solution of potassium dichromate ($K_2Cr_2O_7$) in dilute sulfuric acid is referred to as acidified potassium dichromate and is a strong oxidising agent. When a primary alcohol is refluxed with acidified potassium dichromate the alcohol (RCH_2OH) is oxidised to the corresponding **carboxylic acid** (RCOOH).

$$R{-}CH_2{-}OH + 2[O] \rightarrow R{-}C({=}O){-}OH + H_2O$$

primary alcohol RCH_2OH — carboxylic acid RCOOH

The oxidising agent is a source of oxygen and is represented by the symbol [O] when writing the equation for the reaction. *The equation for the oxidation reveals that the dichromate provides two moles of oxygen atoms (2[O]) for every mole of alcohol oxidised.*

For example, when propan-1-ol is refluxed with acidified potassium dichromate the alcohol is oxidised to propanoic acid, CH_3CH_2COOH

$$CH_3CH_2CH_2OH + 2[O] \rightarrow CH_3CH_2COOH + H_2O$$

propan-1-ol　　　　propanoic acid

There is a change in smell as the orange colour produced by the dichromate ions, $Cr_2O_7^{2-}$ is replaced by the green colour produced by the aqueous chromium(III) ions, $Cr^{3+}(aq)$ that are formed when chromium is reduced from oxidation state +6 in dichromate to +3.

The properties and reactions of propanoic acid are determined by the carboxyl (-COOH) functional group. The systematic names and structures for straight-chain carboxylic acids with up to six carbon atoms are summarised in Table 2.

Exercise 2.6I

1. Ethanol is oxidised when heated with acidified potassium dichromate. Write the equation for the oxidation of ethanol to ethanoic acid using [O] to represent acidified potassium dichromate. *(CCEA June 2005)*

2. The oxidation of ethanol to ethanoic acid is carried out using the following apparatus.

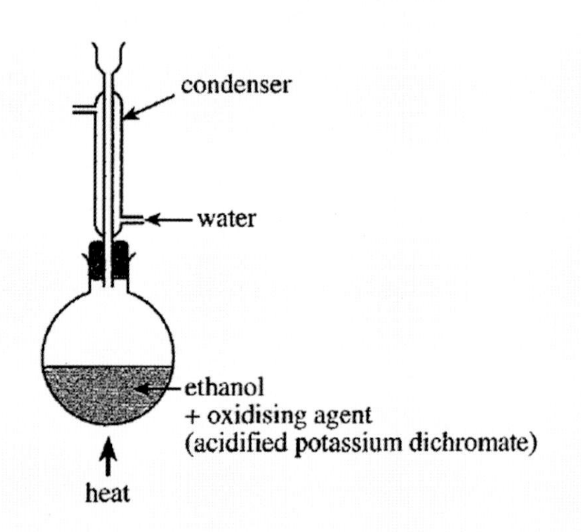

Systematic name and condensed formula	Skeletal formula
HCOOH **methan**oic acid	O, H, OH
CH_3COOH **ethan**oic acid	O, OH or COOH
CH_3CH_2COOH **propan**oic acid	O, OH or COOH
$CH_3(CH_2)_2COOH$ **butan**oic acid	O, OH or COOH
$CH_3(CH_2)_3COOH$ **pentan**oic acid	O, OH or COOH
$CH_3(CH_2)_4COOH$ **hexan**oic acid	O, OH or COOH

Table 2: Systematic names and structures for straight-chain carboxylic acids (RCOOH) with up to six carbon atoms.

The purpose of the condenser is to prevent

A air reacting with the ethanoic acid.
B air reacting with the ethanol.
C the oxidising agent from escaping.
D unreacted ethanol from escaping.

(Adapted from CCEA January 2004)

The oxidation of a primary alcohol to the corresponding carboxylic acid occurs when the alcohol is refluxed with the oxidising agent for an extended period. If the mixture is instead distilled the alcohol is only partially oxidised and the corresponding **aldehyde** (RCHO) is formed.

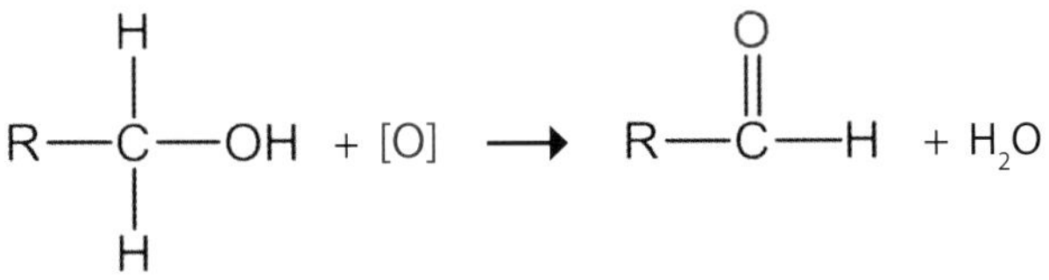

$$R{-}C(H)(H){-}OH + [O] \longrightarrow R{-}C({=}O){-}H + H_2O$$

primary alcohol
RCH_2OH

aldehyde
RCHO

The balanced equation reveals that an aldehyde is formed when dichromate provides one mole of oxygen atoms ([O]) for every mole of alcohol oxidised.

For example, when a mixture of propan-1-ol, $CH_3CH_2CH_2OH$ and acidified potassium dichromate is distilled, the propan-1-ol is oxidised to propanal, CH_3CH_2CHO which distils from the reaction mixture and is collected at its boiling point.

$$CH_3CH_2CH_2OH + [O] \rightarrow CH_3CH_2CHO + H_2O$$

propan-1-ol propanal

Here again, there is a change in smell, and the orange colour produced by the dichromate ions is replaced by the green colour of aqueous chromium(III) ions, $Cr^{3+}(aq)$ as chromium is reduced from oxidation state +6 in dichromate ion to +3.

The properties and reactions of propanal are determined by the aldehyde (-CHO) functional group. The systematic names for aldehydes with up to six carbon atoms are summarised in Table 3.

Systematic name and condensed formula	Skeletal formula
HCHO **methan**al	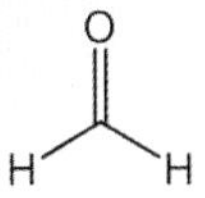 O, H, H
CH_3CHO **ethan**al	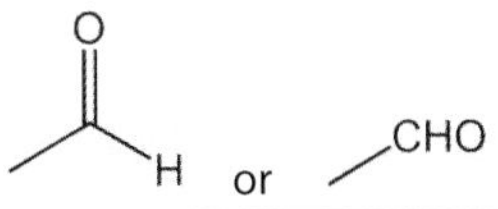 O, H or CHO
CH_3CH_2CHO **propan**al	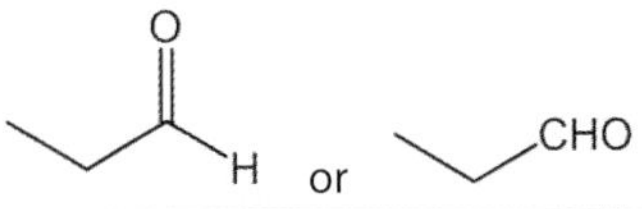 O, H or CHO
$CH_3(CH_2)_2CHO$ **butan**al	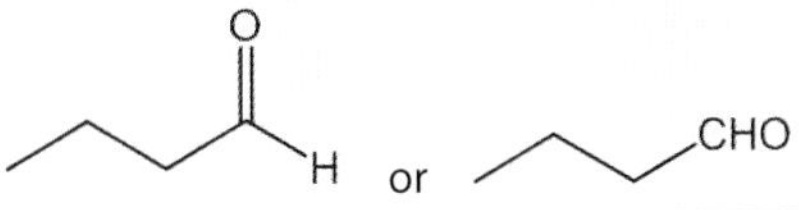 O, H or CHO
$CH_3(CH_2)_3CHO$ **pentan**al	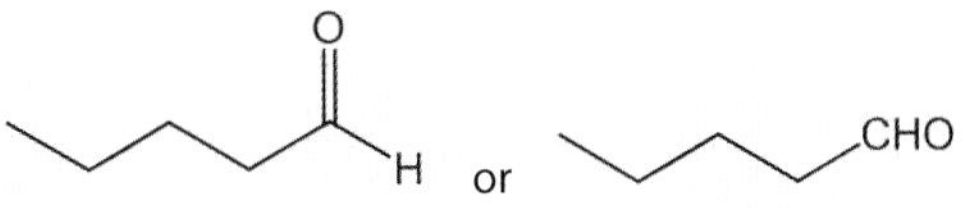 O, H or CHO
$CH_3(CH_2)_4CHO$ **hexan**al	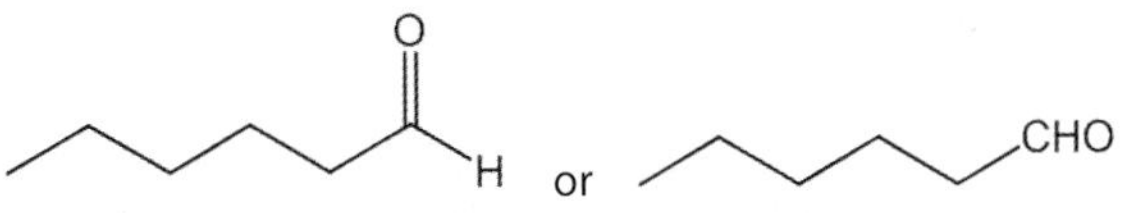 O, H or CHO

Table 3: Systematic names and structures for straight-chain aldehydes (RCHO) with up to six carbon atoms.

Exercise 2.6J

1. Methanol may be oxidised to methanal, HCHO and then to methanoic acid, HCOOH by acidified potassium dichromate. Draw the structures of (a) the aldehyde and (b) the acid showing all bonds present.

(CCEA January 2006)

2. Ethanol can be oxidised to ethanal or ethanoic acid depending on the experimental technique used. (a) Name a suitable oxidising agent. (b) Give the formula of the ion formed by reduction of this oxidising agent. State the experimental techniques required to form (c) ethanal and (d) ethanoic acid.

(CCEA January 2011)

3. Early breathalysers involved the motorist blowing into a tube containing crystals of potassium dichromate mixed with concentrated sulfuric acid. (a) What colour change occurs when acidified potassium dichromate reacts with ethanol? (b) Write an equation for the oxidation of ethanol using [O] to represent the acidified potassium dichromate.

(CCEA January 2014)

4. Ethylene glycol contains primary alcohol groups. When heated under reflux with excess acidified potassium dichromate the solution changes from orange to green. (a) Explain why the alcohol groups in ethylene glycol are classified as primary. (b) Name the type of reaction occurring. (c) Draw the structure of the organic product. (d) Name the functional group present in the organic product.

CH_2OH
|
CH_2OH

ethylene glycol

(CCEA June 2014)

Oxidation of Secondary Alcohols

When a secondary alcohol is refluxed with acidified dichromate the alcohol is oxidised to the corresponding **ketone** (RCOR').

R—C(OH)(H)—R' + [O] → R—C(=O)—R' + H_2O

secondary alcohol RCH(OH)R' → ketone RCOR'

For example, when a mixture of propan-2-ol and acidified potassium dichromate is refluxed, the propan-2-ol is oxidised to propanone, CH_3COCH_3.

$CH_3CH(OH)CH_3 + [O] \rightarrow CH_3COCH_3 + H_2O$

propan-2-ol → propanone

Once again there is a change in smell, and the orange colour produced by the dichromate ions is replaced by the green colour of aqueous chromium(III) ions, $Cr^{3+}{}_{(aq)}$ as chromium is reduced from oxidation state +6 in dichromate ion to +3.

The properties and reactions of propanone are determined by the carbonyl (C=O) functional group. The systematic names and structures of ketones with up to six carbon atoms are summarised in Table 4.

Exercise 2.6K

1. Which one of the following alcohols can be oxidised to form a ketone?
   A Butan-1-ol
   B Butan-2-ol
   C 2-methylpropan-1-ol
   D 2-methylpropan-2-ol

(CCEA June 2013)

2. Which one of the following is formed by the complete oxidation of propan-2-ol?
   A a carboxylic acid
   B a ketone
   C a secondary alcohol
   D an aldehyde

(CCEA January 2013)

Oxidation of Tertiary Alcohols

We have already seen that a primary alcohol can be oxidised to an aldehyde, and then further oxidised to form the corresponding carboxylic acid. In contrast, once a secondary alcohol has been oxidised to the corresponding ketone it cannot be oxidised further, making it more resistant to oxidation than a primary alcohol. Tertiary alcohols are even more resistant to oxidation and will not react when refluxed with oxidising agents such as acidified dichromate.

OH	OH	OH
R—C—H, H	R—C—R', H	R—C—R'', R'
primary alcohol	secondary alcohol	tertiary alcohol

Ease of oxidation decreasing →

Exercise 2.6L

1. Which one of the following molecules would not be oxidised by acidified dichromate?

   A $CH_3CH_2CH_2OH$

   B $CH_3CH(OH)CH_3$

   C $(CH_3)_3COH$

   D $CH_3CH_2CH(OH)CH_3$

   (CCEA June 2003)

2. Which one of the following alcohols is not oxidised by acidified potassium dichromate solution?

   A propan-1-ol
   B propan-2-ol
   C 2-methylpropan-1-ol
   D 2-methylpropan-2-ol

   (CCEA June 2007)

Before moving to the next section, check that you are able to:

- Deduce the structure of the product formed when primary and secondary alcohols are refluxed with acidified dichromate.
- Deduce the structure of the aldehyde formed when a mixture of a primary alcohol and acidified dichromate is distilled.
- Recall that tertiary alcohols are not oxidised by strong oxidising agents such as acidified dichromate.
- Write balanced equations for the oxidation of alcohols using the symbol [O] to represent the oxidising agent.

Parent ketone	Structural isomers	Skeletal formula
C_3H_6O **propan**one	CH_3COCH_3	
C_4H_8O **butan**one	$CH_3CH_2COCH_3$	
$C_5H_{10}O$ **pentan**one	$CH_3CH_2CH_2COCH_3$ pentan-2-one	
	$CH_3CH_2COCH_2CH_3$ pentan-3-one	
$C_6H_{12}O$ **hexan**one	$CH_3CH_2CH_2CH_2COCH_3$ hexan-2-one	
	$CH_3CH_2CH_2COCH_2CH_3$ hexan-3-one	

Table 4: Systematic names and structures for straight-chain ketones (RCOR') with up to six carbon atoms.

Distinguishing Between Alcohols

The colour change that accompanies the oxidation of a primary or secondary alcohol can be used to distinguish primary and secondary alcohols from tertiary alcohols. If an alcohol is oxidised by acidified dichromate, the mixture changes colour from orange to green as dichromate ion ($Cr_2O_7^{2-}$) is reduced to form chromium(III) ions (Cr^{3+}). The colour change that accompanies the oxidation of a primary or secondary alcohol by acidified dichromate is shown in Figure 8.

Figure 8: The colour change that occurs when a primary or secondary alcohol is oxidised by acidified dichromate.

Worked Example 2.6ii

Describe a chemical test which could be used to distinguish between the alcohols butan-1-ol and 2-methylpropan-2-ol.

(CCEA January 2008)

Strategy

Describe the procedure and what would be observed when each alcohol is reacted with acidified potassium dichromate.

Solution

Place samples of the alcohols in separate test tubes. Add an equal volume of acidified potassium dichromate solution to each alcohol and gently heat the mixtures. The mixture containing butan-1-ol will turn from orange to green. The mixture containing 2-methylpropan-2-ol will remain orange.

Exercise 2.6M

1. Which one of the following alcohols will **not** cause the solution to turn from orange to green when heated with acidified potassium dichromate solution?

   A butan-1-ol

   B butan-2-ol

   C 2-methylpropan-1-ol

   D 2-methylpropan-2-ol

   (CCEA June 2010)

2. Explain why the structural isomers of propanol cannot be distinguished using acidified sodium dichromate solution.

3. Explain how acidified potassium dichromate solution can be used to distinguish between the isomers pentan-3-ol and 2-methylbutan-2-ol.

   (Adapted from CCEA June 2005)

4. A mixture of ethanol and gasoline (petrol) is known as Gasohol. Explain how you would carry out a chemical test to show that ethanol was present in a sample of Gasohol.

   (CCEA January 2010)

Before moving to the next section, check that you are able to:

- Describe the use of acidified dichromate to distinguish primary and secondary alcohols from tertiary alcohols.
- Explain the colour change that occurs when an alcohol is oxidised by acidified dichromate in terms of the ions and oxidation states involved.

2.7 Infrared Spectroscopy

CONNECTIONS

- Infrared spectroscopy is used by structural engineers to monitor the decay of plastics and other materials in the environment by detecting changes in chemical bonding within the material.
- Infrared spectroscopy can be used to detect forged documents by analysing the chemical composition of the ink used in the document.

In this section we are learning to:

- Explain how the infrared spectrum of a compound is obtained.
- Use the infrared spectrum of a compound to identify the compound.

Molecular Vibrations

The atoms in molecules are constantly in motion, causing the molecule to vibrate. A molecule of water vibrates at three distinct frequencies: 3756 cm^{-1}, 3657 cm^{-1} and 1595 cm^{-1}, which are expressed in **wavenumbers** (Unit: cm^{-1}) as it is convenient for frequencies of this size. The vibrations at 3756 cm^{-1} and 3657 cm^{-1} primarily involve stretching of the O-H bonds, while the vibration at 1595 cm^{-1} primarily involves a bending motion that varies the angle between the O-H bonds.

O-H stretching at 3756 cm^{-1} | O-H stretching at 3657 cm^{-1} | O-H bending at 1595 cm^{-1}

A molecule will absorb infrared (IR) radiation if the frequency of the radiation is the same as the frequency at which the molecule vibrates. Often, molecular vibrations are associated with individual bonds or groups of atoms in a molecule. As a result, *the absorption of IR radiation with a particular frequency can be used to detect specific bonds or groups of atoms within a molecule.*

For example, stretching of the O-H bond within a hydroxyl (-OH) group gives rise to absorption in the range 3500-3100 cm^{-1}. As a result, the absorption of IR radiation in this wavenumber range can be used to help verify the presence of a hydroxyl group in a compound.

At room temperature most molecular vibrations are in their **ground state**, a term used when referring to *the lowest energy state of the vibration*. When a bond absorbs IR radiation it vibrates more vigorously as it uses the energy from the radiation to move from its ground state to a higher energy state.

The infrared analysis of a substance involves passing a beam of IR radiation, containing a range of frequencies, through a specially prepared sample of the substance. The **infrared (IR) spectrum** for the substance is then obtained by plotting the percentage of the radiation transmitted through the sample as a function of frequency as illustrated by Figure 1. The process of obtaining the IR spectrum of a substance is known as **infrared (IR) spectroscopy**.

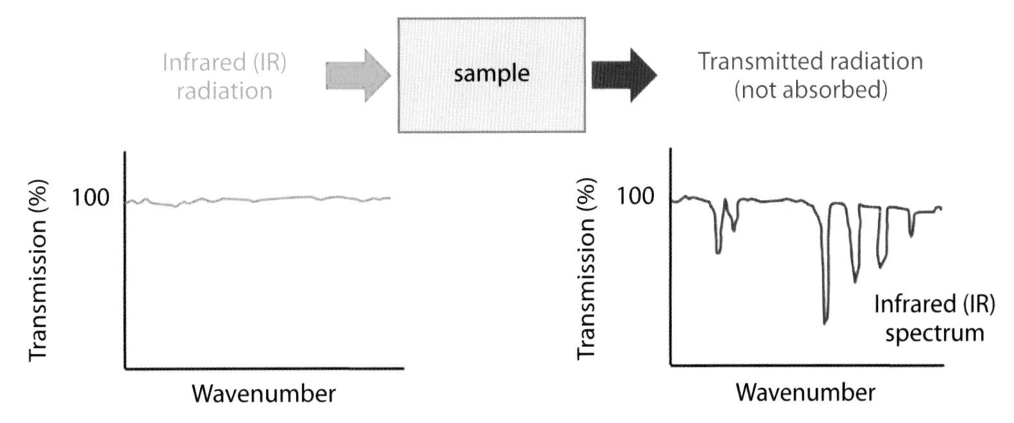

Figure 1: Generating the infrared (IR) spectrum for a substance by measuring the absorption of infrared radiation over a range of frequencies.

Exercise 2.7A

1. Absorption of infrared radiation by molecules is caused by

   A electronic transitions.

   B electronic vibrations.

   C molecular transitions.

   D molecular vibrations.

   (CCEA June 2012)

2. Which one of the following occurs when a molecule absorbs infrared radiation?

   A electrons in the bonds are excited

   B the bonds bend and eventually break

   C the bonds rotate more

   D the bonds vibrate more

   (Adapted from CCEA January 2010)

Before moving to the next section, check that you are able to:

- Explain why molecules absorb IR radiation.
- Describe how the infrared (IR) spectrum of a compound is obtained.

Identifying Functional Groups

The IR spectrum of a compound can be used to quickly detect individual bonds, and in some cases, entire functional groups within the compound. For example, the IR spectrum for butan-2-ol in Figure 2 reveals strong absorption (low %T) over the range 3500-3200 cm^{-1}. Strong absorption within this range confirms the presence of the hydroxyl (-OH) group in butan-2-ol as the hydroxyl groups in other alcohols also absorb strongly within this range. Similarly, strong absorption over the range 3000-2850 cm^{-1} confirms the presence of C-H bonds in butan-2-ol as the C-H bonds in other compounds also absorb within this range.

Worked Example 2.7i

The infrared spectrum for an alcohol with molecular formula C_2H_6O is shown below.

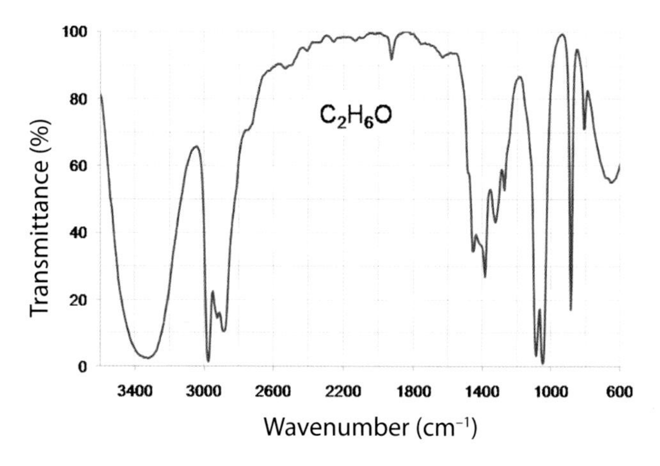

(a) Draw the structural formula for the alcohol showing all bonds present. (b) Explain, using information obtained from the spectrum, how the

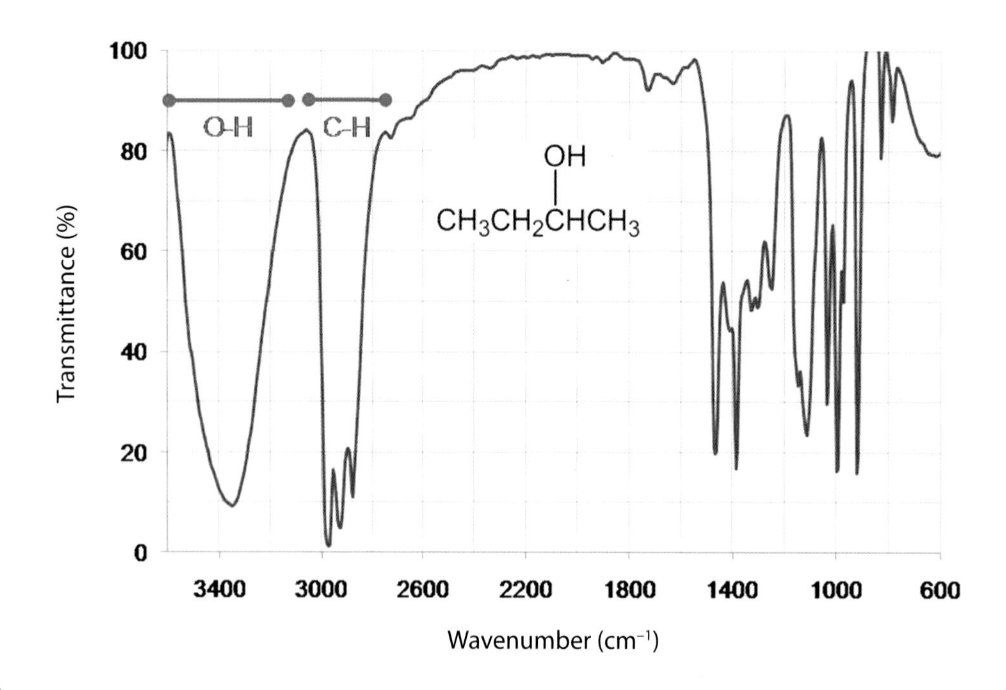

Figure 2: The IR spectrum for butan-2-ol, $CH_3CH_2CH(OH)CH_3$.

infrared spectrum can be used to identify the compound as an alcohol.

Solution

(a)

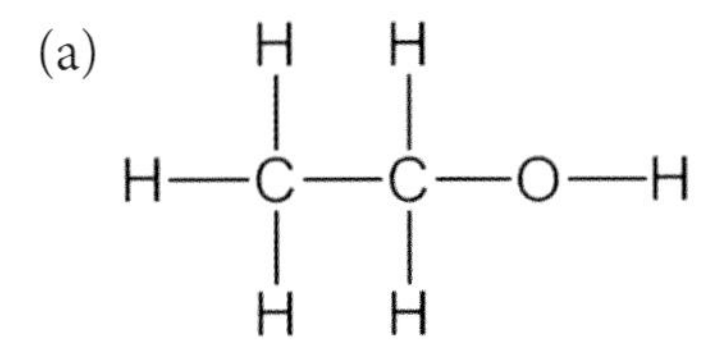

(b) Absorption within the range 3500–3100 cm^{-1} is due to the O-H bond within the hydroxyl (-OH) group. The presence of a hydroxyl group identifies the compound as an alcohol.

The IR spectrum for butanone in Figure 3 again reveals absorption within the range 3000-2850 cm^{-1} resulting from the presence of C-H bonds in the molecule. The IR spectrum for butanone also reveals strong absorption within the range 1750-1650 cm^{-1} resulting from the presence of the C=O bond in the molecule. As would be expected, both features are also present in the IR spectrum of ethyl ethanoate, $CH_3COOCH_2CH_3$ shown in Figure 3.

The consistency with which C-H bonds absorb in the range 3000-2850 cm^{-1} and C=O bonds absorb in the range 1750-1650 cm^{-1} demonstrates that *absorption within a well-defined wavenumber range can be used to reliably determine the presence of specific types of bond in a compound*. The range of wavenumber over which

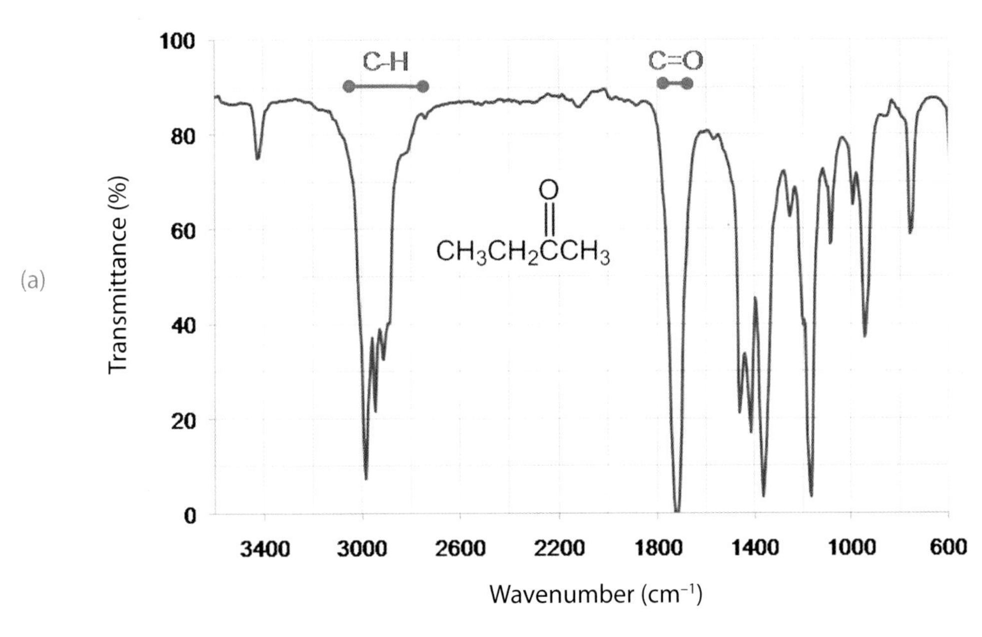

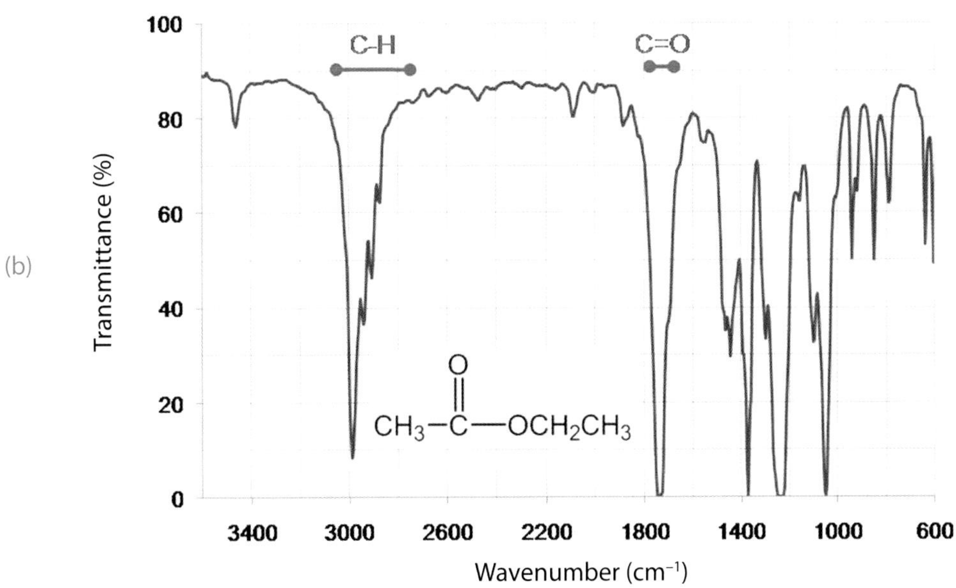

Figure 3: The IR spectrum of (a) butanone, $CH_3CH_2COCH_3$ and (b) ethyl ethanoate, $CH_3COOCH_2CH_3$.

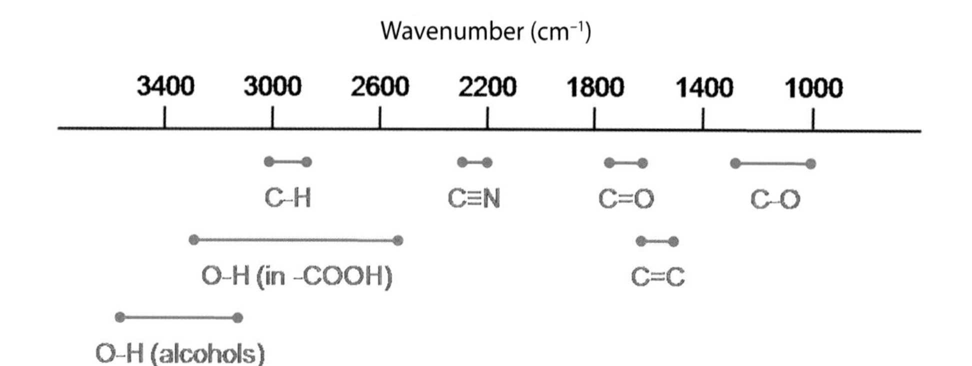

Figure 4: The approximate wavenumber range over which several common types of chemical bond absorb IR radiation.

several common bond types absorb IR radiation is summarised in Figure 4.

The IR spectrum for ethanoic acid, CH_3COOH is shown in Figure 5. As in butanone (Figure 3) and ethyl ethanoate (Figure 3), strong absorption within the range 1750-1650 cm^{-1} signals the presence of the C=O bond within the carboxyl (-COOH) group. The O-H bond in the carboxyl (-COOH) group gives rise to strong absorption within the range 3300-2500 cm^{-1}, and is easily distinguished from absorption by the hydroxyl (-OH) group in an alcohol that occurs within the range 3500-3100 cm^{-1}. In this way absorption within the range 1750-1650 cm^{-1} (C=O) can be combined with broad absorption within the range 3300-2500 cm^{-1} (O-H in -COOH) to detect the presence of a carboxyl (-COOH) group.

Exercise 2.7B

1. Explain how infrared spectroscopy could be used to distinguish between 2-bromo-2-methylpropane and the tertiary alcohol produced when 2-bromo-2-methylpropane reacts with aqueous potassium hydroxide.

(Adapted from CCEA January 2013)

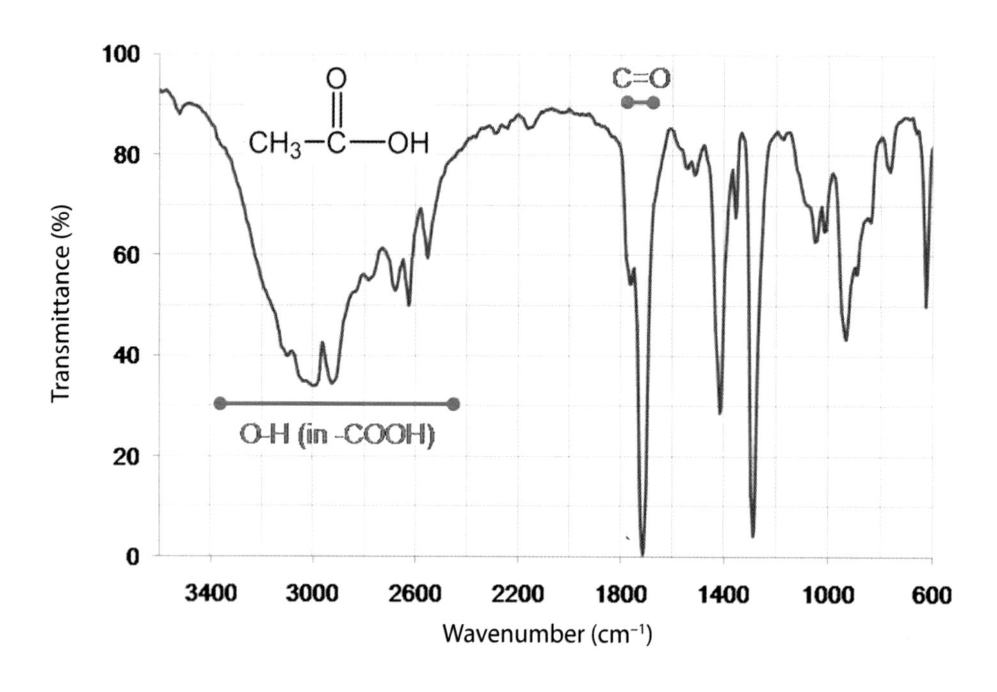

Figure 5: The IR spectrum for ethanoic acid, CH_3COOH.

2. Which one of the following molecules produces the infrared spectrum shown below?

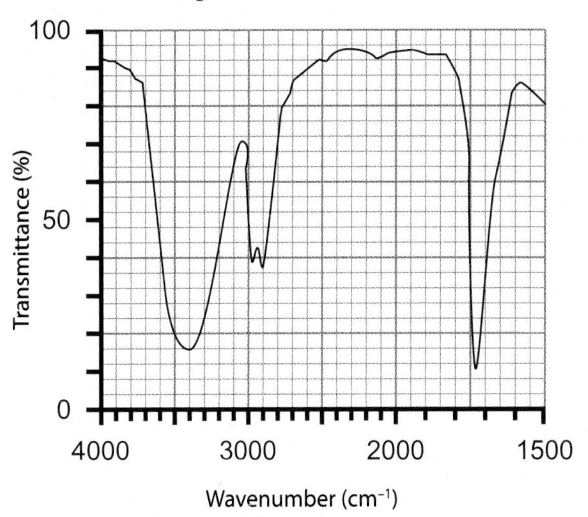

Bond	Wavenumber /cm^{-1}
O-H (alcohols)	3200-3700 strong broad
C-H	2700-3300 medium sharp
O-H (carboxylic acids)	2500-3200 strong broad
C=O	1680-1780 strong sharp

A

```
    H   H
    |   |      O
H — C — C — C //
    |   |      \
    H   H       H
```

B

```
    H   H   H
    |   |   |
H — C — C — C — O — H
    |   |   |
    H   H   H
```

C

```
    H   H
    |   |      O
H — C — C — C //
    |   |      \
    H   H       O — H
```

D

```
        H   H
        |   |      O
H — O — C — C — C //
        |   |      \
        H   H       H
```

(CCEA June 2013)

3. Parts of three infrared spectra A, B and C are given below. The spectra are those of ethanol, ethanoic acid and ethyl ethanoate but not necessarily in that order. Use the following information to help identify each compound. Explain your reasoning.

```
         H
         |
H3C  —  C  — OH
         |
         H
```

ethanol

```
         O
         ||
H3C  —  C  — OH
```

ethanoic acid

```
         O
         ||
H3C  —  C  —  O  — CH2CH3
```

ethyl ethanoate

- O-H bonds produce broad absorptions in the range 2500-3500 cm^{-1}
- C=O bonds produce sharp absorptions in the range 1650-1750 cm^{-1}

A

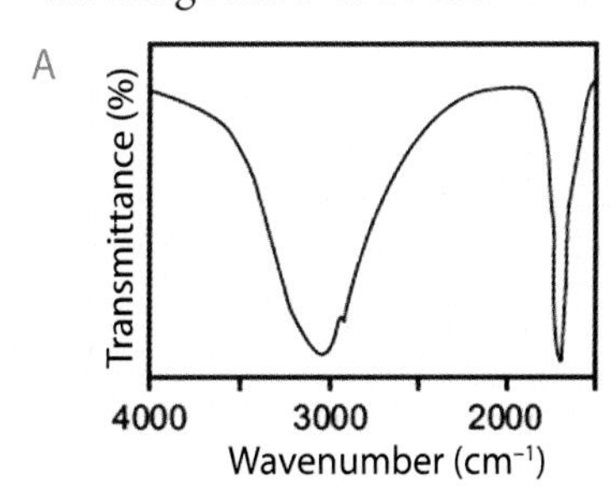

B

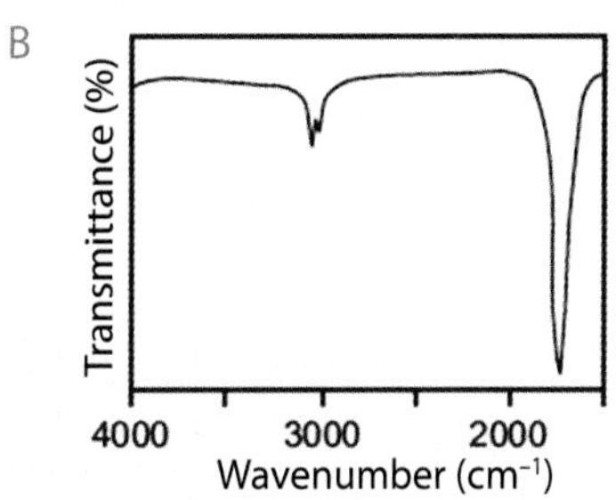

C

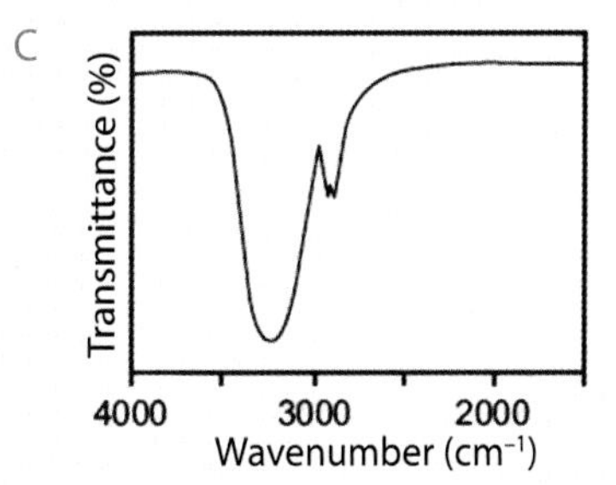

(Adapted from CCEA June 2010)

4. The infrared spectra for ethanol, ethanal and ethanoic acid are shown below, but not necessarily in that order. (a) Explain how the absorption of infrared radiation arises in molecules. (b) Use the data in the table to identify the absorptions at 3000 cm^{-1} and 3400 cm^{-1} in Spectrum A and 1700 cm^{-1} in Spectrum B. (c) Identify which molecules give rise to Spectrum B and Spectrum C.

Bond	Wavenumber /cm^{-1}
C-H	2850–3300
C=C	1620–1680
C=O	1680–1750
C-O	1000–1300
O-H (alcohols)	3230–3550
O-H (acids)	2500–3000

(Adapted from CCEA January 2011)

Spectrum A

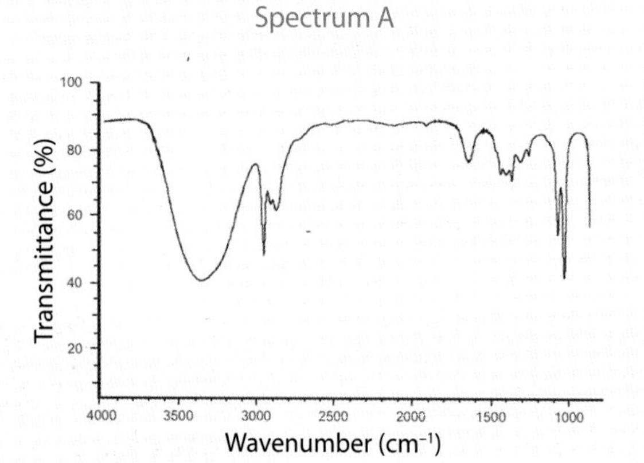

Spectrum B

Transmittance (%)

100 80 60 40 20

4000 3500 3000 2500 2000 1500 1000

Wavenumber (cm^{-1})

Spectrum C

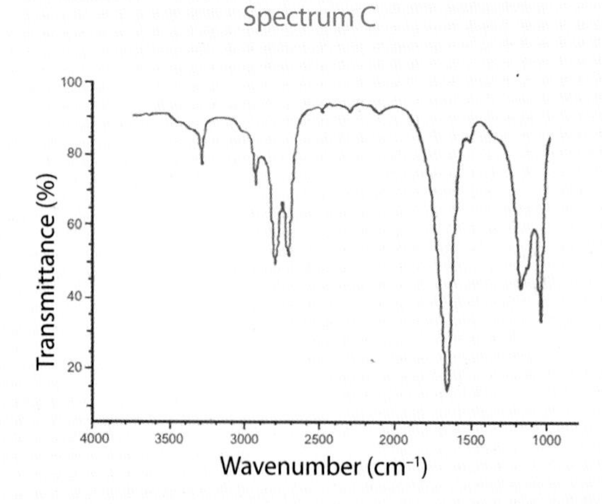

5. In the case of a positive breathalyser test, a motorist is tested a second time using infrared spectroscopy. (a) State and explain the effect of infrared radiation on molecules. (b) The infrared spectrum must be carefully read to distinguish between ethanol and propanone, which is found in the breath of diabetics. Use the data in the following table to determine which spectrum, A or B, is for ethanol. Explain your choice and refer to both spectra in your answer.

Bond	Wavenumber /cm^{-1}
C-H	2850–3300
C=O	1680–1750
C-O	1000–1300
O-H	3230–3550

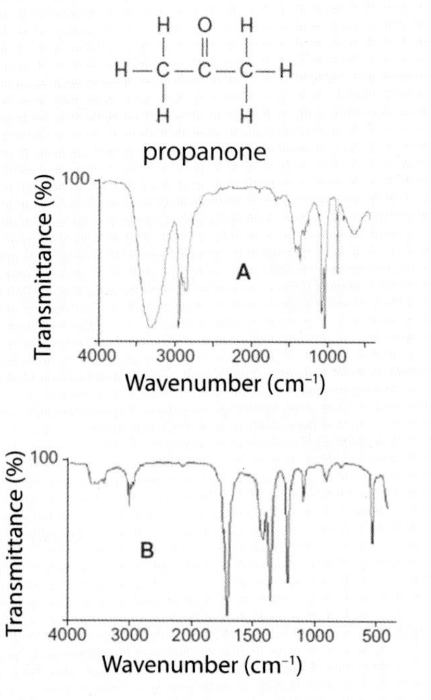

(CCEA January 2014)

6. Ethylene glycol contains primary alcohol groups. When heated under reflux with excess acidified potassium dichromate the solution changes from orange to green. (a) Draw the structure of the organic product. (b) Name the functional group present in the organic product. (c) What would be the most significant difference between the infrared spectrum of the organic product and that of ethylene glycol?

CH_2OH
|
CH_2OH

ethylene glycol

(CCEA June 2014)

Before moving to the next section, check that you are able to:

- Use the IR spectrum of a compound to identify bonds and functional groups within the compound.

Identifying Compounds

The IR spectrum of a compound is unique and can be used to identify an unknown substance by matching the IR spectrum of the unknown with the IR spectrum of a pure compound. If the substance contains impurities the IR spectra will not be an exact match as the impurities may absorb in different wavenumber ranges, or may absorb in regions of the spectrum where the substance absorbs.

Worked Example 2.7ii

Propenonitrile, CH_2=CHCN can be polymerised to form a fibre known as Orlon that is used to make clothes. Propenonitrile is manufactured from propene, ammonia and oxygen. Suggest how you could use infrared spectroscopy to show that no propene was present in the product from the reaction.

(CCEA June 2009)

Solution

The IR spectrum of a compound is unique. If the product is pure and no propene is present, the IR spectrum of the product will exactly match the IR spectrum for propenonitrile.

Exercise 2.7C

1. Today, the identification of compounds is carried out using spectroscopic methods rather than test tube reactions. (a) Explain why molecules absorb infrared radiation. (b) Which group of atoms, present in butanol, is used to identify alcohols in infrared spectroscopy? (c) Explain how you would use the infrared spectra of butanol isomers to identify an unknown butanol.

(Adapted from CCEA June 2015)

2. Suggest how you would use infrared spectroscopy to identify an unknown sample of bromopropane as 1-bromopropane or 2-bromopropane.

(CCEA June 2014)

Before moving to the next section, check that you are able to:

- Describe how IR spectroscopy can be used to verify the identity of a compound, and check for impurities in a mixture.

2.8 Energetics

CONNECTIONS

- An ice cube feels cold to the touch because the water molecules in the ice have less energy than the molecules in contact with the ice cube.
- An explosive such as TNT produces lots of energy when it explodes by forming very stable substances.

Enthalpy

In this section we are learning to:

- Use the terms exothermic and endothermic to describe the energy changes that accompany chemical reactions.
- Construct an enthalpy diagram to explain the nature of the energy change for a chemical reaction.

The term **enthalpy** (Symbol: H) refers to the amount of heat energy contained within a substance. In an **exothermic reaction** *the enthalpy of the products is less than the enthalpy of the reactants* and energy is released from the reaction mixture. The energy released may be a combination of heat, light or sound, and is often sufficient to increase the temperature of the reaction mixture.

The enthalpy diagram in Figure 1a illustrates the relationship between the enthalpy of the reactants ($H_{reactants}$), the enthalpy of the products ($H_{products}$), and the amount of energy released from the reaction mixture (ΔH) for an exothermic reaction.

On defining the **enthalpy of reaction**, ΔH to be *the enthalpy change when the amounts specified in the chemical equation react,* Figure 1a demonstrates that the enthalpy of reaction for an exothermic reaction is negative.

Enthalpy of reaction for an exothermic reaction:
$\Delta H = H_{products} - H_{reactants} < 0$

In contrast, in an **endothermic reaction** *the enthalpy of the products is greater than the enthalpy of the reactants* and energy is needed for the reaction to occur. The enthalpy diagram in Figure 1b reveals that the enthalpy of reaction (ΔH) for an endothermic reaction is positive.

Enthalpy of reaction for an endothermic reaction:
$\Delta H = H_{products} - H_{reactants} > 0$

In most cases the energy needed for an endothermic reaction to occur is taken from the reaction mixture and, as a result, the temperature of the reaction mixture decreases as the reaction proceeds.

Worked Example 2.8i

The partial oxidation of ammonia is an exothermic reaction and is used in the manufacture of nitric acid. Draw a labelled enthalpy diagram for the reaction.

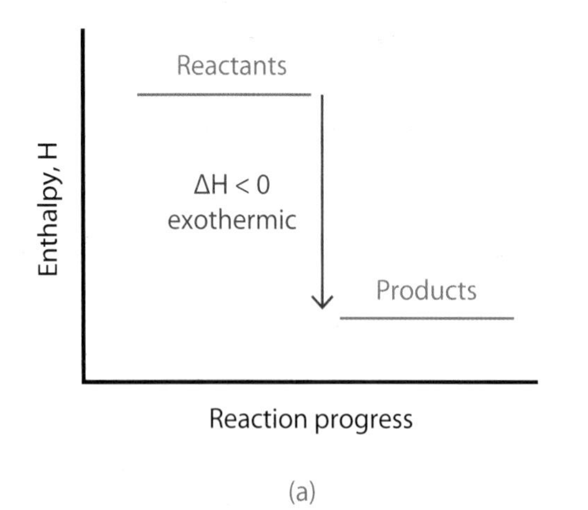

(a)

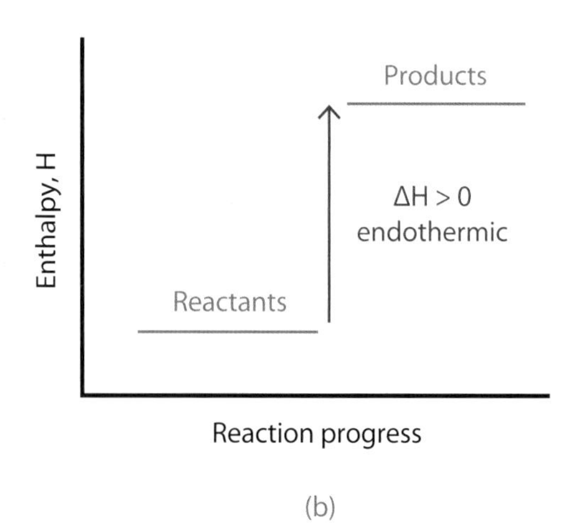

(b)

Figure 1: Enthalpy diagram for (a) an exothermic reaction and (b) an endothermic reaction.

$$4NH_3 + 5O_2 \rightarrow 4NO + 6H_2O$$

(Adapted from CCEA January 2007)

Strategy

Use an arrow labelled ΔH to describe the relationship between the enthalpy of the reactants and products.

Solution

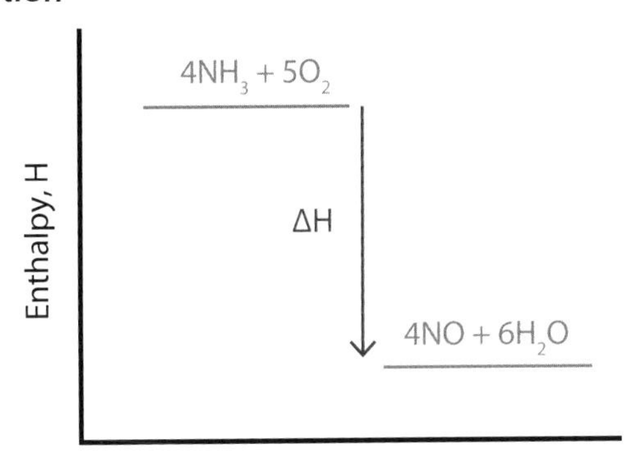

Exercise 2.8A

1. The formation of ice crystals from water is an exothermic reaction. Draw an enthalpy diagram for the freezing of water. Mark the enthalpy change as ΔH.

   (CCEA January 2005)

2. Photosynthesis is an endothermic process used by plants to produce carbohydrates. (a) Explain what is meant by the term *endothermic*. (b) Draw a fully labelled enthalpy diagram for the reaction.

   $6CO_{2(g)} + 6H_2O_{(l)} \rightarrow C_6H_{12}O_{6(s)} + 6O_{2(g)}$

   (CCEA June 2003)

3. Some reusable hand warmers contain concentrated sodium thiosulfate solution, $Na_2S_2O_{3(aq)}$. Squeezing the packet releases a seed crystal which causes sodium thiosulfate crystals, $Na_2S_2O_3.5H_2O_{(s)}$ to be produced. The process of forming sodium thiosulfate crystals is exothermic. (a) Explain what is meant by the term *exothermic*. (b) Draw an enthalpy diagram for the reaction.

   (CCEA January 2003)

Before moving to the next section, check that you are able to:

- Use the terms exothermic and endothermic to describe the relationship between the enthalpy of the reactants and products in a chemical reaction.
- Construct labelled enthalpy diagrams to describe enthalpy changes.

Measuring Enthalpy Changes

In this section we are learning to:

- Use the idea of standard conditions to define the standard enthalpy change for a reaction and the standard state of a substance.
- Use a simple calorimeter to determine the standard enthalpy change for a reaction.
- Define and calculate the standard enthalpy of neutralisation for an acid.

Standard Conditions

The enthalpy change for a chemical reaction depends on the conditions under which the reaction is carried out. As a result enthalpy changes can only be compared if they are obtained under **standard conditions** which *are defined to be 25 °C and a pressure of 100 kPa.*

An enthalpy change that occurs under standard conditions is known as a **standard enthalpy change**, is given the symbol ΔH° and *is equal to the change in heat energy that occurs when the process is carried out at constant pressure under standard conditions*. The standard temperature of 25 °C is often reported as 298 K where we are using the following conversion between degrees Celcius (°C) and Kelvin (K).

Converting temperature in °C to K:
Temperature (K) = Temperature (°C) + 273

Having defined standard conditions, the term **standard state** can then be used to refer to *the physical state of a substance under standard conditions.* For example, the standard state of iron is solid as iron is a solid under standard conditions. Similarly, the standard state of water is liquid as water is a liquid under standard conditions.

Worked Example 2.8ii

Which formula does not represent a substance in its standard state? Explain your answer and write the correct formula.

$CH_{4(g)}$ $CO_{2(g)}$ $C_4H_{10(g)}$ $C_2H_5OH_{(aq)}$

Solution

The formula $C_2H_5OH_{(aq)}$ represents a solution of ethanol in water. The standard state of ethanol is $C_2H_5OH_{(l)}$.

Exercise 2.8B

1. Which one of the following formulas does not represent a substance in its standard state? Explain your answer and write the correct formula.

   $Cl_{2(g)}$ $Br_{2(g)}$ $I_{2(s)}$ $S_{8(s)}$

2. Which one of the following formulas does not represent a substance in its standard state? Explain your answer and write the correct formula.

   $N_{2(g)}$ $Cu_{(s)}$ $H_{(g)}$ $CH_3OH_{(l)}$

Before moving to the next section, check that you are able to:

- Explain what is meant by the term standard conditions.
- Use the concept of standard conditions to define the *standard state* of a substance.

Burning Fuels

The amount of heat produced when a fuel burns can be calculated by measuring the rise in temperature when the heat produced by burning the fuel is used to heat a known mass of liquid.

Liquid fuels such as ethanol can be burnt in a controlled way using a 'spirit burner' of the type shown in Figure 2. The heat produced (q) is related to the change in temperature (ΔT) and the mass of liquid heated (m) by Equation 1.

$$q = m\,c\,\Delta T \qquad \text{Equation 1}$$

The amount of energy produced per mole of fuel burnt is referred to as the **molar enthalpy change**, and is calculated by dividing the amount of energy produced (q) by the moles of fuel burnt (n) as in Equation 2.

$$\Delta H = \frac{-q}{n} \qquad \text{Equation 2}$$

Figure 2: Use of a 'spirit burner' to control the burning of a liquid fuel such as ethanol.

A negative sign is added in Equation 2 as heat is produced ($q > 0$) resulting in an exothermic reaction ($\Delta H < 0$).

The **specific heat capacity** of the liquid being heated (c) is defined as the amount of energy needed to raise the temperature of 1 g of the liquid by 1 °C and has units of $J\ g^{-1}\ °C^{-1}$. If we recall that a change in temperature of one Kelvin (1 K) is the same size as one degree centigrade (1 K = 1 °C) the units for specific heat capacity can also be written as $J\ g^{-1}\ K^{-1}$.

The accuracy of the enthalpy change (ΔH) calculated by this method can be improved by insulating the beaker of liquid, and shielding the flame from draughts, to minimise heat loss to the surroundings. The temperature of the liquid in the beaker will also be more accurate if the liquid is stirred during heating to ensure that the liquid is heated uniformly.

The enthalpy change (ΔH) obtained by this method will, in any event, be less than expected due to incomplete combustion of the fuel.

Worked Example 2.8iii

Burning 0.050 moles of a liquid fuel increased the temperature of 200 g of water by 10 K. Calculate the molar enthalpy of combustion of the fuel. The heat capacity of water is $4.2\ J\ g^{-1}\ K^{-1}$.

(Adapted from CCEA June 2010)

Strategy

- Use $q = m\,c\,\Delta T$ to calculate the enthalpy change for burning 0.050 mol of liquid fuel.
- Calculate the enthalpy change for burning one mol of liquid fuel.

Solution

Mass of water heated, m = 200 g

Specific heat capacity of water, $c = 4.2\ J\ g^{-1}\ K^{-1}$
$= 4.2\ J\ g^{-1}\ {}^{\circ}C^{-1}$

Temperature change, $\Delta T = +\ 10\ K = +\ 10\ {}^{\circ}C$

$q = m\,c\,\Delta T = 200\ g \times 4.2\ J\ g^{-1}\ {}^{\circ}C^{-1} \times 10\ {}^{\circ}C$
$= 8.4 \times 10^{3}\ J = 8.4\ kJ$

Molar enthalpy change, $\Delta H =$

$$= \frac{-\,q}{\text{Moles of fuel}} = \frac{-\,8.4\ kJ}{0.050\ mol} = -\,1.7 \times 10^{2}\ kJ\ mol^{-1}$$

Exercise 2.8C

1. The enthalpy of combustion for ethanol can be obtained using the apparatus shown below. The results of the experiment are summarised in the table. (a) Calculate the heat energy released by the combustion of the ethanol using the equation: heat energy released (J) = mass of water × 4.18 × temperature rise. (b) Calculate the enthalpy of combustion for ethanol in $kJ\ mol^{-1}$.

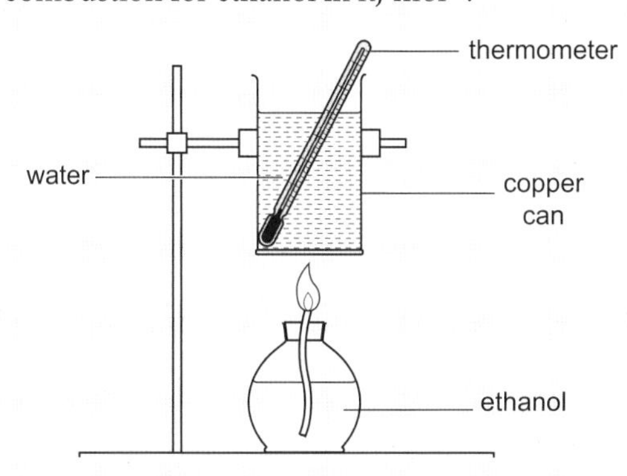

Mass of water in the copper can	200.0 g
Mass of spirit burner + ethanol (before burning)	150.0 g
Mass of spirit burner + ethanol (after burning)	148.6 g
Initial temperature of water	18.0 °C
Final temperature of water	42.0 °C

(CCEA January 2013)

2. Icosane (RMM = 282) is an alkane containing 20 carbon atoms and is a solid at room temperature. A calorimeter is set up as shown in the diagram using a container of icosane with a wick in it. (a) Complete the table. (b) Calculate the number of moles of icosane burned. (c) Calculate the amount of heat energy in Joules transferred to the water (specific heat capacity of water = $4.18\ J\ g^{-1}\ {}^{\circ}C^{-1}$). (d) Calculate the enthalpy of combustion of icosane in $kJ\ mol^{-1}$. (e) Give one practical reason why the experimental value you calculated is significantly different from a value quoted in a data book.

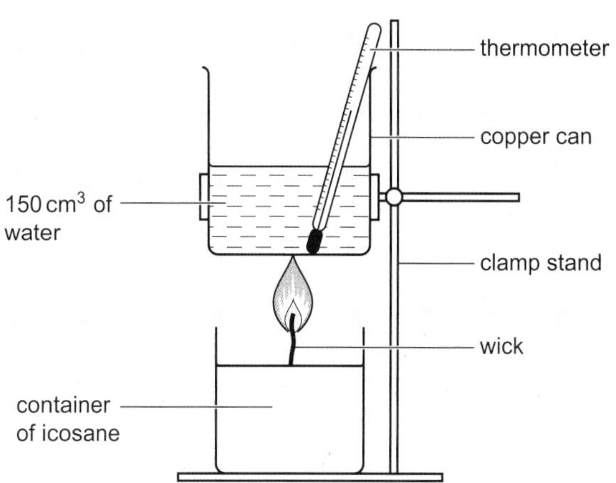

Initial mass of icosane	26.20 g
Final mass of icosane	25.30 g
Mass of icosane burned	
Initial water temperature	19 °C
Final water temperature	80 °C
Temperature change	
Mass of water	150 g

(Adapted from CCEA January 2014)

3. Disposable lighters contain liquid butane. The following experiment was carried out to determine the molar enthalpy of combustion for butane. The results of the experiment are summarised in the table. The specific heat capacity of water is $4.2\ J\ g^{-1}\ {}^{\circ}C^{-1}$. (a) Calculate the heat received by the water. (b) Calculate the molar enthalpy of combustion for butane. (c) A similar experiment was carried out to determine the molar enthalpy of combustion for ethanol. A value of half the theoretical value was obtained. State *three* reasons for the low experimental value. (d) Explain why the molar enthalpy of combustion for butane is theoretically more than that of ethanol.

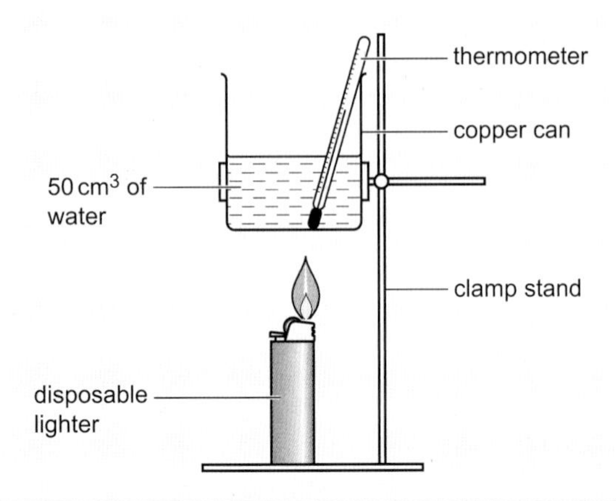

Mass of water in the copper can	50 g
Mass of lighter before ignition	15.00 g
Mass of lighter after burning	14.53 g
Temperature of water before heating	25 °C
Temperature of water after heating	80 °C

(Adapted from CCEA January 2012)

4. 0.47 g of a hydrocarbon was completely burnt in air. The heat produced raised the temperature of 200 g of water by 28.2 °C. The standard enthalpy of combustion of the hydrocarbon is −2220 kJ mol^{-1}. The specific heat capacity of water is 4.2 J g^{-1} $°C^{-1}$. Calculate the molar mass of the hydrocarbon.

(CCEA June 2012)

Before moving to the next section, check that you are able to:

- Describe how to accurately measure the temperature change associated with the burning of a known amount of fuel.
- Use a measured temperature change to calculate the molar enthalpy change for the burning of a fuel.

Reactions in Solution

A simple 'coffee cup calorimeter' of the type shown in Figure 3 can be used to measure temperature changes that accompany reactions in solution. The heat produced by the reaction (q) is again related to the temperature change in the solution (ΔT), and the mass of the solution (m), by the equation: $q = m\,c\,\Delta T$.

If the reaction is exothermic ($\Delta H < 0$) the temperature of the solution increases ($\Delta T > 0$) as heat is produced by the reaction ($q > 0$). Conversely, if the reaction is endothermic ($\Delta H > 0$), the temperature of the solution decreases ($\Delta T < 0$) as heat is absorbed by the reaction ($q < 0$).

Worked Example 2.8iv

Explain how you would carry out an experiment to determine the enthalpy change for dissolving magnesium sulfate, stating the equipment used, the measurements made, and the possible sources of

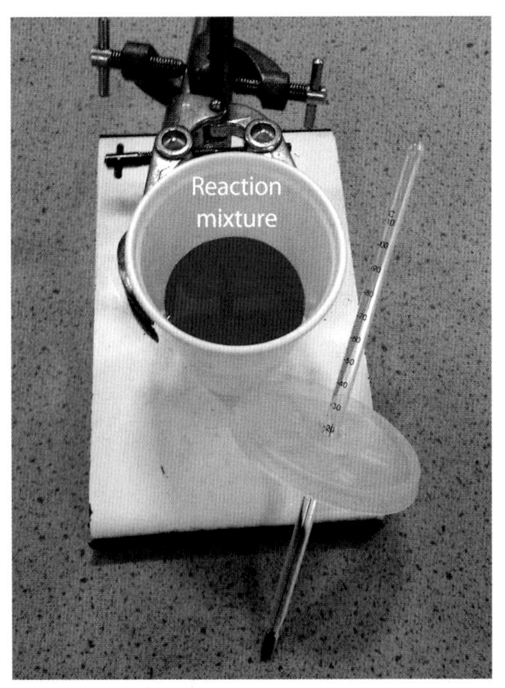

Figure 3: A simple 'coffee cup calorimeter' made from a polystyrene cup.

error. Explain how these errors can be minimised.

(Adapted from CCEA June 2009)

Solution

Add a known amount of water to an insulated container and measure the temperature of the water with a thermometer. Add a known amount of magnesium sulfate to the water and stir with a glass rod to dissolve.

Measure the temperature of the solution at regular intervals, then calculate the biggest temperature change (ΔT) produced by the reaction.

A major source of error is heat loss. Heat loss can be minimised by adding more insulation to the container, and reducing draughts around the apparatus.

If the reaction occurs in aqueous solution it is reasonable to assume that the density of the solution is similar to the density of water ($1\ g\ cm^{-3}$). It is also reasonable to assume that the specific heat capacity of an aqueous solution is similar to the specific heat capacity for water ($4.2\ J\ g^{-1}\ {}^{o}C^{-1}$).

Worked Example 2.8v

A solution of zinc sulfate is formed when zinc powder is added to a solution of copper sulfate. When 6.00 g of zinc powder (in excess) was added to 50.0 cm^3 of a 0.50 mol dm^{-3} solution of copper sulfate in a polystyrene cup, the temperature of the solution increased by 25.3 K. Calculate the energy change for the reaction in units of kJ mol^{-1} (of copper sulfate). The heat capacity and density of the solution is $4.2\ J\ g^{-1}\ K^{-1}$ and $1.0\ g\ cm^{-3}$ respectively.

(Adapted from CCEA June 2009)

Strategy

- Calculate the enthalpy change for the reaction (in kJ).
- Calculate the molar enthalpy change (in kJ mol^{-1}).

Solution

Mass of solution, m = Volume × Density

$= 50.0\ cm^3 \times 1.0\ g\ cm^{-3} = 50\ g$

Specific heat capacity of solution, $c = 4.2\ J\ g^{-1}\ K^{-1}$

$= 4.2\ J\ g^{-1}\ {}^{o}C^{-1}$

Temperature change, $\Delta T = +\ 25.3\ K = +\ 25.3\ {}^{o}C$

$q = m\ c\ \Delta T = 50\ g \times 4.2\ J\ g^{-1}\ {}^{o}C^{-1} \times 25.3\ {}^{o}C$

$= 5.3 \times 10^3\ J$

$$\Delta H = \frac{-\ q}{\text{Moles of } CuSO_4} = \frac{-\ 5.3\ kJ}{0.025\ mol}$$

$= -\ 2.1 \times 10^2\ kJ\ mol^{-1}$

Exercise 2.8D

1. The temperature of the water dropped from 25.00 °C to 24.10 °C when 5.0 g of ammonium nitrate was added to 100 g of water. Calculate (a) the energy change taking place, (b) the molar enthalpy change for dissolving ammonium nitrate in 100 g of water, and (c) the mass of ammonium nitrate needed to decrease the temperature of 120 g of water by 25 °C. The specific heat capacity of water is $4.2\ J\ K^{-1}\ g^{-1}$.

   (Adapted from CCEA January 2011)

2. Hex-1-ene (density = $0.68\ g\ cm^{-3}$) reacts rapidly with bromine in a solvent such as 1,1,1-trichloroethane (TCE). The reaction is exothermic. 100 cm^3 of a solution of bromine in TCE is placed in an insulated glass beaker and its temperature measured. 2.0 cm^3 of hex-1-ene is added to the beaker, the mixture stirred, and the temperature measured again. (a) Calculate the heat received by the 100 cm^3 of TCE (density = $1.33\ g\ cm^{-3}$; heat capacity = $1.30\ J\ g^{-1}\ {}^{o}C^{-1}$) if the temperature of the TCE increases from 24.9 °C to 32.1 °C. (b) Calculate the molar enthalpy of bromination for hex-1-ene.

   (Adapted from June 2015)

Before moving to the next section, check that you are able to:

- Describe how to use a basic calorimeter to accurately measure the temperature change associated with a reaction in solution.
- Use a measured temperature change to calculate the molar enthalpy change for a reaction in solution.

Enthalpy of Neutralisation

The term **standard enthalpy of neutralisation, $\Delta_n H^{\circ}$** refers to *the enthalpy change that occurs when one mole of water is produced by a neutralisation reaction under standard conditions.*

Neutralisation is an exothermic reaction ($\Delta H < 0$), and the associated rise in temperature can be determined using a simple calorimeter of the type shown in Figure 3.

When one mole of hydrochloric acid completely reacts with a strong alkali, one mole of water is formed, and the total enthalpy change is equal to the enthalpy of neutralisation for hydrochloric acid.

$HCl_{(aq)} + NaOH_{(aq)} \rightarrow NaCl_{(aq)} + H_2O_{(l)}$

Ionic equation: $H^+_{(aq)} + OH^-_{(aq)} \rightarrow H_2O_{(l)}$

Enthalpy change = $\Delta_n H^{\circ}$

Similarly when one mole of nitric acid completely reacts with a strong alkali, one mole of water is formed, and the total enthalpy change is equal to the enthalpy of neutralisation for nitric acid.

$HNO_{3(aq)} + NaOH_{(aq)} \rightarrow NaNO_{3(aq)} + H_2O_{(l)}$

Ionic equation: $H^+_{(aq)} + OH^-_{(aq)} \rightarrow H_2O_{(l)}$

Enthalpy change = $\Delta_n H^{\circ}$

In contrast, when one mole of sulfuric acid reacts with a strong alkali, two moles of water are formed, and the total enthalpy change is equal to twice the enthalpy of neutralisation for sulfuric acid.

$H_2SO_{4(aq)} + NaOH_{(aq)} \rightarrow NaHSO_{4(aq)} + H_2O_{(l)}$

$NaHSO_{4(aq)} + NaOH_{(aq)} \rightarrow Na_2SO_{4(aq)} + H_2O_{(l)}$

Total enthalpy change = $2\Delta_n H^{\circ}$

The enthalpies of neutralisation for hydrochloric acid, nitric acid and sulfuric acid are very similar as the other ions present in the reaction mixture do not significantly affect the enthalpy change.

Worked Example 2.8vi

Describe, giving practical details, how the molar enthalpy change for the neutralisation of sulfuric acid with sodium hydroxide could be determined in the laboratory. Measurements and calculations should be described.

(CCEA June 2004)

Solution

Place a known amount of sulfuric acid in an insulated flask. Measure the temperature of the acid in the flask using a thermometer. Use the thermometer to stir the solution while slowly adding a known amount of sodium hydroxide.

Record the temperature of the solution at regular intervals and use the highest temperature recorded to calculate the temperature change (ΔT) produced by the reaction.

Use the temperature change to calculate the heat produced by the reaction (q) using the equation: $q = m\,c\,\Delta T$. Divide the heat produced (q) by the moles of acid neutralised, and then by two, to obtain the enthalpy change per mole of water formed.

Exercise 2.8E

1. (a) Write the equation for the reaction between potassium hydroxide and hydrochloric acid. (b) Describe how the enthalpy of neutralisation for this reaction is determined. Include experimental details, including one potential source of error, and one safety precaution. Details of calculations are not required.

(Adapted from CCEA June 2007)

Before moving to the next section, check that you are able to:

- Explain what is meant by the *standard enthalpy of neutralisation* for an acid.
- Describe how to use a basic calorimeter to accurately measure the temperature change associated with a neutralisation reaction in solution.
- Use a measured temperature change to calculate the enthalpy of neutralisation for an acid.

Calculating Enthalpy Changes

In this section we are learning to:

- Use average bond enthalpies to *estimate* the enthalpy change for a reaction.
- Use enthalpies of formation to *calculate* the enthalpy change for a reaction.

Bond Enthalpy Calculations

The enthalpy change for a chemical reaction can be estimated by calculating the difference between the energy needed to break bonds in the reactants, and the energy released when bonds are formed in the products.

If the energy needed to break bonds in the reactants ($E_{breaking}$) is greater than the energy released when bonds are made in the products (E_{making}) the reaction will be endothermic ($\Delta H > 0$) as illustrated by Figure 4a. Conversely, if the energy needed to break bonds in the reactants ($E_{breaking}$) is less than the energy released when bonds are made in the products (E_{making}) the reaction will be exothermic ($\Delta H < 0$) as illustrated by Figure 4b.

This relationship between the energy involved in breaking and making bonds, and the enthalpy change for the reaction is summarised by Equation 3.

$$\Delta H = E_{breaking} - E_{making} \qquad \text{Equation 3}$$

On defining the **average bond enthalpy** for a particular type of bond to be *the energy needed to break one mole of a particular type of bond averaged over many compounds* the energy changes associated with the making and breaking of bonds in Equation 3 can be *estimated* by combining average bond enthalpies for the bonds in the reactants and products.

Worked Example 2.8vii

Use the following average bond enthalpies to calculate the enthalpy change when one mole of propane is completely burnt:

$C_3H_8 + 5O_2 \rightarrow 3CO_2 + 4H_2O$

Bond:	C-C	C-H	O=O	C=O	O-H
Bond Enthalpy (kJ mol^{-1}):	347	413	498	805	464

Strategy

- Calculate the moles of bonds broken and the moles of bonds formed when one mole of propane is burnt.
- Use the average bond enthalpies to calculate the total energy needed to break bonds and the total energy released by making bonds.

Solution

$CH_3-CH_2-CH_3$ (H–C–C–C–H with all H shown) + 5O=O

→ 3O=C=O + 4H-O-H

Bonds broken (in reactants):

2 moles of C-C bonds + 8 moles of C-H bonds + 5 moles of O=O bonds

Total energy needed to break bonds ($E_{breaking}$):

2 E(C-C) + 8 E(C-H) + 5 E(O=O)

= 2(347) + 8(413) + 5(498) = 6488 kJ

Bonds made (in products):

6 moles of C=O bonds + 8 moles of O-H bonds

Total energy released by making bonds (E_{making}):

6 E(C=O) + 8 E(O-H) = 6(805) + 8(464) = 8542 kJ

Enthalpy change, $\Delta H = E_{breaking} - E_{making}$

= 6488 – 8542 = –2044 kJ mol^{-1}

Exercise 2.8F

1. (a) Explain what is meant by the term *average bond enthalpy*. (b) Use the average bond enthalpy values given below to calculate (i) the total bond enthalpy for the products, and (ii) the enthalpy change for the following reaction, given that the total bond enthalpy of the reactants is 16548 kJ.

$$C_8H_{18(l)} + \frac{25}{2}O_{2(g)} \rightarrow 8CO_{2(g)} + 9H_2O_{(l)}$$

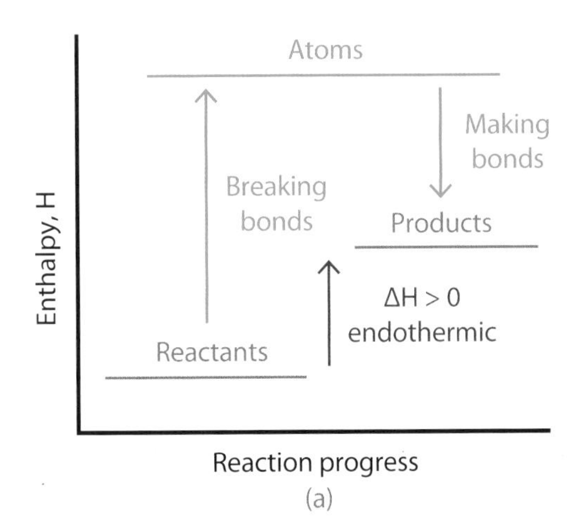

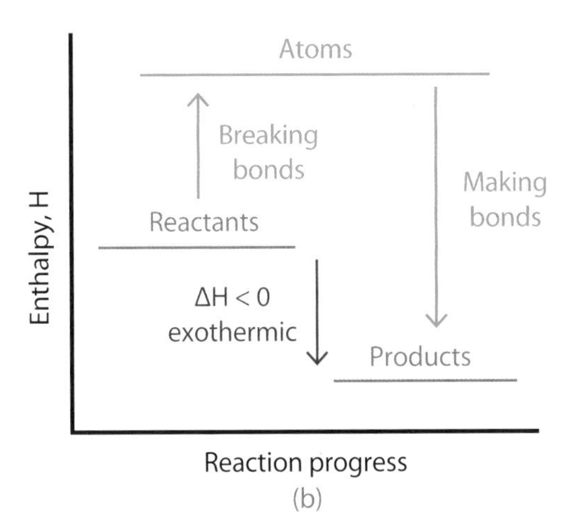

Figure 4: The relationship between the energy involved in breaking and making bonds, and the enthalpy of reaction for (a) an endothermic reaction and (b) an exothermic reaction.

Bond:	C=O	O-H
Bond Enthalpy (kJ mol^{-1}):	750	463

(Adapted from CCEA January 2008)

2. Chloromethane, CH_3Cl is formed in an exothermic reaction between methane and chlorine. (a) Use the following average bond enthalpies to calculate the enthalpy change for the reaction. (b) Draw an enthalpy diagram for the reaction. (c) Explain why the measured enthalpy of reaction is slightly different from the value calculated in part (a).

$$CH_{4(g)} + Cl_{2(g)} \rightarrow CH_3Cl_{(g)} + HCl_{(g)}$$

Bond:	C-H	Cl-Cl	C-Cl	H-Cl
Bond Enthalpy (kJ mol^{-1}):	413	243	346	432

(Adapted from CCEA January 2006)

3. (a) Explain what is meant by the term *average bond enthalpy*. (b) Use the following information to calculate the bond enthalpy for a C-H bond in methane, CH_4.

$$CH_4 + 2O_2 \rightarrow CO_2 + 2H_2O \quad \Delta H = -698 \text{ kJ mol}^{-1}$$

Bond:	C=O	O=O	O-H
Bond Enthalpy (kJ mol^{-1}):	743	496	463

(CCEA June 2010)

4. Use the following bond energies to calculate the enthalpy change for the reaction:

$$C_2H_4 + 3O_2 \rightarrow 2CO_2 + 2H_2O$$

Bond:	C=C	C-H	O-H	O=O	C=O
Bond Enthalpy (kJ mol^{-1}):	612	412	464	497	803

(Adapted from CCEA June 2005)

5. Ethanol can be used as a biofuel. The enthalpy of combustion of ethanol can be calculated using bond enthalpies. (a) Explain what is meant by the term *average bond enthalpy*. (b) Use the average bond enthalpies given in the table below to calculate the enthalpy of combustion of ethanol. (c) Using experimental data the standard enthalpy of combustion of ethanol is found to be −1407 kJ mol^{-1}. Explain the difference between this value and that obtained using average bond enthalpies.

$$CH_3CH_2OH + 3O_2 \rightarrow 2CO_2 + 3H_2O$$

Bond:	C-O	C-H	O-H	C=O	O=O	C-C
Bond Enthalpy (kJ mol^{-1}):	358	413	464	805	498	347

(Adapted from CCEA June 2012)

6. (a) Use the following bond enthalpies to calculate the enthalpy change for the combustion of one mole of ammonia. (b) Explain why the reaction is exothermic.

$$4NH_3 + 3O_2 \rightarrow 2N_2 + 6H_2O$$

Bond:	N-H	O=O	N≡N	O-H
Bond Enthalpy (kJ mol^{-1}):	391	498	945	464

(CCEA January 2007)

7. The reaction between fluorine and diborane, a hydride of boron, produces a large amount of energy. Use the following bond enthalpies to calculate the enthalpy change when one mole of diborane reacts completely with fluorine.

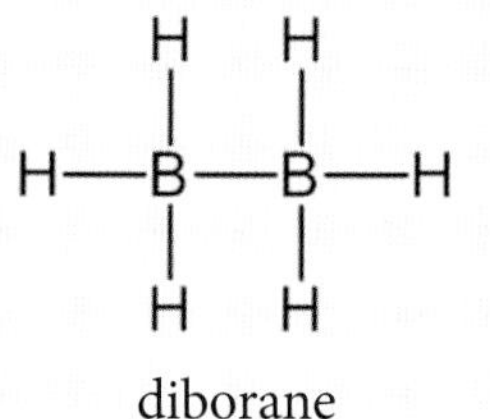

diborane

$$B_2H_6 + 6F_2 \rightarrow 6HF + 2BF_3$$

Bond:	F-F	B-H	B-B	H-F	B-F
Bond Enthalpy (kJ mol^{-1}):	158	389	293	566	627

(Adapted from CCEA June 2006)

Before moving to the next section, check that you are able to:

- Describe the relationship between the energy involved in making and breaking bonds for exothermic and endothermic reactions.
- Explain what is meant by the term *average bond enthalpy*.
- Use bond enthalpies to *estimate* the enthalpy change for a reaction.
- Explain why enthalpy changes calculated using bond enthalpies are estimates.

Enthalpy of Formation

The **standard enthalpy of formation,** $\Delta_f H^\circ$ of a substance is *the enthalpy change when one mole of the substance is formed from its elements under standard conditions.*

For example, the standard enthalpy of formation for propane, $\Delta_f H^\circ(C_3H_8)$ is the standard enthalpy change when one mole of propane is formed from carbon (graphite) and hydrogen gas under standard conditions.

$$3C_{(s)} + 4H_{2\,(g)} \rightarrow C_3H_{8\,(g)} \qquad \Delta H^\circ = \Delta H_f^\circ(C_3H_8)$$

Similarly, the standard enthalpy of formation for sodium chloride, $\Delta_f H^\circ(NaCl)$ is the standard enthalpy change when one mole of sodium chloride is formed from sodium metal and chlorine gas under standard conditions.

$$Na_{(s)} + \frac{1}{2}Cl_{2\,(g)} \rightarrow NaCl_{(s)} \qquad \Delta H^\circ = \Delta H_f^\circ(NaCl)$$

Worked Example 2.8viii

Which one of the following defines the standard enthalpy of formation for water?

A $H_{2(g)} + \frac{1}{2}O_{2(g)} \rightarrow H_2O_{(g)}$

B $2H_{(g)} + O_{(g)} \rightarrow H_2O_{(l)}$

C $2H_{2(g)} + O_{2(g)} \rightarrow 2H_2O_{(l)}$

D $H_{2(g)} + \frac{1}{2}O_{2(g)} \rightarrow H_2O_{(l)}$

Strategy

- Reactions B and D form one mole of water in its standard state.
- In reaction D the reactants are elements in their standard states.

Solution

Answer D.

Exercise 2.8G

1. Ammonia is a gas under standard conditions. Which one of the following defines the standard enthalpy of formation for ammonia?

   A $N_{2(g)} + 3H_{2(g)} \rightarrow 2NH_{3(g)}$

   B $N_{2(g)} + 3H_{2(g)} \rightarrow 2NH_{3(l)}$

   C $\frac{1}{2}N_{2(g)} + \frac{3}{2}H_{2(g)} \rightarrow NH_{3(l)}$

   D $\frac{1}{2}N_{2(g)} + \frac{3}{2}H_{2(g)} \rightarrow NH_{3(g)}$

2. Which one of the following represents the standard enthalpy of formation for potassium bromide?

   A $2K_{(s)} + Br_{2(g)} \rightarrow 2KBr_{(s)}$

   B $2K_{(s)} + Br_{2(l)} \rightarrow 2KBr_{(s)}$

   C $K_{(s)} + \frac{1}{2}Br_{2(g)} \rightarrow KBr_{(s)}$

   D $K_{(s)} + \frac{1}{2}Br_{2(l)} \rightarrow KBr_{(s)}$

3. Which one of the following represents the standard enthalpy change for the formation of ethanol?

   A $2C_{(g)} + 6H_{(g)} + O_{(g)} \rightarrow C_2H_5OH_{(g)}$

   B $2C_{(s)} + 3H_{2(g)} + O_{(g)} \rightarrow C_2H_5OH_{(l)}$

   C $2C_{(s)} + 3H_{2(g)} + \frac{1}{2}O_{2(g)} \rightarrow C_2H_5OH_{(g)}$

   D $2C_{(s)} + 3H_{2(g)} + \frac{1}{2}O_{2(g)} \rightarrow C_2H_5OH_{(l)}$

   (CCEA January 2011)

4. Which one of the following equations corresponds to a standard enthalpy of formation?

   A $2NO_{(g)} \rightarrow N_{2(g)} + O_{2(g)}$

   B $2H_{2(g)} + O_{2(g)} \rightarrow 2H_2O_{(g)}$

   C $Na_{(s)} + Cl_{(g)} \rightarrow NaCl_{(s)}$

   D $Mg_{(s)} + Br_{2(l)} \rightarrow MgBr_{2(s)}$

   (CCEA June 2009)

5. (a) Use the following bond enthalpies to calculate the standard enthalpy of formation for hydrogen fluoride, HF. (b) Draw an enthalpy diagram for the reaction.

Bond:	H-H	F-F	H-F
Bond Enthalpy (kJ mol^{-1}):	436	158	568

(CCEA January 2004)

Having defined standard enthalpies of formation in this way it follows that *the standard enthalpy of formation for an element is zero*, and the standard enthalpy change for a reaction can be calculated by subtracting the total enthalpy of formation for the reactants from the total enthalpy of formation for the products (Equation 4).

$\Delta H^\circ = \Delta_f H^\circ(\text{products}) - \Delta_f H^\circ(\text{reactants})$ Equation 4

Worked Example 2.8ix

A highly explosive mixture of hydrazine, N_2H_4 and hydrogen peroxide, H_2O_2 was used to power the first jet aircraft. Use the following enthalpies of formation to calculate the enthalpy change for the reaction between hydrazine and hydrogen peroxide.

$N_2H_4 + 2H_2O_2 \rightarrow N_2 + 4H_2O$

	N_2H_4	H_2O_2	H_2O
Standard enthalpy of formation (kJ mol^{-1}):	+50	−191	−286

(CCEA January 2005)

Solution

$\Delta_f H^\circ(\text{products}) = \Delta_f H^\circ(N_2) + 4\,\Delta_f H^\circ(H_2O)$

$= 0 + 4(-286) = -1144$ kJ

$\Delta_f H^\circ(\text{reactants}) = \Delta_f H^\circ(N_2H_4) + 2\,\Delta_f H^\circ(H_2O_2)$

$= 50 + 2(-191) = -332$ kJ

Enthalpy change per mole of hydrazine reacted:

$\Delta H^\circ = \Delta_f H^\circ(\text{products}) - \Delta_f H^\circ(\text{reactants})$

$= -1144 - (-332) = -812$ kJ mol^{-1}

Exercise 2.8H

1. The standard enthalpy of formation of water is −286 kJ mol^{-1}. Calculate the enthalpy change when hydrogen reacts with oxygen under standard conditions to form 1.0 g of water.

   (Adapted from CCEA January 2012)

2. Ammonia is a gas under standard conditions. The standard enthalpy change for the formation of ammonia is −46.2 kJ mol^{-1}. What is the enthalpy change (in kJ) for the reaction:

   $2NH_{3(g)} \rightarrow N_{2(g)} + 3H_{2(g)}$

   (Adapted from CCEA January 2007)

Before moving to the next section, check that you are able to:

- Explain what is meant by the *standard enthalpy of formation* for a substance.
- Use standard enthalpies of formation to calculate the standard enthalpy change for a reaction.

Enthalpy of Combustion

The **standard enthalpy of combustion,** $\Delta_c H^\circ$ of a substance is *the enthalpy change when one mole of the substance is completely burnt in oxygen under standard conditions.*

For example, the standard enthalpy of combustion for butane is the enthalpy change that occurs when one mole of butane, C_4H_{10} burns to form carbon dioxide and water under standard conditions.

$$C_4H_{10(g)} + \frac{13}{2}O_{2(g)} \rightarrow 4CO_{2(g)} + 5H_2O_{(l)}$$

$$\Delta H^\circ = \Delta_c H^\circ(C_4H_{10})$$

Carbon dioxide and water are very stable compounds and a great deal of energy is released when they are formed. As a result, the combustion of hydrocarbons is highly exothermic, and many are good fuels.

Worked Example 2.8x

Burning glucose releases energy:

$C_6H_{12}O_{6(s)} + 6O_{2(g)} \rightarrow 6CO_{2(g)} + 6H_2O_{(l)}$

The standard enthalpy of combustion for glucose may be measured directly, or calculated using standard enthalpies of formation. Use the following standard enthalpies of formation to calculate the standard enthalpy of combustion for glucose.

Substance:	$CO_{2(g)}$	$H_2O_{(l)}$	$C_6H_{12}O_{6(s)}$
$\Delta_f H^\circ$ (kJ mol^{-1}):	−394	−286	−1273

(CCEA June 2005)

Solution

$\Delta_f H^\circ(\text{products}) = 6\,\Delta_f H^\circ(CO_2) + 6\,\Delta_f H^\circ(H_2O)$
$= 6(-394) + 6(-286) = -4080$ kJ

$\Delta_f H^\circ(\text{reactants}) = \Delta_f H^\circ(C_6H_{12}O_6) + 6\,\Delta_f H^\circ(O_2)$
$= -1273 + 6(0) = -1273$ kJ

Enthalpy change per mole of glucose burnt:

$\Delta H_c^{\ominus} = \Delta_f H^{\ominus}(\text{products}) - \Delta_f H^{\ominus}(\text{reactants})$
$= -4080 - (-1273) = -2807 \text{ kJ mol}^{-1}$

Exercise 2.8I

1. Butane is sold commercially as 'bottled gas'. The standard enthalpy of combustion for butane is $-2876.5 \text{ kJ mol}^{-1}$. Calculate the amount of energy released for every kilogram of carbon dioxide produced.

   $2C_4H_{10} + 13O_2 \rightarrow 8CO_2 + 10H_2O$

   (CCEA June 2014)

2. (a) Define the term *standard enthalpy of combustion*. (b) Use the following standard enthalpies of formation to calculate the enthalpy change for the complete combustion of methane:

   $CH_4 + 2O_2 \rightarrow CO_2 + 2H_2O$.

Substance:	$CO_{2(g)}$	$H_2O_{(l)}$	$CH_{4(g)}$
$\Delta_f H^{\ominus}$ (kJ mol^{-1}):	−394	−286	−75

(CCEA June 2010)

3. (a) Define the term *standard enthalpy of combustion*. (b) Use the standard enthalpies of formation given in the table to calculate the standard enthalpy of combustion of ethanol.

   $C_2H_5OH_{(l)} + 3O_{2(g)} \rightarrow 2CO_{2(g)} + 3H_2O_{(l)}$

Compound	$\Delta_f H^{\ominus}$ (kJ mol^{-1})
$C_2H_5OH_{(l)}$	−277
$CO_{2(g)}$	−394
$H_2O_{(l)}$	−286

(CCEA January 2013)

4. (a) Use the standard enthalpies of formation given in the table to calculate the standard enthalpy of combustion for butan-1-ol. (b) State the conditions used for measuring standard enthalpies of formation. (c) Explain why no value is given for the standard enthalpy of formation of oxygen.

   $C_4H_9OH_{(l)} + 6O_{2(g)} \rightarrow 4CO_{2(g)} + 5H_2O_{(l)}$

Compound	$\Delta_f H^{\ominus}$ (kJ mol^{-1})
$C_4H_9OH_{(l)}$	−327
$CO_{2(g)}$	−394
$H_2O_{(l)}$	−286

(Adapted from CCEA June 2012)

5. Kerosine is a mixture that contains the compound dodecane, $C_{12}H_{26}$. (a) Write an equation for the complete combustion of dodecane. (b) Draw an enthalpy diagram for the combustion of dodecane. (c) Explain, in terms of enthalpy, why the combustion of dodecane is exothermic.

   (CCEA January 2008)

6. (a) Define the term *standard enthalpy of formation*. (b) Use the following standard enthalpies of formation to calculate the standard enthalpy change for photosynthesis:

   $6CO_{2(g)} + 6H_2O_{(l)} \rightarrow C_6H_{12}O_{6(s)} + 6O_{2(g)}$

	$CO_{2(g)}$	$H_2O_{(l)}$	$C_6H_{12}O_{6(s)}$
Standard enthalpy of formation (kJ mol^{-1})	−394	−286	−1273

(CCEA June 2003)

Before moving to the next section, check that you are able to:

- Explain what is meant by the *standard enthalpy of combustion* for a substance.
- Use standard enthalpies of formation to calculate the standard enthalpy of combustion for a reaction.

Hess's Law

In this section we are learning to:

- Use Hess's law to calculate enthalpy changes that are difficult to measure directly in the laboratory.

Many of the enthalpy changes that accompany chemical reactions can be measured using a calorimeter. Enthalpy changes that are too difficult to measure under normal laboratory conditions must instead be calculated using Hess's law. For example, the enthalpy change when anhydrous copper(II) sulfate dissolves in water (ΔH_1), and the enthalpy change when the hydrate $CuSO_4.5H_2O_{(s)}$ dissolves in water (ΔH_2), can be measured in the laboratory using a simple calorimeter.

$CuSO_{4(s)} \rightarrow CuSO_{4(aq)}$ $\quad\quad \Delta H = \Delta H_1$

$CuSO_4.5H_2O_{(s)} \rightarrow CuSO_{4(aq)}$ $\quad\quad \Delta H = \Delta H_2$

In contrast, the enthalpy change that occurs when anhydrous copper(II) sulfate, $CuSO_{4(s)}$ reacts with water to form the hydrate $CuSO_4.5H_2O_{(s)}$ (ΔH_3) is difficult to measure, and must instead be calculated from ΔH_1 and ΔH_2 using Hess's law.

$CuSO_{4(s)} + 5H_2O_{(l)} \rightarrow CuSO_4.5H_2O_{(s)}$ $\quad \Delta H = \Delta H_3$

According to **Hess's law** *the enthalpy change for a chemical reaction depends on the state of the reactants and products and does not depend on the way in which the reactants are converted to the products.* Thus, according to Hess's law the enthalpy change that occurs when anhydrous copper(II) sulfate dissolves (ΔH_1) must be equivalent to the total change in enthalpy when the hydrate $CuSO_4.5H_2O$ is formed from anhydrous copper(II) sulfate (ΔH_3), and then dissolved to form a solution of copper(II) sulfate (ΔH_2). This relationship between the enthalpy changes can be used to construct the enthalpy cycle in Figure 5, and can also be expressed in the form of Equation 5.

$\Delta H_1 = \Delta H_3 + \Delta H_2$ $\quad\quad$ Equation 5

In this way Hess's law is a statement of the principle that energy is conserved, being neither created or destroyed, as it is transformed from one form to another during a chemical reaction.

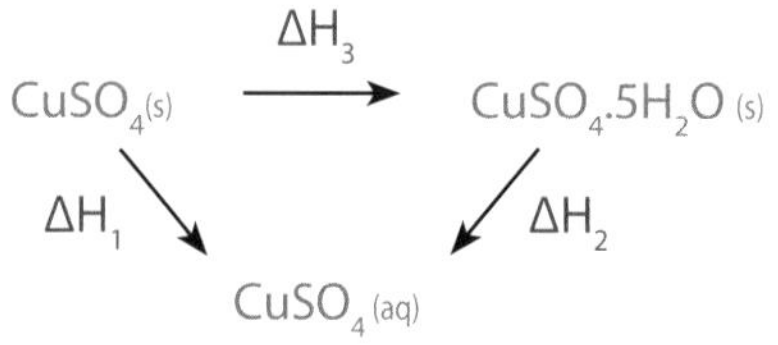

ΔH_3: $CuSO_{4(s)} \rightarrow CuSO_4.5H_2O_{(s)}$
ΔH_1: $CuSO_{4(s)} \rightarrow CuSO_{4(aq)}$
ΔH_2: $CuSO_4.5H_2O_{(s)} \rightarrow CuSO_{4(aq)}$

Figure 5: Enthalpy cycle to calculate the enthalpy change for the reaction of anhydrous copper(II) sulfate with water to form the hydrate $CuSO_4.5H_2O$.

Worked Example 2.8xi

The following diagram shows the enthalpy changes that are used to determine the enthalpy change for the reaction between anhydrous copper(II) sulfate and water to form the hydrate $CuSO_4.5H_2O_{(s)}$.

ΔH_1: $CuSO_{4(s)} \rightarrow CuSO_{4(aq)}$
ΔH: $CuSO_{4(s)} \rightarrow CuSO_4.5H_2O_{(s)}$
ΔH_2: $CuSO_4.5H_2O_{(s)} \rightarrow CuSO_{4(aq)}$

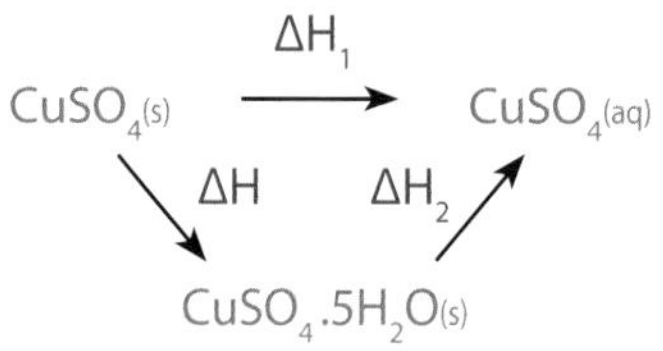

(a) Dissolving 8.0 g of anhydrous copper(II) sulfate, $CuSO_{4(s)}$ in 100 g of water produces a rise in temperature of 1.2 °C. Calculate the enthalpy change ΔH_1. The specific heat capacity of water is 4.2 J °C^{-1} g^{-1}.

(b) When 12.5 g of hydrated copper(II) sulfate, $CuSO_4.5H_2O_{(s)}$ dissolves in 100 g of water the temperature of the solution drops by 1.4 °C. Calculate the enthalpy change ΔH_2. The specific heat capacity of water is 4.2 J °C^{-1} g^{-1}.

(c) Use the values for ΔH_1 and ΔH_2 to calculate the molar enthalpy change ΔH.

(Adapted from CCEA June 2011)

Solution

(a) $q = m\,c\,\Delta T = 100\text{ g} \times 4.2\text{ J °C}^{-1}\text{ g}^{-1} \times 1.2\text{ °C}$
$= 5.0 \times 10^2\text{ J} = 0.50\text{ kJ}$

$$\Delta H_1 = \frac{-q}{\text{Moles of } CuSO_4} = \frac{-0.50\text{ kJ}}{0.050\text{ mol}} = -10\text{ kJ mol}^{-1}$$

(b) $q = m\,c\,\Delta T = 100\text{ g} \times 4.2\text{ J }^{\circ}\text{C}^{-1}\text{ g}^{-1} \times (-1.4\ ^{\circ}\text{C})$
$= -5.9 \times 10^2\text{ J} = -0.59\text{ kJ}$

$$\Delta H_2 = \frac{-q}{\text{Moles of } CuSO_4.5H_2O} = \frac{0.59\text{ kJ}}{0.050\text{ mol}}$$

$= 12\text{ kJ mol}^{-1}$

(c) Applying Hess's law gives: $\Delta H_1 = \Delta H + \Delta H_2$
$\Delta H = \Delta H_1 - \Delta H_2 = (-10) - 12 = -22\text{ kJ mol}^{-1}$

Exercise 2.8J

1. The enthalpy change for the combustion of graphite is –393.5 kJ mol^{-1} and the enthalpy change for the combustion of diamond is –395.4 kJ mol^{-1}. Calculate the enthalpy change for the reaction: C(graphite) → C(diamond).

   (CCEA June 2011)

2. When 0.10 mol of anhydrous $MgSO_4$ was dissolved in 100 g of water the temperature rose by 9.0 °C. When 0.10 mol of $MgSO_4.7H_2O$ was dissolved in 100 g of water, the temperature dropped by 3.0 °C. The specific heat capacity for water is 4.2 J °C^{-1} g^{-1}. Use this information to calculate the enthalpy change for the hydration of magnesium sulfate:

   $MgSO_{4(s)} + 7H_2O_{(l)} \rightarrow MgSO_4.7H_2O_{(s)}$

   (Adapted from CCEA June 2009)

3. (a) Explain what is meant by the term *standard enthalpy of combustion*. (b) State Hess's law. (c) Use the following enthalpy changes to calculate the enthalpy of combustion for octane, C_8H_{18}.

   $C_8H_{18(l)} + \frac{25}{2}O_{2(g)} \rightarrow 8CO_{2(g)} + 9H_2O_{(l)}$

   $C_{(s)} + O_{2(g)} \rightarrow CO_{2(g)}$ $\quad \Delta H^{\circ} = -393.5\text{ kJ mol}^{-1}$

   $8C_{(s)} + 9H_{2(g)} \rightarrow C_8H_{18(l)}$ $\quad \Delta H^{\circ} = -250.0\text{ kJ mol}^{-1}$

   $H_{2(g)} + \frac{1}{2}O_{2(g)} \rightarrow H_2O_{(l)}$ $\quad \Delta H^{\circ} = -286.0\text{ kJ mol}^{-1}$

   (CCEA January 2008)

4. The standard enthalpies of combustion for carbon, hydrogen and ethyne, C_2H_2 are given below. Use this data to calculate the standard enthalpy of formation for ethyne.

   $C_{(s)} + O_{2(g)} \rightarrow CO_{2(g)}$ $\quad \Delta_cH^{\circ} = -394\text{ kJ mol}^{-1}$

   $H_{2(g)} + \frac{1}{2}O_{2(g)} \rightarrow H_2O_{(l)}$ $\quad \Delta_cH^{\circ} = -286\text{ kJ mol}^{-1}$

   $C_2H_{2(g)} + \frac{5}{2}O_{2(g)} \rightarrow 2CO_{2(g)} + H_2O_{(l)}$

   $\Delta_cH^{\circ} = -1300\text{ kJ mol}^{-1}$

   (Adapted from CCEA January 2011)

5. Butane is sold commercially as 'bottled gas'. (a) Write an equation, with state symbols, which represents the standard enthalpy of formation for butane. (b) Use the following enthalpies of combustion to calculate the standard enthalpy of formation for butane, C_4H_{10}.

	C	H_2	C_4H_{10}
Standard enthalpy of combustion (kJ mol^{-1})	–393.5	–285.8	–2876.5

   (Adapted from CCEA June 2014)

6. The standard enthalpies of combustion for carbon (C), hydrogen (H_2), and propane (C_3H_8) are –394, –286, and –2219 kJ mol^{-1} respectively. Calculate the standard enthalpy of formation for propane in kJ mol^{-1}.

   (Adapted from CCEA June 2013)

7. Use the following standard enthalpies of combustion to calculate the standard enthalpy of formation for ethanoic acid, $CH_3COOH_{(l)}$.

	$C_{(s)}$	$H_{2(g)}$	$CH_3COOH_{(l)}$
Standard enthalpy of combustion (kJ mol^{-1})	–393	–286	–487

   (Adapted from CCEA June 2007)

8. The enthalpy change for the thermal decomposition of sodium hydrogencarbonate (ΔH_1) can be determined using Hess's law. (a) State Hess's law. (b) Use the following enthalpy changes and energy cycle to calculate the enthalpy change ΔH_1.

$$2NaHCO_{3(s)} \xrightarrow{\Delta H_1} Na_2CO_{3(s)} + H_2O_{(l)} + CO_{2(g)}$$

ΔH_3 (from $2NaHCO_{3(s)}$), ΔH_2 (from $Na_2CO_{3(s)} + H_2O_{(l)} + CO_{2(g)}$), both to:

$$2NaCl_{(aq)} + 2H_2O_{(l)} + 2CO_{2(g)}$$

$NaHCO_{3(s)} + HCl_{(aq)} \rightarrow NaCl_{(aq)} + H_2O_{(l)} + CO_{2(g)}$
$\Delta H = +16$ kJ

$Na_2CO_{3(s)} + 2HCl_{(aq)} \rightarrow 2NaCl_{(aq)} + H_2O_{(l)} + CO_{2(g)}$
$\Delta H = -21$ kJ

(Adapted from CCEA June 2004)

Before moving to the next section, check that you are able to:

- Recall what is meant by *Hess's law* and explain why Hess's law can be considered an example of the principle that energy must be conserved.
- Use Hess's law to determine enthalpy changes.

2.9 Equilibrium

CONNECTIONS

- The transport of oxygen around the human body depends on the ability of oxygen to bind reversibly to the protein haemoglobin in red blood cells.
- The pH of blood is maintained in the narrow range 7.35-7.45 by an equilibrium involving dissolved carbon dioxide and hydrogencarbonate ions.

Dynamic Equilibrium

In this section we are learning to:

- Explain how a reversible reaction can give rise to a state of dynamic equilibrium in a reaction mixture.

The term **reversible reaction** describes *a reaction that can operate in the forward direction (reactants → products) and the reverse direction (products → reactants)*. The melting of ice is an example of a reversible reaction as a block of ice will melt if its temperature is above 0 °C, and will reform if the water formed by the melting is cooled to below 0 °C.

Water freezes below 0°C	Heat →	Ice melts above 0°C
$H_2O(l) \rightarrow H_2O(s)$	← Cool	$H_2O(s) \rightarrow H_2O(l)$

If a mixture of ice and water is maintained at exactly 0 °C a **dynamic equilibrium** is established in which ice is converted to water at the same rate as water is converted to ice, and the amount of ice and water in the mixture remains constant. In this context the term *dynamic* indicates that *the forward reaction is operating at the same rate as the reverse reaction,* and the term *equilibrium* indicates that *the amount of each reactant and product in the mixture remains constant.*

Figure 1: The dynamic equilibrium between ice and water at 0 °C.

The dynamic equilibrium between ice and water at 0 °C is shown in Figure 1, and is represented by the equation: $H_2O(s) \leftrightharpoons H_2O(l)$, where the double-arrow ($\leftrightharpoons$) is used to indicate a state of dynamic equilibrium.

Many industrial processes that produce chemicals on a large scale involve reactions operating in a state of dynamic equilibrium. For example, ammonia (NH_3) is used in the manufacture of fertilisers, and is produced on an industrial scale by reacting a mixture of nitrogen and hydrogen at a high temperature and pressure in the presence of a catalyst.

Formation of ammonia (forward reaction):

$N_2 + 3H_2 \rightarrow 2NH_3$

The reaction between nitrogen and hydrogen is reversible, and the ammonia formed in the reaction immediately begins to decompose and reform nitrogen and hydrogen.

Decomposition of ammonia (reverse reaction):

$2NH_3 \rightarrow N_2 + 3H_2$

The rate at which ammonia decomposes increases as the amount of ammonia in the reaction mixture increases. Eventually a state of dynamic equilibrium is established in which the rate at which ammonia is formed (the forward reaction) is equal to the rate at which ammonia decomposes (the reverse reaction), and the composition of the reaction mixture remains constant.

The changes in concentration that occur when a mixture of nitrogen and hydrogen react are illustrated in Figure 2. The shaded region denotes a state of dynamic equilibrium.

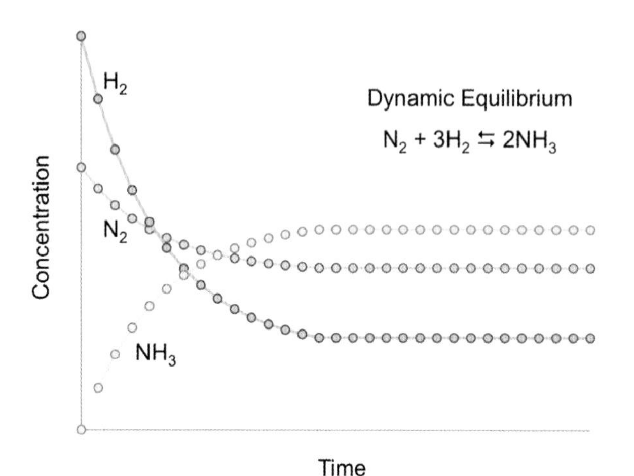

Figure 2: The changes in concentration that occur when a mixture of nitrogen and hydrogen react and establish a dynamic equilibrium (shown shaded).

Exercise 2.9A

1. Nitrogen dioxide, NO_2 exists in dynamic equilibrium with dinitrogen tetroxide, N_2O_4. Explain the term *dynamic equilibrium*.

   $N_2O_{4(g)} \leftrightharpoons 2NO_{2(g)}$ $\Delta H = +56$ kJ

   (CCEA January 2007)

Before moving to the next section, check that you are able to:

- Explain what is meant by the term *reversible reaction.*
- Describe how a reversible reaction can give rise to a state of dynamic equilibrium and define the term *dynamic equilibrium.*

Factors Affecting Equilibrium

In this section we are learning to:

- Explain how changing the temperature, pressure and composition of a dynamic equilibrium affects the amount of product formed at equilibrium.
- Recall the effect of a catalyst on the composition of an equilibrium.

Changes in Composition

The composition of a dynamic equilibrium depends on the properties of the reactants and products, and the conditions under which the equilibrium is established. If the composition of a dynamic equilibrium is suddenly changed, the mixture is no longer at equilibrium, and the composition of the mixture adjusts until a new dynamic equilibrium is established as shown in Figure 3.

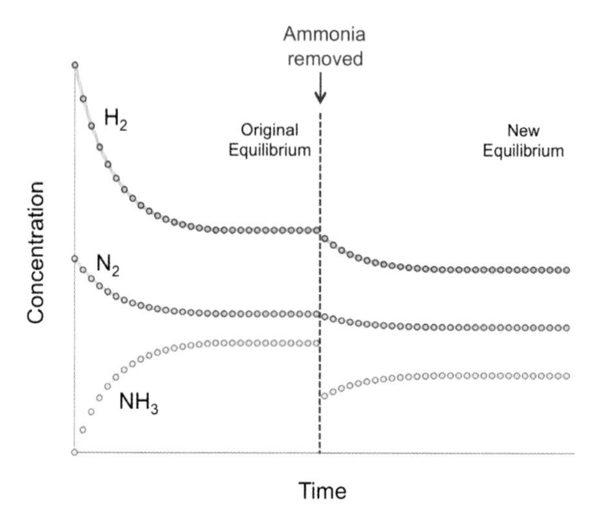

Figure 3: The effect of removing ammonia on the composition of the dynamic equilibrium $N_2 + 3H_2 \leftrightharpoons 2NH_3$.

The concentrations in Figure 3 reveal that, on removing ammonia, the mixture attempts to re-establish a state of dynamic equilibrium by using the forward reaction to increase the amount of ammonia in the mixture.

Making more ammonia decreases the amount of nitrogen and hydrogen in the reaction mixture. As a result, the amount of nitrogen and hydrogen present when equilibrium is re-established is lower than in the original equilibrium. In this way removing ammonia shifts the **position of equilibrium** to the right in favour of ammonia.

Adding nitrogen or hydrogen to the dynamic equilibrium will also shift the position of equilibrium to the right. Conversely, the position of the equilibrium can be shifted to the left by adding ammonia, or removing some of the reactants.

Add reactants or remove products

$N_2 + 3H_2 \leftrightharpoons 2NH_3$ $N_2 + 3H_2 \leftrightharpoons 2NH_3$

Left Position of equilibrium Right

Add products or remove reactants

Together these observations show that *when the composition of a dynamic equilibrium is changed, the mixture attempts to re-establish equilibrium by responding in a way that reverses the change.*

Worked Example 2.9i

The reaction between bromine and water is represented by the following equation. Which one of the following reagents would move the equilibrium position to the right?

$$Br_2 + H_2O \leftrightarrows Br^- + OBr^- + 2H^+$$

A nitric acid
B sodium carbonate
C sodium bromide
D sulfuric acid

(CCEA June 2011)

Strategy

- Substances that increase the concentration of Br^-, OBr^- or H^+ will shift the position of equilibrium to the left.
- Substances that decrease the concentration of Br^-, OBr^- or H^+ will shift the position of equilibrium to the right.

Solution

Answer B

Explanation

- Sodium carbonate (B) forms an alkali when it dissolves in water. The hydroxide ions in the alkali react with H^+ ions in the equilibrium to form water: $H^+ + OH^- \rightarrow H_2O$. The position of equilibrium then shifts to the right to replace the H^+ ions.
- Adding nitric acid (A) or sulfuric acid (D) increases the concentration of H^+ and shifts the position of equilibrium to the left.
- Adding sodium bromide (C) increases the concentration of Br^- and shifts the position of equilibrium to the left.

Exercise 2.9B

1. The salt ammonium ethanoate, CH_3COONH_4 partially dissociates on heating to form ammonia and ethanoic acid. (a) Write an equilibrium equation for the dissociation. (b) Explain why heating ammonium ethanoate in the presence of ethanoic acid prevents dissociation.

   (Adapted from CCEA January 2012)

2. Hydrogen cyanide is manufactured by passing a mixture of ammonia and methane over a platinum catalyst. Suggest how the yield of hydrogen cyanide is affected by removing hydrogen cyanide from the reaction mixture as the reaction proceeds.

   $NH_{3(g)} + CH_{4(g)} \leftrightarrows HCN_{(g)} + 3H_{2(g)}$

   (Adapted from CCEA January 2010)

3. The following equilibrium is established when excess silver chloride is added to aqueous ammonia. Which one of the following occurs when aqueous sodium chloride is added?

   $AgCl_{(s)} + 2NH_{3(aq)} \leftrightarrows Ag(NH_3)_2^+{}_{(aq)} + Cl^-{}_{(aq)}$

   A $AgCl_{(s)}$ dissolves
   B More $Ag(NH_3)_2^+{}_{(aq)}$ is formed
   C More $AgCl_{(s)}$ is precipitated
   D $NH_4Cl_{(s)}$ is formed

   (CCEA May 2014)

4. Bromine reacts with water according to the following equation. Which one of the following will move the equilibrium to the right?

   $Br_2 + H_2O \leftrightarrows HOBr + HBr$

   A Adding bromide ions
   B Adding hydrogen ions
   C Decreasing the concentration of bromine
   D Increasing the concentration of hydroxide ions

   (CCEA May 2015)

Before moving to the next section, check that you are able to:

- Predict and explain the effect of changing the composition of a dynamic equilibrium in terms of the position of equilibrium.

Adding a Catalyst

In industry the reaction between nitrogen and hydrogen is carried out in the presence of a catalyst. The catalyst increases the rate of the forward and reverse reactions by equal amounts. As a result *adding a catalyst increases the rate at which equilibrium is established, and does not affect the composition of the equilibrium mixture as the rate of the forward reaction and the rate of the reverse reaction increase by an equal amount.*

The effect of adding a catalyst to a mixture of hydrogen and nitrogen as it reacts to form ammonia is illustrated by Figure 4.

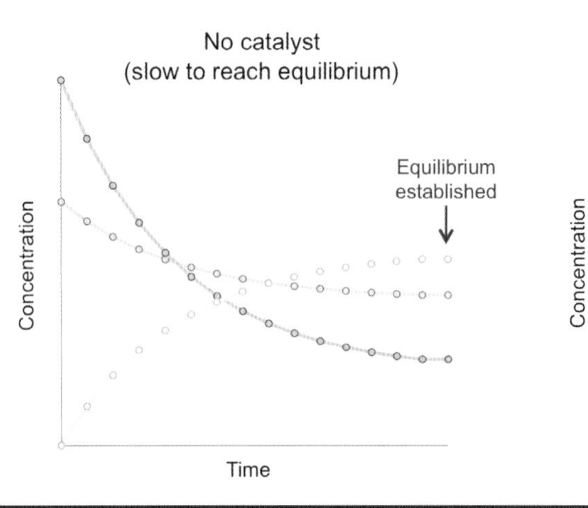

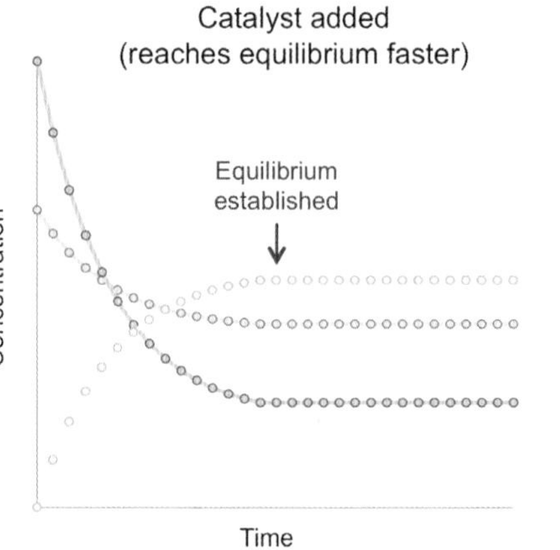

Figure 4: The effect of adding a catalyst to a mixture of hydrogen and nitrogen as it reacts to form ammonia $N_2 + 3H_2 \rightarrow 2NH_3$.

Before moving to the next section, check that you are able to:

- Explain how a catalyst affects the rate at which a dynamic equilibrium is established, and the composition of the dynamic equilibrium.

Changes in Pressure

The effect of increasing the pressure in a dynamic equilibrium formed from a mixture of hydrogen and nitrogen is illustrated in Figure 5. As the pressure is increased the rate of the forward reaction increases in order to reduce the amount of gas in the mixture, and the position of equilibrium shifts to the right until a new dynamic equilibrium is established.

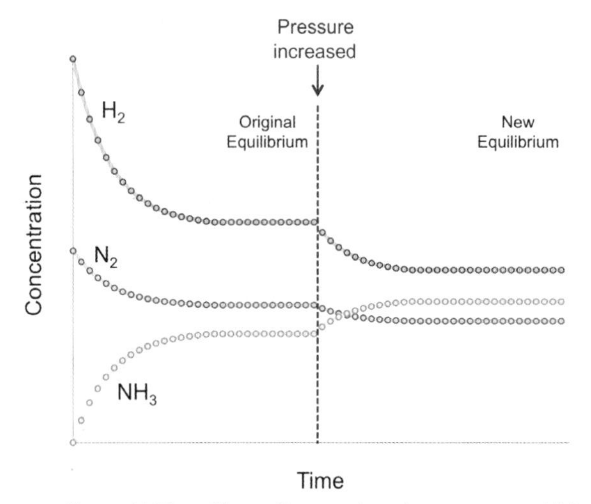

Figure 5: The effect of increasing the pressure within the dynamic equilibrium $N_2 + 3H_2 \leftrightarrows 2NH_3$.

Conversely, if the pressure in the dynamic equilibrium is reduced, the rate of the reverse reaction increases in an attempt to increase the amount of gas in the mixture, and the position of equilibrium shifts to the left until a new dynamic equilibrium is established.

Increase pressure (forms less molecules)

$N_2 + 3H_2 \leftrightarrows 2NH_3$ | Left | Position of equilibrium | Right | $N_2 + 3H_2 \leftrightarrows 2NH_3$

Decrease pressure (forms more molecules)

Together these observations show that *when the pressure in a dynamic equilibrium is changed, the mixture attempts to re-establish equilibrium by responding in a way that reverses the change.*

Worked Example 2.9ii

Hydrogen cyanide is manufactured by passing a mixture of ammonia and methane over a platinum catalyst. Explain if a high pressure should be used in the manufacture of hydrogen cyanide.

$$NH_{3(g)} + CH_{4(g)} \leftrightarrows HCN_{(g)} + 3H_{2(g)}$$

(CCEA January 2010)

Solution

A high pressure would shift the position of equilibrium to the left as there are more molecules on the right side of the equilibrium. As a result a high pressure should not be used as it would decrease the yield of hydrogen cyanide.

Exercise 2.9C

1. Nitrogen dioxide, NO_2 is a brown gas that reacts to form the colourless gas dinitrogen tetroxide, N_2O_4. State and explain the effect of reducing the total pressure on the equilibrium.

   $N_2O_{4(g)} \leftrightarrows 2NO_{2(g)}$ $\quad \Delta H = +58 \text{ kJ mol}^{-1}$

   (Adapted from CCEA May 2008)

2. Hydrogen gas for the Haber process is produced by reacting methane with steam. Explain how the equilibrium concentration of hydrogen gas would be affected by operating the process at a higher pressure.

   $CH_4(g) + H_2O(g) \leftrightarrows CO(g) + 3H_2(g)$

Before moving to the next section, check that you are able to:

- Predict and explain the effect of changing the pressure in a dynamic equilibrium in terms of the position of equilibrium.

Changes in Temperature

The effect of increasing the temperature in a dynamic equilibrium formed from a mixture of hydrogen and nitrogen is illustrated in Figure 6.

In this equilibrium the forward reaction ($N_2 + 3H_2 \rightarrow 2NH_3$) is exothermic. As a result the reverse reaction ($2NH_3 \rightarrow N_2 + 3H_2$) is endothermic, and increasing the temperature increases the rate of the reverse reaction in an attempt to reduce the amount of heat energy in the mixture. In this way the position of equilibrium shifts to the left until a new dynamic equilibrium is established.

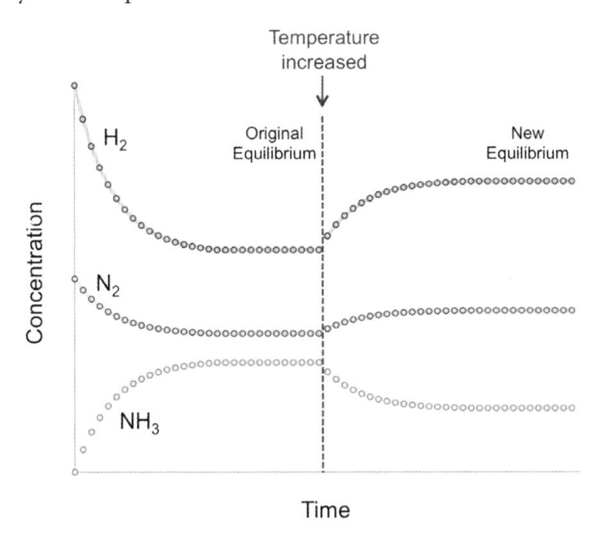

Figure 6: The effect of increasing the temperature within the dynamic equilibrium $N_2 + 3H_2 \leftrightarrows 2NH_3$.

Conversely, if the temperature in the equilibrium is reduced, the rate of the forward reaction increases in an attempt to increase the amount of heat energy in the mixture, and the position of equilibrium shifts to the right until a new dynamic equilibrium is established.

Decrease temperature (exothermic reaction)

$N_2 + 3H_2 \leftrightarrows 2NH_3$ Left — Position of equilibrium — Right $N_2 + 3H_2 \leftrightarrows 2NH_3$

Increase temperature (endothermic reaction)

Together these observations show that *when the temperature in a dynamic equilibrium is changed, the mixture attempts to re-establish equilibrium by responding in a way that reverses the change.*

Worked Example 2.9iii

Hydrogen cyanide is manufactured by passing a mixture of ammonia and methane over a platinum catalyst. The reaction is endothermic. Suggest why the reaction is carried out at 1000 °C.

$NH_3(g) + CH_4(g) \leftrightarrows HCN(g) + 3H_2(g)$

(CCEA January 2010)

Solution

The forward reaction is endothermic therefore a high temperature would increase the yield of hydrogen cyanide by shifting the position of equilibrium to the right. The high temperature would also increase the rate at which hydrogen cyanide is formed.

Exercise 2.9D

1. Sulfuryl chloride, SO_2Cl_2 dissociates at high temperatures. State and explain the effect of increasing the temperature on the extent of dissociation.

   $SO_2Cl_2(g) \leftrightarrows SO_2(g) + Cl_2(g)$ $\Delta H = +93\ kJ\ mol^{-1}$

   (CCEA May 2010)

2. Methanol is manufactured by the reaction of carbon monoxide with hydrogen. The reaction is carried out at 250 °C and 50 atmospheres pressure. Describe the effect of increasing temperature on the equilibrium yield of methanol.

   $CO(g) + 2H_2(g) \leftrightarrows CH_3OH(g)$ $\Delta H = -91\ kJ\ mol^{-1}$

   (Adapted from CCEA May 2010)

3. When heated, ammonium cyanide decomposes to form ammonia and hydrogen cyanide. An equilibrium is formed in a sealed test tube. Explain if the reverse reaction is exothermic or endothermic.

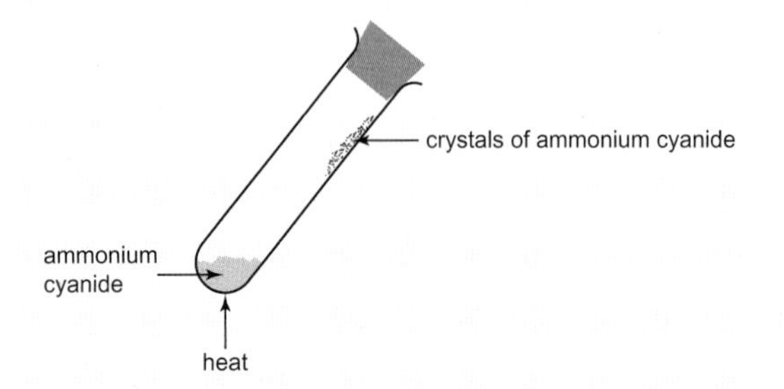

(CCEA May 2014)

4. Methanol, CH_3OH is produced by the reaction between carbon dioxide and hydrogen. If the reaction is carried out at a higher temperature the equilibrium yield of methanol decreases. Determine the sign of the enthalpy change for the forward reaction and explain your reasoning.

$$CO_{2(g)} + 3H_{2(g)} \leftrightarrows CH_3OH_{(g)} + H_2O_{(g)}$$

(Adapted from CCEA January 2008)

5. Hydrogen is produced on an industrial scale by the reaction between methane and steam. The reaction is endothermic and is carried out at a pressure of 40 atmospheres in the presence of a nickel catalyst. Explain why the reaction is also carried out at 1023 K.

$$CH_{4(g)} + 2H_2O_{(g)} \leftrightarrows CO_{2(g)} + 4H_{2(g)}$$

(CCEA June 2009)

6. When solid ammonium chloride is heated, it dissociates to form a mixture of gases:

$$NH_4Cl_{(s)} \leftrightarrows NH_{3(g)} + HCl_{(g)}$$

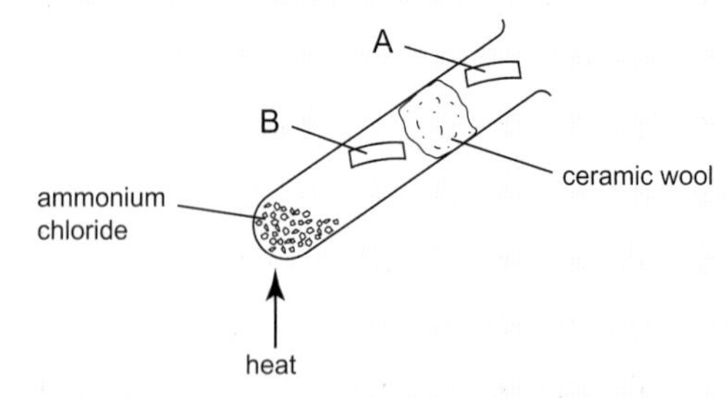

(a) Explain why the pH paper changes to a blue colour at A. (b) Explain why the pH paper turns a red colour at B. (c) Suggest the purpose of the ceramic wool. (d) Determine if the forward reaction is exothermic or endothermic and explain your reasoning.

(CCEA June 2011)

Before moving to the next section, check that you are able to:

- Predict and explain the effect of changing the temperature in a dynamic equilibrium in terms of the position of equilibrium.

Additional Problems

Exercise 2.9E

1. Phosphorous pentachloride, PCl_5 vaporises and decomposes to form an equilibrium mixture. State and explain the effect of (a) increasing the pressure and (b) increasing the temperature on the equilibrium.

$$PCl_{5(g)} \leftrightarrows PCl_{3(g)} + Cl_{2(g)} \qquad \Delta H = +91 \text{ kJ mol}^{-1}$$

(CCEA May 2009)

2. The polymer Teflon, polytetrafluoroethene, is made by polymerising tetrafluoroethene which itself is produced from chlorodifluoromethane according to the equation:

$$2CHClF_{2(g)} \leftrightarrows C_2F_{4(g)} + 2HCl_{(g)} \qquad \Delta H = +128 \text{ kJ mol}^{-1}$$

State and explain the effect of (a) increasing the overall pressure and (b) increasing the temperature on the yield of tetrafluoroethene.

(CCEA January 2010)

3. Phosgene, $COCl_2$ is a colourless gas which condenses readily to a liquid at 8 °C. It is manufactured by passing carbon monoxide and chlorine through a bed of porous carbon which acts as a catalyst. The reaction is exothermic and is carried out at a temperature between 50 °C and 150 °C. (a) Explain the effect of increased pressure on the equilibrium. (b) Explain why the temperature is in the range 50–150 °C. (c) Suggest how phosgene can be removed from the reaction mixture.

$$CO_{(g)} + Cl_{2(g)} \leftrightarrows COCl_{2(g)}$$

(CCEA January 2008)

4. Which one of the following statements regarding the following chemical equilibrium is correct?

   $2SO_{2(g)} + O_{2(g)} \leftrightarrows 2SO_{3(g)}$ $\Delta H = -192$ kJ

   A Increasing the concentration of oxygen, O_2 increases the amount of sulfur dioxide, SO_2 at equilibrium.

   B Decreasing the pressure increases the concentration of sulfur trioxide, SO_3 at equilibrium.

   C Increasing the temperature increases the concentration of sulfur dioxide, SO_2 at equilibrium.

   D Removing sulfur trioxide, SO_3 from the reaction increases the concentration of SO_2 at equilibrium.

   (Adapted from CCEA June 2006)

5. The following equilibrium reaction takes place in a catalytic converter. (a) State and explain the effect of a decrease in temperature on the above equilibrium. (b) State and explain the effect of an increase in pressure on the above equilibrium.

   $2CO + 2NO \leftrightarrows 2CO_2 + N_2$ $\Delta H = -745$ kJ mol^{-1}

   (CCEA January 2014)

6. The second stage in the production of nitric acid involves the reaction of nitrogen monoxide with oxygen to form nitrogen dioxide:

   $2NO_{(g)} + O_{2(g)} \leftrightarrows 2NO_{2(g)}$ $\Delta H = -116$ kJ mol^{-1}

   (a) Explain what is meant by the term *dynamic equilibrium*. State and explain how (b) increasing the pressure, (c) increasing the temperature and (d) adding a catalyst will affect the equilibrium yield of nitrogen dioxide.

   (CCEA June 2010)

7. The first step in the Ostwald Process for the manufacture of nitric acid is the oxidation of ammonia. The reaction is carried out at approximately 900 °C. (a) Explain how the yield of nitrogen(II) oxide is affected by increasing the temperature to 1500 °C. (b) Explain how increasing the pressure would affect the yield of nitrogen(II) oxide. (c) Explain the effect, if any, the catalyst has on the yield of the nitrogen(II) oxide.

   $4NH_{3(g)} + 5O_{2(g)} \leftrightarrows 4NO_{(g)} + 6H_2O_{(g)}$
   $\Delta H = -950$ kJ mol^{-1}

   (CCEA June 2012)

8. The Birkeland-Eyde process for the manufacture of nitric acid was developed in 1903. In the first step of the process, nitrogen and oxygen react to form nitrogen(II) oxide. Explain the effect, if any, of each of the following changes on the yield of nitrogen(II) oxide.

   $N_{2(g)} + O_{2(g)} \leftrightarrows 2NO_{(g)}$ $\Delta H = +180$ kJ mol^{-1}

   (a) Increasing the temperature.
   (b) Adding more nitrogen.
   (c) Increasing the pressure.
   (d) Adding a catalyst.

   (CCEA January 2011)

9. Ethylene glycol, CH_2OHCH_2OH is produced by passing a mixture of ethene and air over silver metal at high temperature. The reaction occurs in two steps.

   Step 1 $CH_2{=}CH_2 + \frac{1}{2}O_2 \leftrightarrows CH_2OCH_2$
   ethylene oxide

   Step 2 $CH_2OCH_2 + H_2O \rightarrow CH_2OHCH_2OH$

   (a) Suggest the purpose of the silver. Using an equilibrium argument explain why (b) a high pressure and (c) a high temperature is used.

   (CCEA January 2005)

10. The hydrogen needed for the production of ammonia is obtained from methane or naphtha. Previously hydrogen was obtained from coke and steam using a two stage process known as the water gas reaction.

    Stage 1 $C_{(s)} + H_2O_{(g)} \leftrightarrows CO_{(g)} + H_{2(g)}$
    $\Delta H = +122$ kJ

    Stage 2 $CO_{(g)} + H_2O_{(g)} \leftrightarrows CO_{2(g)} + H_{2(g)}$
    $\Delta H = -41$ kJ

    Explain the effect, if any, of (a) increasing the temperature, (b) increasing the pressure and (c) adding a catalyst on the position of equilibrium in stage 2.

    (CCEA January 2004)

Describing Equilibrium

In this section we are learning to:

- Use the concept of an equilibrium constant to describe the composition of an equilibrium mixture and the extent to which a reaction has occurred.

Equilibrium Constants

When potassium chromate(VI), K_2CrO_4 is dissolved in water the following equilibrium between chromate(VI) ions, CrO_4^{2-} and dichromate(VI) ions, $Cr_2O_7^{2-}$ is established.

$$2CrO_4^{2-}(aq) + 2H^+(aq) \leftrightharpoons Cr_2O_7^{2-}(aq) + H_2O(l)$$

On adding alkali the position of equilibrium moves to the left and the yellow colour of chromate(VI) ion, CrO_4^{2-} dominates as shown in Figure 7a. Conversely, on adding acid, the position of the equilibrium moves to the right and the orange colour of dichromate(VI) ion, $Cr_2O_7^{2-}$ (orange) dominates as shown in Figure 7b.

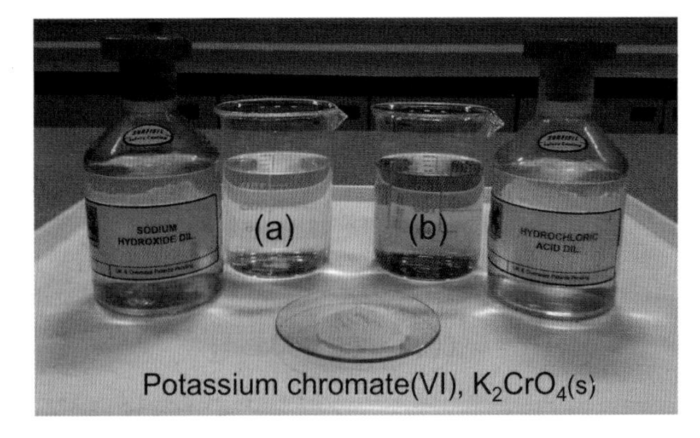

Figure 7: (a) The yellow colour of chromate(VI) ion, CrO_4^{2-} dominates in alkaline solution. (b) The orange colour of dichromate(VI) ion, $Cr_2O_7^{2-}$ dominates in acidic solution.

The composition of the equilibrium mixtures in Figure 7 is described by the **equilibrium constant, K_c** for the reaction; *a positive number that relates the concentrations of the species in the equilibrium, and does not change unless the temperature in the equilibrium changes.*

The equilibrium constant, K_c for the chromate-dichromate equilibrium is calculated by inserting the concentration of chromate(VI) ion, $[CrO_4^{2-}]$ the concentration of dichromate(VI) ion, $[Cr_2O_7^{2-}]$ and the concentration of hydrogen ion, $[H^+]$ into the equation:

$$K_c = \frac{[Cr_2O_7^{2-}]}{[CrO_4^{2-}]^2[H^+]^2}$$

The concentration of water is not included in the expression for K_c as water is the solvent for the reaction and its concentration will not change significantly if the amounts of the other species in the equilibrium change.

If we had instead defined the equilibrium by the equation:

$$Cr_2O_7^{2-}(aq) + H_2O(l) \leftrightharpoons 2CrO_4^{2-}(aq) + 2H^+(aq)$$

the equilibrium constant would have been written:

$$K_c' = \frac{[CrO_4^{2-}]^2[H^+]^2}{[Cr_2O_7^{2-}]} = \frac{1}{K_c}$$

In this way the expression for the equilibrium constant, K_c depends on how the chemical equation is written, and it becomes necessary to specify how the chemical equation is written when talking about the equilibrium constant for a chemical reaction.

Before moving to the next section, check that you are able to:

- Explain what is meant by the *equilibrium constant* for a reaction.
- Explain why solvents are not included in the expression for K_c.
- Relate the value of K_c for a reaction to the value of K_c for the reverse reaction.

Defining K_c

The examples in Table 1 demonstrate how the following relationship can be used to define K_c for a **homogeneous reaction**; a reaction in which *the reactants and products are all in the same physical state.*

$$aA + bB \leftrightharpoons cC + dD \qquad K_c = \frac{[C]^c[D]^d}{[A]^a[B]^b}$$

The following worked example demonstrates how the expression for K_c can be used to deduce the units for K_c. The relationship between the expression for K_c and the units of K_c is further illustrated by the examples in Table 1.

Chemical Reaction	Equilibrium Constant	Units for K_c
$N_2O_4(g) \leftrightarrows 2NO_2(g)$	$K_c = \frac{[NO_2]^2}{[N_2O_4]}$	$\frac{(mol\ dm^{-3})^2}{mol\ dm^{-3}} = mol\ dm^{-3}$
$H_2(g) + I_2(g) \leftrightarrows 2HI(g)$	$K_c = \frac{[HI]^2}{[H_2]\ [I_2]}$	$\frac{(mol\ dm^{-3})^2}{(mol\ dm^{-3})^2}$ = no units
$2SO_2(g) + O_2(g) \leftrightarrows 2SO_3(g)$	$K_c = \frac{[SO_3]^2}{[SO_2]^2\ [O_2]}$	$\frac{(mol\ dm^{-3})^2}{(mol\ dm^{-3})^3} = mol^{-1}\ dm^3$

Table 1: The equilibrium constant, K_c for several homogeneous reactions.

Worked Example 2.9iv

Deduce the units of K_c for the equilibrium $N_{2(g)} + 3H_{2(g)} \leftrightarrows 2NH_{3(g)}$.

(Adapted from CCEA January 2011)

Strategy

- Write the expression for K_c.
- Each concentration has units of mol dm^{-3}.

Solution

The equilibrium constant, $K_c = \frac{[NH_3]^2}{[N_2]\ [H_2]^3}$

The units for K_c are

$$\frac{(mol\ dm^{-3})^2}{(mol\ dm^{-3})^4} = \frac{1}{(mol\ dm^{-3})^2} = mol^{-2}\ dm^6.$$

A different procedure is used to define K_c for a **heterogeneous reaction** in which *the reactants and products are not all in the same physical state.*

Exercise 2.9F

1. The solubility of iodine in water is increased by adding potassium iodide as aqueous iodine combines with iodide ions to form I_3^- ions. (a) Write an expression for the equilibrium constant K_c. (b) Deduce the units for K_c.

   $I_{2(aq)} + I^-_{(aq)} \leftrightarrows I_3^-{}_{(aq)}$

   (CCEA June 2004)

2. A catalyst containing metals such as platinum is used to speed up the conversion of carbon monoxide to carbon dioxide in car exhaust gases. The process can be represented by the following equilibrium. (a) Write an expression for the equilibrium constant K_c. (b) Deduce the units of K_c for this process.

   $2CO_{(g)} + O_{2(g)} \leftrightarrows 2CO_{2(g)}$

3. (a) Write an expression for K_c, and state the units of K_c, for each of the following homogeneous reactions. (b) Explain what is meant by the term *homogeneous reaction.*

   A $CO_{(g)} + H_2O_{(g)} \leftrightarrows CO_{2(g)} + H_{2(g)}$

   B $2H_{2(g)} + O_{2(g)} \leftrightarrows 2H_2O_{(g)}$

   C $4NH_{3(g)} + 5O_{2(g)} \leftrightarrows 4NO_{(g)} + 6H_2O_{(g)}$

Before moving to the next section, check that you are able to:

- Explain what is meant by *homogeneous reaction* and *heterogeneous reaction.*
- Write an expression for the equilibrium constant of a homogeneous reaction.
- Use the expression for K_c to deduce the units for K_c.

Interpreting K_c

The value of K_c for an equilibrium can be used to estimate the extent to which a reaction has occurred. For most reactions a value for K_c in the range 10^{-5} to 10^5 indicates that there are significant amounts of reactants and products in the equilibrium.

In contrast, a value for K_c larger than 10^5 indicates that essentially all of the reactants have been converted to products, and the double arrow (⇆) can be replaced by a single arrow (→) to indicate that the reaction has gone to completion.

Similarly, a value for K_c smaller than 10^{-5} indicates that essentially none of the reactants have been converted to products and no reaction has occurred.

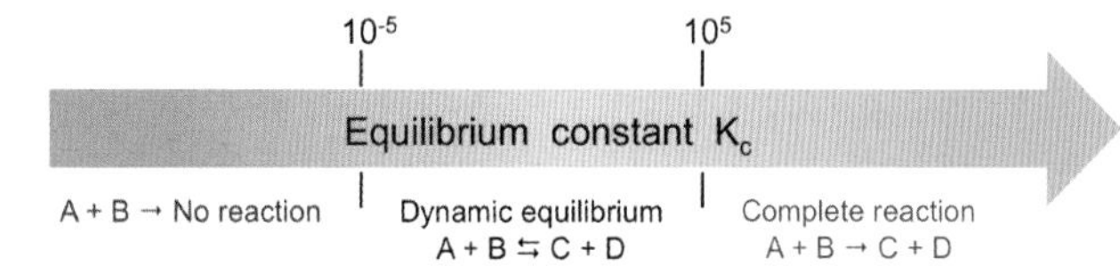

Exercise 2.9G

1. The ester pentyl ethanoate, $CH_3COOC_5H_{11}$ is formed from pentan-1-ol and ethanoic acid.

   $CH_3COOH + C_5H_{11}OH$

   $\leftrightarrows CH_3COOC_5H_{11} + H_2O$

   (a) Write an expression for the equilibrium constant, K_c. (b) State the units for K_c. (c) The value of K_c is 6.47 at 330 K and 8.31 and 350 K. State if the reaction is exothermic or endothermic. Explain your answer.

   (Adapted from CCEA May 2013)

2. The formation of ammonium carbonate is being investigated as a possible method to remove carbon dioxide from combustion processes.

   $CO_2 + 2NH_3 + H_2O \leftrightarrows (NH_4)_2CO_3$

   The equilibrium constant for the reaction at 20 °C is 2.5×10^4. At 80 °C the equilibrium constant is 0.12. Explain the drop in the value of K and state how this would affect the CO_2 removal process.

   (Adapted from CCEA January 2012)

3. Ozone is 30 % dissociated at equilibrium. (a) Explain the effect of increasing pressure on the equilibrium. (b) What effect would an increase in pressure have on the value of K_c?

   $2O_{3(g)} \leftrightarrows 3O_{2(g)}$

   (CCEA May 2012)

4. Which one of the following statements about the equilibrium constant K_c is correct for the reaction shown below?

   $2SO_{2(g)} + O_{2(g)} \leftrightarrows 2SO_{3(g)}\ \Delta H = -114\ kJ\ mol^{-1}$

   A K_c increases on adding a catalyst.

   B K_c has units of mol dm^{-3}.

   C K_c decreases on increasing the temperature.

   D K_c increases on increasing the pressure.

   (Adapted from CCEA January 2002)

Before moving to the next section, check that you are able to:

- Recall that the value of K_c increases as the position of equilibrium moves to the right.
- Use the value of K_c for a reaction to deduce the extent of the reaction.

Applications of Equilibrium

In this section we are learning to:

- Recall the conditions used to produce ammonia by the Haber-Bosch process and the conditions used to make sulfuric acid by the Contact process.
- Use equilibrium arguments to account for the reaction conditions used in each process.

Production of Ammonia

In industry ammonia is produced by the Haber-Bosch process. A mixture of nitrogen and hydrogen react in the presence of a granulated iron catalyst at a temperature of 450 °C and a total pressure of 250 atmospheres.

The Haber-Bosch process:

$N_{2(g)} + 3H_{2(g)} \leftrightarrows 2NH_{3(g)} \qquad \Delta H = -92\ kJ$

The reaction conditions are chosen to maximise profits by producing ammonia as quickly and as cheaply as possible. This is achieved by *using a catalyst to increase the rate at which equilibrium is established, and removing ammonia from the reaction mixture as it forms.*

The effect of temperature and pressure on the amount of ammonia at equilibrium is shown in Figure 8. At higher pressures the yield of ammonia increases as the position of equilibrium moves to the right in order to reduce the number of molecules in the equilibrium. As a result we might expect the process to be more profitable at high pressure. However, the increased cost of maintaining equipment capable of operating at high pressure reduces profits to the extent that it becomes necessary to operate the process at a pressure of 250 atmospheres. In this way *an operating pressure of 250 atmospheres represents a compromise between a better yield of ammonia at higher pressure, and the increased cost of operating the process at higher pressures.*

A similar compromise is involved in choosing to operate the process at a temperature of 450 °C. The effect of temperature on the equilibrium yield of ammonia in Figure 6 reveals the extent to which the reverse reaction ($2NH_3 \rightarrow N_2 + 3H_2$) is used to absorb heat, shifting the position of equilibrium to the left, as the operating temperature increases.

A higher yield of ammonia at lower temperatures is not desirable as reducing the temperature reduces the rate at which equilibrium is established and, as a result, the rate at which ammonia is produced. In this way *an operating temperature of 450 °C represents a compromise between maintaining a satisfactory rate of reaction and keeping the temperature low enough to obtain a reasonable yield of ammonia in the equilibrium.*

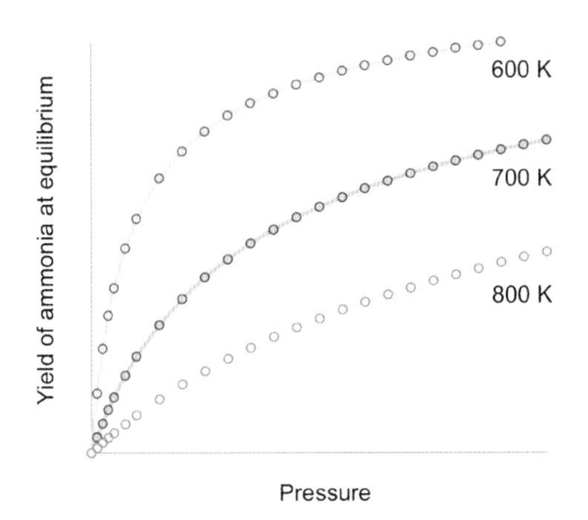

Figure 8: The effect of temperature and pressure on the equilibrium yield of ammonia obtained from the Haber-Bosch process.

Exercise 2.9H

1. (a) State the temperature and pressure used in the production of ammonia by the Haber process. (b) Why is the temperature used for the Haber process described as a compromise temperature?

   (CCEA June 2010)

2. The Haber process for the manufacture of ammonia involves the equilibrium reaction between nitrogen and hydrogen. (a) Write the equation for the reaction. (b) Name the catalyst used in the Haber process. (c) Explain why a combination of high pressure and low temperature would maximise the yield of ammonia.

   (CCEA January 2009)

3. The graph shows how the equilibrium yield of ammonia varies with temperature for the production of ammonia by the Haber-Bosch process. Use the graph to determine if the reaction is exothermic or endothermic.

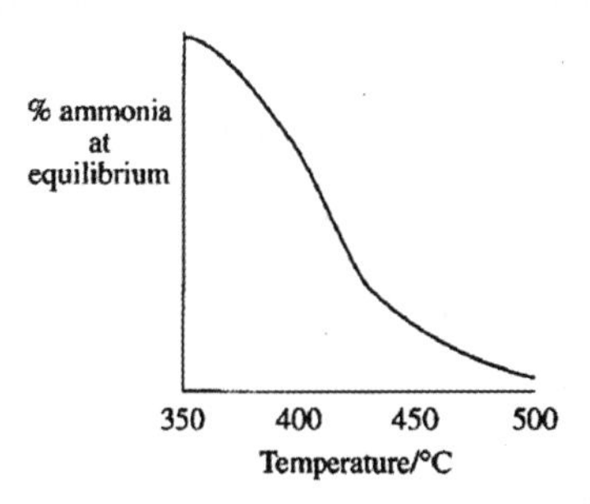

(Adapted from CCEA June 2005)

Before moving to the next section, check that you are able to:

- Recall the conditions used for the Haber-Bosch process.
- Use equilibrium arguments to explain why the temperature and pressure used in the Haber-Bosch process represent a compromise.

Manufacture of Sulfuric Acid

Sulfuric acid is used to make a wide range of useful materials including fertiliser, paints, detergents and pharmaceuticals. It is produced on an industrial scale by a process known as the Contact process. In the first stage of the process sulfur dioxide, SO_2 is produced by burning sulfur in air.

Step 1 $S_{(s)} + O_{2(g)} \rightarrow SO_{2(g)}$

The sulfur dioxide is then mixed with more air and heated to 450 °C before being passed over a solid vanadium(V) oxide, V_2O_5 catalyst. In the presence of the catalyst sulfur dioxide reacts with oxygen to form sulfur trioxide, SO_3 and a dynamic equilibrium is established.

Step 2 $2SO_{2(g)} + O_{2(g)} \leftrightarrows 2SO_{3(g)}$ $\Delta H = -197$ kJ

In the final stage of the process sulfuric acid is produced by dissolving the sulfur trioxide formed in Step 2 in sulfuric acid.

Step 3 $SO_{3(g)} + H_2SO_{4(aq)} + H_2O(l) \rightarrow 2H_2SO_{4(aq)}$

The reaction between sulfur dioxide and oxygen in Step 2 is exothermic. Increasing the operating temperature increases the rate at which the equilibrium is established, but decreases the amount of sulfur trioxide formed as the position of equilibrium shifts to the left in an effort to absorb heat. As a result, *choosing to operate the process at 450 °C represents a compromise between increasing the rate of reaction to establish the equilibrium more quickly, and higher yields of sulfur trioxide at lower temperatures.*

Operating Step 2 at high pressure would also be expected to increase the yield of sulfur trioxide as there are fewer gas molecules on the right side of the equilibrium. However, the additional cost of operating the process at high pressure cannot be justified as the yield of sulfur trioxide is already quite high when the process is operated at a slightly elevated pressure of 1-2 atmospheres.

Exercise 2.9I

1. Which set of conditions are used in the Contact Process?

Conditions	Catalyst	Pressure (atm)	Temp (°C)
A	Iron	1–2	250
B	Iron	200–1000	450
C	vanadium(V) oxide	1–2	450
D	vanadium(V) oxide	200–1000	250

(CCEA June 2008)

2. The Contact Process for the production of sulfuric acid involves the following equilibrium. (a) State and explain the effect of increasing the total pressure on the yield of sulfur trioxide. (b) Explain why the reaction is carried out at a moderate temperature.

$2SO_{2(g)} + O_{2(g)} \leftrightarrows 2SO_{3(g)}$ $\Delta H = -197$ kJ

(Adapted from CCEA June 2001)

Before moving to the next section, check that you are able to:

- Recall the conditions used for the Contact process.
- Explain why the temperature used in the Contact process represents a compromise and why a pressure of 1–2 atmospheres is sufficient for the Contact process.

2.10 Chemical Kinetics

CONNECTIONS

- Ahmed Zewail received the Nobel Prize for Chemistry in 1999 for his pioneering use of lasers to study bond making and breaking that occurs in femtoseconds during chemical reactions. One femtosecond, 1fs = 1×10^{-15} s.
- Many chemical reactions that occur in humans require an enzyme to make the reaction occur at a reasonable speed. The enzyme acts as a catalyst.

Rate of Reaction

In this section we are learning to:

- Use the concept of rate of reaction to suggest how the speed of a chemical reaction could be measured.

The speed or 'rate' of a chemical reaction can be measured using a wide variety of techniques. The technique chosen to measure the rate of a particular reaction depends on the nature of the reactants and products.

If a reaction produces a gas, the rate of the reaction can be determined by using a gas syringe to collect and measure the volume of gas produced at regular intervals during the reaction. The use of a gas syringe to measure the volume of hydrogen gas produced when magnesium reacts with dilute hydrochloric acid is shown in Figure 1.

$$Mg_{(s)} + 2HCl_{(aq)} \rightarrow MgCl_{2(aq)} + H_{2(g)}$$

If the total volume of hydrogen is measured every minute, the rate of the reaction can be expressed as the volume of hydrogen produced per minute. As the reaction proceeds the volume of gas produced every minute decreases, and becomes zero when the reaction is complete.

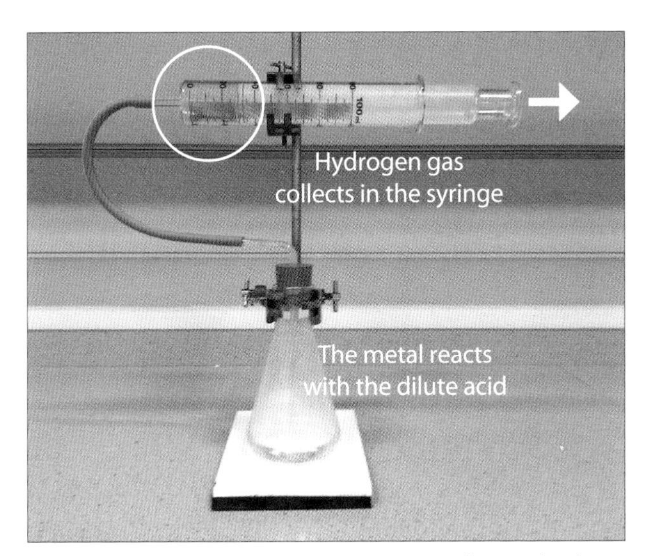

Figure 1: Using a gas syringe to measure the total volume of hydrogen gas produced at regular intervals when magnesium reacts with dilute hydrochloric acid.

In this way the **rate of reaction** can be defined as *the amount of substance that reacts, or the amount of product formed, per unit time.*

Alternatively, if a reaction produces a colour change, the rate of reaction can be calculated by measuring the time taken for the colour change to occur. In the reaction between aqueous sodium thiosulfate and dilute hydrochloric acid, the rate of reaction is calculated by measuring the time taken for the reaction to produce enough sulfur to make the solution too cloudy to see through.

$$Na_2S_2O_{3(aq)} + 2HCl_{(aq)} \rightarrow 2NaCl_{(aq)} + H_2O_{(l)} + SO_{2(g)} + S_{(s)}$$

A rate of reaction measured in this way is known as the **average rate of reaction** as it refers to *the average amount of a substance reacted or formed per unit time during the reaction*, and is calculated using the equation:

$$\text{Average rate} = \frac{1}{\text{Time for reaction}}$$

Before moving to the next section, check that you are able to:

- Explain what is meant by the term *rate of reaction*.
- Recall experimental techniques to measure the rate of a reaction.
- Explain how *the average rate* of a reaction differs from *the rate* of a reaction.

Factors Affecting Rate

In this section we are learning to:

- Use collision theory to describe how the reaction conditions affect the rate of a chemical reaction.

Collision Theory

The particles in liquids and gases are constantly moving and colliding with each other. The particles gain or lose energy when they collide and, as a result, are constantly changing speed and direction.

If particles collide with sufficient energy and the correct orientation they will react. A collision that results in a reaction is referred to as a **successful collision**. The role of orientation in determining if a collision will be successful is illustrated in Figure 2.

The rate of a chemical reaction is proportional to the number of successful collisions per second between reacting particles. As a result, if changing the reaction conditions increases the number of successful collisions per second, the rate of reaction will increase. Conversely, if changing the reaction conditions reduces the number of successful collisions per second, the rate of reaction will decrease.

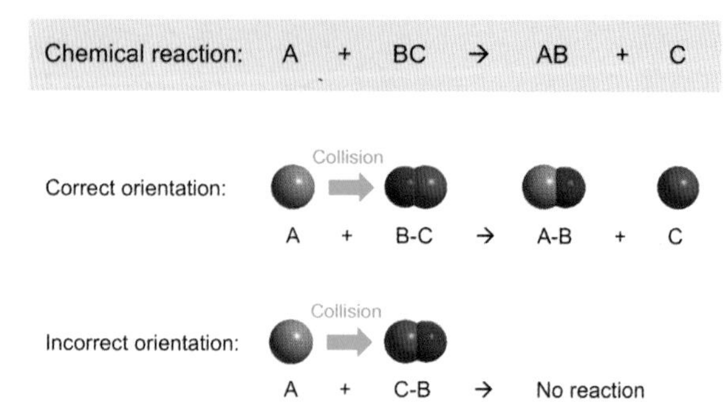

Figure 2: The role of orientation in determining if a collision is successful.

Effect of Concentration

In a mixture of gases the number of collisions per second can be increased by increasing the concentration of the gas particles in the mixture. This can be achieved by confining the mixture to a smaller volume or adding more gas to the mixture as illustrated in Figure 3.

Increasing the concentration does not affect the energy of the particles. As a result, when the concentration of the particles is increased, the particles have the same average speed and a smaller average separation (d_{avg}). *A smaller average separation results in more collisions per second and, if the particles react, more successful collisions per second.*

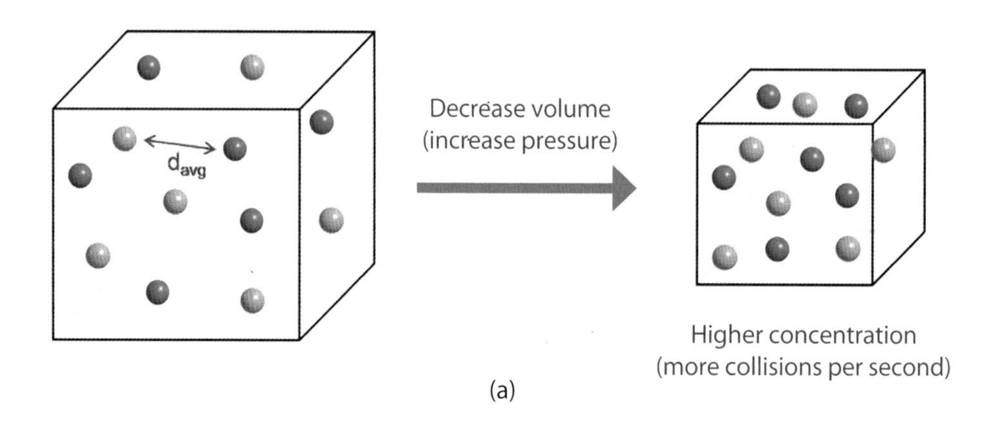

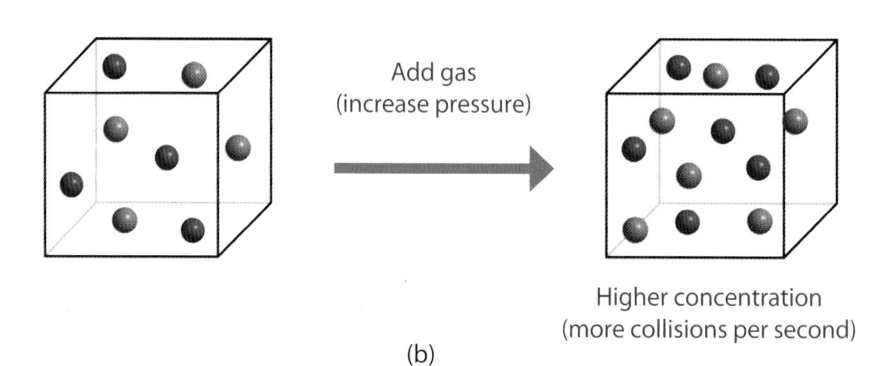

Figure 3: Increasing the concentration of particles in a mixture of gases by (a) reducing the volume of the mixture or (b) adding more of the mixture to the same volume. In each case the average separation of the particles, d_{avg} becomes smaller as the concentration increases.

Exercise 2.10A

1. The first step in the Ostwald Process for the manufacture of nitric acid is the oxidation of ammonia. Explain how increasing the pressure would affect the rate of the reaction.

   $4NH_{3(g)} + 5O_{2(g)} \leftrightarrows 4NO_{(g)} + 6H_2O_{(g)}$

   $\Delta H = -950\ kJ\ mol^{-1}$

   (CCEA June 2012)

Before moving to the next section, check that you are able to:

- Explain, in terms of successful collisions per second, why increasing the concentration of reacting particles increases the rate of reaction.

Effect of Temperature

The particles in a mixture of liquids or gases have a range of energies. The **Maxwell-Boltzmann distribution** in Figure 4 is a function that describes the number of particles in the mixture with a particular energy. As a result, *the total area under the distribution curve represents the total number of particles in the mixture.*

The form of the distribution curve reveals that all of the particles in the mixture have energy and are moving. The distribution also reveals that while a few particles have lots of energy and are moving very fast, most have a moderate amount of energy and move relatively slowly.

On defining the **activation energy, E_a** for a reaction to be *the minimum amount of energy needed for the reaction to occur*, the shaded area between the

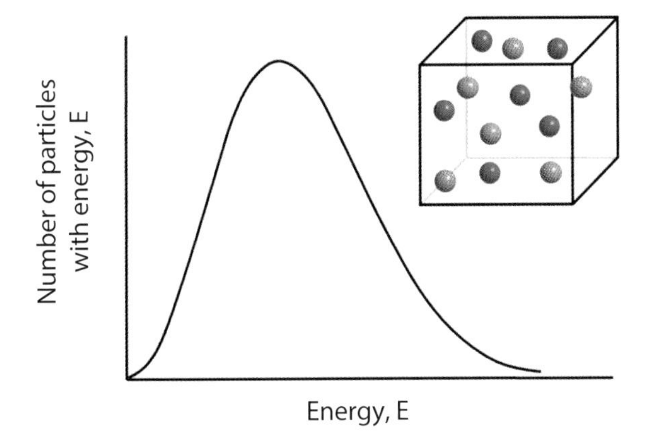

Figure 4: A Maxwell-Boltzmann distribution showing how energy is distributed amongst the particles in a mixture of liquids or gases.

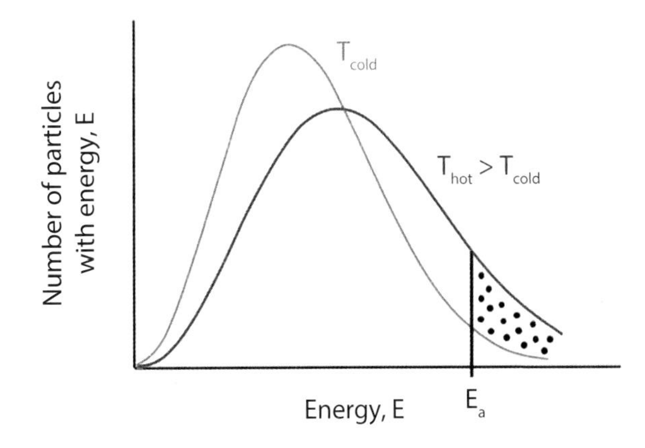

Figure 5: Maxwell-Boltzmann distributions for a mixture of liquids or gases at a temperature T_{cold} and the same mixture at a temperature T_{hot} where $T_{hot} > T_{cold}$. The shaded area represents the additional number of particles with energy greater than E_a as the temperature increases from T_{cold} to T_{hot}.

Maxwell-Boltzmann distributions in Figure 5 represents the additional number of particles with energy greater than E_a as the temperature in the mixture increases from T_{cold} to T_{hot}.

Increasing the number of particles with energy greater than E_a results in more collisions per second involving particles with enough energy to react. This in turn results in more successful collisions per second, and explains why the rate of reaction increases when the temperature of the reaction mixture is increased.

Exercise 2.10B

1. Which one of the following graphs most accurately represents the distribution of molecular energies in a gas at 500 K if the dotted curve represents the distribution for the same gas at 300 K? *(CCEA January 2010)*

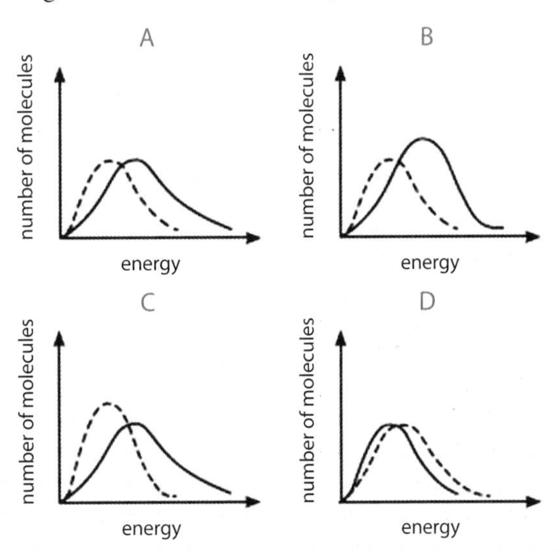

2. The energy distribution for the molecules in a mixture of nitrogen and oxygen at 298 K is shown below. (a) Label the axes and state the significance of the area under the curve. (b) Sketch the distribution of molecular energies for the mixture at a higher temperature on the same axes.

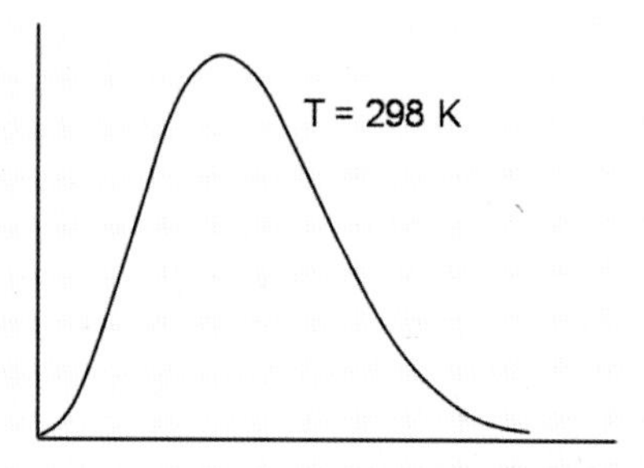

(CCEA January 2003)

3. The Maxwell-Boltzmann distribution for a reaction mixture is shown below. N is the number of molecules with the most probable energy and E_A is the activation energy. Which one of the following shows the effect of increasing the temperature on E_A and on N?

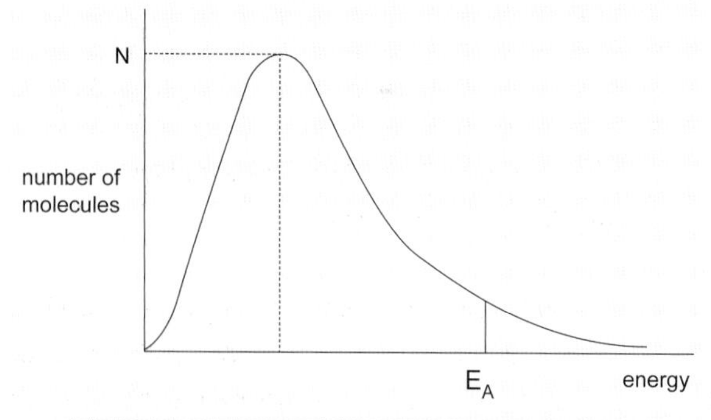

	E_A	N
A	constant	decreases
B	constant	increases
C	decreases	increases
D	decreases	increases

(Adapted from CCEA June 2014)

4. Chloroethane, CH_3CH_2Cl reacts with nucleophiles such as ammonia. The distribution of molecular energies in a gaseous mixture of chloroethane and ammonia at 20 °C is shown below. (a) Explain the significance of the shaded area. (b) Sketch the distribution for the same mixture at 30 °C and use the distributions to explain the difference between the rate of reaction at 20 °C and 30 °C.

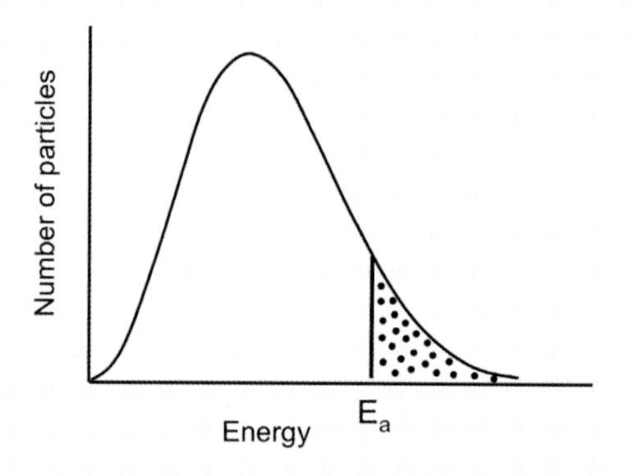

(Adapted from CCEA June 2003)

5. The Maxwell-Boltzmann distribution for a mixture of SO_2 and O_2 at 450 K is shown below. (a) Label the axes on the diagram and suggest why the curve starts at the origin. (b) Draw a second curve on the same axes to describe the mixture at 440 K. (c) With reference to the two curves state and explain the effect of reducing the temperature on the rate of reaction between SO_2 and O_2 to form SO_3.

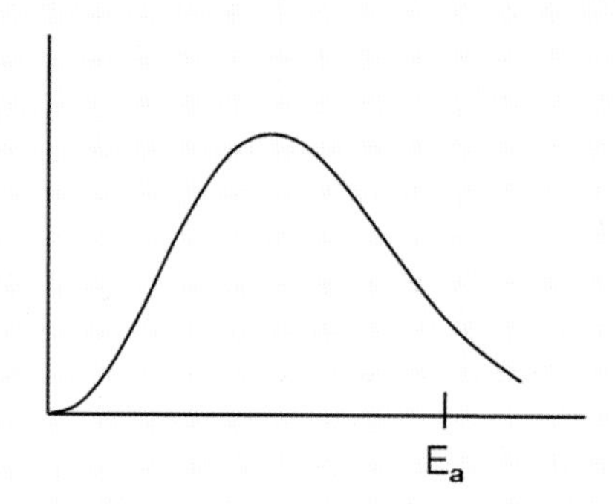

(CCEA June 2010)

6. The distribution of molecular kinetic energies in a mixture of CO and NO with a temperature T_1 is shown on the next page. The activation energy, E_a is indicated on the diagram. (a) Explain the term *activation energy*. (b) Explain why most collisions between molecules of CO and NO do not result in a reaction. (c) On the diagram draw the distribution of molecular kinetic energies at a higher temperature and label it T_2. (d) Using the distribution curves explain why the reaction between CO and NO is faster at the higher temperature.

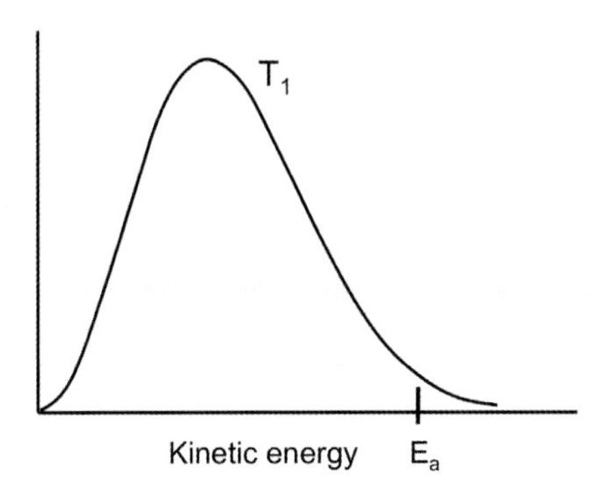

(Adapted from CCEA January 2008)

7. The distribution of energy amongst the molecules in a mixture of SO_2 and O_2 is shown below. Which one of the following changes increases the proportion of molecules with enough energy to react and form SO_3?

$2SO_2 + O_2 \leftrightarrows 2SO_3$ $\Delta H = -92$ kJ

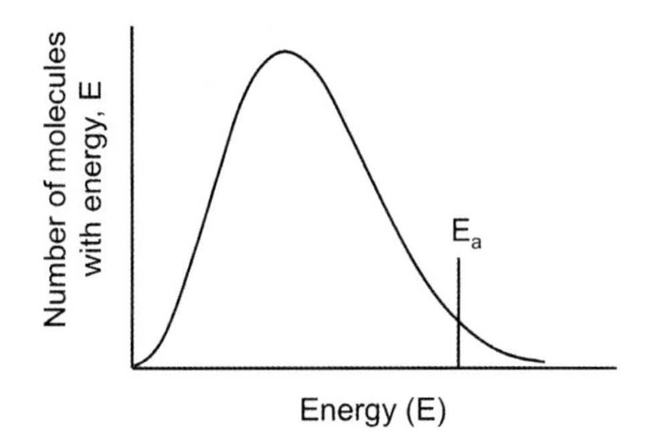

A Decreasing the pressure of the mixture.

B Decreasing the temperature of the mixture.

C Increasing the pressure of the mixture.

D Increasing the temperature of the mixture.

(CCEA January 2009)

8. The graph shows the effect of thiosulfate concentration on the rate of reaction between sodium thiosulfate solution and hydrochloric acid at 25 °C. Sketch the graph of rate against concentration for a similar series of experiments carried out at 35 °C. Use collision theory to explain the difference in the graphs.

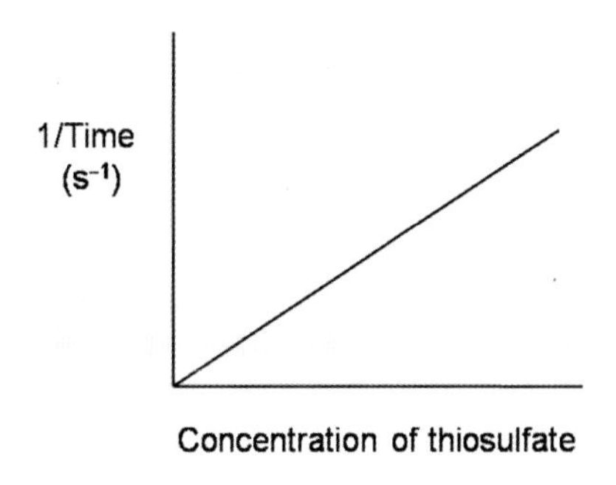

(Adapted from CCEA June 2003)

Before moving to the next section, check that you are able to:

- Use the Maxwell-Boltzmann distribution to describe the distribution of energy amongst the particles in a mixture of liquids or gases.
- Explain what is meant by the term *activation energy*.
- Use the concept of activation energy to explain why the rate of reaction increases when the temperature of the reaction mixture is increased.

Catalysis

In this section we are learning to:

- Use collision theory to explain how a catalyst increases the rate of a reaction.

Adding a small amount of solid manganese dioxide, MnO_2 to an aqueous solution of hydrogen peroxide, $H_2O_{2(aq)}$ greatly increases the rate at which the hydrogen peroxide decomposes to form water and oxygen:

$$2H_2O_{2(aq)} \rightarrow 2H_2O_{(l)} + O_{2(g)}$$

The manganese dioxide acts as a **catalyst** as it *increases the rate of reaction by providing an alternative pathway for the reaction with a lower activation energy, and is not consumed by the reaction*. The effect of a catalyst on the pathway for a reaction is illustrated in Figure 6.

In order to lower the activation energy *the catalyst must participate in the reaction and be changed as it reacts with other species in the reaction mixture*. The catalyst is then regenerated in its original form as the reaction proceeds, and does not appear in the chemical equation for the reaction as it is not consumed by the reaction.

The Maxwell-Boltzmann distributions in Figure 7

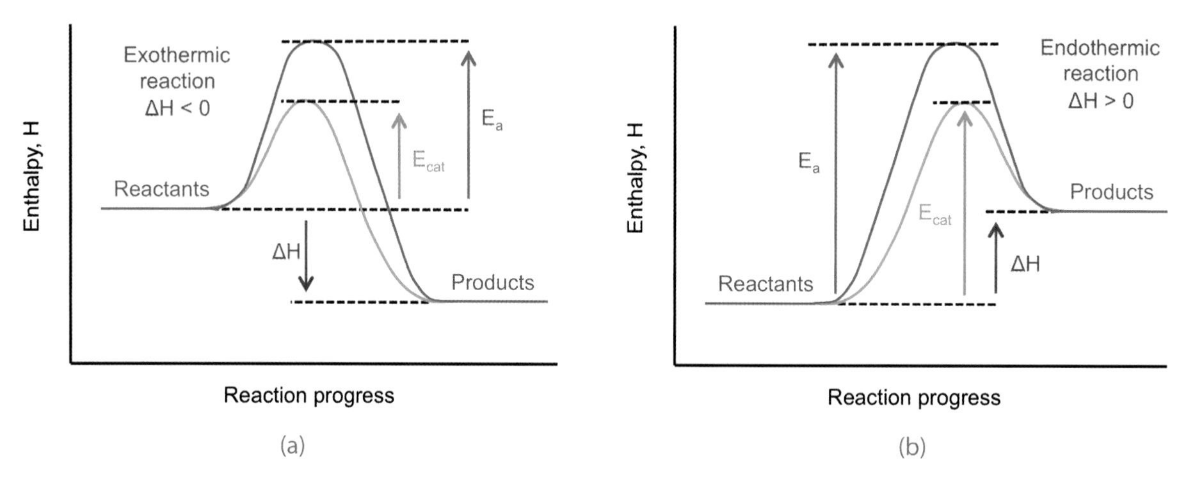

Figure 6: Using a catalyst to generate a reaction pathway with a lower activation energy, E_{cat} for (a) an exothermic reaction and (b) an endothermic reaction. The pathway for the reaction without a catalyst is drawn in blue.

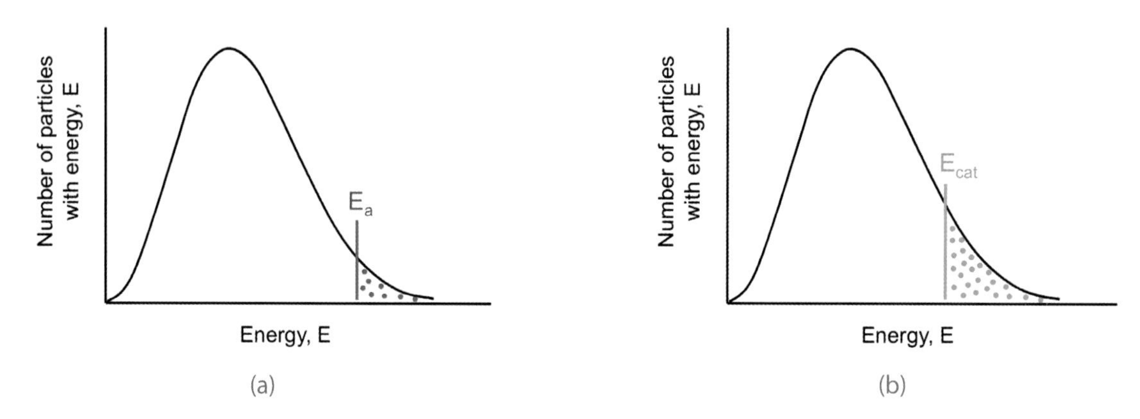

Figure 7: The fraction of particles in a mixture of liquids or gases with enough energy to react (shown shaded) when (a) the reaction is carried out without a catalyst and (b) a catalyst is added to the reaction mixture.

show how reducing the activation energy (from E_a to E_{cat}) increases the number of particles in the reaction mixture with enough energy to react. By increasing the number of particles with enough energy to react, more collisions involve particles with enough energy to react, and the number of successful collisions per second increases. In this way adding a catalyst increases the rate of reaction.

Exercise 2.10C

1. Which one of the following represents a correctly labelled enthalpy diagram?

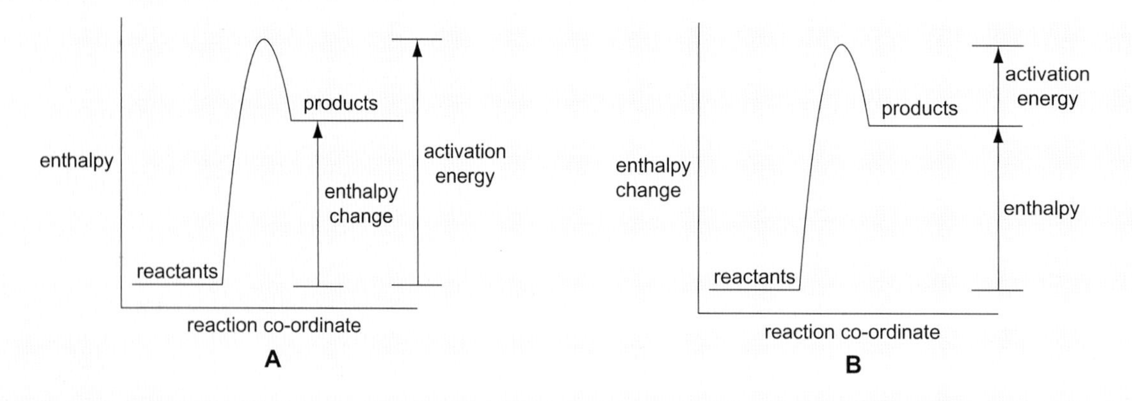

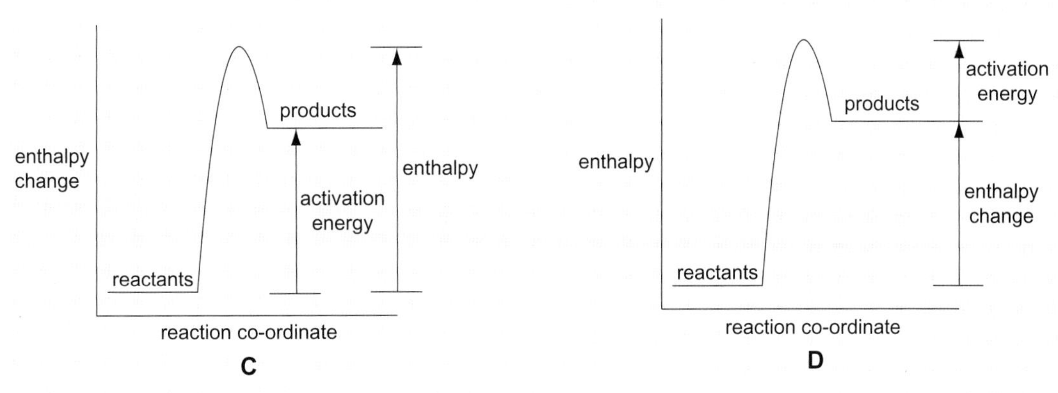

(CCEA June 2012)

2. Which one of the following reaction profiles represents a reaction for which the enthalpy change is numerically greater than the activation energy?

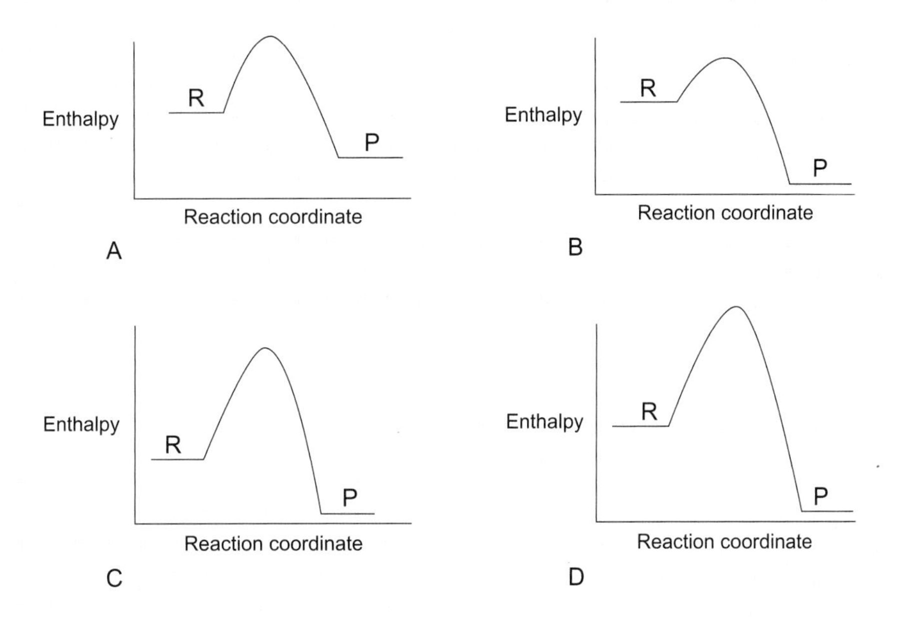

(CCEA June 2013)

3. Which one of the following pairs of enthalpy changes shows how a catalyst works in an endothermic reaction?

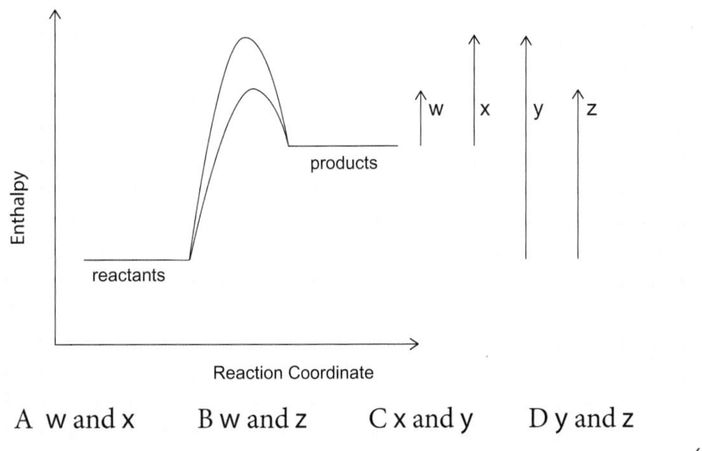

A w and x   B w and z   C x and y   D y and z

(CCEA January 2013)

4. A catalytic converter is a device in a car exhaust system in which a catalyst is spread thinly over a honeycomb structure. Explain, in terms of energy, what is meant by the term *catalyst.*

(CCEA January 2014)

5. A mixture of ethanol and gasoline (petrol) is known as gasohol. Toxic gases produced by the combustion of gasohol can be eliminated by passing them through a catalytic converter. Explain the role of the catalyst in the catalytic converter with the help of a simple labelled enthalpy diagram. Assume the reactions catalysed are exothermic.

(Adapted from CCEA January 2010)

6. Disposable lighters contain liquid butane which forms a gas when it escapes from the lighter. A spark is used to ignite the gas, which then burns to produce heat. State and explain whether the spark acts as a catalyst.

(Adapted from CCEA January 2012)

7. Sketch the distribution of molecular kinetic energies for the gas molecules in a reaction mixture. Mark the activation energy, E_a for the uncatalysed reaction and the activation energy, E_{cat} for the catalysed reaction on the distribution.

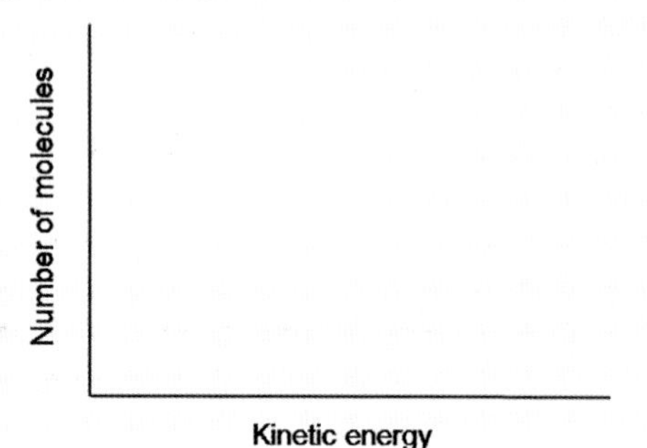

(Adapted from CCEA June 2008)

8. The diagram shows the distribution curve for the energies of the gas molecules in a reaction mixture. E_a is the activation energy for a particular reaction. (a) Explain why the curve starts at the origin. (b) Why does the curve not meet the horizontal axis at high energies? (c) What does the shaded area represent? (d) With reference to the distribution curve explain the effect of a catalyst on the rate of a chemical reaction.

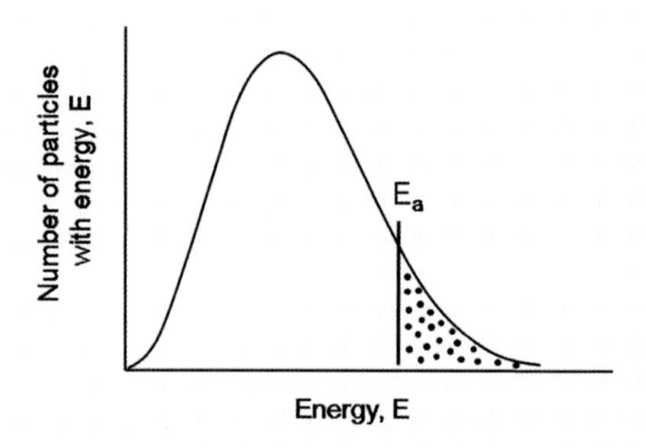

(Adapted from CCEA June 2009)

9. The first step in the Ostwald Process for the manufacture of nitric acid is the oxidation of ammonia.

$$4NH_{3(g)} + 5O_{2(g)} \leftrightarrows 4NO_{(g)} + 6H_2O_{(g)} \quad \Delta H = -950 \text{ kJ mol}^{-1}$$

The diagram shows the distribution of molecular energies in the reaction mixture at 900 °C. Explain, referring to the diagram, how a catalyst containing rhodium and platinum increases the rate of the reaction.

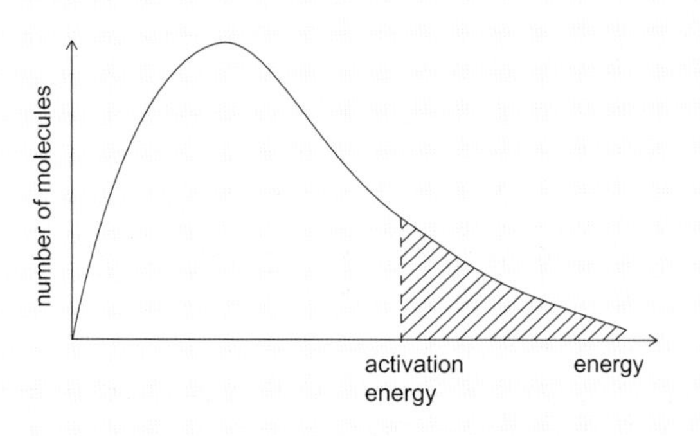

(CCEA June 2012)

10. The diagram shows a Maxwell-Boltzmann distribution curve for a mixture of sulfur dioxide and oxygen at a temperature, T. The symbol E_a represents the activation energy. (a) Use the distribution curve to explain the role of a catalyst. (b) Explain why the reaction between sulfur dioxide and oxygen is faster in the presence of a catalyst.

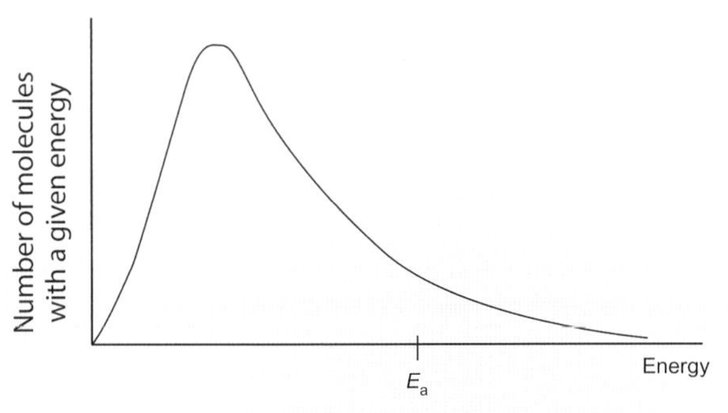

(Adapted from CCEA June 2013)

Before moving to the next section, check that you are able to:

- Recall that a catalyst speeds up a reaction by providing an alternative pathway for the reaction with lower activation energy.
- Construct reaction pathways to show the effect of adding a catalyst.
- Recall that a catalyst does not appear in the chemical equation for a reaction as it is regenerated by the reaction.
- Explain, with the aid of a Maxwell-Boltzmann distribution, why adding a catalyst increases the rate of reaction.

Applications of Catalysts

In this section we are learning to:

- Explain the effect of a catalyst on the composition of a chemical equilibrium and the rate at which equilibrium is established.
- Explain why catalysts are more effective when the surface area of the catalyst particles is increased.

Catalysts are widely used in the chemical industry to speed up the production of valuable chemicals such as sulfuric acid and ammonia. In the Haber-Bosch process ammonia is produced by reacting a mixture of nitrogen and hydrogen at high temperature and pressure in the presence of a granulated iron catalyst.

The Haber-Bosch process:

$N_2 + 3H_2 \leftrightarrows 2NH_3$

The granulated iron catalyst increases the rate at which equilibrium is established by lowering the activation energy for the forward and reverse reactions (E_a and E_a') by the same amount as illustrated in Figure 8.

The composition of the reaction mixture at equilibrium is determined by the enthalpy change for the reaction (ΔH), and is not affected by the size of the activation energy for the forward and reverse reactions (E_a and E_a'). In this way *adding a catalyst increases the rate at which equilibrium is established but does not affect the composition of the equilibrium mixture.*

Exercise 2.10D

1. Which letter (A–D) represents the activation energy for the back reaction?

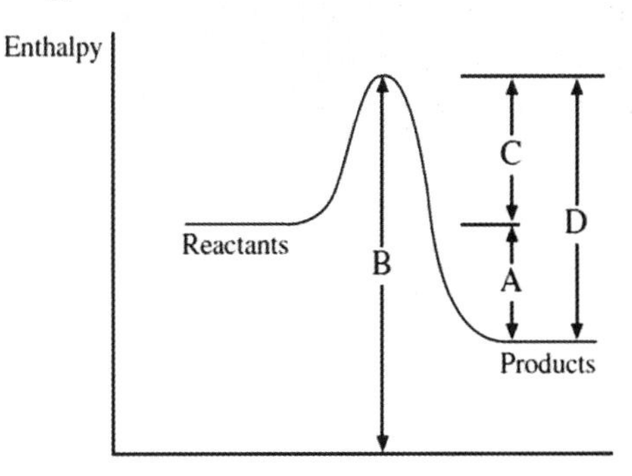

(CCEA June 2006)

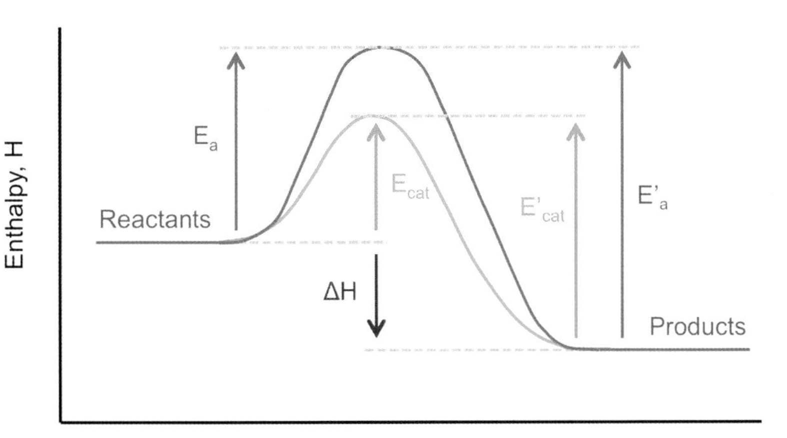

Figure 8: Adding a catalyst alters the reaction pathway for an equilibrium reaction by lowering the activation energy for the forward reaction (E_a) and the reverse reaction (E_a') by the same amount.

2. The energy level diagram for a reversible reaction is shown below. Which one of the following statements is correct?

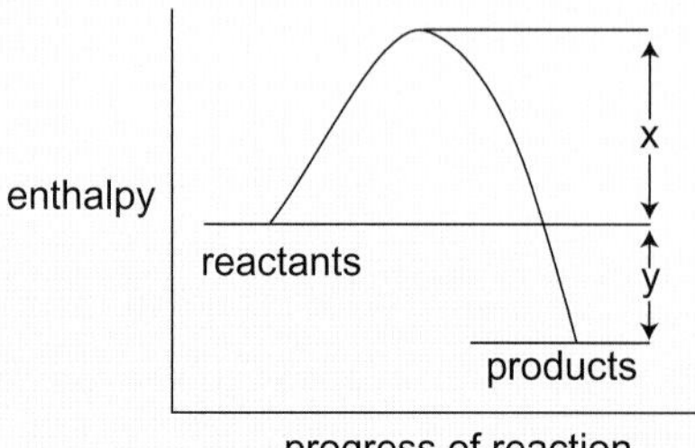

A The forward reaction is endothermic.
B The activation energy for the reverse reaction is y.
C The activation energy for the forward reaction is x.
D The enthalpy change for the reverse reaction is y-x.

(Adapted from CCEA June 2011)

3. The enthalpy diagram shows reaction pathways for a catalysed and uncatalysed reversible reaction. Which one of the following is the activation energy for the reverse reaction without a catalyst?

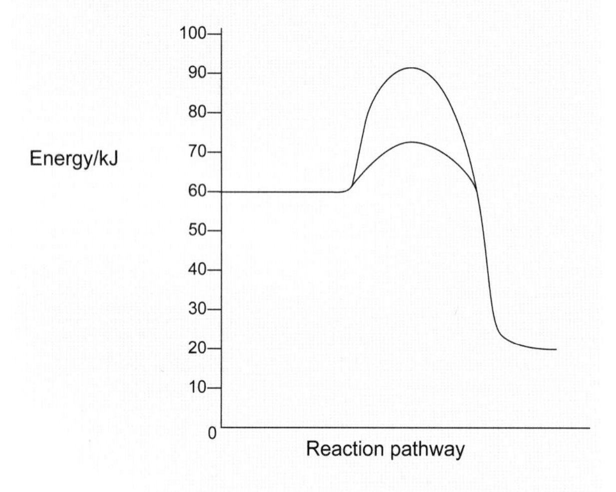

A 10 kJ B 30 kJ C 40 kJ D 70 kJ

(CCEA June 2015)

4. The diagram represents the enthalpy of a reversible reaction plotted against the reaction coordinate. Which one of the following combined enthalpy changes would be most affected by the use of a catalyst?

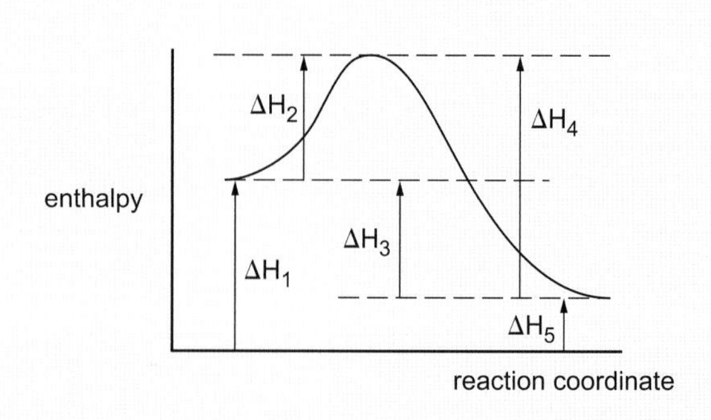

A $\Delta H_1 + \Delta H_5$

B $\Delta H_2 + \Delta H_3$

C $\Delta H_2 + \Delta H_4$

D $\Delta H_3 + \Delta H_4$

(Adapted from CCEA January 2012)

The hydrogenation of vegetable fats and oils (C=C → C-C) for use in the manufacture of margarine is carried out using a finely divided nickel catalyst where the term **finely divided** refers to a compound with a large surface to volume ratio. The large surface area of a finely divided catalyst *increases the efficiency of the reaction as it provides a large surface on which the reactants can react.*

Finely divided catalysts are also used to reduce toxic emissions from motor vehicles. When the exhaust gases from a motor vehicle are passed through a catalytic converter, toxic gases such as carbon monoxide and nitrogen oxides are converted into less toxic gases such as carbon dioxide and nitrogen.

The catalytic converter contains a finely divided catalyst consisting of small metal particles supported on an inert material with a honeycomb structure. The large surface created by the honeycomb structure increases the efficiency of the catalyst by increasing the number of metal particles that come into contact with the exhaust gases. *The small size of the metal particles further increases the efficiency of the catalyst by increasing the area on which the exhaust gases can bind to the metal and react.*

Exercise 2.10E

1. The inside surfaces of a catalytic converter are coated with finely divided metal particles. Suggest why the particles are finely divided.

   (CCEA January 2009)

2. Ammonia forms nitrogen(II) oxide and steam when it is mixed with air and passed through a heated platinum gauze.

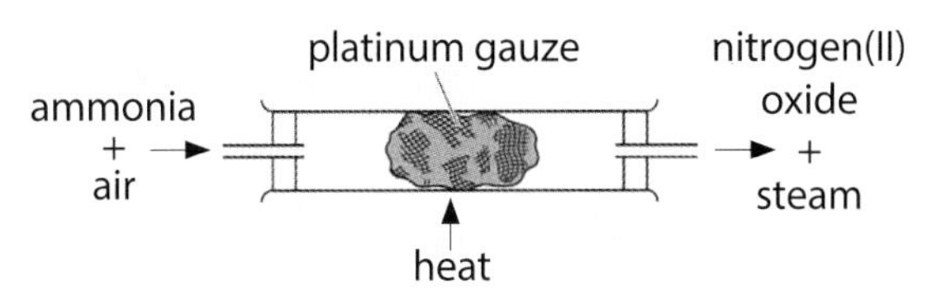

   $4NH_3 + 5O_2 \rightarrow 4NO + 6H_2O$

   (a) Suggest the purpose of the platinum gauze. (b) Explain why the gauze is used rather than a pile of platinum powder. (c) Explain why the laws of chemical equilibrium cannot be applied to this particular mixture of ammonia and oxygen.

   (CCEA January 2012)

Before moving to the next section, check that you are able to:

- Explain why adding a catalyst increases the rate at which equilibrium is established but does not affect the composition of the equilibrium mixture.
- Construct reaction pathways to illustrate the effect of a catalyst on a dynamic equilibrium.
- Recall what is meant by the term *finely divided* and explain why finely divided catalysts are efficient.

2.11 Group II: The Alkaline Earth Metals

CONNECTIONS

- Group II metals and their salts are used in emergency flares and fireworks.
- Magnesium is used as a reducing agent in the industrial scale production of less reactive metals such as titanium and uranium.
- A 'barium meal' containing barium sulfate is used to improve the quality of CT scans showing the interior of the human body.

Properties

In this section we are learning to:

- Recall the appearance and physical properties of the Group II metals.
- Explain trends in the atomic radius and first ionisation energy of the Group II metals and make comparisons with Group I metals.

The elements in Group II are hard, silvery metals known as the *Alkaline Earth Metals*. They are less reactive than the Group I metals, but too reactive to be found in nature. Most Group II metals are found in the form of carbonates and are referred to as **s-block elements** as *their outermost electrons are located in an s-type subshell*.

Atomic Properties

The electron configurations for the Group II metals in Table 1 reveal that each has a pair of electrons in its outermost shell. The pair of electrons in the outermost shell determines many of the properties of the Group II metals. For example, when a Group II metal forms an ionic compound, it satisfies the Octet rule by losing the pair of electrons in its outermost shell to form the corresponding 2+ ion (M^{2+}).

Period	Element	Ground State Configuration	Atomic Radius (pm)	First Ionisation Energy (kJ mol^{-1})
2	Be	[He] $2s^2$	112	900
3	Mg	[Ne] $3s^2$	145	738
4	Ca	[Ar] $4s^2$	194	590
5	Sr	[Kr] $5s^2$	219	550
6	Ba	[Xe] $6s^2$	253	503

Table 1: Selected properties of Group II metal atoms.

Notes:
- The shorthand [He], [Ne], ... is used to represent the ground state electron configurations of the Noble gases.
- 1 pm = 1×10^{-12} m

The analysis in Figure 1 reveals that *the first ionisation energies of the Group I and II metals become smaller as atomic radius increases down the group*. This relationship can be explained by noting that the electrons in the outermost s-type subshell are further from the nucleus and are better shielded as the number of filled shells increases down the group. As a result, the electrons in the outermost s-subshell become easier to remove, and the first ionisation energy becomes smaller, as the atoms get bigger down the group.

The trends in Figure 1 also reveal that *each Group II metal has a smaller atomic radius, and a larger first ionisation energy, than the adjacent Group I metal*. This is explained by noting that the electron configuration for a Group II element is obtained by adding one electron to the electron configuration for the adjacent Group I element.

... $3s^2\ 3p^6\ 4s^1$ → ... $3s^2\ 3p^6\ 4s^2$

Potassium atom (Group 1) → Calcium atom (Group 2)

The additional electron forms a pair of electrons in the outermost s-type subshell, and is shielded by the same amount as the outermost electron in the adjacent Group I element. As a result, the electrons in the outermost s-type subshell of a Group II metal experience a greater nuclear charge and are held more tightly than the electron in the outermost s-type subshell of the adjacent Group I element.

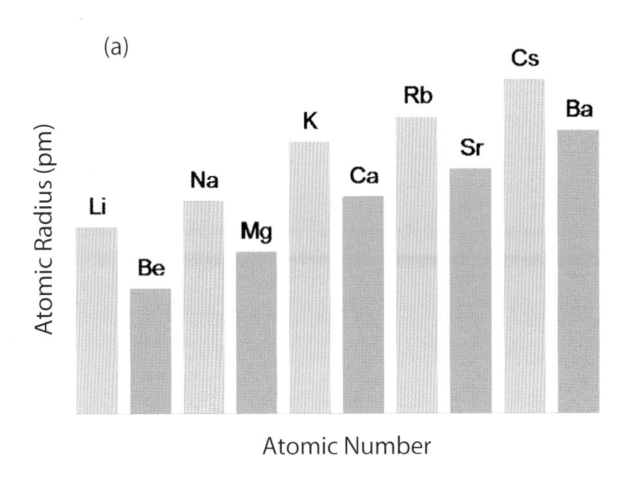

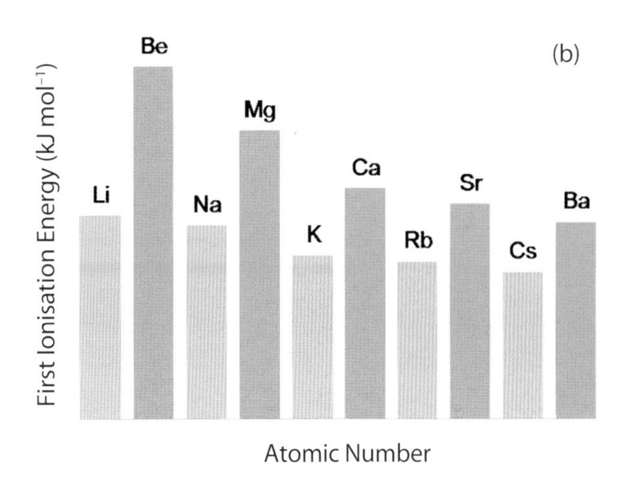

Figure 1: The trend in (a) atomic radius and (b) first ionisation energy as atomic number increases down Groups I and II.

Exercise 2.11A

1. (a) Explain why the Group II elements are regarded as s-block elements. (b) Write an equation, including state symbols, for the first ionisation energy of magnesium. (c) State and explain the change in the value of the first ionisation energy from magnesium to barium.

   (CCEA January 2011)

2. The table contains information about the Group II elements.

	magnesium	calcium	strontium	barium
Atomic number	12	20	38	56
Atomic radius (nm)	0.160	0.197	0.215	
Density (g cm^{-3})	1.74	1.54	2.6	3.5

   (a) Explain why none of the Group II elements are found as free elements. (b) Write the electronic configuration for calcium. (c) Explain why the atomic radius of the elements increases down the group. (d) Suggest a relationship between the atomic radius of the Group II elements and the corresponding elements in Group I. (e) Predict the atomic radius of barium. (f) Use the information in the table to explain why barium has the highest density of all the Group II elements.

   (CCEA June 2011)

Before moving to the next section, check that you are able to:

- Recall the appearance of the Group II elements and explain why they are found in nature as carbonates and other compounds.
- Write the ground state electron configuration for a Group II element and explain why the Group II elements are referred to as s-block elements.
- Explain why first ionisation energies of the Group II metals decrease as the atomic radius of the Group II metals increases down the group.
- Explain why a Group II metal has a higher first ionisation energy than the adjacent Group I metal.

Reactions

In this section we are learning to:

- Account for the reactions of Group II metals with oxygen, water and dilute acids in terms of the relative reactivity of the Group II metals.

Reaction with Oxygen

The Group II metals form the corresponding metal oxide when heated in air. The reaction is exothermic and the metal emits visible light with a colour that is characteristic of the metal.

$2Mg_{(s)} + O_{2(g)} \rightarrow 2MgO_{(s)}$
Observations: *white powder formed, intense white light.*

$2Ca_{(s)} + O_{2(g)} \rightarrow 2CaO_{(s)}$
Observations: *white powder formed, brick red flame.*

$2Sr_{(s)} + O_{2(g)} \rightarrow 2SrO_{(s)}$
Observations: *white powder formed, crimson flame.*

$2Ba_{(s)} + O_{2(g)} \rightarrow 2BaO_{(s)}$
Observations: *white powder formed, green flame.*

Heating the heavier Group II metals in an oxygen rich atmosphere also produces the corresponding metal peroxide, MO_2.

$Ba_{(s)} + O_{2(g)} \rightarrow BaO_{2(s)}$

Worked Example 2.11i

Strontium burns with a crimson flame when heated in air. (a) Write an equation for the reaction. Include state symbols. (b) Explain, using a labelled diagram, how you would safely heat strontium granules in air. (c) With reference to the diagram, suggest one way to ensure that all of the strontium reacts when heated.

Solution

(a) $2Sr_{(s)} + O_{2(g)} \rightarrow 2SrO_{(s)}$

(b)

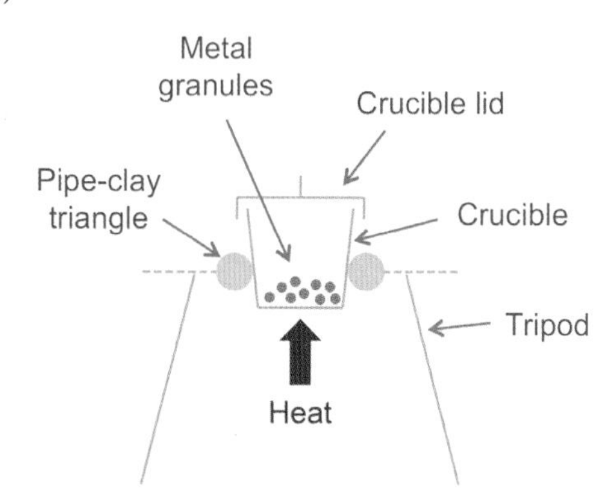

(c) The lid of the crucible is lifted at intervals during heating to allow oxygen into the crucible.

Exercise 2.11B

1. A white solid containing a single compound is formed when calcium is heated in air. (a) Write an equation for the reaction. Include state symbols. (b) Calculate the mass of product formed when 2.0 g of calcium is heated in air, assuming a percentage yield of 76%.
2. A white solid containing a single compound is formed when barium is heated in air. If barium is instead heated in an oxygen-rich atmosphere, the solid product is found to be a mixture of two compounds. (a) Write the name and formula for the compound formed when barium is heated in air. (b) Write equations to show how each product is formed when barium is heated in an oxygen-rich atmosphere.

Before moving to the next section, check that you are able to:

- Recall what is observed when a Group II metal reacts with air and oxygen.
- Write equations for the reactions of Group II metals with air and oxygen.

Reaction with Water

Group II metals react with water to form the corresponding metal hydroxide. For example, granules of calcium metal react vigorously with cold water to form a solution of calcium hydroxide as shown in Figure 2a.

$Ca_{(s)} + 2H_2O_{(l)} \rightarrow Ca(OH)_{2(aq)} + H_{2(g)}$

Observations: metal sinks then rises, metal disappears, bubbles of colourless gas, reaction mixture warms up.

Calcium is more dense than water and initially sinks to the bottom of the beaker. The calcium granules then begin to rise as the heat produced by the reaction warms the surrounding water, causing it to rise. Once the granules reach the surface, the heat generated by the reaction results in a fizzing sound and the production of steamy fumes.

The solution becomes increasingly alkaline as the reaction proceeds and is strongly alkaline within a few seconds. Calcium hydroxide is only sparingly soluble in water and the solution quickly becomes saturated. Once the solution has become saturated a suspension of solid calcium hydroxide forms giving the solution a cloudy white appearance as shown in Figure 2b.

Strontium (Sr) and barium (Ba) also react vigorously with cold water. In each case the resulting solution is colourless as the hydroxides of strontium and barium are more soluble than calcium hydroxide.

The following Worked Example demonstrates how the hydrogen gas produced by the reaction can be collected by *the downward displacement of water.*

(a)

(b)

Figure 2: (a) The reaction between calcium metal and cold water. (b) A suspension of calcium hydroxide forms once the solution has become saturated.

Worked Example 2.11ii

Strontium reacts vigorously with cold water to form a strongly alkaline solution. (a) Write an equation for the reaction. Include state symbols. (b) Explain, using a labelled diagram, how you would collect a sample of the gas produced by the reaction. (c) Describe a test to verify that the resulting solution is strongly alkaline. (d) Suggest two observations that could be used to indicate that the reaction is exothermic.

Solution

(a) $Sr_{(s)} + 2H_2O_{(l)} \rightarrow Sr(OH)_{2(aq)} + H_{2(g)}$

(b)

Hydrogen gas
Clamp
Beaker of water
Inverted boiling tube
Metal granules
Inverted glass funnel

(c) The solution turns purple on adding a few drops of Universal Indicator.

(e) The reaction mixture warms up and steamy fumes are produced as the reaction mixture 'fizzes'.

The reaction between magnesium turnings and cold water again produces a solution of the corresponding metal hydroxide.

$Mg_{(s)} + 2H_2O_{(l)} \rightarrow Mg(OH)_{2(aq)} + H_{2(g)}$

Observations: metal surface becomes dull, colourless solution formed, bubbles of colourless gas produced slowly.

The reaction occurs very slowly and can be monitored over several hours by placing a few drops of universal indicator in the solution. The increase in pH as magnesium hydroxide forms is accompanied by a gradual colour change from green to blue as shown in Figure 3.

Figure 3: Using universal indicator to monitor the increase in pH as magnesium reacts with cold water.

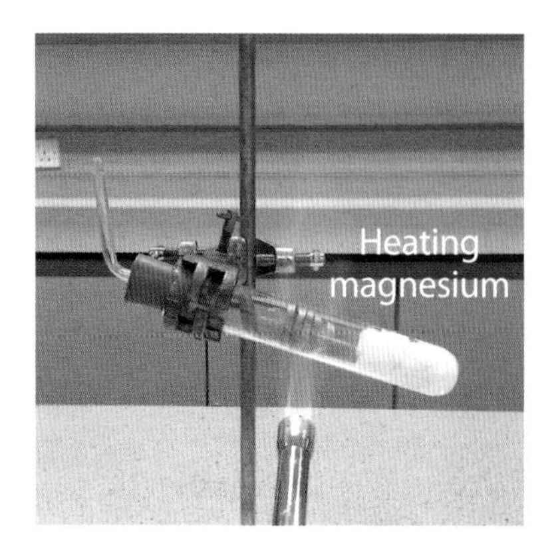

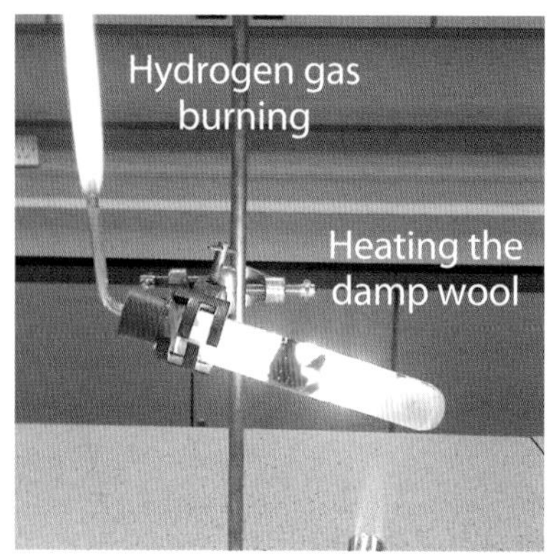

(a)

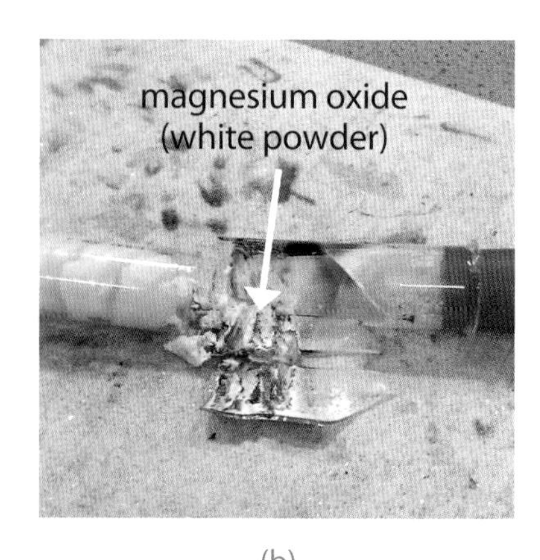

(b)

Figure 4: (a) The reaction between magnesium and steam.
(b) Magnesium oxide formed in the reaction between magnesium and steam.

The rate at which calcium and magnesium react with cold water demonstrates that *Group II metals react more vigorously with water as the metals become more reactive down the group*.

Less reactive Group II metals such as magnesium react much more vigorously when heated in the presence of steam. The reaction between magnesium and steam is very exothermic and produces intense white light as shown in Figure 4.

$Mg_{(s)} + H_2O_{(g)} \rightarrow MgO_{(s)} + H_{2(g)}$

Observations: *intense white light, white solid formed, heat produced.*

Worked Example 2.11iii

Magnesium reacts vigorously when heated in the presence of steam. (a) Write an equation for the reaction. Include state symbols. (b) Explain, using a labelled diagram, how you would react magnesium with steam and collect the gas produced by the reaction. (c) Name the experimental technique used to collect the gas. (d) Describe a test to verify the identity of the gas produced by the reaction.

Solution

(a) $Mg_{(s)} + H_2O_{(g)} \rightarrow MgO_{(s)} + H_{2(g)}$

(b)

Magnesium ribbon
Damp cotton wool
Clamp
Delivery tube
Hydrogen gas
Inverted boiling tube
Beehive shelf
Heat
Heat
Water trough

(c) Downward displacement of water.

(d) The gas is hydrogen if it burns with a 'pop' when a lit splint is placed in an inverted boiling tube filled with the gas.

Exercise 2.11C

1. (a) Does calcium float or sink in water? Explain your answer. (b) State two further observations when calcium reacts with water. (c) Write the equation for the reaction of calcium with water. (d) Draw a labelled diagram to show how the gas given off when calcium reacts with water can be collected using a test tube and a beaker.

(CCEA June 2011)

2. Magnesium ribbon and calcium turnings were added to water in separate boiling tubes as shown in the diagram below. A lit splint was held above each boiling tube. (a) Describe what is observed in each boiling tube. (b) Describe what happens around the lit splint for each boiling tube. (c) Describe and explain what happens when Universal Indicator is added to the boiling tubes after thirty minutes.

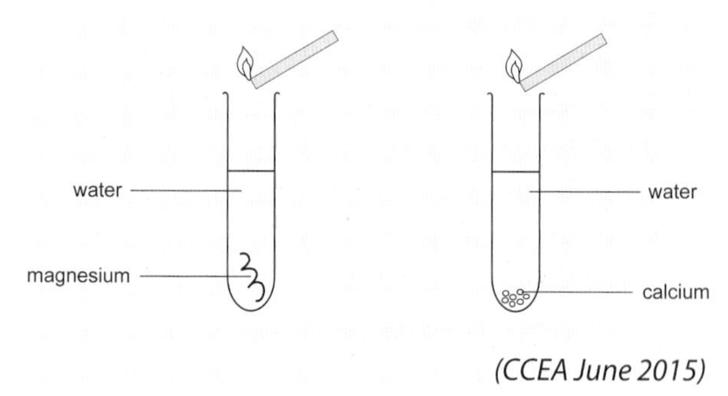

(CCEA June 2015)

3. (a) State the trend in reactivity with water of the elements from magnesium to barium. (b) Write the equation for the reaction of barium with water.

(CCEA June 2013)

4. The reaction between magnesium and steam is very exothermic. (a) State two further observations for the reaction between magnesium and steam. (b) Write an equation for the reaction. Include state symbols. (c) Zinc is less reactive than magnesium. Suggest two ways in which the reaction between zinc and steam would differ from the reaction between magnesium and steam.

5. Which of the following will **not** occur when calcium reacts with water?

   A The metal sinks and then rises to the surface.

   B A colourless gas is produced.

   C A colourless solution is formed.

   D The metal disappears.

Before moving to the next section, check that you are able to:

- Recall what is observed when calcium and magnesium react with cold water and write equations for the reactions that occur.
- Relate what is observed when calcium and magnesium react with cold water to the reactivity of the metal.
- Recall what is observed when magnesium reacts with steam and write an equation for the reaction.
- Describe, using a diagram, the collection of hydrogen gas by the downward displacement of water.

Reaction with Dilute Acids

The Group II metals react with dilute acids to form hydrogen gas and an aqueous solution of the corresponding metal salt. The reaction becomes much more vigorous as the reactivity of the metal increases down the group. Only the lighter Group II elements (Be, Mg and Ca) react safely with dilute acids.

$$Mg_{(s)} + 2HCl_{(aq)} \rightarrow MgCl_{2(aq)} + H_{2(g)}$$

$$Mg_{(s)} + H_2SO_{4(aq)} \rightarrow MgSO_{4(aq)} + H_{2(g)}$$

$$Ca_{(s)} + 2HNO_{3(aq)} \rightarrow Ca(NO_3)_{2(aq)} + H_{2(g)}$$

Observations: fizzing, metal disappears, reaction mixture warms up.

The reaction between calcium and sulfuric acid does not go to completion. The calcium sulfate formed in the reaction is insoluble, and stops the remaining calcium from reacting with the acid by forming an unreactive layer over the surface of the calcium.

Exercise 2.11D

1. Magnesium reacts vigorously with hydrochloric and sulfuric acids to form solutions of magnesium chloride and magnesium sulfate. (a) Write the equation for the reaction of magnesium with hydrochloric acid. (b) State two observations when this reaction is carried out.

(CCEA June 2013)

2. Compare the chemistry of calcium with that of magnesium using the headings: (a) Combustion, (b) Reaction with water and (c) Reaction with dilute hydrochloric acid. Include observations.

(CCEA January 2005)

Before moving to the next section, check that you are able to:

- Write equations to describe the reaction of Group II metals with dilute acids.
- Recall that the reaction of Group II metals with dilute acids becomes more vigorous as the metals become more reactive down the group.

Properties of Group II Compounds

In this section we are learning to:

- Account for the appearance and properties of Group II compounds.
- Explain how to determine the solubility curve for a compound.
- Describe the reactions of Group II oxides, hydroxides and carbonates.

Structure and Bonding

When a Group II metal (M) forms a compound it loses two electrons to form the corresponding M^{2+} cation. As a result compounds such as MgO, $CaCO_3$, $SrCl_2$ and $BaSO_4$ are ionic compounds in which the M^{2+} ions form part of an ionic lattice that is held together by strong attractive forces between oppositely charged ions.

The M^{2+} ions in a Group II compound have a full outer shell and do not absorb visible light. As a result, *Group II metal compounds are white crystalline solids, and have a high melting point that results from strong ionic bonding in the lattice.*

Acid-Base Character

Salts of Group II metals such as $BaCl_2$, $CaBr_2$ and $MgSO_4$ are formed when a strong acid reacts with a strong base, and dissolve to form an aqueous solution with a pH close to 7.

$CaBr_{2(s)} \rightarrow Ca^{2+}_{(aq)} + 2Br^{-}_{(aq)}$ $\quad pH \approx 7$

$MgSO_{4(s)} \rightarrow Mg^{2+}_{(aq)} + SO_4^{2-}{}_{(aq)}$ $\quad pH \approx 7$

In contrast, Group II oxides, hydroxides and carbonates are bases, and may dissolve or react with water to form an alkali.

$Ca(OH)_{2(s)} \rightarrow Ca^{2+}_{(aq)} + 2OH^{-}_{(aq)}$ $\quad pH > 7$

$BaO_{(s)} + H_2O_{(l)} \rightarrow Ba(OH)_{2(aq)}$ $\quad pH > 7$

Trends in Solubility

The hydroxides formed by Group II metals become more soluble down the group. The hydroxides of the lighter Group II metals such as $Mg(OH)_2$ and $Ca(OH)_2$ are only sparingly soluble in water, while barium hydroxide, $Ba(OH)_2$ readily dissolves to form a colourless solution.

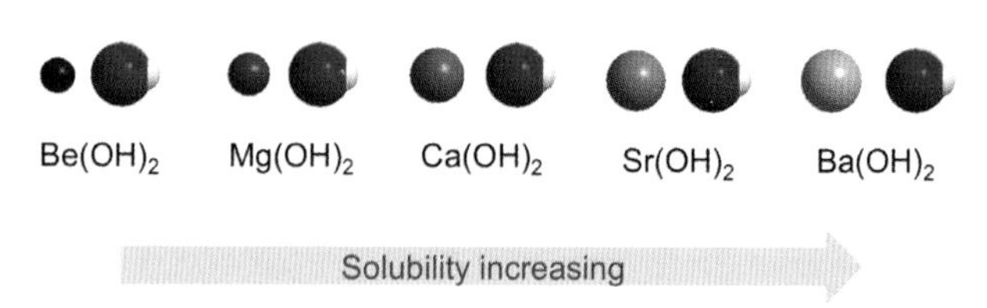

In contrast the sulfates formed by Group II metals become less soluble down the group to the extent that calcium sulfate, $CaSO_4$ is sparingly soluble, and barium sulfate, $BaSO_4$ is insoluble in water at room temperature.

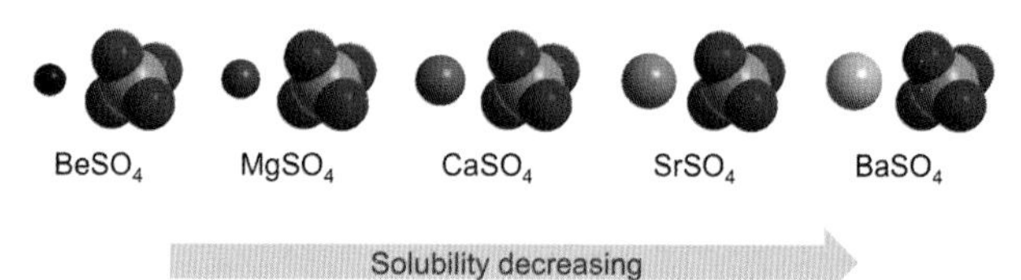

Determining Solubility

If a Group II compound dissolves in water, the following procedure can be used to determine its solubility at a particular temperature using the apparatus in Figure 5.

A known mass of solid is added to a measured volume of water in a boiling tube. The mixture is then placed in a water bath and stirred gently until the solid dissolves. The solution is then removed from the water bath and stirred as it cools until crystals of the solid begin to form. *The formation of crystals indicates that the solution in the boiling tube is saturated as it contains the maximum amount of dissolved solid at that temperature.*

The solubility is then the mass of solute in the saturated solution expressed in units of grams per

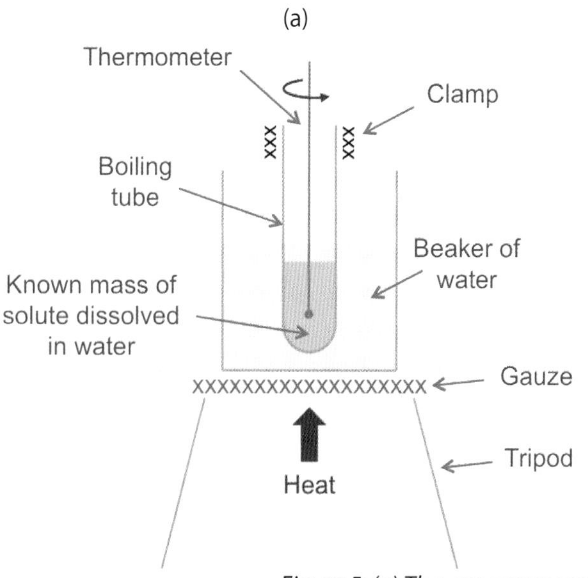

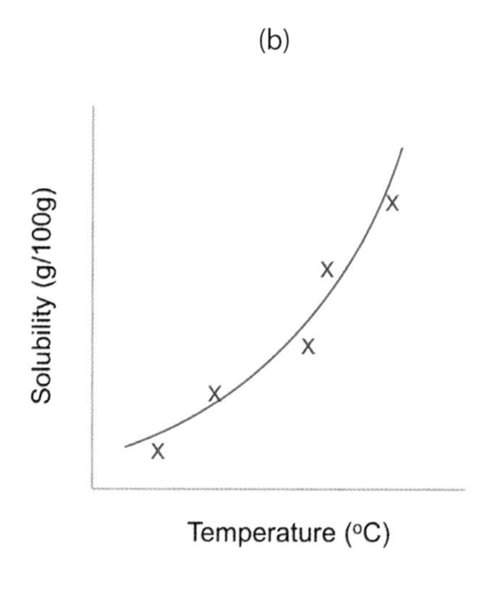

Figure 5: (a) The apparatus used to determine the solubility of a solid at a single temperature. (b) A solubility curve obtained by plotting the solubility of a solid at a range of temperatures.

100 g of solvent. For example, if crystals form when a solution containing 4 g of solid dissolved in 10 g of water is cooled to 50 °C, the solubility of the solid at 50 °C is 4 g/10 g × 10 = 40 g/100 g of water. In this way the **solubility** of the solid becomes *the maximum mass of solid that will dissolve in 100 g of solvent at a given temperature*.

Having noted the temperature at which crystals begin to form, the solubility at a second temperature can then be obtained by adding a small volume of water to the solution and repeating the experiment.

The solubility values obtained in this way can then be used to construct a solubility curve, with each saturated solution contributing one data point to the curve as shown in Figure 5.

In contrast, the solubility of Group II hydroxides and other sparingly soluble bases can be determined by titrating a saturated solution of the compound against a strong acid using the procedure in the following example.

Worked Example 2.11iv

The hydroxides of the Group II metals, magnesium to barium, are white ionic solids. They are sparingly soluble in water, the solubility rising with atomic number. Suggest how you could determine the solubility of strontium hydroxide at 25 °C. Give full practical details.

Compound:	$Mg(OH)_2$	$Ca(OH)_2$	$Sr(OH)_2$	$Ba(OH)_2$
Solubility: (g/100g at 25 °C)	0.01	0.15	0.89	3.32

(CCEA January 2007)

Solution

Place a measured volume of water in a boiling tube. Place the boiling tube in a water bath set at 25 °C. Add a small amount of solid strontium hydroxide to the water and stir to dissolve. Repeat until no more dissolves. Filter the solution to remove any undissolved solid, then determine the amount of strontium hydroxide in the solution by titrating a portion of the solution against a standard acid.

Exercise 2.11E

1. Which one of the following lists both sets of compounds in order of increasing solubility (least soluble first)?

	Sulfates	Hydroxides
A	$MgSO_4$ $CaSO_4$ $SrSO_4$ $BaSO_4$	$Ba(OH)_2$ $Sr(OH)_2$ $Ca(OH)_2$ $Mg(OH)_2$
B	$MgSO_4$ $CaSO_4$ $SrSO_4$ $BaSO_4$	$Mg(OH)_2$ $Ca(OH)_2$ $Sr(OH)_2$ $Ba(OH)_2$
C	$BaSO_4$ $SrSO_4$ $CaSO_4$ $MgSO_4$	$Mg(OH)_2$ $Ca(OH)_2$ $Sr(OH)_2$ $Ba(OH)_2$
D	$BaSO_4$ $SrSO_4$ $CaSO_4$ $MgSO_4$	$Ba(OH)_2$ $Sr(OH)_2$ $Ca(OH)_2$ $Mg(OH)_2$

(CCEA June 2010)

2. Explain the difference in pH when 0.1 mole of magnesium hydroxide and 0.1 mole of barium hydroxide are stirred with separate 100 cm^3 portions of water.

(CCEA January 2011)

3. Barium hydroxide reacts with dilute nitric acid to form a solution of barium nitrate. (a) Write an equation for the reaction and (b) explain why barium hydroxide does not dissolve in dilute sulfuric acid.

(CCEA January 2007)

4. The table shows the solubility of strontium sulfate at different temperatures. (a) Explain whether the dissolving of strontium sulfate is an exothermic or endothermic process. (b) Compare the solubility of strontium sulfate with that of calcium and barium sulfates.

Solubility (g dm^{-3})	Temperature (°C)
0.0111	18
0.0135	25

(CCEA January 2012)

5. Which one of the following decreases as Group II is descended from magnesium to barium?
   A Atomic radius
   B First ionisation energy
   C Reactivity with water
   D Solubility of the hydroxides

(CCEA June 2012)

6. Which one of the following chlorides with the formula MCl_2 is the chloride of a Group II element?

   A White solid. Melting point 280 °C. Boiling point 304 °C. Fairly soluble in water to give a colourless neutral solution with poor electrical conductivity.

   B White solid. Melting point 815 °C. Readily soluble in water to give a green-blue solution with good electrical conductivity.

   C White solid. Melting point 875 °C. Readily soluble in water to give a colourless neutral solution with good electrical conductivity.

   D White solid. Melting point 672 °C. Dissolves to give a pale green solution with good electrical conductivity.

   (CCEA January 2010)

Before moving to the next section, check that you are able to:

- Account for the appearance of Group II compounds and the pH of solutions formed by Group II compounds.
- Recall trends in the solubility of Group II hydroxides and sulfates.
- Describe how to determine the solubility of a Group II compound, and construct a solubility curve.

Reactions of Group II Compounds

In this section we are learning to:

- Describe the reactions of Group II oxides, hydroxides and carbonates.

Reaction with Water

Group II oxides such as MgO and CaO react with water to form the corresponding hydroxide. The reaction is very exothermic and must be carried out by the controlled addition of water to the solid oxide.

$$CaO_{(s)} + H_2O_{(l)} \rightarrow Ca(OH)_{2(aq)}$$

A solution of calcium hydroxide, $Ca(OH)_{2(aq)}$ is known as limewater. Observing the formation of a cloudy white suspension when carbon dioxide is bubbled through limewater is a positive test for carbon dioxide.

$$Ca(OH)_{2(aq)} + CO_{2(aq)} \rightarrow CaCO_{3(s)} + H_2O_{(l)}$$

Observations: colourless solution turns milky.

Worked Example 2.11v

(a) A saturated solution of calcium hydroxide is known as limewater. Describe how you would prepare limewater and use it to test for carbon dioxide. State the result of a positive test. (b) Write the equation for the reaction of aqueous calcium hydroxide with carbon dioxide. Include state symbols. (c) When 5.0 dm^3 of polluted air containing an excess of carbon dioxide was passed through limewater 0.050 g of calcium carbonate was precipitated. Calculate the % carbon dioxide in the air sample. All measurements were carried out at 20 °C and 1 atmosphere pressure.

(Adapted from CCEA January 2010)

Solution

(a) Add solid calcium hydroxide to water with stirring until no more dissolves then filter the solution. The solution is colourless and turns milky when carbon dioxide is bubbled through the solution.

(b) $Ca(OH)_{2(aq)} + CO_{2(g)} \rightarrow CaCO_{3(s)} + H_2O_{(l)}$

(c) Moles of $CaCO_3 = 5.0 \times 10^{-4}$ mol

Volume of CO_2
$= 5.0 \times 10^{-4} \text{ mol} \times 24 \text{ dm}^3 \text{ mol}^{-1} = 0.012 \text{ dm}^3$

$$\% \, CO_2 = \frac{0.012 \text{ dm}^3}{5.0 \text{ dm}^3} \times 100 = 0.24 \, \%$$

Reaction with Dilute Acid

Group II oxides and hydroxides are bases and react with acids to form a solution of the corresponding salt and water. Group II carbonates are also bases and react with acids to form a solution of the corresponding salt, water and carbon dioxide.

$$CaO_{(s)} + 2HCl_{(aq)} \rightarrow CaCl_{2(aq)} + H_2O_{(l)}$$

$$Ca(OH)_{2(s)} + 2HCl_{(aq)} \rightarrow CaCl_{2(aq)} + 2H_2O_{(l)}$$

$$CaCO_{3(s)} + 2HCl_{(aq)} \rightarrow CaCl_{2(aq)} + H_2O_{(l)} + CO_{2(g)}$$

If we consider a **base** to be *a substance that accepts a hydrogen ion (H^+) from another substance*, the ionic equations for the reaction of CaO, $Ca(OH)_2$ and $CaCO_3$ with dilute acid show that, in each case, the calcium compound acts as a base by accepting hydrogen ions from the acid to form the products of the reaction.

$CaO_{(s)} + 2H^+_{(aq)} \rightarrow Ca^{2+}_{(aq)} + H_2O_{(l)}$

$Ca(OH)_{2(s)} + 2H^+_{(aq)} \rightarrow Ca^{2+}_{(aq)} + 2H_2O_{(l)}$

$CaCO_{3(s)} + 2H^+_{(aq)} \rightarrow Ca^{2+}_{(aq)} + H_2O_{(l)} + CO_{2(g)}$

A number of indigestion remedies contain bases such as $Mg(OH)_2$ or $CaCO_3$ that reduce indigestion by neutralising excess acid in the stomach. Bases such as calcium carbonate are also added to toothpaste as they help reduce tooth decay by neutralising acid deposited on the surface of teeth by bacteria.

Thermal Decomposition

When heated, Group II carbonates and hydroxides decompose to form the corresponding oxide.

$CaCO_{3(s)} \rightarrow CaO_{(s)} + CO_{2(g)}$

$Ca(OH)_{2(s)} \rightarrow CaO_{(s)} + H_2O_{(l)}$

The decomposition of a Group II carbonate or hydroxide is an example of **thermal decomposition** as it is *a process in which a substance is broken down into two or more simpler substances by the action of heat*.

In industry the thermal decomposition of limestone, which is mostly $CaCO_3$, is carried out on a large scale using rotary kilns such as those in Figure 6. The calcium oxide formed by the thermal decomposition is known as quick lime, and is used in the manufacture of important building materials such as cement and concrete.

Figure 6: Rotary lime kilns used for the thermal decomposition of limestone.

Exercise 2.11F

1. Calcium oxide is basic and reacts with water to form calcium hydroxide. (a) Explain the term *basic*. (b) Write the equation for the reaction of calcium oxide with water. (c) If the solubility of calcium hydroxide is 0.021 mol dm^{-3} at 20 °C, calculate the mass of calcium hydroxide that could dissolve in 250 cm^3 of water at 20 °C. (d) Write the equation for the reaction of calcium hydroxide with hydrochloric acid.

(Adapted from CCEA June 2003)

2. Barium hydroxide can be formed from barium oxide. (a) How would you convert barium oxide into barium hydroxide? (b) Write an equation for the reaction between barium hydroxide and sulfuric acid. (c) What colour is the solid product formed in the reaction?

(Adapted from CCEA January 2014)

3. (a) Write an equation for the reaction of barium carbonate with hydrochloric acid. (b) Describe what is observed when the gas produced is bubbled through limewater. (c) Calculate the volume of carbon dioxide produced at 20 °C and 1 atmosphere when 0.66 g of barium carbonate is reacted with an excess of acid.

(CCEA January 2005)

4. (a) Write the equation for the formation of strontium chloride from strontium oxide and hydrochloric acid. (b) Describe how a pure, dry sample of strontium chloride could be obtained from the reaction mixture.

(CCEA January 2003)

5. (a) Write the equation for the reaction of calcium carbonate with hydrochloric acid. (b) An impure sample of limestone is 95% calcium carbonate. Calculate the volume of carbon dioxide produced at 20 °C and 1 atmosphere pressure when 15.00 g of impure limestone is added to excess hydrochloric acid.

(CCEA June 2005)

6. The mineral Dolomite has the formula $CaCO_3$.$MgCO_3$. (a) State one observation when dolomite is treated with dilute hydrochloric acid. (b) Write an equation for the reaction. (c) State and explain the flame colour expected when a sample of dolomite is used in a flame test.

(Adapted from CCEA June 2007)

7. Magnesium is obtained from calcined (heated) dolomite, $CaCO_3.MgCO_3$. The calcined dolomite is slaked (reacted with water) to form the hydroxides. Seawater, which contains magnesium chloride, can be treated with an excess of slaked calcined dolomite. The magnesium hydroxide formed is then filtered off and calcined.

$$MgCl_2 + Ca(OH)_2.Mg(OH)_2 \rightarrow 2Mg(OH)_2 + CaCl_2$$

(a) Write the equation for the reaction of calcined dolomite with water. (b) Magnesium hydroxide is precipitated. What is the solubility trend of the Group II hydroxides? (c) Calculate the atom economy of $Mg(OH)_2$ from the reaction of magnesium chloride with slaked calcined dolomite. (d) Write an equation for the calcination of magnesium hydroxide.

(CCEA June 2015)

Before moving to the next section, check that you are able to:

- Describe, with the aid of equations, the reactions of Group II oxides, hydroxides and carbonates with water and dilute acids.
- Explain, in terms of hydrogen ions, why Group II oxides, hydroxides and carbonates are bases.
- Explain what is meant by the term *thermal decomposition*.
- Describe, with the aid of equations, the thermal decomposition of Group II hydroxides and carbonates.
- Recall the use of Group II compounds in indigestion remedies, toothpaste, and the manufacture of concrete and cement.

Thermal Stability of Group II Compounds

In this section we are learning to:

- Account for trends in the thermal stability of Group II carbonates and hydroxides.

Group II carbonates (MCO_3) undergo thermal decomposition to form the corresponding metal oxide (MO) and carbon dioxide. The temperature above which a Group II carbonate decomposes increases down the group.

$$MgCO_{3(s)} \rightarrow MgO_{(s)} + CO_{2(g)}$$

Decomposes above 540 °C

$$BaCO_{3(s)} \rightarrow BaO_{(s)} + CO_{2(g)}$$

Decomposes above 1360 °C

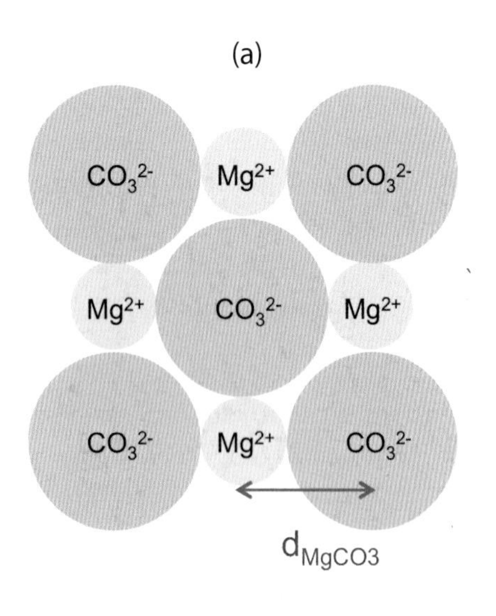

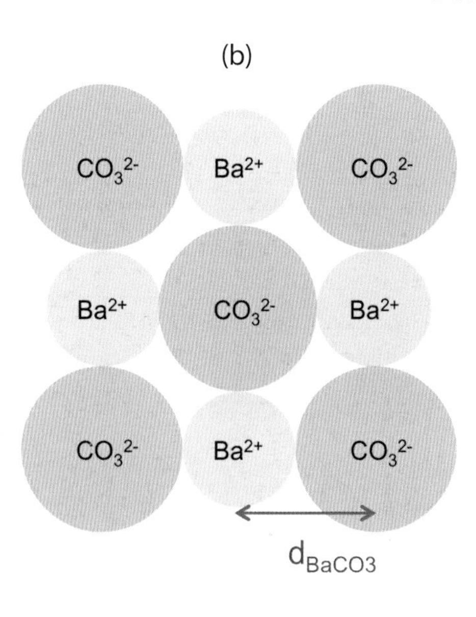

Figure 7: The ionic lattice in (a) $MgCO_3$ and (b) $BaCO_3$. The distance between neighbouring metal ions and carbonate ions, d is greater in $BaCO_3$ as Ba^{2+} ion is bigger than Mg^{2+} ion.

Figure 8: (a) The polarisation of a C-O bond within a carbonate ion by a neighbouring Mg^{2+} ion. (b) Decomposition of a carbonate ion in $MgCO_3$ to form magnesium oxide, MgO and carbon dioxide, CO_2.

The ionic lattices in Figure 7 reveal that the bonding in $BaCO_3$ is weaker than in $MgCO_3$ as the ions are less tightly packed. As a result the greater **thermal stability** of $BaCO_3$ cannot be due to stronger ionic bonding, and must instead be explained in terms of how the metal ions in the lattice affect the neighbouring carbonate ions.

As the Group II metal ions (M^{2+}) get smaller towards the top of the group, the positive charge becomes more concentrated, and the density of positive charge within the ion increases. The higher **charge density** within the metal ion results in greater attraction between the metal ion and the electrons in neighbouring carbonate ions, and increases the polarity of C-O bonds adjacent to the metal ion as shown in Figure 8a.

Polarising a C-O bond reduces the extent to which electrons are shared in the bond, and by so doing reduces the strength of the C-O bond. As a result, less energy is needed to break the C-O bond and the carbonate ion becomes less resistant to the effects of heating. The thermal decomposition of a polarised carbonate ion to form an oxide ion (O^{2-}) and carbon dioxide (CO_2) is illustrated in Figure 8b.

In this way the thermal stability of Group II carbonates increases down the group *as the charge density on the metal decreases, and the metal ion becomes less able to polarise C-O bonds within neighbouring carbonate ions.*

The Group II hydroxides also become more thermally stable down the group as the charge density on the metal decreases, and the metal ions become less able to polarise O-H bonds within the neighbouring hydroxide ions.

Worked Example 2.11vi

Currently 280 million tonnes of calcium oxide is produced annually by heating limestone (calcium carbonate) to a temperature of 1200 °C. The heat needed to sustain the reaction is provided by the combustion of fossil fuels.

$$CaCO_3 \rightarrow CaO + CO_2$$

(a) Compare the thermal stability of calcium carbonate with other Group II metal carbonates. (b) Explain how the thermal stability of a Group II carbonate is related to the charge and size of the cation.

(CCEA January 2010)

Solution

(a) Calcium carbonate is more stable than $BeCO_3$ and $MgCO_3$ and less stable than $SrCO_3$ and $BaCO_3$.

(b) The Group II carbonates become more resistant to heating as the charge density within the metal ion decreases down the group.

Exercise 2.11G

1. Which one of the following (A-D) lists the Group II carbonates and hydroxides in order of increasing thermal stability?

	Carbonates	Hydroxides
A	$MgCO_3$ $CaCO_3$ $SrCO_3$ $BaCO_3$	$Ba(OH)_2$ $Sr(OH)_2$ $Ca(OH)_2$ $Mg(OH)_2$
B	$BaCO_3$ $SrCO_3$ $CaCO_3$ $MgCO_3$	$Mg(OH)_2$ $Ca(OH)_2$ $Sr(OH)_2$ $Ba(OH)_2$
C	$MgCO_3$ $CaCO_3$ $SrCO_3$ $BaCO_3$	$Mg(OH)_2$ $Ca(OH)_2$ $Sr(OH)_2$ $Ba(OH)_2$
D	$BaCO_3$ $SrCO_3$ $CaCO_3$ $MgCO_3$	$Ba(OH)_2$ $Sr(OH)_2$ $Ca(OH)_2$ $Mg(OH)_2$

(CCEA January 2008)

2. Barium oxide is formed by the thermal decomposition of barium carbonate. (a) Write an equation for the decomposition of barium carbonate. (b) Explain why calcium carbonate will decompose at a lower temperature than barium carbonate.

(CCEA January 2014)

3. (a) Write the equation for the thermal decomposition of calcium carbonate. (b) Explain why the decomposition temperature for magnesium carbonate would be expected to be greater than or less than the decomposition temperature for calcium carbonate.

(CCEA June 2005)

4. (a) Write an equation for the decomposition of calcium hydroxide. (b) Compare and explain the thermal stability of magnesium hydroxide with barium hydroxide.

(CCEA January 2011)

5. Magnesium is obtained from calcined (heated) dolomite, $CaCO_3.MgCO_3$. When calcined dolomite decomposes it forms the metal oxides. (a) Write the equation for the decomposition of dolomite. (b) State the order of thermal stability of the Group II metal carbonates starting with the least stable. (c) Explain the thermal stability of the Group II carbonates in terms of the cations.

(CCEA June 2015)

6. (a) Calcium sulfate occurs as gypsum, $CaSO_4.2H_2O$. Calculate the percentage yield of $CaSO_4$ when 34.4 g of gypsum is heated to form 26.0 g of anhydrous calcium sulfate. (b) Heating an anhydrous Group II sulfate produces the corresponding Group II oxide and sulfur trioxide. Write the equation for the decomposition of anhydrous calcium sulfate. (c) The thermal stability of Group II sulfates can be explained in a similar way as the stability of the Group II carbonates. Explain the relative stability of the Group II sulfates with reference to the cations involved.

(Adapted from CCEA June 2009)

7. (a) Write the equation for the decomposition of barium carbonate. (b) Suggest why the thermal stability of barium carbonate is higher than that of beryllium carbonate. (c) The decomposition of barium carbonate occurs at a much lower temperature if it is heated with carbon to form barium oxide and carbon monoxide. Write the equation for the reaction.

(CCEA January 2009)

8. Group II carbonates and nitrates decompose when heated. Magnesium nitrate forms magnesium oxide, nitrogen(IV) oxide and oxygen. (a) Write an equation for the decomposition of magnesium nitrate. (b) Suggest whether or not strontium nitrate would be more or less stable than magnesium nitrate when heated. Explain your reasoning.

(CCEA January 2013)

9. Which one of the following describes a trend down Group II from beryllium to barium?

   A The reactivity of the metal decreases.

   B The solubility of the hydroxide decreases.

   C The thermal stability of the carbonates increases.

   D The solubility of the sulfate increases.

(CCEA June 2004)

Before moving to the next section, check that you are able to:

- Explain what is meant by the term *thermal stability.*
- Account for trends in the thermal stability of Group II carbonates and hydroxides in terms of the size of the metal ion.

Unit AS 3:

Practical Assessment

Practical Assessment

Introduction

The AS Practical Assessment consists of two parts. The first part of the assessment is laboratory-based, and is focussed on making and recording observations. The second part is timetabled for a different day, and is focussed on the analysis of experimental data and the evaluation of experimental techniques.

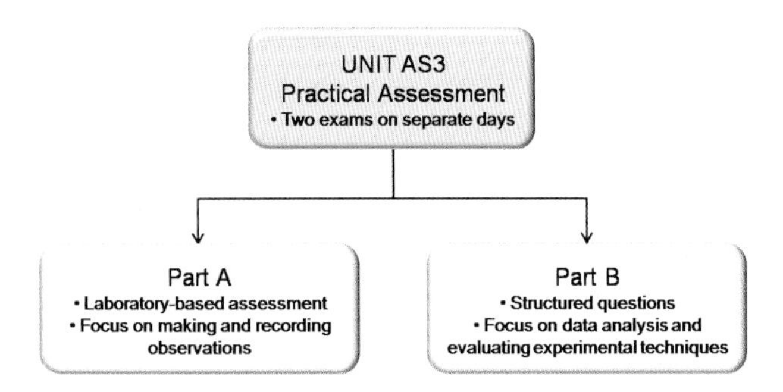

The tasks in Part A afford candidates an opportunity to demonstrate their knowledge of practical chemistry, including the tasks listed in the AS3 Section of the specification. The Worked Examples and Problems in the following sections exemplify the level of detail expected when making observations, analysing data, and evaluating experimental techniques, and should be used in conjunction with the Specimen Assessment Materials (SAMs) published by CCEA to gain insight into the likely format and scope of questions in both parts of the practical assessment.

Volumetric Analysis

In this section we are learning to:

- Identify the knowledge and skills required to answer volumetric analysis components of the practical assessment.

In the first part of the practical assessment (Paper A) a candidate may be asked to perform a titration, record their results in a suitable table, and report any other observations that could be made during the experiment. An outline of the method will be given, and candidates will be expected to demonstrate knowledge of how the method can be used to generate accurate results as in Worked Example 3i.

Worked Example 3i

You are required to titrate a vinegar solution using sodium hydroxide solution and phenolphthalein indicator.

You are provided with:

Sodium hydroxide solution of concentration 0.10 mol dm^{-3}.
Vinegar (ethanoic acid) of unknown concentration.
Phenolphthalein indicator.

Carry out the titration as follows:

- Rinse and fill a burette with the sodium hydroxide solution.
- Use a pipette filler to rinse the pipette and transfer 25.0 cm^3 of the vinegar solution to a conical flask.
- Add three drops of phenolphthalein indicator to the conical flask and titrate until the end point is reached.

Present your results in a suitable table and calculate the average titre.

	Initial Burette Reading (cm^3)	Final Burette Reading (cm^3)	Titre (cm^3)
Rough	0.0	22.0	22.0
First Accurate	21.7	43.1	21.4
Second Accurate	0.0	21.3	21.3

$$\text{Average titre} = \frac{21.4 + 21.3}{2} = 21.35\ cm^3$$

State the colour change at the end point of your titration.

Colourless to pink.

(Adapted from CCEA June 2009)

In the second part of the assessment (Paper B) a candidate may be asked to analyse the results of a titration, and to describe relevant procedures such as: how to prepare a solution, how to perform a titration, or how to ensure that the results of a titration experiment are accurate. The level of detail expected in Part B of the practical assessment is demonstrated in Worked Example 3ii.

Worked Example 3ii

The following results were obtained when 25 cm^3 portions of a vinegar solution of unknown concentration were titrated using 0.10 mol dm^{-3} sodium hydroxide solution and phenolphthalein indicator.

	Initial Burette Reading (cm^3)	Final Burette Reading (cm^3)	Titre (cm^3)
Rough	0.0	22.0	22.0
First Accurate	21.7	43.1	21.4
Second Accurate	0.0	21.3	21.3

(a) Give details of the procedure used.

Rinse the pipette with the vinegar solution. Use the pipette and a pipette filler to transfer 25 cm^3 of vinegar solution to a conical flask. Add 2-3 drops of phenolphthalein to the conical flask. Rinse the burette with sodium hydroxide solution before filling the burette with sodium hydroxide solution. Perform a rough titration by adding sodium hydroxide in 1 cm^3 amounts until the solution changes from colourless to pink at the end point. Perform an accurate titration, adding sodium hydroxide dropwise near the end point. Repeat to obtain accurate titres that are in close agreement.

(b) Write the equation, including state symbols, for the reaction of sodium hydroxide with the ethanoic acid present in vinegar.

$CH_3COOH_{(aq)} + NaOH_{(aq)} \rightarrow CH_3COONa_{(aq)} + H_2O_{(l)}$

(c) Calculate the number of moles of sodium hydroxide used in the titration.

Moles of NaOH used = 0.10 mol dm^{-3} × 2.135 × 10^{-2} dm^{-3} = 2.1 × 10^{-3} mol

Calculate the number of moles of ethanoic acid neutralised in the titration.

Moles of ethanoic acid = moles of NaOH used = 2.1×10^{-3} mol

Calculate the concentration (in mol dm^{-3}) of the ethanoic acid in the vinegar.

$$\text{Molarity of ethanoic acid} = \frac{2.1 \times 10^{-3}\ \text{mol}}{0.025\ \text{dm}^3} = 0.084\ \text{mol dm}^{-3}$$

Calculate the concentration (in g dm^{-3}) of the ethanoic acid in the vinegar.

0.084 mol $dm^{-3} \times 60$ g $mol^{-1} = 5.0$ g

(Adapted from CCEA June 2009)

Exercise 3A

1. Assuming all the apparatus is clean and dry (a) suggest **two** ways in which the accuracy of a titration can be increased, and (b) suggest **two** ways in which the reliability of a titration can be increased.

   (Adapted from CCEA May 2016)

2. The formula of an organic acid, RCOOH can be determined by titrating a solution of the acid against sodium hydroxide solution. (a) Describe how you would prepare 250 cm^3 of a 7.4 g dm^{-3} solution of the organic acid. The density of the acid is 0.80 g cm^{-3}. (b) An average titre of 24.8 cm^3 is obtained when 25 cm^3 portions of the organic acid solution are titrated using 0.10 mol dm^{-3} sodium hydroxide solution and phenolphthalein indicator. (i) Write the equation for the titration reaction. (ii) State the colour change at the end point. (iii) Calculate the formula of the acid, RCOOH using the following headings to structure your work.

   - Molarity of the organic acid solution.
   - Relative molecular mass of the acid.
   - Relative formula mass of the alkyl group, R.
   - Formula of the acid, RCOOH.

   (Adapted from CCEA May 2012)

3. Washing soda crystals ($Na_2CO_3.xH_2O$) lose water of crystallisation when left in the air. (a) Describe how you would prepare 250 cm^3 of a 11.6 g dm^{-3} washing soda solution from the solid. (b) An average titre of 19.8 cm^3 is obtained when 25 cm^3 portions of the washing soda solution are titrated using 0.10 mol dm^{-3} hydrochloric acid and methyl orange indicator. (i) Write the equation for the titration reaction. (ii) State the colour change at the end point. (iii) Calculate the value of x in the formula $Na_2CO_3.xH_2O$ using the following headings to structure your work.

   - Moles of sodium carbonate in 25 cm^3 of solution.
   - Molarity of the sodium carbonate solution.
   - Mass of sodium carbonate in 1 dm^3 of solution.
   - Mass of water of crystallisation in 11.6 g of washing soda.
   - The value of x.

   (Adapted from CCEA May 2013)

Before moving to the next section, check that you are able to:

- Describe the methods used to prepare and titrate solutions accurately.
- Use the average titre to calculate amounts of solute being titrated.

Observation-Deduction

In this section we are learning to:

- Record appropriate observations and deductions when performing chemical tests on mixtures of compounds.

In the first part of the assessment (Paper A) candidates may be asked to perform chemical tests on a substance, or a mixture of substances, and record observations using appropriate terminology. The following exercises demonstrate the level of detail expected in the laboratory-based observation tasks, and the nature of the deductions that can be discerned from observations in the second part of the assessment (Paper B).

Testing for Ions

If a mixture is coloured it likely contains a transition metal compound. Conversely, if the mixture is a white solid it likely contains compounds of Group I and II metals, or ammonium salts.

Examples of observation-deduction based on appearance:

Experiment	Observations	Deductions
Describe the mixture.	Blue solid	The mixture contains a transition metal compound.

Experiment	Observations	Deductions
Describe the mixture.	White solid	The mixture contains Group I, II or ammonium compounds.

The presence of individual ions in a mixture is confirmed by performing chemical tests for specific ions. *In an observation-deduction exercise the presence of an ion can only be confirmed if the observations indicate a positive test for the ion.*

Often, a positive test for an ion can only be obtained after it has been determined that other ions are not present in the mixture. For example, in Worked Example 3iii, the sequence of observations and deductions in Experiment 3 demonstrates that the presence of chloride ion can only be confirmed once dilute acid has been used to confirm that the mixture does not contain carbonate ions.

Worked Example 3iii

The following table details the results of tests conducted on a mixture of two salts, labelled X, which have a common cation. Record appropriate deductions in the table and identify both salts in the mixture.

Experiment	Observations	Deductions
1. Describe X.	White solid.	Group I, II or ammonium compound.
2. (a) Fill a test tube one quarter full of water and record the temperature.	20°C	
(b) Add three spatula measures of X to the test tube, stir and record the temperature.	18°C	
(c) Record the temperature change. Keep the contents of this test tube for experiments 3 and 4.	-2°C	Reaction is endothermic.

3. (a) Add 1-2 cm^3 of the solution formed in experiment 2 to a test tube, acidify with 1 cm^3 of dilute nitric acid, and then add 1 cm^3 of silver nitrate solution.	No effervescence. White precipitate formed.	Carbonate ion not present. Chloride ion present.
(b) Add 5 cm^3 of dilute ammonia solution to the test tube.	Precipitate dissolves to form a colourless solution.	Confirms the presence of chloride ion.
4. Add 1-2 cm^3 of the solution formed in experiment 2 to a test tube, acidify with 3 drops of dilute nitric acid, and then add 3 drops of barium chloride solution.	White precipitate formed.	Sulfate ion present.
5. Add a spatula measure of X to a test tube one third full of dilute sodium hydroxide and warm gently, testing any gas evolved with a glass rod dipped in concentrated hydrochloric acid.	Pungent smell. White fumes of solid ammonium chloride.	 Ammonia gas. The mixture contains an ammonium salt.

Name the two salts present in X:

Ammonium chloride and ammonium sulfate.

(Adapted from CCEA May 2011)

In Worked Example 3iii the observations and deductions confirm that carbonate ion was not present in the mixture. The observations and deductions needed to confirm the presence of carbonate ion, and the presence of a hydrated salt, are illustrated in Worked Example 3iv.

Worked Example 3iv

You are provided with an unknown solid B. Record appropriate deductions in the table and identify the compound B.

Test	Observations	Deductions
1. Describe the appearance of B.	White solid.	Group I, II or ammonium compound.
2. (a) Add half a spatula measure of B to a test tube one quarter filled with dilute ethanoic acid.	Effervescence. Colourless gas formed.	Carbonate ion present.
(b) Use limewater to test any gas that is produced.	Cloudy white suspension turns limewater milky.	The gas is carbon dioxide. Confirms carbonate ion present.
3. Add a spatula measure of B to a dry boiling tube and heat.	Colourless liquid forms on the side of the boiling tube.	Liquid water formed. B is a hydrated salt.
4. Dip a clean nichrome wire loop into concentrated hydrochloric acid, touch sample B with the wire, then hold it in a blue Bunsen flame.	Yellow/orange flame.	Sodium ion present.

Name the compound B:

Hydrated sodium carbonate

(Adapted from CCEA May 2015)

In Worked Example 3v a mixture of potassium iodide and potassium sulfate is identified by testing for the presence of potassium ions, halide ions and sulfate ions. The presence of each ion is again confirmed from observations that indicate a positive test for the ion. The reaction between potassium iodide and concentrated sulfuric acid is also included in the scheme, and is used to provide initial evidence for the presence of iodide ions in the mixture.

Worked Example 3v

The following table details the results of tests conducted on a mixture of two salts, labelled B, which have a common cation. Record appropriate deductions in the table and identify both salts in the mixture.

Test	Observations	Deductions
1. Describe the appearance of B.	White solid.	Group I, II or ammonium compound.
2. Dip a wire loop in concentrated hydrochloric acid, touch sample B with the wire, then hold it in a blue Bunsen flame.	Lilac flame (pink through cobalt glass).	Potassium ion present.
In a fume cupboard: 3. Add about 1 cm^3 of concentrated sulfuric acid to a half spatula measure of B in a test tube. Heat the test tube gently.	Steamy fumes. Purple vapour. Grey-black solid formed.	 Iodine formed. Confirms iodine formed. Iodide ion present.
4. Make up a solution of B by dissolving a half spatula-measure of B in a test tube half-full of water. Put 1 cm^3 of the solution into each of two separate test tubes.	Colourless solution formed.	
(a) (i) Add a few drops of silver nitrate solution to the first test tube.	Yellow precipitate formed.	Iodide ion present.
(ii) Add about 2 cm^3 of concentrated ammonia to the first test tube.	Yellow precipitate remains.	Confirms iodide ion present.
(b) Add a few drops of barium chloride solution to the second test tube and then add 2 cm^3 of dilute nitric acid.	White precipitate formed. No effervescence with acid.	Sulfate ion present. Carbonate ion not present.

Name the two salts present in B:

Potassium iodide and potassium sulfate.

(Adapted from CCEA June 2009)

Exercise 3B

1. The following table details the results of tests conducted on a mixture of two salts, labelled A, which have a common cation. Record appropriate deductions in the table and identify both salts in the mixture.

Experiment	Observations	Deductions
1. Describe the appearance of A.	White solid.	
2. Dip a wire loop in concentrated hydrochloric acid, touch sample A with the wire, then hold it in a blue Bunsen flame.	Yellow/orange flame.	
In a fume cupboard: 3. Add about 1 cm^3 of concentrated sulfuric acid to a half spatula measure of A in a test tube. Test the gas evolved using a glass rod dipped in concentrated ammonia solution.	Steamy fumes. White fumes of solid ammonium chloride.	
4. Make up a solution of A by dissolving a half spatula-measure of A in a test tube half-full of dilute nitric acid. Put 1 cm^3 of the solution into each of two separate test tubes. (a) (i) Add a few drops of silver nitrate solution to the first test tube. (ii) Add about 1 cm^3 of concentrated ammonia into the first test tube. (b) Add a few drops of barium chloride solution into the second test tube.	No effervescence. Colourless solution formed. White precipitate formed. White precipitate dissolves. White precipitate formed.	

(Adapted from CCEA June 2009)

> Before moving to the next section, check that you are able to:
>
> - Record appropriate observations and deductions when testing for the presence of cations and anions in mixtures of ionic compounds.

Testing for Functional Groups

If the unknown is an organic compound the observation-deduction exercise may involve:

- describing the physical properties of the unknown,
- burning a sample of the unknown,
- using bromine water to test for a C=C bond,
- using PCl_5 to test for a hydroxyl (–OH) group,
- using acidified dichromate to oxidise the unknown, or
- using a base to test for acidity.

The deductions may then be used in conjunction with the IR spectrum of the compound to deduce its structure. Worked Example 3vi illustrates a sequence of tests to detect and then confirm the presence of a C=C functional group.

Worked Example 3vi

The table details the results of tests conducted on an organic liquid labelled Y. Record appropriate deductions in the table and identify the functional group present in the liquid.

Experiment	Observations	Deductions
1. Place 10 drops of Y in a test tube and add 1 cm^3 of water.	The liquids form two layers. Liquid Y floats on water.	Y cannot hydrogen bond effectively with water. Y is less dense than water.
2. Place 10 drops of Y on a watch glass placed on a heat proof mat and ignite it using a splint.	Y burns with a smoky yellow flame.	Y has a high carbon content.
3. Add approximately 10 drops of Y to a test tube one quarter full of bromine water and mix well.	The bromine water is decolourised.	Y contains one or more C=C bonds.
4. Add 10 drops of Y to 2 cm^3 of acidified potassium dichromate solution in a test tube. Warm the mixture gently.	The orange colour of dichromate remains.	Y is not oxidised. Y is not a primary or secondary alcohol.

Based on the above tests suggest a functional group which is present in Y:

A carbon-carbon double bond (a C=C bond).

(Adapted from CCEA May 2010)

Exercise 3C

1. Write appropriate deductions for the following sequence of experiments carried out on an organic liquid B.

Experiment	Observations	Deductions
1. Place 10 drops of B in a test tube and add 1 cm^3 of water.	B mixes completely with water to form a colourless solution.	
2. Place 10 drops of B on a watch glass placed on a heat proof mat and ignite it using a splint.	B burns with a clean blue flame.	
3. Add approximately 10 drops of B to a test tube one quarter full of bromine water and mix well.	The orange colour of bromine remains.	
4. Add 10 drops of B to 2 cm^3 of acidified potassium dichromate solution in a test tube. Warm the mixture gently.	The mixture changes from orange to green and there is a change in smell.	

(Adapted from CCEA June 2010)

2. Write appropriate deductions for the following sequence of experiments carried out on an aqueous solution of an organic liquid B.

Experiment	Observations	Deductions
1. Describe the solution and place a few drops on a piece of Universal Indicator paper.	Colourless solution. Turns Universal Indicator paper green.	
In a fume cupboard: 2. Shake a small volume of the solution with bromine water.	The orange colour of bromine remains.	
3. Gently heat 2 cm^3 of the solution with 2 cm^3 of acidified potassium dichromate solution.	The mixture changes from orange to green and there is a change in smell.	

(Adapted from CCEA June 2009)

The presence of C-Cl, C-Br and C-I bonds in halogenoalkanes can be detected by forming the corresponding silver halide precipitate as demonstrated by the hydrolysis reactions in Worked Example 3vii.

Worked Example 3vii

The halogenoalkanes 1-chlorobutane, 1-bromobutane and 1-iodobutane are colourless liquids. A sample of each liquid is provided in containers labelled X, Y and Z. Identify the samples by writing appropriate deductions for the following experiments carried out on X, Y and Z.

Experiment	Observations	Deductions
Place 1 cm^3 of X, Y and Z separately into three test tubes. Label the test tubes with their contents. Add 1 cm^3 of ethanol and 1 cm^3 of silver nitrate solution to each test tube. Place the three test tubes in a beaker of water heated to 50-60 °C. Leave for 5 minutes and note the relative rate of reaction.	X Yellow precipitate formed. First precipitate formed.	X Silver iodide precipitate. Fastest reaction. X contains a C-I bond.
	Y White precipitate formed. Last precipitate formed.	Y Silver chloride precipitate. Slowest reaction. Y contains a C-Cl bond.
	Z Cream precipitate formed. Second precipitate formed.	Z Silver bromide precipitate. Z contains a C-Br bond.

Based on the above tests:

Identify X 1-iodobutane
Identify Y 1-chlorobutane

(Adapted from CCEA June 2011)

Exercise 3D

1. The following observations were recorded for tests carried out on an organic liquid labelled Z. Complete the deductions.

Experiment	Observations	Deductions
1. Place 1 cm^3 of Z in a test tube and add 1 cm^3 of water. Add a bung and shake the test tube.	Two layers formed.	
2. Place 10 drops of Z on a watch glass placed on a heatproof mat and ignite it using a burning splint.	Z burns with a smoky yellow flame.	
3. In a fume cupboard add approximately 0.5 cm^3 of Z to a test tube one quarter full of bromine water and mix well.	The bromine water is decolourised.	
4. Place 1 cm^3 of Z in a test tube. Add 1 cm^3 of ethanol and 1 cm^3 of silver nitrate solution before placing the test tube in a beaker of hot water for 5 minutes.	A yellow precipitate forms.	

(Adapted from CCEA May 2014)

In Worked Example 3viii an aqueous solution of ethanoic acid (vinegar) is quickly identified by its characteristic smell. Subsequent tests are then conducted to confirm the presence of a carboxyl group (-COOH) by demonstrating that the organic compound behaves as an acid when mixed with bases and metals.

Worked Example 3viii

Write appropriate deductions for the following sequence of experiments carried out on an aqueous solution of an organic liquid Y.

Experiment	Observations	Deductions
1. Describe the smell of solution Y.	Vinegar smell.	Y may be an aqueous solution of ethanoic acid.
2. Using a glass rod place a drop of Y onto Universal Indicator paper.	The Universal Indicator paper turns orange.	Y is a weak acid. Confirms that Y is a carboxylic acid.
3. Add a spatula measure of anhydrous sodium carbonate to a test tube one quarter full of solution Y and identify the gas evolved using a suitable reagent.	Effervescence. The gas produced turns limewater milky. The sodium carbonate reacts to form a colourless solution.	Carbon dioxide gas is produced. Confirms that Y is an acid.
4. Add 1 cm^3 of Y to a test tube and then add a 2 cm length of magnesium ribbon.	Effervescence. The mixture warms up. Magnesium reacts to form a colourless solution.	 The reaction is exothermic. Confirms that Y is an acid.

Based on the above tests suggest a functional group which is present in Y:

A carboxyl (-COOH) group.

Liquid Y contains one functional group and two carbon atoms. Write an equation for the reaction occurring in experiment 4.

$2CH_3COOH + Mg \rightarrow (CH_3COO)_2Mg + H_2$

(CCEA May 2011)

Before moving to the next section, check that you are able to:

- Record appropriate observations and deductions when testing for functional groups in organic compounds.

Structured Questions

In this section we are learning to:

- Use practical knowledge to analyse experimental data, evaluate the techniques used to acquire data, and account for the techniques used to prepare and purify compounds.

The structured questions in the second part of the practical assessment (Paper B) test understanding of how practical techniques such as reflux and distillation are used in the preparation and purification of compounds. The questions may also test understanding of the techniques used to collect and analyse data when determining reacting amounts, or physical quantities such as: the rate of a reaction, the solubility of a compound, and enthalpy changes.

Determining Reacting Amounts

The following examples test understanding of practical methods used to determine reacting amounts, and how the data obtained can be used to deduce the formula of a compound.

Worked Example 3ix

You are required to plan an experiment to determine the degree of hydration in a sample of sodium carbonate. If the sample of hydrated sodium carbonate is heated in a crucible to constant mass and appropriate masses measured, the value of x in the formula $Na_2CO_3.xH_2O$ can be found.

(a) Explain the meaning of the term hydrated sodium carbonate.

Solid sodium carbonate that contains water of crystallisation.

Draw a labelled diagram to show the apparatus which could be used to heat the hydrated sodium carbonate.

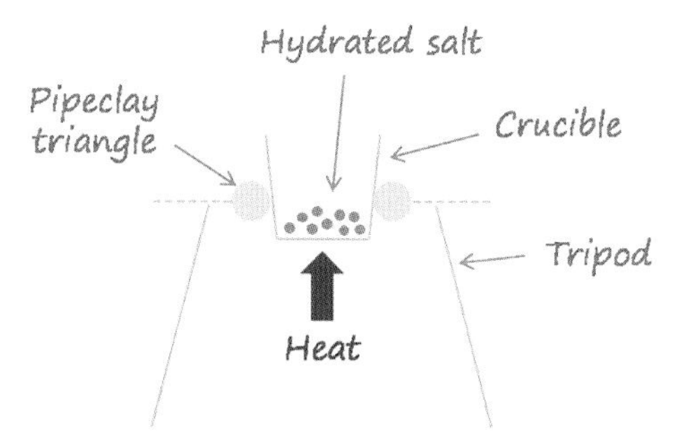

(b) What masses should be recorded before heating the hydrated carbonate?

The mass of the empty crucible and the mass of the crucible when it contains the sample of hydrated sodium carbonate.

The hydrated sodium carbonate is heated to remove all the water. What steps would you take to ensure that it had all been removed?

Heat then weigh the crucible containing the sample. Repeat until the mass of the crucible containing the sample stops decreasing.

State one safety precaution which should be followed after the sample is heated and before weighing.

Use tongs to lift the crucible after it has been heated.

(c) When 11.44 g of hydrated sodium carbonate was heated 4.24 g of anhydrous sodium carbonate was formed.

What is the mass of water lost?

11.44 – 4.24 = 7.20 g

What is the number of moles of water lost?

$$\frac{7.20\ g}{18\ g\ mol^{-1}} = 0.400\ mol$$

What is the number of moles of anhydrous sodium carbonate formed?

$$\frac{4.24\ g}{106\ g\ mol^{-1}} = 0.0400\ mol$$

Calculate the value of x in $Na_2CO_3.xH_2O$.

$$X = \frac{0.400\ mol\ of\ water}{0.0400\ mol\ of\ hydrate} = 10$$

(CCEA June 2009)

Exercise 3E

1. The formula of hydrated zinc sulfate is $ZnSO_4.\ xH_2O$. The value of x can be found by heating a sample of hydrated zinc sulfate to constant mass. (a) Explain the meaning of the term *hydrated zinc sulfate.* (b) Draw a labelled diagram to show the apparatus which could be used to heat the hydrated zinc sulfate. (c) What masses should be recorded before heating the hydrated zinc sulfate? (d) The hydrated zinc sulfate is heated to remove all the water. What steps would you take to ensure that it had all been removed? (e) State one safety precaution which should be followed after heating and before weighing. (f) Heating 8.63 g of hydrated zinc sulfate produced 4.85 g of anhydrous zinc sulfate. Calculate (i) the mass of water lost, (ii) the moles of water lost, (iii) the moles of anhydrous salt formed, and (iv) the formula of hydrated zinc sulfate.

(CCEA June 2009)

2. Heating a Group II carbonate, MCO_3 to constant mass can be used to identify the Group II element, M. The equation for the action of heat on the metal carbonate is:

$$MCO_3 \rightarrow MO + CO_2$$

(a) Draw a diagram of the apparatus which could be used to heat the carbonate to constant mass. (b) Apart from wearing safety glasses, state one safety precaution which should be taken. (c) What weighings must be recorded before heating the sample? (d) Explain the term *heated to constant mass.* (e) 2.33 g of metal oxide, MO is obtained on heating 4.90 g of the metal carbonate, MCO_3 to constant mass. Calculate (i) the mass of carbon dioxide evolved, (ii) the moles of carbon dioxide evolved, (iii) the moles of metal oxide formed, (iv) the RAM of the Group II element, and (v) the identity of M.

(Adapted from CCEA May 2005)

Worked Example 3x

(a) The empirical formula of an oxide of copper can be found by reducing the oxide in the following apparatus.

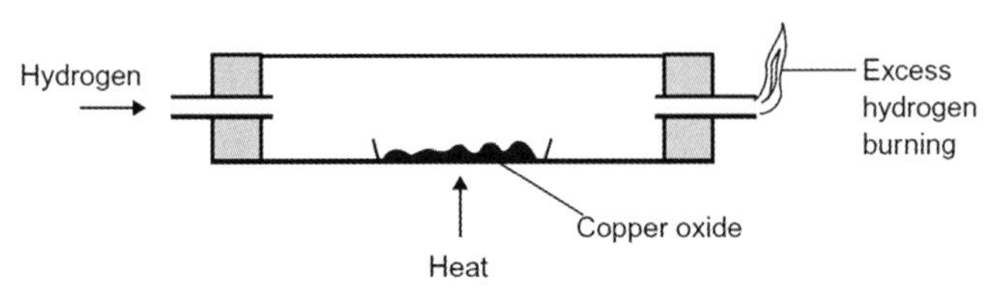

What weighings should be taken before heating?

The mass of the container and the mass of the container when filled with copper oxide.

In addition to wearing safety glasses, suggest and explain one other safety precaution which should be taken.

Flush the apparatus with hydrogen gas as mixtures of hydrogen and oxygen are explosive and should not be heated.

What steps would you take to ensure that all of the oxygen has been removed from the copper?

The sample is heated in its container. The sample is then weighed in its container and the process repeated until the mass does not change.

Explain why the hydrogen continues to be passed through the apparatus after all the copper oxide has been reduced.

To allow the apparatus to cool and to prevent the copper reacting with oxygen in the air.

(b) When 2.16 g of the copper oxide was reduced, 1.92 g of copper was formed.

What mass of oxygen was present in the copper oxide?

2.16 – 1.92 = 0.24 g

How many moles of oxygen were present in the copper oxide?

$$\frac{0.24 \text{ g}}{16 \text{ g mol}^{-1}} = 0.015 \text{ mol}$$

How many moles of copper were formed?

$$\frac{1.92 \text{ g}}{64 \text{ g mol}^{-1}} = 0.0300 \text{ mol}$$

Calculate the empirical formula of the copper oxide.

1 mol of O reacts with

$$\frac{0.0300 \text{ mol}}{0.015 \text{ mol}} = 2 \text{ mol of Cu.}$$

The formula of the copper oxide is Cu_2O.

(CCEA June 2010)

Exercise 3F

1. The empirical formula of an oxide of lead can be found by reducing the lead oxide using the apparatus below.

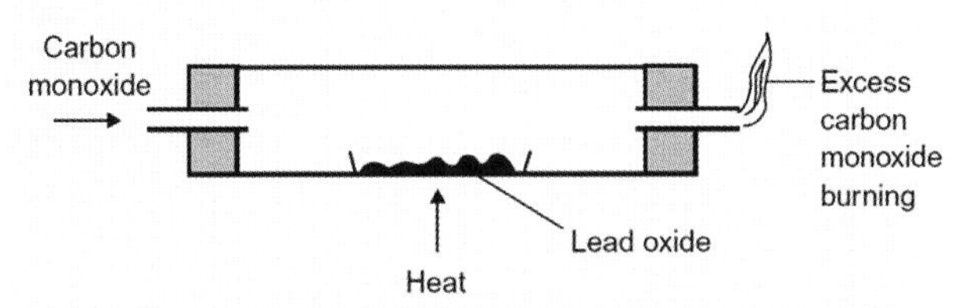

(a) What weighings should be taken before heating? (b) In addition to wearing safety glasses, suggest and explain one other safety precaution which should be taken. (c) What steps would you take to ensure that all of the oxygen has been removed from the lead? (d) Explain why the carbon monoxide continues to be passed through the apparatus after all the lead oxide has been reduced. (e) When 1.39 g of the lead oxide was reduced, 1.26 g of lead was formed. (i) What mass of oxygen was present in the lead oxide? (ii) How many moles of oxygen were present in the lead oxide? (iii) How many moles of lead were formed? (iv) Calculate the empirical formula of the lead oxide.

(CCEA June 2010)

Before moving to the next section, check that you are able to:

- Explain the application of laboratory techniques to the study of reactions.
- Use experimental data to calculate reacting amounts.

Synthesis and Purification of Compounds

The following examples test understanding of the laboratory methods used to synthesise and purify compounds.

Exercise 3G

1. Heating a reacting mixture under reflux is an important practical technique in organic chemistry. Draw a labelled diagram of the apparatus used to reflux a reaction mixture.

(CCEA June 2009)

2. Ethanoic acid can be prepared in the laboratory by the following method.

Place 50 cm³ of water in a round-bottomed flask with some anti-bump granules. Carefully add concentrated sulfuric acid to the flask with swirling and cooling. Add 50 cm³ of potassium dichromate solution. Slowly add a mixture of 15 cm³ of ethanol in 50 cm³ of water, shake the flask and cool if a vigorous reaction occurs. Heat the solution under reflux for twenty minutes. Allow the flask to cool.

Rearrange the apparatus and collect the product by distillation.

(a) State the purpose of the anti-bump granules. (b) Suggest why the flask is cooled when concentrated sulfuric acid is added. (c) Explain what is meant by the term *reflux*. (d) State the function of the acidified potassium dichromate solution. (e) Name a suitable drying agent for the distillate and suggest how it may be removed. (f) Suggest two reasons why the percentage yield obtained is less than 100%.

(Adapted from CCEA May 2016)

3. Separating a product from a mixture by distillation is an important practical technique in organic chemistry. Draw a labelled diagram of the apparatus used to carry out a distillation.

(CCEA June 2009)

4. Diethyl ether, C_2H_5-O-C_2H_5 is a highly flammable, colourless liquid with a boiling point of 35 °C. It can be prepared by the slow addition of ethanol to concentrated sulfuric acid followed by heating in a distillation apparatus to 155 °C. The crude distillate is cooled in an ice bath as it is collected.

$2C_2H_5OH \rightarrow C_2H_5\text{-O-}C_2H_5 + H_2O$

(a) Suggest the role of the concentrated sulfuric acid in this preparation. (b) Suggest why the diethyl ether collected must be cooled in an ice bath. (c) Suggest an organic impurity which would be present in the crude distillate. (d) The crude distillate is purified by shaking with a 10% sodium carbonate solution in a separating funnel. Once separated the organic layer is further treated with anhydrous magnesium sulfate. (i) State the purpose of adding sodium carbonate solution to the crude distillate. (ii) Explain why diethyl ether does not mix with water. (iii) State the purpose of the anhydrous magnesium sulfate. (iv) How would the magnesium sulfate be removed from the organic layer. (e) The diethyl ether is finally purified by distillation. Explain why distillation achieves a satisfactory separation.

(CCEA May 2011)

5. Iodoethane can be prepared in the laboratory by the following method.

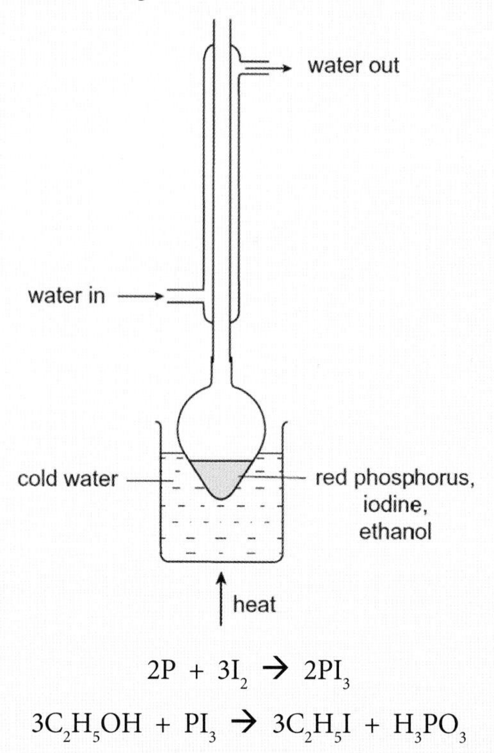

$$2P + 3I_2 \rightarrow 2PI_3$$

$$3C_2H_5OH + PI_3 \rightarrow 3C_2H_5I + H_3PO_3$$

Place 0.5 g of red phosphorus (in excess) and 5 g of iodine in a flask. Immerse the flask in a beaker of cold water and, using a dropping pipette, add 5 cm³ of ethanol, in 1 cm³ portions, down the reflux condenser. When all the ethanol has been added, slowly bring the water in the beaker to the boil. Allow the contents of the flask to reflux for an hour.

Allow the apparatus to cool and adjust the condenser for distillation. Bring the water in the bath gently to the boil and maintain at this temperature until no more oily drops of impure distillate are obtained.

Purify, dry and redistill the iodoethane. Collect the fraction boiling in the range 68-73 °C.

(a) Explain what is meant by the term *reflux*. (b) Suggest why the flask is kept in a beaker of cold water as the ethanol is added. (c) Name **two** inorganic impurities which will be present after refluxing. (d) Describe, giving practical details, how phosphoric(V) acid may be removed from the distillate using a separating funnel. (e) Name a suitable reagent for drying the impure distillate and suggest how it may be removed. (f) Suggest why a water bath can be used to heat the mixture during the refluxing. (g) State **two** reasons why the range 68-73 °C is used to collect the distillate.

(Adapted from CCEA May 2013)

6. A student used 13.7 cm^3 of butan-1-ol (density = 0.81 g cm^{-3} and RFM = 74) to produce 10.28 g of 1-bromobutane (RFM = 137). Calculate the percentage yield using the following headings to structure your work.
   - The mass of butan-1-ol used.
   - The number of moles of butan-1-ol used.
   - The theoretical yield of 1-bromobutane in moles.
   - The actual yield of 1-bromobutane in moles.
   - The percentage yield of product.

(Adapted from CCEA June 2009)

Before moving to the next section, check that you are able to:

- Explain the application of laboratory techniques to the preparation and purification of compounds.
- Use experimental data to calculate the extent of reaction as measured by the atom economy and percentage yield of the reaction.

Answers

Exercise 1.1A

1. (a) Nitrogen, bromine and iodine are diatomic.
   (b) Nitrogen is a gas.
   (c) Iodine, sulfur and silver are solids.

Exercise 1.1B

1. (a) $NaNO_3$ and $MgBr_2$ are ionic compounds.
   (b) NO_2 and CCl_4 are molecular.
   (c) SiO_2 has a giant structure.
2. Answer B
3. Answer B

Exercise 1.1C

1. (a) LiF (b) KCl (c) $MgBr_2$ (d) MgS (e) Al_2O_3

Exercise 1.1D

1. (a) PBr_3 (b) NO_2 (c) SF_4 (d) $HgCl_2$
2. (a) SO_2 (b) PCl_5 (c) MnO_2 (d) XeF_2

Exercise 1.1E

1. (a) calcium chloride (b) aluminium oxide (c) silver bromide (d) lithium hydride (e) zinc sulfide
2. (a) zinc carbonate (b) zinc sulfate (c) magnesium nitrate (d) magnesium nitride (e) potassium permanganate
3. (a) silver nitrate (b) ammonium nitrate (c) sodium nitrite (d) aluminium sulfate (e) sodium sulfite
4. (a) sodium oxide (b) sodium peroxide (c) calcium hydrogencarbonate (d) sodium hypochlorite (e) potassium chromate

Exercise 1.1F

1. (a) sulfur tetrafluoride (b) sulfur trioxide (c) silicon tetrachloride (d) phosphorus tribromide (e) xenon tetrafluoride

Exercise 1.1G

1. (a) copper(II) oxide (b) copper(II) sulfate (c) copper(I) oxide (d) copper(II) chloride (e) mercury(II) chloride (f) mercury(I) chloride

Exercise 1.1H

1. (a) sulfur(IV) fluoride (b) tin(II) chloride (c) silicon(IV) chloride (d) phosphorus(III) bromide (e) xenon(IV) fluoride (f) lead(IV) oxide

Exercise 1.1I

1. (a) $C + O_2 \rightarrow CO_2$
   (b) $2C + O_2 \rightarrow 2CO$
2. (a) $2K + BeCl_2 \rightarrow 2KCl + Be$
   (b) $Be + Cl_2 \rightarrow BeCl_2$
   $Be + 2HCl \rightarrow BeCl_2 + H_2$
3. $3Mg + N_2 \rightarrow Mg_3N_2$
4. $4HF + SiO_2 \rightarrow SiF_4 + 2H_2O$
5. (a) $P_4 + 10Cl_2 \rightarrow 4PCl_5$
   (b) $P_4 + 6Cl_2 \rightarrow 4PCl_3$

Exercise 1.1J

1. (a) $2Na_{(s)} + Cl_{2(g)} \rightarrow 2NaCl_{(s)}$
   (b) $H_{2(g)} + Cl_{2(g)} \rightarrow 2HCl_{(g)}$
   (c) $2Ca_{(s)} + O_{2(g)} \rightarrow 2CaO_{(s)}$
   (d) $Mg_{(s)} + Br_{2(l)} \rightarrow MgBr_{2(s)}$
2. (a) $Fe_{(s)} + S_{(s)} \rightarrow FeS_{(s)}$
   (b) $CaCO_{3(s)} \rightarrow CaO_{(s)} + CO_{2(g)}$
   (c) $2Fe_{(s)} + 3Br_{2(l)} \rightarrow 2FeBr_{3(s)}$
   (d) $2Al_{(s)} + 3I_{2(s)} \rightarrow 2AlI_{3(s)}$

Exercise 1.1K

1. (a) and (c)
2. They are all soluble.
3. (b) and (d)
4. (c) and (d)

Exercise 1.1L

1. Ethanoic acid dissociates to produce hydrogen ions: $C_2H_4O_2 \rightarrow H^+ + C_2H_3O_2^-$
2. (a) Phosphoric acid dissociates to produce hydrogen ions: $H_3PO_4 \rightarrow H^+ + H_2PO_4^-$ (b) Some of the H_3PO_4 molecules do not dissociate.
3. (a) The solution is strongly acidic and turns the indicator red. (b) The solution is neutral and turns the indicator green.
4. The reaction shows zinc oxide reacting with an acid to form a salt and water.
5. Carbonate ions in the solution react with water to form hydroxide ions:
   $CO_3^{2-}{}_{(aq)} + H_2O_{(l)} \rightarrow HCO_3^-{}_{(aq)} + OH^-{}_{(aq)}$

Exercise 1.1M

1. $Ag^+{}_{(aq)} + I^-{}_{(aq)} \rightarrow AgI_{(s)}$
2. $Ca^{2+}{}_{(aq)} + CO_3^{2-}{}_{(aq)} \rightarrow CaCO_{3(s)}$

Exercise 1.1N

1. (a) $CuO + 2HCl \rightarrow CuCl_2 + H_2O$
   (b) $CuO + 2H^+ \rightarrow Cu^{2+} + H_2O$
2. (a) $Na_2CO_3 + 2HCl \rightarrow 2NaCl + H_2O + CO_2$
   (b) $CO_3^{2-} + 2H^+ \rightarrow H_2O + CO_2$

Exercise 1.1O

1. Answer A

Exercise 1.1P

1. Atoms of Cu
 $= 1.56 \times 10^{-2}$ mol $\times 6.02 \times 10^{23}$ mol^{-1}
 $= 9.39 \times 10^{21}$
2. Mass of Na = 0.143 mol × 23 g mol^{-1} = 3.29 g
3. (a) 182 (b) 182 g
 (c) Mass of one molecule $= \dfrac{182 \text{ g}}{6.02 \times 10^{23}}$
 $= 3.02 \times 10^{-22}$ g
 (d) 1.81×10^{-20} g
4. (a) Molar mass $= \dfrac{2.30 \text{ g}}{0.0500 \text{ mol}} = 46.0$ g mol^{-1}
 (b) NO_2
5. (a) $2Na + O_2 \rightarrow Na_2O_2$
 (b) Na_2O
6. Atoms in 0.25 mole = 1.5×10^{23} molecules × 45 atoms per molecule. Answer A.

Exercise 1.1Q

1. (a) Moles of salicylic acid used = 3.62 mol
 Mass of acetic anhydride needed
 = 3.62 mol × 102 g mol^{-1} = 369 g
 (b) Mass of aspirin formed
 = 3.62 mol × 180 g mol^{-1} = 652 g
2. Moles of Li_2O used = 16.7 mol
 Mass of water removed
 = 16.7 mol × 18 g mol^{-1} = 301 g
3. (a) Moles of HgO decomposed = 0.046 mol
 Mass of Hg formed
 = 0.046 mol × 201 g mol^{-1} = 9.2 g
 (b) Need 1 mol of HgO = 217 g
4. Mass of $KClO_3$ = 0.33 mol × 122.5 g mol^{-1} = 40 g
5. Moles of CO_2 consumed = 6.82×10^3 mol
 Mass of O_2 produced
 = 6.82×10^3 mol × 0.032 kg mol^{-1} = 218 kg
6. Moles of Fe in 100 kg = 1.79×10^3 mol
 Mass of CO_2 produced
 = 2.69×10^3 mol × 0.044 kg mol^{-1} = 118 kg
7. (a) Moles of $NaHCO_3$ = Moles of NaCl
 = 4.00×10^3 mol
 (b) Mass of Na_2CO_3
 = 2.00×10^3 mol × 0.106 kg mol^{-1} = 212 kg

Exercise 1.1R

1. Moles of H_2SO_4 in 1 tonne = 1.02×10^4 mol
 Mass of S needed = 1.02×10^4 mol × 32 g mol^{-1}
 = 3.26×10^5 g
 Cost of S = 0.326 tonnes × £160 = £52

Exercise 1.1S

1. Moles of Fe reacted = 7.5×10^{-3} mol
 Mass of Cu formed = 7.5×10^{-3} mol × 64 g mol^{-1}
 = 0.48 g
2. Moles of Fe_2O_3 reacted = 20.0 mol.
 Fe_2O_3 is limiting.
 Moles of Fe formed = 40.0 mol × 0.056 kg mol^{-1}
 = 2.24 kg
3. Moles of Mg reacted = 2.1×10^6 mol.
 Mg is limiting.
 Mass of U formed
 = 1.05×10^6 mol × 0.238 kg mol^{-1} = 2.50×10^5 kg
4. Moles of $Ca_3(PO_4)_2$ = 1.9×10^2 mol.
 $Ca_3(PO_4)_2$ is limiting.
 Mass of H_3PO_4 formed
 = 3.8×10^2 mol × 0.098 kg mol^{-1} = 37 kg
5. (a) $2P + 3Br_2 \rightarrow 2PBr_3$
 (b) Moles of $Br_2 = \dfrac{24.8 \text{ g}}{160 \text{ g mol}^{-1}} = 0.155$ mol
 Mass of PBr_3 = 0.103 mol × 271 g mol^{-1} = 27.9 g
 (c) $PBr_3 + 3H_2O \rightarrow 3HBr + H_3PO_3$
6. Moles of $TiCl_4$ formed = 1.0×10^2 mol.
 $TiCl_4$ is limiting.
 Mass of Ti formed
 = 1.0×10^2 mol × 0.048 kg mol^{-1} = 4.8 kg

Exercise 1.1T

1. 62.9 %
2. (a) $C_{12}H_{22}O_{11} \rightarrow 12C + 11H_2O$
 (b) A compound is hydrated if it contains water of crystallisation. Water of crystallisation is water that is chemically bonded within the structure of the crystals.
 (c) Cane sugar is not hydrated as it does not contain water of crystallisation.

Exercise 1.1U

1. $CoCl_2.7H_2O$
2. x = 2

Exercise 1.2A

1.

	Relative mass	Relative charge
Proton	1	1+
Neutron	1	0
Electron	1/1840	1–

Exercise 1.2B

1. Answer B
2. Answer C
3. Answer C

Exercise 1.2C

1.

	neutrons	electrons	protons
^{43}Ca	43 – 20 = 23	20	20

2. (a) The number of protons in the nucleus of the atom.
   (b) The total number of protons and neutrons in an atom of the isotope
3. Isotopes are atoms with the same atomic number that have different mass numbers.
4. (a)

isotope	protons	neutrons	electrons
^{54}Fe	26	54 – 26 = 28	26
^{56}Fe	26	56 – 26 = 30	26
^{57}Fe	26	57 – 26 = 31	26

   (b) The isotopes have the same electron configuration and therefore have the same chemical properties.
5. (a) An atom of ^{23}Na has 11 protons, 11 electrons, and 12 neutrons.
   (b) ^{23}Na and ^{24}Na are both sodium atoms and have a different number of neutrons.
6.

ion	protons	neutrons	electrons
$^{24}Mg^{2+}$	12	24 – 12 = 12	10
$^{35}Cl^{-}$	17	35 – 17 = 18	18

7. (a) The new element contains 7 electrons, 7 protons and 7 neutrons.
   (b) Nitrogen.
   (c) They have the same atomic number and different mass numbers.
8. (a) 1
   (b) Total mass of reactants = 197 + 18 = 215
   Total mass of products = 210 + 5 = 215
   (c) The total mass of the electrons in each isotope is negligible.

Exercise 1.2D

1. Avogadro's constant is the number of atoms in exactly 12 g of carbon-12.
2. Atomic masses are measured on a scale where one atom of carbon-12 has a mass of exactly 12.

Exercise 1.2E

1. RAM of Na $= \dfrac{(24 \times 2.00) + (23 \times 98.00)}{100}$

   $= \dfrac{(48.0 \times 2254)}{100} = 23.02$

2. RAM of C $= \dfrac{(12 \times 98.89) + (13 \times 1.11)}{100}$

   $= \dfrac{1187 + 14.4}{100} = 12.01$

3. Answer B

Exercise 1.2F

1. RAM of I

   $= \dfrac{(127 \times 95.91) + (129 \times 2.49) + (131 \times 1.60)}{100}$

   $= \dfrac{12180 + 321 + 210}{100} = 127.1$

2. RAM of Fe

   $= \dfrac{(54 \times 5.8) + (56 \times 91.6) + (57 \times 2.6)}{100}$

   $= \dfrac{310 + 5130 + 150}{100} = 55.9$

3. (a) Isotopes are atoms with the same atomic number that have different mass numbers..
   (b) RAM of Ca

   $= \dfrac{(40 \times 96.9) + (42 \times 0.6) + (43 \times 0.2) + (44 \times 2.3)}{100}$

   $= \dfrac{3880 + 30 + 10 + 100}{100} = 40.20$

4. RAM of Xe =

   $\dfrac{(129 \times 27) + (131 \times 23) + (132 \times 28) + (134 \times 12) + (136 \times 10)}{100}$

   $= \dfrac{3500 + 3000 + 3700 + 1600 + 1400}{100} = 132$

Exercise 1.2G

1. Answer C
2. To determine the RAM of elements and the RMM of molecules.

Exercise 1.2H

1. Answer D

Exercise 1.2I

1. A Ca atom $1s^2\,2s^2\,2p^6\,3s^2\,3p^6\,4s^2$ loses two electrons to form a Ca^{2+} ion $1s^2\,2s^2\,2p^6\,3s^2\,3p^6$ and an O atom $1s^2\,2s^2\,2p^4$ gains two electrons to form an O^{2-} ion $1s^2\,2s^2\,2p^6$.
2. (a) $1s^2\,2s^2\,2p^6$ (b) Al^{3+} and F^-
3. Answer D

Exercise 1.2J

1. (a) Mn $1s^2\ 2s^2\ 2p^6\ 3s^2\ 3p^6\ 4s^2\ 3d^5$
   (b) Mn^{2+} $1s^2\ 2s^2\ 2p^6\ 3s^2\ 3p^6\ 3d^5$
   (c) Zn^{2+} $1s^2\ 2s^2\ 2p^6\ 3s^2\ 3p^6\ 3d^{10}$
   (d) Ni $1s^2\ 2s^2\ 2p^6\ 3s^2\ 3p^6\ 4s^2\ 3d^8$
2. (a) Sc $1s^2\ 2s^2\ 2p^6\ 3s^2\ 3p^6\ 4s^2\ 3d^1$
   (b) V $1s^2\ 2s^2\ 2p^6\ 3s^2\ 3p^6\ 4s^2\ 3d^3$
   (c) V^{3+} $1s^2\ 2s^2\ 2p^6\ 3s^2\ 3p^6\ 3d^2$
   (d) Ca $1s^2\ 2s^2\ 2p^6\ 3s^2\ 3p^6\ 4s^2$

Exercise 1.2K

1. Answer C
2. Answer C
3.

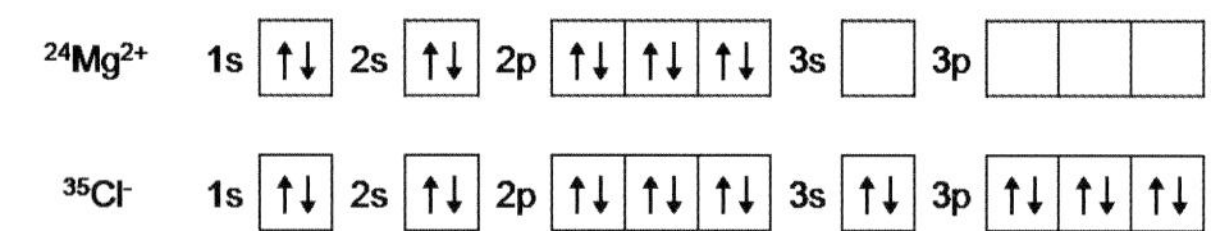

4. Answer D
5. Answer B

Exercise 1.2L

1. (a) $1s^2\ 2s^2\ 2p^6\ 3s^2\ 3p^6\ 3d^6$
   (b) An iron(III) ion is more stable as the ground state of an iron(III) ion has a half-filled d-subshell ($3d^5$).

Exercise 1.2M

1. Answer C
2. Answer C

Exercise 1.2N

1. Answer C
2. The outermost electrons in a calcium atom are further from the nucleus and better shielded than the outermost electrons in a magnesium atom.

Exercise 1.2O

1. (a) The electrons in an atom occupy the lowest energy subshells available to produce the electron configuration with the lowest energy.
   (b)
   1s ↑↓ 2s ↑↓ 2p ↑↓ ↑↓ ↑↓ 3s ↑↓ 3p ↑ ↑ ↑
   (c) The electrons with the highest energy are in a half-filled subshell.
2. Answer C
3. (a) The increase in shielding on moving one element to the right is not sufficient to offset the increased attraction to the nucleus. As a result, the outermost electrons become harder to remove.
   (b) IE1 decreases from 12 to 13 (Mg to Al) because the outermost 3p electron in Al is further from the nucleus and better shielded than the outermost 3s electrons in Mg. *Alternative answer:* It is harder to remove a 3s electron from Mg than it is to remove the 3p electron from Al as the 3s subshell in Mg is filled.
   (c) The paired 3p electrons in S repel and are easier to remove than the unpaired 3p electrons in P. *Alternative answer:* The 3p electrons in P are harder to remove as the 3p subshell is half-filled.

Exercise 1.2P

1.

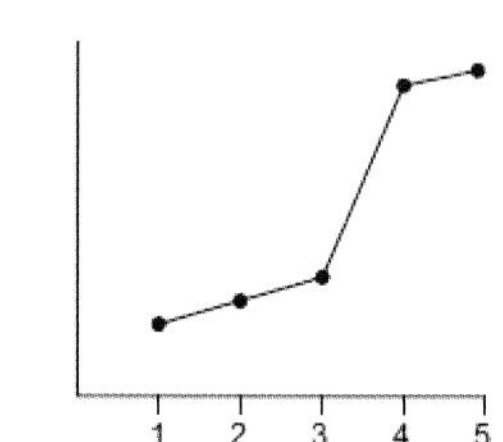

2.

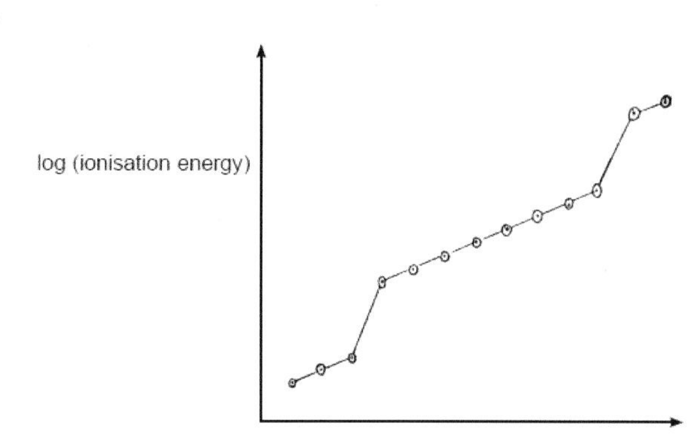

3. (a) Nitrogen.
   (b) The shielding of the outermost electrons decreases as electrons are removed. As a result the outermost electrons are more strongly attracted to the nucleus.
   (c) The sixth electron is removed from the 1s subshell. It is closer to the nucleus and much less shielded than an electron in the 2s subshell.
4. (a) IE1(K) < IE1(Na) because the outermost electron in potassium is further from the nucleus and better shielded by the inner shells of electrons.
   (b) IE2 >> IE1 because removing a second electron from the atom involves taking an electron from a full shell.
5. (a) $Ca^{+}{}_{(g)} \rightarrow Ca^{2+}{}_{(g)} + e^{-}{}_{(g)}$
   (b) $0.20\ mol \times (IE1 + IE2) = 3.5 \times 10^2\ kJ$
6. (a) The energy needed to remove one mole of electrons from one mole of gas phase ions with a 1+ charge to form one mole of gas phase ions with

a 2+ charge.

(b) $Mg^{+}(g) \rightarrow Mg^{2+}(g) + e^{-}(g)$

(c) The third electron is removed from a full 2p subshell, it is also closer to the nucleus and less shielded than the electrons removed previously from the 3s subshell.

7. (a) $1s^2\ 2s^2\ 2p^6\ 3s^2\ 3p^6$

   (b) The 3p electrons in calcium are held more tightly as calcium has a greater positive charge in its nucleus.
8. Answer B
9. Answer D
10. Answer B

Exercise 1.2Q

1. Argon belongs to the p-block as its outermost electrons are in a p-subshell.
2. Answer A
3. MgF_2 (QR_2)
4. (a) Atomic number.

   (b) The outermost electrons are in a d-subshell.
5. (a) The outermost electrons are in an s-subshell.

   (b) The number of filled subshells increases.
6. X (= vanadium, V) belongs to the d-block.

   Y (= sulfur, S) belongs to the p-block.

   Z (= argon, Ar) belongs to the p-block.
7. Answer B
8. Answer A
9. Answer B
10. Answer C
11. (a)

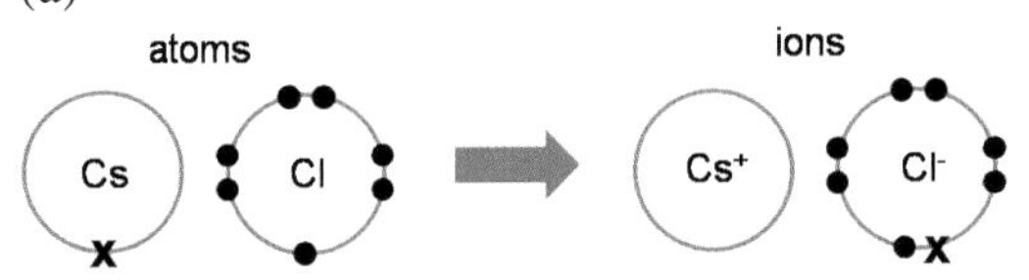

   (b) Cs^{+} $1s^2\ 2s^2\ 2p^6\ 3s^2\ 3p^6\ 4s^2\ 3d^{10}\ 4p^6\ 5s^2\ 4d^{10}\ 5p^6$

   Cl^{-} $1s^2\ 2s^2\ 2p^6\ 3s^2\ 3p^6$
12. X_2Y_3
13. EuH_2

Exercise 1.3A

1. (a) $2Na(s) + F_2(g) \rightarrow 2NaF(s)$

   (b)

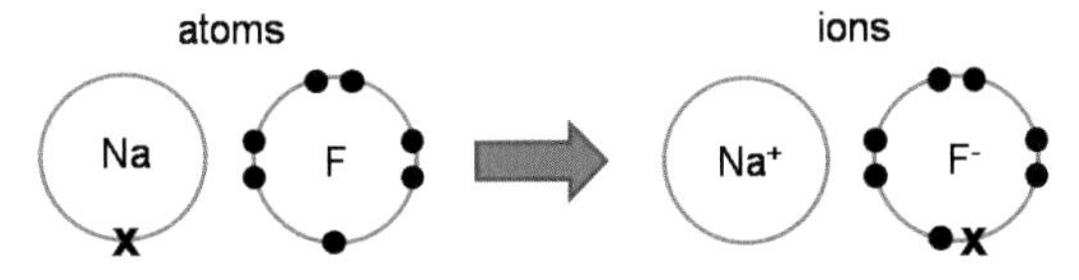

2.

3. (a)

   (b) Ca^{2+} $1s^2\ 2s^2\ 2p^6\ 3s^2\ 3p^6$

   Br^{-} $1s^2\ 2s^2\ 2p^6\ 3s^2\ 3p^6\ 4s^2\ 3d^{10}\ 4p^6$
4. (a)

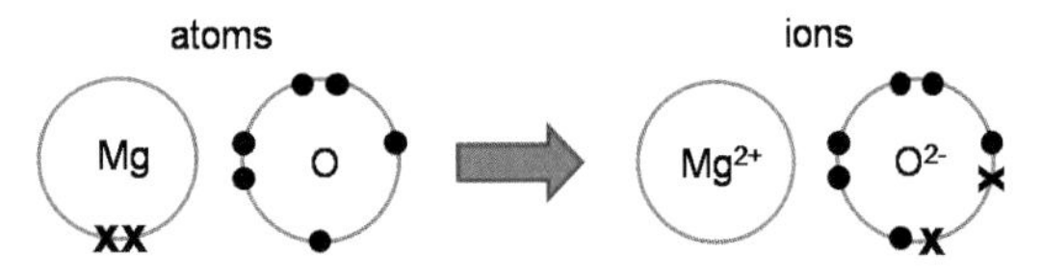

   (b) Ionic bonding

   (c) Magnesium oxide has a high melting point and conducts electricity when molten.
5. (a)

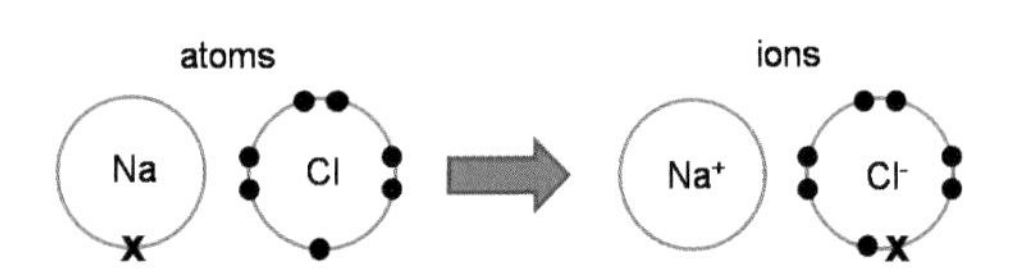

   (b) Ionic bonding

   (c) A regular arrangement of particles such as the ions in sodium chloride.

   (d) Sodium chloride has a high melting point, is soluble in water, and conducts electricity when molten or dissolved in water.
6. (a) $(NH_4)_3PO_4$

   (b) Ammonium phosphate would be a brittle solid as the ions in the compound are arranged in a lattice. It would also have a high melting point as a result of strong electrostatic attraction between oppositely charged ions in the lattice. It would also conduct electricity when molten or dissolved in water as the ions are free to move when the solid melts or dissolves.

Exercise 1.3B

1. (a) A covalent bond is the attractive force between the negatively charged electrons in a shared pair of electrons and the positively charged nucleus of each atom sharing the electrons.

   (b) The octet rule states that atoms will attempt to gain, lose or share electrons to form compounds in which they have a full outer shell containing eight electrons.

2. (a) $P_4 + 6Br_2 \rightarrow 4PBr_3$
   (b)

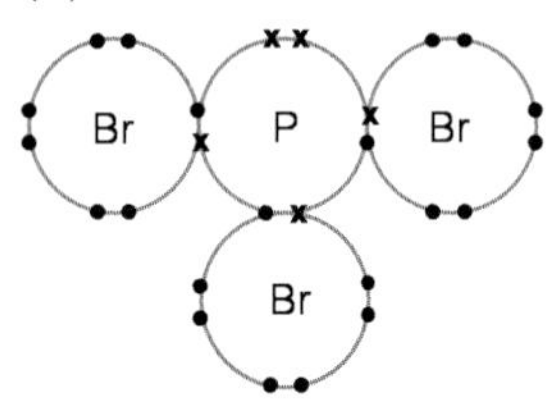

   (c) The phosphorus atom obeys the octet rule by sharing electrons to obtain a full outer shell containing eight electrons.

3. Answer A

Exercise 1.3C

1.

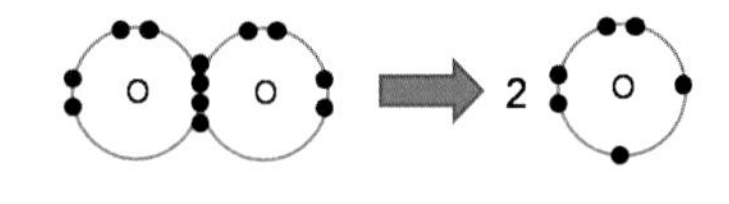

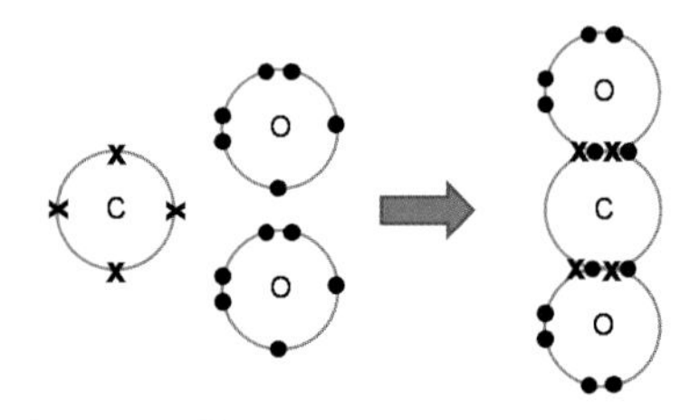

2. Answer C

Exercise 1.3D

1.

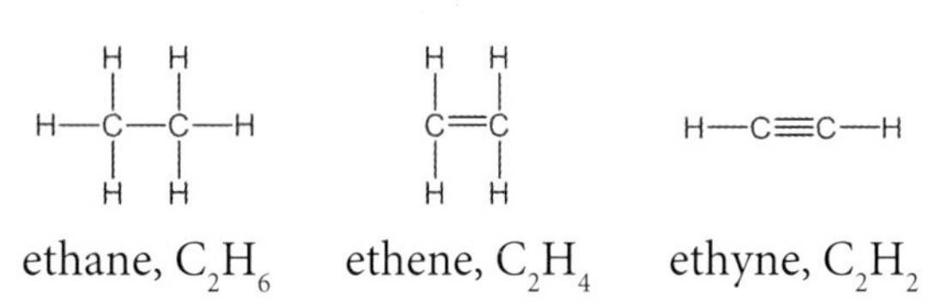

ethane, C_2H_6 ethene, C_2H_4 ethyne, C_2H_2

Exercise 1.3E

1. (a) A coordinate bond.
2. (b) A pair of electrons on the nitrogen atom is shared with the boron atom.
3. Answer B

Exercise 1.3F

1. (a) Beryllium obeys the octet rule by sharing electrons to obtain a full outer shell containing eight electrons.
   (b) Chlorine obeys the octet rule by sharing electrons to obtain a full outer shell containing eight electrons.

2. (a)

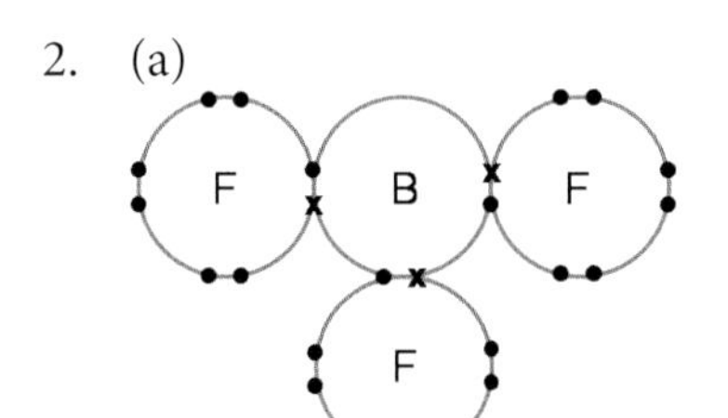

   (b) The octet rule states that atoms will attempt to gain, lose or share electrons to form compounds in which they have a full outer shell containing eight electrons.
   (c) Fluorine obeys the octet rule by sharing electrons to obtain a full outer shell containing eight electrons. Boron does not obey the octet rule as its outer shell contains only six electrons.

3. (a) $BeCl_2$ and H_2O.
   (b) All except $BeCl_2$.

4. (a)

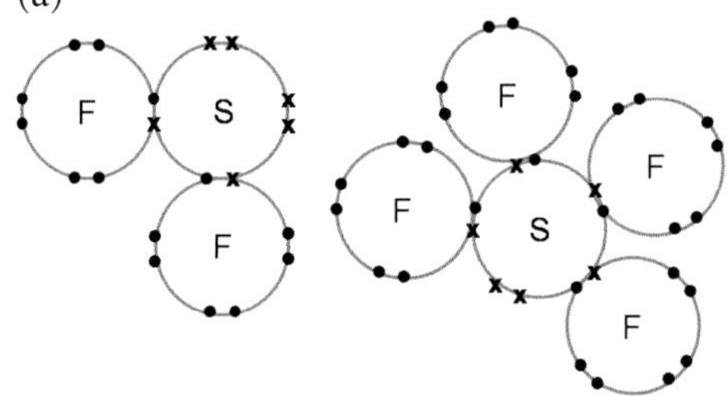

   (b) The octet rule states that atoms will attempt to gain, lose or share electrons to form compounds in which they have a full outer shell containing eight electrons.
   (c) In SF_2 sulfur obeys the octet rule by sharing electrons to obtain a full outer shell containing eight electrons. In SF_4 sulfur does not obey the octet rule as its outer shell contains ten electrons.

5. (a)

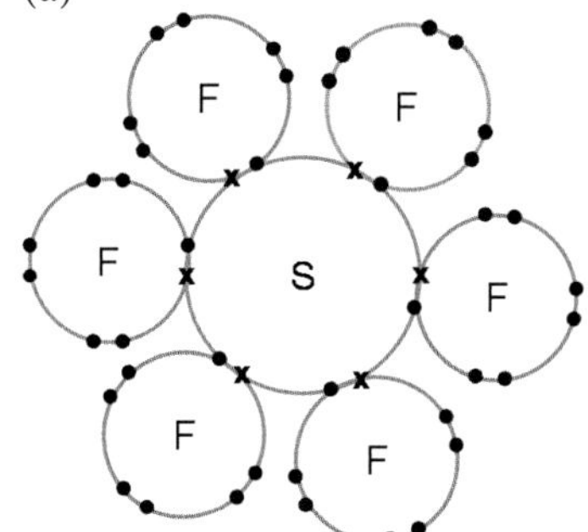

   (b) Each fluorine atom has eight electrons in its outer shell and satisfies the octet rule. The sulfur atom has twelve electrons in its outer shell and does not satisfy the octet rule. In this way the compound SF_6 does not satisfy the octet rule.

Exercise 1.3G

1. (a) The electronegativity of an element is the extent to which an atom of the element attracts the shared electrons in a covalent bond to itself when forming covalent bonds with other atoms.
   (b) Electronegativity increases from left to right across a period.
2. Answer B
3. (a) The electronegativity of an element is the extent to which an atom of the element attracts the shared electrons in a covalent bond to itself when forming covalent bonds with other atoms.
   (b) The electronegativity values for beryllium and chlorine are more similar than the electronegativity values for barium and chlorine.
4. Answer A
5. HF
6. Answer A

Exercise 1.3H

1. Carbon dioxide does not have a permanent dipole as the molecule is symmetric

Exercise 1.3I

1. *Electrical conductor:* The electrons involved in bonding the layers together are free to move and carry charge through the solid. *Use as lubricant:* Graphite contains layers of carbon atoms The bonding between the layers is weak allowing the layers to slide over each other.
2. (a) A covalent bond is the attractive force between the negatively charged electrons in a shared pair of electrons and the positively charged nucleus of each atom sharing the electrons.
   (b) The atoms in diamond form a covalent bond with each of four neighbouring atoms to form a giant structure. The bonds about each carbon atom have a tetrahedral arrangement. The atoms in graphite form a covalent bond with each of three neighbouring atoms to form layers of atoms. The layers of atoms are held together by weak bonds between the layers.
   (c) The layers in graphite are held together by electrons that are free to move between the layers.
   (d) The carbon atoms are held in fixed positions by a network of strong covalent bonds that each require a lot of energy to break.
3. (a) Diamond.
   (b) Diamond is hard because the carbon atoms are held in fixed positions by a network of strong covalent bonds that each require a lot of energy to break.
   (c) Diamond does not conduct electricity because the outer shell electrons on each atom are used to form covalent bonds and are not able to move through the material.
4. Answer D
5. Diamond has a high melting point because a large amount of energy is needed to break the network of strong covalent bonds between the atoms.

Exercise 1.3J

1. (a) Metallic bonding refers to the attraction between the array of positive metal ions and the delocalised electrons in a metal.

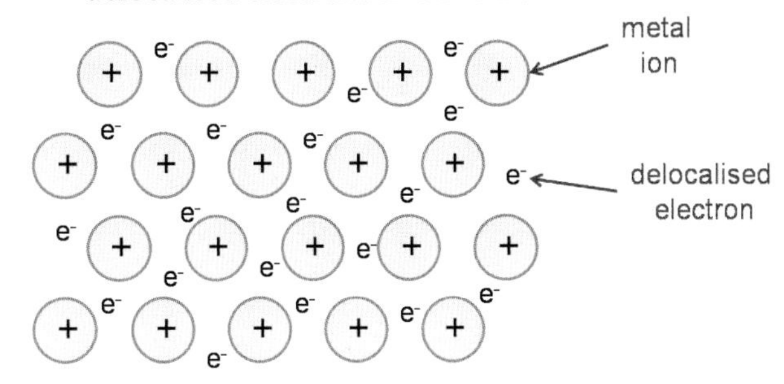

   (b) The attraction between the ions and the delocalised electrons decreases as the metal ions get bigger.
   (c) Calcium has two delocalised electrons for every metal ion. Potassium has only one delocalised electron for every metal ion.
2. (a) The attraction between the metal ions and the delocalised electrons is weak as the metal ions are large.
   (b) The delocalised electrons are less strongly attracted to the metal ions and are able to move more freely as the metal ions are large.
3. (a) The layers of ions can move past each other without disturbing the bonding in the metal.
   (b) The outermost electrons are delocalised and can move freely carrying a charge through the metal.
4. The bonding in a metal results from the attraction between the positively charged metal ions and the delocalised electrons in the metal. Metals are ductile because the metal ions can move past each other without disturbing the bonding in the metal. Metals conduct electricity because the delocalised electrons are able to move and carry a charge

through the metal.

5. Strontium metal is a good conductor because the delocalised electrons in the metal are free to move and carry a charge through the metal. Strontium fluoride is a poor conductor because the ions in the solid are not free to move.

Exercise 1.4A

1.

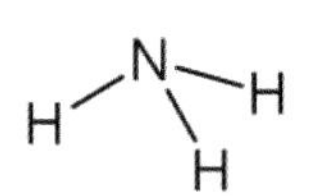

The four pairs of electrons in the outer shell of the nitrogen atom repel to give ammonia a pyramidal shape.

2. (a)

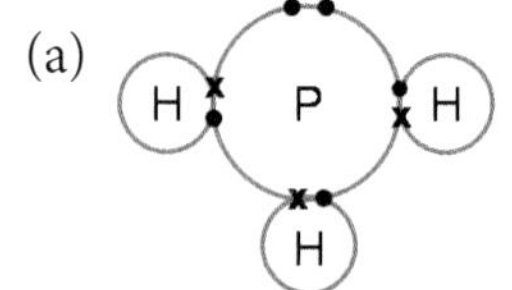

(b)

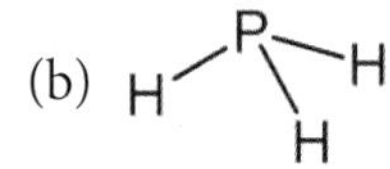

Shape: pyramidal

3. (a) 104.5°

(b) The repulsion between the two bonding pairs and two lone pairs in the outer shell of the oxygen atom gives water a bent shape.

(c) The repulsion between the lone pairs and bonding pairs in the outer shell of oxygen is greater than the repulsion between the bonding pairs in the outer shell of carbon.

4.

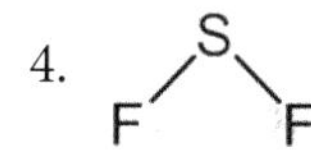

Shape: Bent *F-S-F angle:* 105° – 6° = 99°

5. The C, S and N atoms have four electron pairs in their outer shell that repel and move as far from each other as possible. Methane has a tetrahedral shape with a H-C-H angle of 109.5°. Ammonia has a pyramidal shape in which repulsion between the lone pair and three bonding pairs on N produces an H-N-H angle of 107°. Hydrogen sulfide has a bent shape in which repulsion between two lone pairs and two bonding pairs on S produces an H-S-H angle of 104.5°.

6. (a)

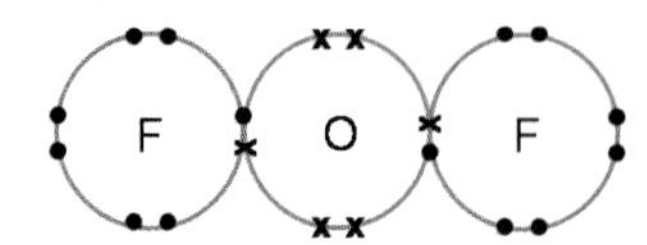

(b) The four pairs of electrons in the outer shell of the oxygen atom repel and move as far from each other as possible to give the molecule a bent shape similar to water.

7. (a)

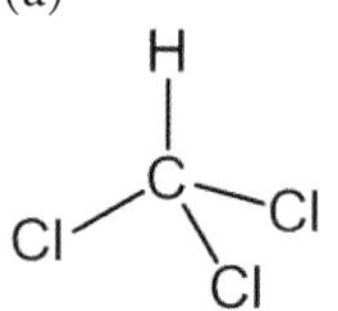

(b)

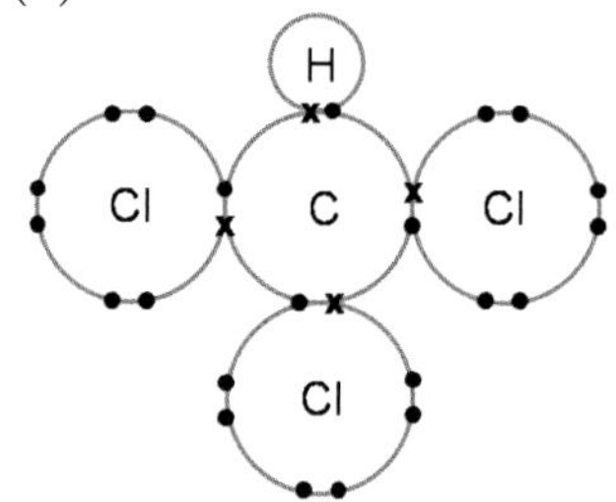

(c) The four pairs of electrons in the outer shell of the carbon atom repel and move as far from each other as possible giving chloroform a tetrahedral shape.

8. (a)

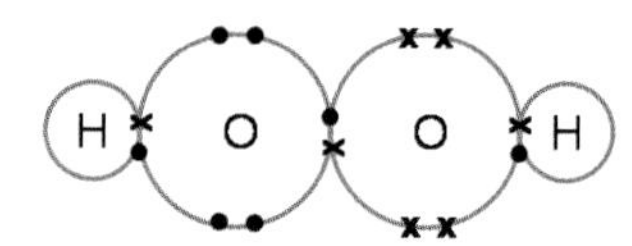

(b) The four pairs of electrons in the outer shell of each oxygen atom repel and move as far from each other as possible to give the molecule a bent shape with an H-O-O angle of around 105°.

9. (a) NH_4^+

(b)

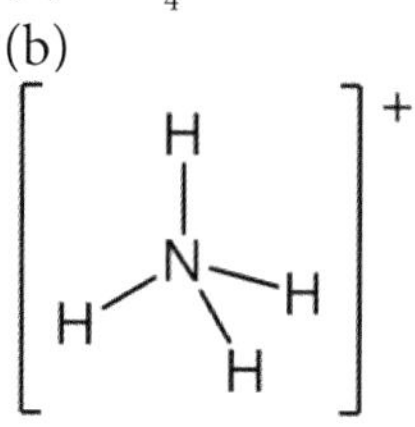

Shape: tetrahedral

(c) 109.5°

(d) $(NH_4)_3PO_4$

10. (a)

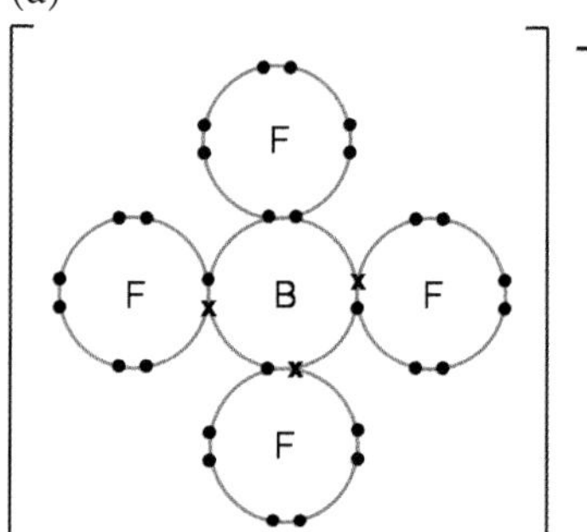

Shape: Tetrahedral *F-B-F angle:* 109.5°

(b) Coordinate bond

11. (a) $2Na + 2NH_3 \rightarrow 2NaNH_2 + H_2$

(b)

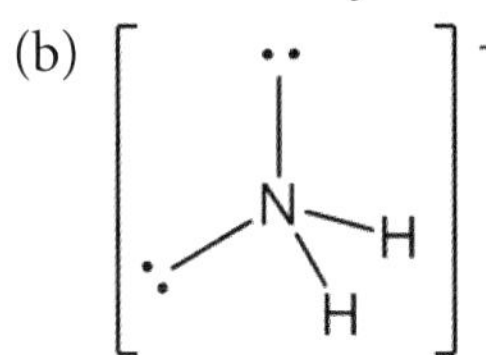

(c) Bent

(d) In NH_2^- the N-H bonding pairs are repelled by two lone pairs. In NH_3 the N-H bonding pairs are repelled by one lone pair. The H-N-H angle in NH_2^- is further reduced by repulsion between the lone pairs on nitrogen which is greater than the repulsion between a lone pair and a bonding pair.

Exercise 1.4B

1. (a) $Be + 2HCl \rightarrow BeCl_2 + H_2$

(b)

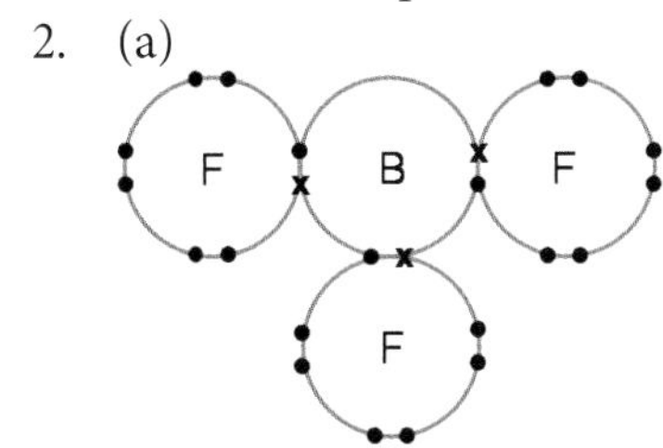

(c) The chlorine atoms have eight electrons in their outer shell and obey the octet rule. The beryllium atom has only four electrons in its outer shell and does not obey the octet rule.

(d) Cl-Be-Cl *Shape:* linear

(e) The two pairs of electrons in the outer shell of the beryllium atom repel and move as far from each other as possible.

2. (a)

(b) The octet rule asserts that atoms will lose gain or share electrons to achieve a full outer shell containing eight electrons. In BF_3 each fluorine atom has eight electrons in its outer shell and obeys the octet rule. The boron atom has six electrons in its outer shell and does not obey the octet rule.

(c)

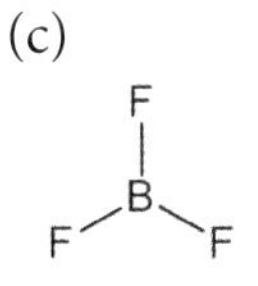

The molecule has a trigonal planar shape as the three pairs of electrons in the outer shell of the boron atom repel and get as far from each other as possible.

3. (a)

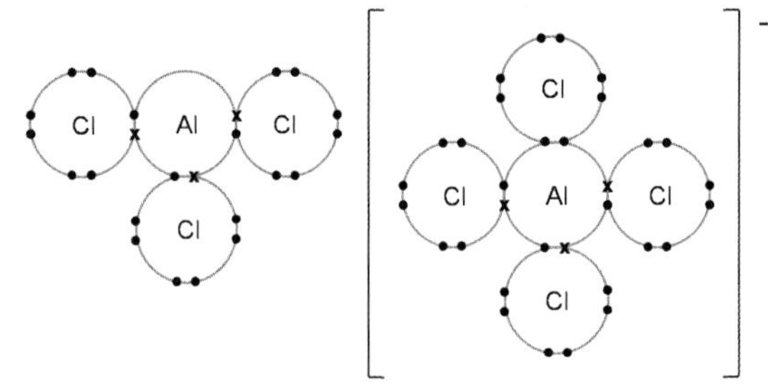

(b) Coordinate bond

(c)

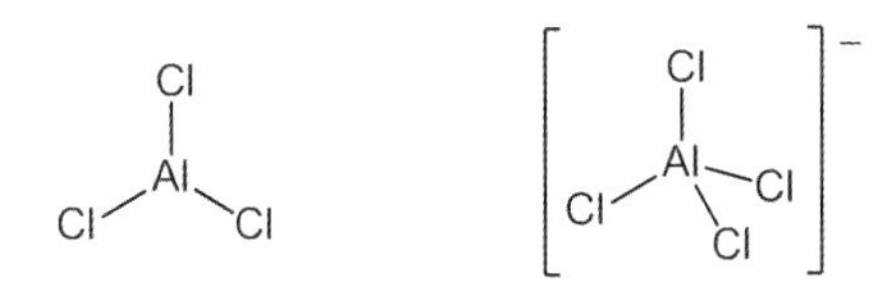

Shape: trigonal planar *Shape:* tetrahedral

4. Molecules A-C do not have a dipole. Answer D.

Exercise 1.4C

1. The silicon atom in $SiCl_4$ has four pairs of electrons in its outer shell that repel and get as far from each other as possible giving $SiCl_4$ a tetrahedral shape. The sulfur atom in SF_4 has five pairs of electrons in its outer shell that repel and get as far from each other as possible giving SF_4 a see-saw shape with a lone pair in the trigonal plane.
2. Boron has three pairs of electrons in its outer shell that repel and get as far from each other as possible giving BF_3 a trigonal planar shape. Phosphorus has four pairs of electrons in its outer shell that repel and get as far from each other as possible giving PF_3 a pyramidal shape. Chlorine has five pairs of electrons that repel and get as far from each other as possible making ClF_3 T-shaped with two lone pairs in the trigonal plane.
3. Answer D

Exercise 1.4D

1. The tetrahedral shape of the CCl_4 molecule allows the C-Cl bond dipoles to cancel. Answer B.
2. (a)

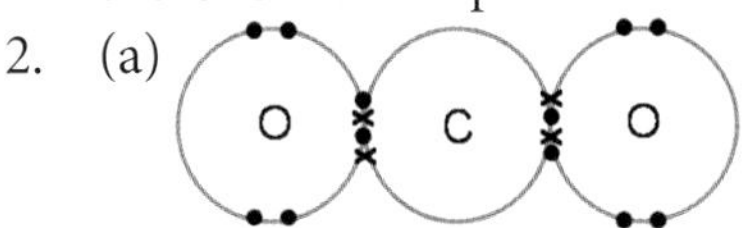

(b) O=C=O *Shape:* linear

(c) The electrons in one C=O bond repel the

electrons in the other C=O bond pushing the C=O bonds as far apart as possible.

(d) The linear shape of the CO_2 molecule allows the C=O bond dipoles to cancel each other.

3. (a) The H-F bond is a polar covalent bond.
   (b) The trigonal planar shape of the BF_3 molecule allows the B-F bond dipoles to cancel each other.
4. The O-F bond dipoles in OF_2 do not cancel as the molecule has a bent shape. Answer C.
5. (a)
   The six pairs of electrons in the outer shell of sulfur repel and move as far from each other as possible to give SF_6 an octahedral shape.
   (b) The octahedral shape of the SF_6 molecule allows the S-F bond dipoles to cancel each other.

Exercise 1.5A

1. Xenon has the greatest number of electrons per atom. Answer D.
2. The S_8 molecules in sulfur have more electrons, and therefore experience greater van der Waals attraction, than the P_4 molecules in phosphorus. The melting point of silicon is much higher as silicon has a giant covalent structure and much more energy is needed to break the strong covalent bonds between the silicon atoms.

Exercise 1.5B

1. (a) Sulfur tetrafluoride.
   (b) The van der Waals attraction between molecules is greater in sulfur hexafluoride as it has a greater RMM.
   (c) The boiling point of sulfur tetrafluoride is higher as the combination of van der Waals attraction and dipole-dipole attraction between molecules in sulfur tetrafluoride is stronger than the van der Waals attraction between molecules in sulfur hexafluoride.
2. (a) A water molecule has a dipole. The negative end of the dipole in each water molecule is attracted to the positively charged rod.
   (b)

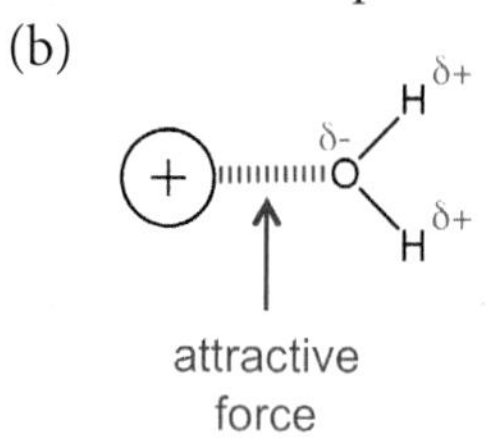

3. Only $CHCl_3$ has a dipole. Answer B.

Exercise 1.5C

1. Answer A
2.

liquid	van der Waals	permanent dipole	hydrogen bonding
water	✓	×	✓
ammonia	✓	×	✓
xenon	✓	×	×
hydrogen chloride	✓	✓	×

3. (a) Hydrogen bonding. (b) Dipole forces. (c) Van der Waals forces.
4. Answer B
5. (a) Ammonia forms hydrogen bonds. Phosphine is not able to form hydrogen bonds.
   (b) The van der Waal attraction between SbH_3 molecules is stronger because antimony has more electrons than arsenic.
6. The hydrogen bonds between water molecules are stronger than the dipole forces between hydrogen sulfide molecules.
7. (a) The van der Waals attraction between molecules increases from HCl to HI as the number of electrons in the molecule increases.
   (b) Hydrogen bonds between HF molecules require an additional amount of energy to overcome.
8. Answer A
9. (a) Dipole forces and van der Waals forces.
   (b) An oxygen atom on one molecule uses a lone pair to bond with a hydrogen atom on a neighbouring water molecule.
   (c) Hydrogen bonding in ice allows the water molecules to form a more open structure with greater distances between the water molecules.
10. The H-F bond is more polar than the O-H bonds in water. As a result the hydrogen bonding in HF is stronger than in water and is harder to break.
11. The molecules have enough energy to overcome the hydrogen bonds between the molecules.
12. (a)
    (b) Ammonia reacts with a hydrogen ion to form

an ammonium ion, NH_4^+. The nitrogen atom in an ammonium ion does not have a lone pair.

13. Answer D
14. Boron trifluoride is not able to form hydrogen bonds. Molecules of boron trifluoride are held together by much weaker van der Waal forces.

Exercise 1.5D

1. The hydrogen atoms in ammonia molecules can hydrogen bond with the oxygen atoms in water molecules. The hydrogen atoms in water molecules can also form hydrogen bonds with the nitrogen atoms in ammonia molecules.
2. The solubility of carbon dioxide is due to dipole forces between carbon ($\delta+$) and the oxygen atoms in water ($\delta-$), and hydrogen bonding between the oxygen atoms in carbon dioxide and the hydrogen atoms in water.

Exercise 1.6A

1. (a) Oxidation is the loss of electrons.
   (b) Reduction is a decrease in oxidation state.
2. (a) $Zn_{(s)} + Cu^{2+}_{(aq)} \rightarrow Zn^{2+}_{(aq)} + Cu_{(s)}$
   (b) Zinc atoms are oxidised to form zinc(II) ions and copper(II) ions are reduced to form copper atoms.
   (c) $Zn_{(s)} \rightarrow Zn^{2+}_{(aq)} + 2e^-$
   (d) Copper(II) ion
3. (a) $Cu_{(s)} + 2Ag^+_{(aq)} \rightarrow Cu^{2+}_{(aq)} + 2Ag_{(s)}$
   (b) Copper atoms are oxidised to form copper(II) ions and silver(I) ions are reduced to form silver atoms.
   (c) $Ag^+_{(aq)} + e^- \rightarrow Ag_{(s)}$
   (d) Copper

Exercise 1.6B

1. (a) $2K_{(s)} + H_{2(g)} \rightarrow 2KH_{(s)}$
   (b) The oxidation number of potassium increases from 0 in K to +1 in KH. The oxidation number of hydrogen decreases from 0 in H_2 to −1 in KH. Potassium is oxidised and hydrogen is reduced.
   (c) Hydrogen is an oxidising agent as it gains electrons by oxidising potassium.
2. (a) $Li_{(s)} + N_{2(g)} \rightarrow Li_3N_{(s)}$
   (b) Ionic lattice containing Li^+ ions and N^{3-} ions.
   (c) The oxidation number of lithium increases from 0 in Li to +1 in Li_3N. The oxidation number of nitrogen decreases from 0 in N_2 to −3 in Li_3N.
   (d) Lithium is a reducing agent as it loses electrons by reducing nitrogen.
3. Uranium has an oxidation number of +4 in UF_4 and +6 in UF_6. ClF_3 is an oxidising agent as it gains electrons by oxidising uranium.

Exercise 1.6C

1. The oxidation number of hydrogen increases from 0 in H_2 to +1 in HCl. The oxidation number of chlorine decreases from 0 in Cl_2 to −1 in HCl. Hydrogen is oxidised and chlorine is reduced.
2. Copper has an oxidation number of +2 in CuO and 0 in Cu. Hydrogen has an oxidation number of 0 in H_2 and +1 in H_2O. Hydrogen is oxidised and copper is reduced.
3. Nitrogen has an oxidation number of −3 in NH_3 and 0 in N_2. Copper has an oxidation number of +2 in CuO and 0 in Cu. Nitrogen is oxidised and copper is reduced.

Exercise 1.6D

1. Chlorine has an oxidation number of 0 in Cl_2, −1 in NaCl and +1 in NaOCl. Chlorine is oxidised to form OCl^- and reduced to form Cl^-.
2. (a) +5 (b) +2
3. Silver has an oxidation number of 0 in Ag and +1 in $AgNO_3$. Nitrogen has an oxidation number of +5 in HNO_3 and +2 in NO. Silver is oxidised and nitrogen is reduced.
4. +4
5. K +1 and Mn +7.
6. K +1 and Cr +6.

Exercise 1.6E

1. (a) K +1, Cl +5, O −2 and S 0
   (b) K +1, Cl −1, O −2 and S +4
   (c) The oxidation number of sulfur increases from 0 in S to +4 in SO_2. The oxidation number of chlorine decreases from +5 in ClO_3^- to −1 in KCl. Sulfur is oxidised and chlorine is reduced.
   (d) P_2S_3
   (e) P_2S_3 is covalently bonded as it is a nonmetal compound.
   (f) Phosphorus oxide and sulfur oxide.
2. The oxidation number of iodine increases from −1 in KI to 0 in I_2. The oxidation number of manganese decreases from +4 in MnO_2 to +2 in $MnSO_4$. Iodine is oxidised and manganese is reduced.
3. Answer A. Nitrogen is oxidised and hydrogen is reduced.

4. Answer A. Bromine is oxidised and sulfur is reduced.
5. Answer D. The reaction in D is a neutralisation reaction.
6. Answer A. Hydrogen peroxide reduces silver.
7. Answer C. Hydrogen peroxide reduces manganese.

Exercise 1.6F

1. Oxidation of iodide: $2I^- \rightarrow I_2 + 2e^-$
   $2HNO_3 + 6H^+ + 6I^- \rightarrow 2NO + 4H_2O + 3I_2$
2. $2HOCl_{(aq)} + 2H^+_{(aq)} + 2Fe^{2+}_{(aq)}$
   $\rightarrow Cl_{2(aq)} + 2H_2O_{(l)} + 2Fe^{3+}_{(aq)}$
3. (a) $O_2 + 4H^+ + 4I^- \rightarrow 2I_2 + 2H_2O$
   (b) Equation 1 is oxidation as iodine loses electrons.
   (c) Equation 2 is reduction as oxygen gains electrons.
4. $2Fe^{2+} + Cl_2 \rightarrow 2Fe^{3+} + 2Cl^-$
5. (a) $Fr \rightarrow Fr^+ + e^-$
   (b) $Cl_2 + 2e^- \rightarrow 2Cl^-$
   (c) $2Fr + Cl_2 \rightarrow 2FrCl$
6. $2MnO_4^- + 16H^+ + 5C_2O_4^{2-}$
   $\rightarrow 2Mn^{2+} + 8H_2O + 10CO_2$
7. (a) $N_2H_4 + 2H_2O_2 \rightarrow N_2 + 4H_2O$
   (b) The oxidation number of nitrogen increases from –2 in N_2H_4 to 0 in N_2. The oxidation number of oxygen decreases from –1 in H_2O_2 to –2 in H_2O. Nitrogen is oxidised and oxygen is reduced.

Exercise 1.6G

1. Balancing oxygen gives z = 6.
   Balancing hydrogen gives x = 12.
   y = 10 as each bromine gains five electrons.
2. (a) Balancing oxygen gives z = 4.
   Balancing hydrogen gives x = 8.
   y = 5 as Mn gains five electrons.
   (b) $MnO_4^- + 8H^+ + 5Fe^{2+} \rightarrow 5Fe^{3+} + Mn^{2+} + 4H_2O$
3. (a) $2CsAu + 2H_2O \rightarrow 2Au + 2CsOH + H_2$
   (b) The oxidation number of gold increases from –1 in Au^- to 0 in Au. The oxidation number of hydrogen decreases from +1 in H_2O to 0 in H_2. Gold is oxidised and hydrogen is reduced.
   (c) Oxidation of gold: $Au^- \rightarrow Au + e^-$
   Reduction of hydrogen: $2H_2O + 2e^- \rightarrow 2OH^- + H_2$

Exercise 1.6H

1. (a) +6 in $HXeO_4^-$, 0 in Xe and +8 in XeO_6^{4-}
   (b) Xenon is oxidised to form XeO_6^{4-} and reduced to form Xe.
2. (a) +5 in $NaClO_3$ and –1 in NaCl
   (b) Chlorine is oxidised to form ClO_3^- and reduced to form Cl^-.
3. (a) $3I_{2(aq)} + 6NaOH_{(aq)}$
   $\rightarrow 5NaI_{(aq)} + NaIO_{3(aq)} + 3H_2O_{(l)}$
   (b) +1 in NaIO, –1 in NaI and +5 in $NaIO_3$. Iodine is oxidised to form IO_3^- and reduced to form I^-.
4. (a) +4 in N_2O_4, +5 in HNO_3 and +2 in NO. Nitrogen is oxidised to form HNO_3 and reduced to form NO.
   (b) $3N_2O_4 + 2H_2O \rightarrow 4HNO_3 + 2NO$

Exercise 1.7A

1.

Halogen	Colour	Physical State
Chlorine	green	gas
Bromine	red-brown	liquid
Iodine	grey-black	solid

2. Answer D
3. Solid iodine has a molecular covalent structure. A crystal of iodine contains diatomic molecules held together by van der Waals attractions. Water is a polar liquid. Iodine is a nonpolar substance and is more soluble in nonpolar liquids such as hexane.
4. The van der Waals attraction between halogen molecules increases as the number of electrons in the halogen molecules increases down the group.

Property	Result for Astatine
molecular formula	At_2
physical state at room temperature	solid
colour at room temperature	grey-black
colour of vapour	purple
solubility in water (answer yes or no)	no
solubility in hexane (answer yes or no)	yes

Exercise 1.7B

1. (a) $Fe + I_2 \rightarrow FeI_2$
   (b) The oxidation number of iron increases from 0 in Fe to +2 in FeI_2. The oxidation number of iodine decreases from 0 in I_2 to –1 in FeI_2. Iron is oxidised and iodine is reduced.
   (c) Iodine is not able to oxidise iron to iron(III) as the oxidising ability of the halogens decreases down the group.
2. (a) $P_4 + 10Cl_2 \rightarrow 4PCl_5$

(b) The oxidation number of phosphorus increases from 0 in P_4 to +5 in PCl_5. The oxidation number of chlorine decreases from 0 in Cl_2 to −1 in PCl_5. Phosphorus is oxidised and chlorine is reduced.
(c) The oxidising ability of the halogens decreases down the group. As a result Cl forms PCl_5 in which P has an oxidation number of +5 and I forms PI_3 in which P has an oxidation number of +3.

Exercise 1.7C

1. (a) $Br_2 + 2I^- \rightarrow 2Br^- + I_2$
(b) The orange colour of bromine water is replaced with the brown colour of aqueous iodine.
2. (a) $Cl_2 + 2NaBr \rightarrow 2NaCl + Br_2$
(b) Both solutions are colourless. The orange colour of bromine appears on mixing the solutions.
3. The solution is colourless. The brown colour of aqueous iodine appears when chlorine is bubbled through the solution: $Cl_2 + 2I^- \rightarrow 2Cl^- + I_2$
4. $I_2 + 2NaAt \rightarrow 2NaI + At_2$
5. Answer B

Exercise 1.7D

1. (a) $Cl_2 + H_2O \rightarrow HOCl + HCl$
(b) Chlorine is oxidised as its oxidation number increases from 0 in Cl_2 to +1 in HOCl, and reduced as its oxidation number decreases from 0 in Cl_2 to −1 in HCl. Chlorine is simultaneously oxidised and reduced.
(c) *Advantages:* Significant quantities can be stored as chlorine gas can be compressed. *Disadvantages:* The concentration of chlorine in the water must be low as chlorine is toxic. Adding chlorine to water could be opposed on the grounds it is mass medication of the population.
2. (a) $Br_2 + H_2O \rightarrow HOBr + HBr$
(b) Bromine is oxidised as its oxidation number increases from 0 in Br_2 to +1 in HOBr, and reduced as its oxidation number decreases from 0 in Br_2 to −1 in HBr. Bromine is simultaneously oxidised and reduced.
3. (a) $Cl_2 + 2NaOH \rightarrow NaCl + NaOCl + H_2O$
(b) Disproportionation refers to the oxidation and reduction of the same element in a reaction.
4. (a) The orange colour of bromine is replaced by a colourless solution.
(b) Bromine is oxidised as its oxidation number increases from 0 in Br_2 to +1 in BrO^-, and reduced as its oxidation number decreases from 0 in Br_2 to −1 in Br^-. Bromine is simultaneously oxidised and reduced.
5. (a) $3Cl_2 + 6NaOH \rightarrow 5NaCl + NaClO_3 + 3H_2O$
(b) Disproportionation
6. (a) $3Cl_2 + 6KOH \rightarrow 5KCl + KClO_3 + 3H_2O$
(b) Bubble chlorine through hot concentrated potassium hydroxide solution.
7. $3Br_2 + 6OH^- \rightarrow 5Br^- + BrO_3^- + 3H_2O$
8. Answer C
9. Answer D

Exercise 1.7E

1. Answer D
2. Answer C
3. Answer B
4. Sodium hydrogensulfate, hydrogen bromide, sulfur dioxide, water.
5. Answer B
6. Answer B
7. Answer D
8. (a) $H_2SO_4 + 8H^+ + 8e^- \rightarrow H_2S + 4H_2O$
(b) $H_2SO_4 + 8HI \rightarrow 4I_2 + H_2S + 4H_2O$
(c) Smell of rotten eggs.
(d) Sulfur dioxide and sulfur.
(e) The electrons in the outermost shell of an iodide ion can be removed more easily as they are further from the nucleus and are better shielded than the outermost electrons in a chloride ion.
9. The reaction produces: a grey-black solid that forms purple fumes, a yellow solid, the smell of rotten eggs, and a choking colourless gas (any 2).
10. Compound O is KBr and reacts to form bromine (P), hydrogen bromide and sulfur dioxide (Q and R).

Exercise 1.8A

1. Rinse a pipette with oven cleaner. Use the pipette and a pipette filler to add 25 cm^3 of oven cleaner to a 500 cm^3 volumetric flask. Add distilled water until the meniscus lies on the fill line. Stopper and invert the volumetric flask to mix the contents. Rinse the pipette with the diluted solution then use the pipette and a pipette filler to transfer 25 cm^3 of the diluted solution to a conical flask.

Exercise 1.8B

1. Molarity of $FeCl_2 = \frac{0.54 \text{ mol}}{0.100 \text{ dm}^3} = 5.4 \text{ mol dm}^{-3}$
2. Molarity of $Na_2CO_3 = \frac{0.025 \text{ mol}}{0.250 \text{ dm}^3}$
   $= 0.10 \text{ mol dm}^{-3}$
3. Molarity of $Na_2CO_3 = 0.200 \text{ mol dm}^{-3}$. Answer D.
4. Molarity of SO_4^{2-} ion = Molarity of K_2SO_4
   $= 0.200 \text{ mol dm}^{-3}$
5. Molarity of K^+ ion $= 2 \times$ Molarity of K_2SO_4
   $= 1.00 \text{ mol dm}^{-3}$
6. Molarity of $Br_2 = \frac{4.0 \times 10^{-3} \text{ g}}{160 \text{ g mol}^{-1}}$
   $= 2.5 \times 10^{-5} \text{ mol dm}^{-3}$
7. Molarity of $HCl = \frac{0.552 \text{ mol}}{0.250 \text{ dm}^3} = 2.21 \text{ mol dm}^{-3}$.
   Answer C.

Exercise 1.8C

1. Mass of fluoride $= 7.3 \times 10^{-2}$ g
   Moles of fluoride $= 3.8 \times 10^{-3}$ mol
   Moles of NaF = Moles of fluoride $= 3.8 \times 10^{-3}$ mol
   Volume in tube $= \frac{50 \text{ g}}{1.6 \text{ g cm}^{-3}} = 31 \text{ cm}^3$
   Concentration of NaF $= \frac{3.8 \times 10^{-3} \text{ mol}}{31 \text{ cm}^3}$
   $= 1.2 \times 10^{-4} \text{ mol cm}^{-3}$

Exercise 1.8D

1. Molarity of $HCl = \frac{3.0 \text{ mol dm}^{-3} \times 50 \text{ cm}^3}{200 \text{ cm}^3}$
   $= 0.75 \text{ mol dm}^{-3}$

Exercise 1.8E

1. (a) Volume of $HNO_3 = \frac{0.50 \text{ mol}}{2.0 \text{ mol dm}^{-3}} = 0.25 \text{ dm}^3$
   (b) Mass of $Mg(NO_3)_2$
   $= 0.25 \text{ mol} \times 148 \text{ g mol}^{-1} = 37$ g
2. Moles of HCl reacted = 0.015 mol
   Mass of $ZnCl_2$ formed
   $= 7.5 \times 10^{-3} \text{ mol} \times 136 \text{ g mol}^{-1} = 1.0$ g
3. Moles of H_2SO_4 used = 0.010 mol
   Volume of KOH needed $= \frac{0.020 \text{ mol}}{0.20 \text{ mol dm}^{-3}}$
   $= 0.10 \text{ dm}^3$
4. (a) Moles of Na_2CO_3 = Moles of H_2SO_4
   $= 5.0 \times 10^{-3}$ mol
   (b) Moles in 250 cm³ $= 5.0 \times 10^{-3} \text{ mol} \times \frac{250 \text{ cm}^3}{25 \text{ cm}^3}$
   $= 5.0 \times 10^{-2}$ mol
   (c) Molar mass of hydrate $= \frac{6.0 \text{ g}}{5.0 \times 10^{-2} \text{ mol}} = 120$
   (d) x = 1
5. Moles of HCl reacted = 0.020 mol
   Molar mass of $MCO_3 = \frac{0.84 \text{ g}}{0.010 \text{ mol}} = 84 \text{ g mol}^{-1}$
   The metal (M) is Mg.
6. Moles of HCl reacted = 0.010 mol
   Molar mass of MO $= \frac{0.28 \text{ g}}{5.0 \times 10^{-3} \text{ mol}}$
   $= 56 \text{ g mol}^{-1}$
   The RAM of M is 40.
7. Moles of chloride in 1 dm³ $= 3.0 \times 10^{-2}$ mol
   Molar mass of $MCl_2 = \frac{3.12 \text{ g}}{1.5 \times 10^{-2} \text{ mol}}$
   $= 208 \text{ g mol}^{-1}$
   The metal (M) is Ba.

Exercise 1.8F

1. (a) A stable solution of known concentration.
   (b) Phenolphthalein OR methyl orange.
   (c) *Phenolphthalein*: colourless to pink.
   Methyl orange: red to yellow.
2. (a) A stable solution of known concentration.
   (b) $CH_3COOH + NaOH \rightarrow CH_3COONa + H_2O$
   (c) *Phenolphthalein*: colourless to pink.
3. (a) $Na_2CO_3 + 2HCl \rightarrow 2NaCl + H_2O + CO_2$
   (b) *Methyl orange*: yellow to red.
4. Answer B

Exercise 1.8G

1. (a) Rinse the pipette with the vinegar solution. Use the pipette and a pipette filler to transfer 25 cm³ of vinegar solution to a conical flask. Add 2–3 drops of phenolphthalein to the conical flask. Rinse the burette with sodium hydroxide solution. Fill the burette with sodium hydroxide solution and record the initial burette reading. Perform a rough titration, swirling the contents of the flask after each addition. Perform an accurate titration by adding sodium hydroxide solution dropwise near the end point. Repeat to obtain accurate titres that are in close agreement. Use the accurate titres to calculate an average titre.
   (b) *Colour change*: colourless to pink.
   $NaOH_{(aq)} + CH_3COOH_{(aq)}$
   $\rightarrow CH_3COONa_{(aq)} + H_2O_{(l)}$

Exercise 1.8H

1. (a) *Phenolphthalein*: colourless to pink.
   (b) $CH_3COOH + NaOH \rightarrow CH_3COONa + H_2O$
   (c) *Titres (in table)*: 21.4 cm³ and 21.3 cm³.
   Average titre = 21.35 cm³.

(d) Moles of NaOH used = Average Titre × Molarity = 2.135×10^{-3} mol

(e) Molarity = $\frac{2.135 \times 10^{-3} \text{ mol}}{0.025 \text{ dm}^3}$

= $0.0854 \text{ mol dm}^{-3}$

(f) Molarity = $0.0854 \text{ mol dm}^{-3} \times \frac{250 \text{ cm}^3}{25 \text{ cm}^3}$

= $0.854 \text{ mol dm}^{-3}$

2. (a) Titre × Molarity of NaOH = 3.0×10^{-3} mol

(b) $3.0 \times 10^{-3} \text{ mol} \times \frac{250 \text{ cm}^3}{25 \text{ cm}^3} = 3.0 \times 10^{-2}$ mol

(c) $3.0 \times 10^{-2} \text{ mol} \times 60 \text{ g mol}^{-1} = 1.8$ g

(d) Mass of wood vinegar

= $25.0 \text{ cm}^3 \times 1.02 \text{ g cm}^{-3} = 25.5$ g

% by mass = $\frac{1.8 \text{ g}}{25.5 \text{ g}} \times 100 = 7.1$ %

Exercise 1.8I

1. Moles of Na_2CO_3 reacted = 5.00×10^{-4} mol

Moles of Na_2CO_3 in 1 dm^3

= $5.00 \times 10^{-4} \text{ mol} \times \frac{1000 \text{ cm}^3}{25 \text{ cm}^3} = 0.0200$ mol

Molar mass of hydrate = $\frac{4.64 \text{ g}}{0.0200 \text{ mol}} = 232 \text{ g mol}^{-1}$

x = 7

2. (a) $BaCl_2 + 2AgNO_3 \rightarrow 2AgCl + Ba(NO_3)_2$

(b) $Ag^+ + Cl^- \rightarrow AgCl$

(c) Moles of chloride in 1 dm^3 = 0.100 mol

Molar mass of hydrate = $\frac{12.2 \text{ g}}{5.00 \times 10^{-2} \text{ mol}}$

= 244 g mol^{-1}

x = 2

Exercise 1.9A

1. A glowing splint will relight when it comes into contact with oxygen gas.
2. White fumes of solid ammonium chloride are formed when the ammonia vapour contacts the fumes from a glass rod dipped in concentrated hydrochloric acid.
3.

Gas	Test	Observation
Hydrogen	Burning splint	Burns with a 'pop'
Hydrogen chloride	Glass rod dipped in concentrated ammonia	White fumes of ammonium chloride

Exercise 1.9B

1. (a) The acid (W) is concentrated hydrochloric acid, the wire (X) is nichrome wire, the flame (Y) is blue and the flame (Z) is blue-green.

(b) To clean the wire and to make the solid stick to the wire.

2. Wash the end of a nichrome wire in concentrated hydrochloric acid. Dip the end of the wire in the solid. Then place the end of the wire in a blue Bunsen flame. The lithium ions in the solid turn the flame a crimson colour.
3. Wash the end of a nichrome wire in concentrated hydrochloric acid. Dip the end of the wire in the solid. Then place the end of the wire in a blue Bunsen flame. If the solid contains calcium ions the flame will be a brick red colour. If the solid contains barium ions the flame will be a green colour.
4. Answer D
5. Answer B

Exercise 1.9C

1. Add dilute sodium hydroxide solution and gently heat the mixture. White fumes form when the ammonia gas produced by the reaction contacts the fumes from a glass rod dipped in concentrated hydrochloric acid.

Exercise 1.9D

1. (a) Aqueous barium chloride.

(b) Colourless solution formed with magnesium chloride. White precipitate formed with magnesium sulfate.

(c) With magnesium sulfate:

$Ba^{2+}{(aq)} + SO_4^{2-}{(aq)} \rightarrow BaSO_{4(s)}$

2. Compound A is copper sulfate, compound B is sulfuric acid, and compound C is potassium sulfate.

Exercise 1.9E

1. Add dilute hydrochloric acid to the solid. Collect the gas evolved using a pipette and bubble the gas through limewater. If a cloudy white suspension turns the limewater milky the gas is carbon dioxide and the solid contains carbonate ions.
2. (a) The presence of calcium ions in the solid is confirmed by a brick red flame when the solid is subject to a flame test.

(b) The presence of carbonate ions in the solid is

confirmed by effervescence and the production of a colourless gas that turns limewater milky when the solid reacts with dilute acid.

Exercise 1.9F

1.

Silver halide	Formula of halide	Colour of halide	Type of bonding	Soluble in dilute ammonia	Soluble in concentrated ammonia
silver fluoride	AgF	white	ionic	yes	yes
silver chloride	AgCl	white	ionic	yes	yes
silver bromide	AgBr	cream	ionic	no	yes
silver iodide	AgI	yellow	ionic	no	no

2. A white precipitate is formed on adding acidified silver nitrate solution. The precipitate dissolves on adding dilute ammonia solution.
3. A cream precipitate is formed on adding acidified silver nitrate solution. The precipitate dissolves on adding concentrated ammonia but does not dissolve on adding dilute ammonia to excess.
4. Add dilute nitric acid followed by several drops of silver nitrate solution to an aqueous solution of the solid. Formation of a white precipitate confirms that chloride is present. A yellow precipitate confirms that iodide is present.
5. Silver chloride will dissolve in dilute ammonia and concentrated ammonia to form a colourless solution. Silver iodide is insoluble in dilute ammonia and concentrated ammonia.
6. Add silver nitrate solution to solutions A, B and C in separate test tubes. Chloride ion forms a white precipitate that dissolves on adding dilute ammonia. Bromide ion forms a cream precipitate that dissolves on adding concentrated ammonia. Iodide ion forms a yellow precipitate that does not dissolve on adding concentrated ammonia.

Exercise 2.1A

1. Theoretical yield = 0.150 mol × 137 g mol^{-1} = 20.6 g
   Percentage yield = $\frac{12.3 \text{ g}}{20.6 \text{ g}} \times 100 = 59.7$ %
2. Theoretical yield = 0.34 mol × 92.5 g mol^{-1} = 31 g
   Percentage yield = $\frac{28 \text{ g}}{31 \text{ g}} \times 100 = 90$ %
3. Ammonia is in excess.
   Theoretical yield = 8.24 × 10^{-3} mol × 300 g mol^{-1} = 2.47 g
   Percentage yield = $\frac{2.08 \text{ g}}{2.47 \text{ g}} \times 100 = 84.2$ %
4. Salicylic acid is limiting.
   Theoretical yield = 2.17 × 10^{-2} mol × 180 g mol^{-1} = 3.91 g
   Percentage yield = $\frac{3.08 \text{ g}}{3.91 \text{ g}} \times 100 = 78.8$ %

Exercise 2.1B

1. Isopentyl alcohol is limiting.
   Actual yield = 5.40 × 10^{-2} mol × $\frac{\text{Percentage yield}}{100}$ = 0.0243 mol
   Mass formed = 0.0243 mol × 130 g mol^{-1} = 3.16 g
2. (a) Percentage yield = $\frac{\text{actual yield}}{\text{theoretical yield}} \times 100\%$
   (b) Actual yield of C_4H_9Br = 0.0400 mol
   Theoretical yield = 0.0400 mol × $\frac{100}{\text{Percentage yield}}$ = 0.10 mol
   Mass needed = 0.10 mol × 74 g mol^{-1} = 7.4 g
3. Actual yield of C_2H_4 = 3.6 × 10^4 mol
   Theoretical yield = 3.6 × 10^4 mol × $\frac{100}{\text{Percentage yield}}$ = 3.8 × 10^4 mol
   Mass needed = 3.8 × 10^4 mol × 142 g mol^{-1} = 5.4 × 10^6 g = 5.4 tonnes

Exercise 2.1C

1. Atom economy = $\frac{153 \text{ g}}{153 \text{ g} + 44 \text{ g}} \times 100$ % = 77.7 %
2. Atom economy = $\frac{114 \text{ g}}{114 \text{ g} + 28 \text{ g}} \times 100$ % = 80.3 %
3. Atom economy = $\frac{112 \text{ g}}{112 \text{ g} + 102 \text{ g}} \times 100$ % = 52.3 %
4. Atom economy = $\frac{154 \text{ g}}{154 \text{ g} + 146 \text{ g}} \times 100$ % = 51.3 %
5. (a) Atom economy = $\frac{48 \text{ g}}{48 \text{ g} + 80 \text{ g}} \times 100$ % = 37.5 %
   (b) Atom economy = $\frac{48 \text{ g}}{48 \text{ g} + 32 \text{ g}} \times 100$ % = 60.0 %
6. (a) Atom economy = $\frac{\text{Mass of useful products}}{\text{Mass of all products}} \times 100$ %
   (b) Atom economy = $\frac{137 \text{ g}}{137 \text{ g} + 18 \text{ g}} \times 100$ % = 88.4 %
7. (a) Atom economy is the mass of useful product formed as a percentage of the total mass of all

products formed.

(b) Atom economy = $\frac{180\text{ g}}{180\text{ g} + 60\text{ g}} \times 100\ \%$
= 75.0 %

Exercise 2.1D

1. Volume of oxygen = 2.50×10^{-2} mol × 24 $dm^3\ mol^{-1}$
= 0.600 dm^3
2. Volume of gas = 0.125 mol × 24 $dm^3\ mol^{-1}$
= 3.00 dm^3
3. Volume of ammonia = 0.695 mol × 24 $dm^3\ mol^{-1}$
= 16.7 dm^3
4. Volume of hydrogen = 0.333 mol × 24 $dm^3\ mol^{-1}$
= 8.00 dm^3
5. Moles of nitrogen = $\frac{50\text{ dm}^3}{24\text{ dm}^3\text{ mol}^{-1}}$ = 2.1 mol
Mass of azide = 1.4 mol × 65 g mol^{-1} = 91 g
6. Volume of hydrogen = 0.050 mol × 24 $dm^3\ mol^{-1}$
= 1.2 dm^3
7. Volume of oxygen = 0.050 mol × 24 $dm^3\ mol^{-1}$
= 1.2 dm^3
8. Moles of NO_2 = $\frac{720\text{ dm}^3}{24\text{ dm}^3\text{ mol}^{-1}}$ = 30.0 mol

Volume = $\frac{30.0\text{ mol}}{2.0\text{ mol dm}^{-3}}$ = 15 dm^3

Exercise 2.1E

1. (a) Mass of H_2 in 1 cm^3
= 4.2×10^{-5} mol × 2 g mol^{-1} = 8.4×10^{-5} g cm^{-3}
(b) Mass of O_2 in 1 cm^3
= 4.2×10^{-5} mol × 32 g mol^{-1} = 1.3×10^{-3} g cm^{-3}
(c) Mass of CO_2 in 1 cm^3
= 4.2×10^{-5} mol × 44 g mol^{-1} = 1.8×10^{-3} g cm^{-3}
2. Weigh the flasks. The flask containing nitrogen is heavier because the flasks contain the same number of moles and nitrogen has a greater molar mass.

Exercise 2.1F

1. (a) 81.8% C, 18.2% H
(b) 83.7% C, 16.3% H
2. 40.0% C, 6.7% H, 53.3% O

Exercise 2.1G

1. (a) P_2O_5
(b) P_4O_{10}
(c) $P_4O_{10} + 12NaOH \rightarrow 4Na_3PO_4 + 6H_2O$
2. (a) The simplest whole number ratio of atoms of each element in one molecule.
(b) The number of atoms of each element in one molecule.
3. (a) $C_2H_3O_2$
(b) $C_4H_6O_4$
4. CF_2Cl
5. $C_4H_8Br_2$
6. $C_8H_{20}Pb$
7. CH_2Cl

Exercise 2.2A

1. (a) Construct a space-filling model of each compound.
(b) Isomers are compounds with the same molecular formula that have a different arrangement of atoms.
(c) A compound that contains only carbon and hydrogen.
(d) Ethanol contains carbon, hydrogen and oxygen.
2. (a) A family of compounds with similar chemical properties, whose formulas derive from the same general formula and differ by CH_2 when arranged by mass, and whose physical properties vary smoothly as the molecules get bigger.
(b) C_nH_{2n+2}
(c) C_6H_{14}
3. Answer C
4. Answer C
5. Answer C
6. CH_3OH C_2H_5OH C_3H_7OH C_4H_9OH

Exercise 2.2B

1. (a) The formulas CH_3 and CO are used to describe the bonding between the atoms in different parts of a propanone molecule.
(b) The structural formula for a molecule describes how the atoms are bonded and does not describe how the atoms are arranged in space.

Exercise 2.2C

1. A group of atoms in a molecule that together determine the reactions of the compound.
2. (a) $C_3H_6O_3$
(b)

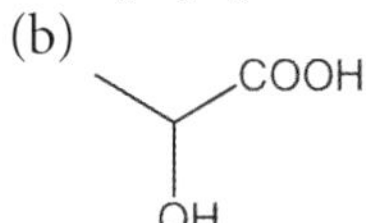

3. (a) $C_6H_{12}O_6$
(b) CH_2O

(c)

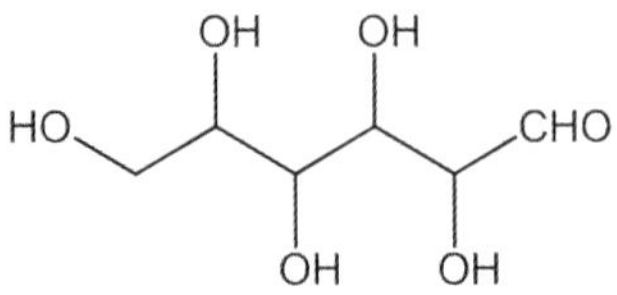

OH OH

HO CHO

OH OH

4. (a) $C_9H_8O_4$
   (b) $C_9H_8O_4$
   (c)

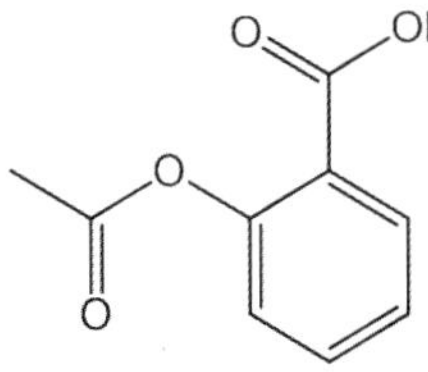

O OH

O

O

Exercise 2.3A

1. (a) C_nH_{2n+2}
   (b) $C_{20}H_{42}$
   (c) The compound does not contain C=C or C≡C bonds.
   (d) The van der Waals attraction between icosane molecules is greater as icosane molecules contain more electrons than ethane molecules.

Exercise 2.3B

1. Answer C
2. Answer B
3. Answer D
4. Answer B

Exercise 2.3C

1. (a) C_nH_{2n+2}
   (b) Structural isomers are molecules with the same molecular formula that have different structural formulas.
   (c) Isopentane is 2-methylbutane. Neopentane is 2,2-dimethylpropane.
   (d) The van der Waals attraction between molecules is greatest between straight-chain alkanes. More branched isomers have lower boiling points due to weaker van der Waals attraction between molecules.
2. Two
3. Answer B

Exercise 2.3D

1. Answer A
2. (a) 3-ethyl-2-methylpentane
   (b) 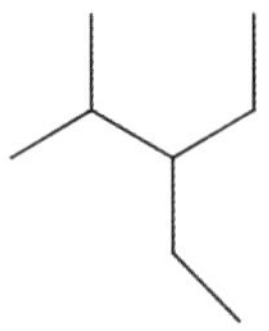
3. (a) 4-ethyl-2,2-dimethylhexane
   (b) 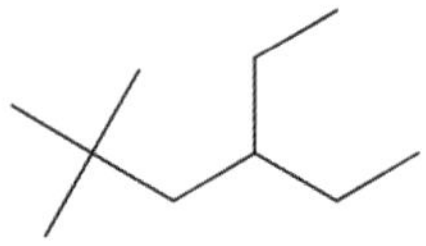
4. (a) C_9H_{20}
   (b) Compound A is 3-ethyl-3-methylhexane.
   Compound B is 4-ethyl-2-methylhexane.
   (c) A and B have the same molecular formula.
5. Answer C

Exercise 2.3E

1. (a)

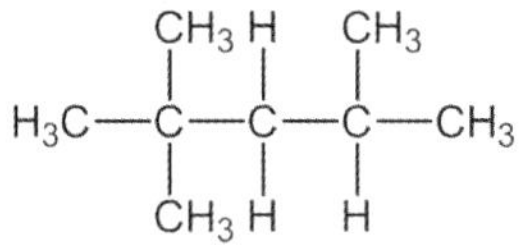

```
      CH3 H  CH3
       |  |   |
H3C—C—C—C—CH3
       |  |   |
      CH3 H   H
```

   (b) $C_8H_{18} + \frac{25}{2}O_2 \rightarrow 8CO_2 + 9H_2O$

   or $2C_8H_{18} + 25O_2 \rightarrow 16CO_2 + 18H_2O$

2. $C_{20}H_{42} + \frac{61}{2}O_2 \rightarrow 20CO_2 + 21H_2O$

   or $2C_{20}H_{42} + 61O_2 \rightarrow 40CO_2 + 42H_2O$

3. Answer C
4. Answer C
5. C_3H_6

Exercise 2.3F

1. (a) $C_5H_{12} + 8O_2 \rightarrow 5CO_2 + 6H_2O$
   (b) $C_5H_{12} + \frac{11}{2}O_2 \rightarrow 5CO + 6H_2O$
   or $2C_5H_{12} + 11O_2 \rightarrow 10CO + 12H_2O$
2. $C_{17}H_{36} + \frac{35}{2}O_2 \rightarrow 17CO + 18H_2O$
   or $2C_{17}H_{36} + 35O_2 \rightarrow 34CO + 36H_2O$
3. (a) Carbon monoxide and water.
   (b) A limited supply of oxygen during combustion.
4. (a) 2,5-dimethylhexane
   (b) Structural isomers are molecules with the same molecular formula that have different structural formulas.
   (c) $C_8H_{18} + \frac{17}{2}O_2 \rightarrow 8CO + 9H_2O$.
   (d) There is insufficient oxygen to react with all of

the carbon.

5. (a) The volume of one mole of gas at a specified temperature and pressure.
   (b) Moles of CO produced $= \dfrac{6000 \text{ dm}^3}{30 \text{ dm}^3 \text{ mol}^{-1}}$
   $= 200$ mol
   Moles of O_2 used $= 2 \times 200$ mol $= 400$ mol
   (c) Mass of $C_3H_8 = 100$ mol $\times 0.044$ kg mol^{-1}
   $= 4.40$ kg
   (d) $6CH_4 + 11O_2 \rightarrow 4CO_2 + 2CO + 12H_2O$

Exercise 2.3G

1. The combustion of hydrocarbon-based fuels produces large quantities of carbon dioxide that contribute to global warming. Repeated exposure to nitrogen oxides, hydrocarbons and particulates (mostly soot) from the combustion of gasoline causes respiratory problems. Carbon monoxide is toxic and is produced by the incomplete combustion of hydrocarbons. Brief exposure to carbon monoxide produces acute breathing difficulties.
2. (a) Carbon dioxide. (b) Carbon dioxide and water. (c) Nitrogen.
3. $2CO + O_2 \rightarrow 2CO_2$
   $2NO \rightarrow N_2 + O_2$
4. (a) Chemisorption refers to the formation of a chemical bond between a substance and a surface, and the weakening of bonds within the substance when it bonds with the surface.
   (b) The honeycomb structure increases the efficiency of the catalyst by providing a large surface over which the catalyst can be dispersed.
5. (a) To increase the efficiency of the catalyst by increasing the surface area on which chemisorption can occur.
   (b) Molecules in the exhaust gases form a bond with the metal surface. The bonds within the molecules weaken as the molecules bond with the surface, making it easier for the molecules to react. Chemisorption also brings the reacting molecules closer together making it easier for them to react.
   (c) Lead produced by the combustion of 'leaded' petrol reduces the efficiency of the catalyst by bonding to its surface and preventing the chemisorption of exhaust gases.
6. (a) $2CO + 2NO \rightarrow N_2 + 2CO_2$
   (b) Any one of: platinum, rhodium or palladium.
   (c) To increase the efficiency of the catalyst by increasing the surface area on which molecules can chemisorb.
   (d) The exhaust gases and the metal catalyst are in different physical states.

Exercise 2.3H

1. Answer A
2. Answer C
3. (a) A free radical is an atom, molecule or ion with an unpaired electron.
   (b) $CH_4 + Cl_2 \rightarrow CH_3Cl + HCl$
   (c) (i) Initiation $Cl_2 \rightarrow 2Cl\bullet$
   (ii) Propagation $CH_4 + Cl\bullet \rightarrow \bullet CH_3 + HCl$
   $\bullet CH_3 + Cl_2 \rightarrow CH_3Cl + Cl\bullet$
   (iii) Termination $Cl\bullet + \bullet CH_3 \rightarrow CH_3Cl$
   or $\bullet CH_3 + \bullet CH_3 \rightarrow C_2H_6$
4. (a) Ultraviolet radiation provides the energy for homolytic fission of the Cl-Cl bond in chlorine.
   (b)

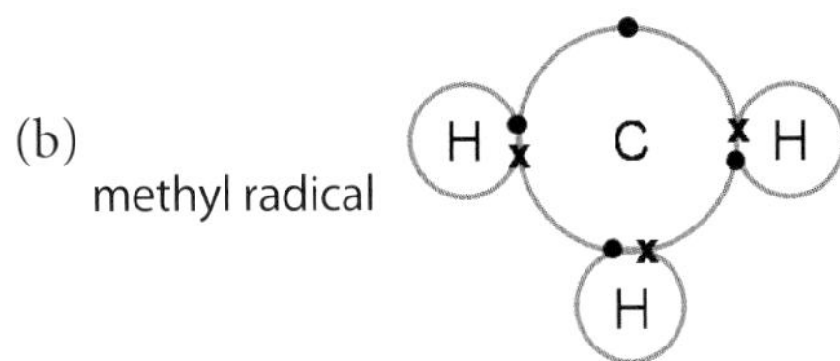

5. Answer C
6. A catalyst speeds up a reaction but is not consumed by the reaction. Light cannot be considered a catalyst as it is consumed by the reaction.
7. (a) $CH_3\bullet$ $C_2H_5\bullet$ $H\bullet$
   (b) Initiation: P Propagation: Q, R, S
   Termination: T U
8. (a) Ultraviolet light.
   (b) $C_2H_6 + Cl\bullet \rightarrow \bullet C_2H_5 + HCl$
   $\bullet C_2H_5 + Cl_2 \rightarrow C_2H_5Cl + Cl\bullet$
   (c) Butane.

Exercise 2.4A

1. Answer C
2. Answer D
3. (a) The molecular formula for hex-1-ene is obtained by setting n = 6 in the general formula for an alkene, C_nH_{2n}.
   (b) The water molecules in bromine water form hydrogen bonds with each other. Molecules of hex-1-ene cannot hydrogen bond with the water molecules in bromine water.

Exercise 2.4B

1. Answer B
2. Answer C

3. (a) A structural isomer has the same molecular formula but a different structural formula.

   (b)

   $CH_2{=}CHCH_2CH_3$ — but-1-ene

   $CH_3CH{=}CHCH_3$ — but-2-ene

   $CH_3C(CH_3){=}CH_2$ — 2-methylpropene

Exercise 2.4C

1. Answer B
2. One

Exercise 2.4D

1. (a) A C=C bond is stronger than the σ-bond in a C-C bond as it consists of a σ-bond and a π-bond. A C=C bond is also shorter as the atoms are held together more tightly by a stronger bond.

   (b) The π-bond within a C=C bond is weaker than the σ-bond in a C-C bond. It is also an electron rich region and will attract electrophiles.
2. (a) See answer to 1(a).

   (b) See answer to 1(b).

Exercise 2.4E

1. Answer C
2. Answer B
3. Either:

   (a) $H_3C{-}CH{=}C(CH_3){-}CH_2{-}CH_3$

   (b) 3-methylpent-2-ene

   (c) H and CH_3 / H_3C and CH_2CH_3 on C=C; H and CH_2CH_3 / H_3C and CH_3 on C=C

   Or:

   (a) $H_3C{-}CH{=}CH{-}CH(CH_3){-}CH_3$

   (b) 4-methylpent-2-ene

   (c) H and H / H_3C and $CH(CH_3)_2$ on C=C; H and $CH(CH_3)_2$ / H_3C and H on C=C

Exercise 2.4F

1. (a) E isomer: H_3C and H / H and CH_3 on C=C. Z isomer: H_3C and CH_3 / H and H on C=C.

   (b) The molecule cannot rotate about the C=C bond and each carbon atom forming the C=C bond is attached to two different groups of atoms.
2. (a) Geometric isomers have the same structural formula but a different arrangement of atoms in space as a result of the structure being unable to rotate about one or more C=C bonds.

   (b) See answer to 1b.

   (c) H_3C and CH_3 / H and CH_2CH_3 on C=C
3. (a)

   E isomer: CH_3CH_2 and CH_2CH_3 / H and F on C=C. Z isomer: CH_3CH_2 and F / H and CH_2CH_3 on C=C.

   (b) At one end of the C=C bond the -F group has a higher priority than the $-CH_2CH_3$ group. At the other end the $-CH_2CH_3$ group is assigned a higher priority than the -H group. In the Z-isomer the high priority groups are on the same side of the C=C bond.
4. E isomer: Z isomer:

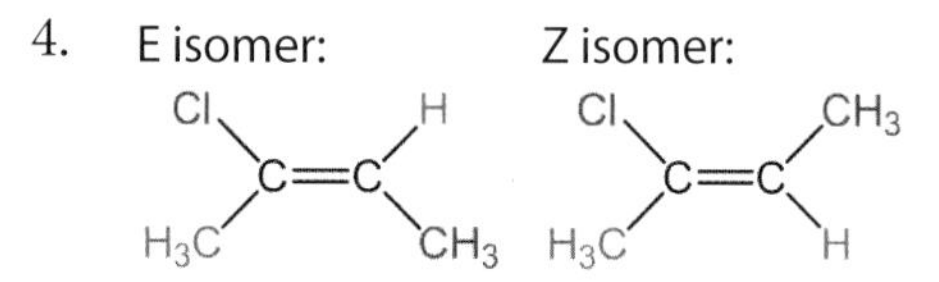

5. E isomer: H_3C and H / H and CH_2OH on C=C. Z isomer: H_3C and CH_2OH / H and H on C=C.
6. E isomer: Cl and CH_3 / H and CH_2OH on C=C. Z isomer: Cl and CH_2OH / H and CH_3 on C=C.
7.

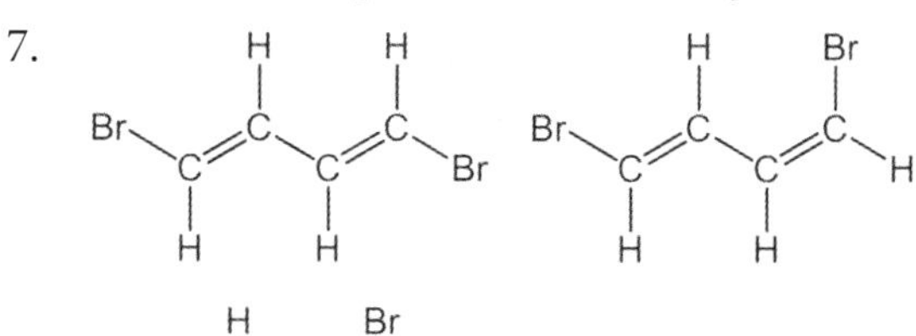

   H, Br, H, C, C, H, C, C, Br, H

Exercise 2.4G

1. (a)

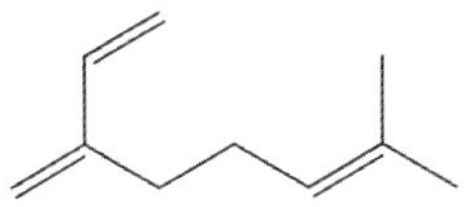

(b) Zero geometric isomers. The carbon atom at one end of each C=C bond is attached to two identical groups.

2. (a)

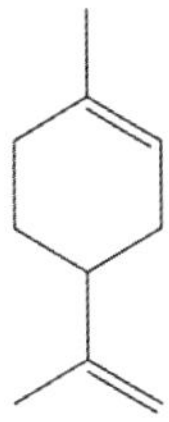

(b) The ring cannot be completed if the arrangement of groups about the C=C bond in the ring is changed. The C=C bond attached to the ring has two hydrogen atoms attached to one end of the bond.

3. (a) E isomer:

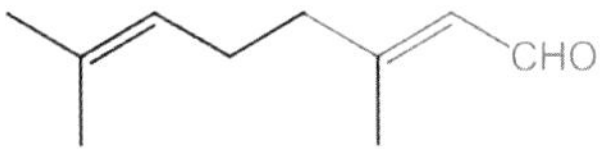

CHO

Z isomer:

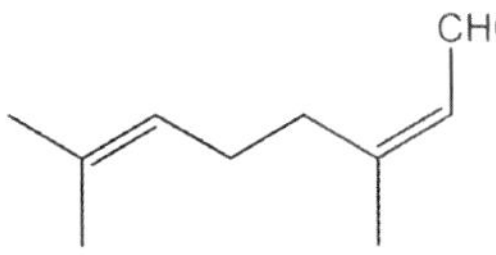

CHO

(b) The molecules have the same molecular formula but are unable to rotate about the C=C bond attached to the -CHO group.

4.

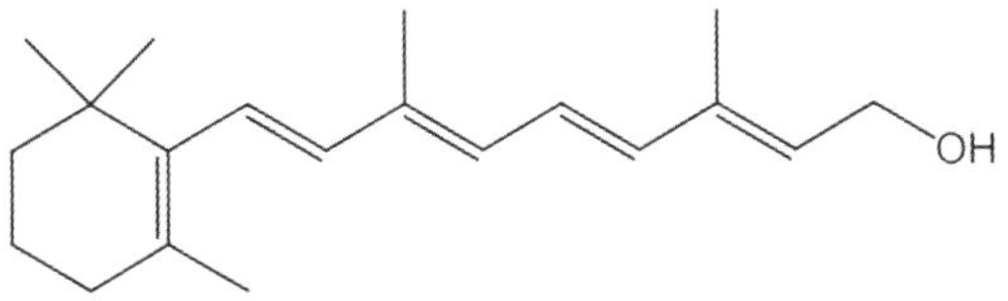

OH

Exercise 2.4H

1. (a) $(H)_2C=C(CH_3)_2$

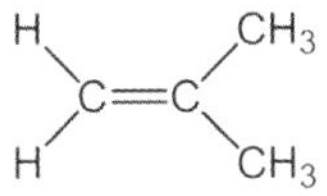

(b) An unsaturated hydrocarbon is a compound that contains only carbon and hydrogen, and contains one or more C=C or C≡C bonds.

(c) Moles of $C_4H_8 = \dfrac{4.2\text{ g}}{56\text{ g mol}^{-1}} = 7.5 \times 10^{-2}$ mol

Volume of H_2 needed
$= 7.5 \times 10^{-2}\text{ mol} \times 24\text{ dm}^3\text{ mol}^{-1} = 1.8\text{ dm}^3$

2. (a) Nickel in the form of a finely divided solid.
(b) $C_6H_{12} + H_2 \rightarrow C_6H_{14}$

The reaction does not produce any non-useful 'waste' products. Atom economy = 100 %.

3. (a) H—C(H)=C(H)—C(H)(H)—C(H)=C(H)—H

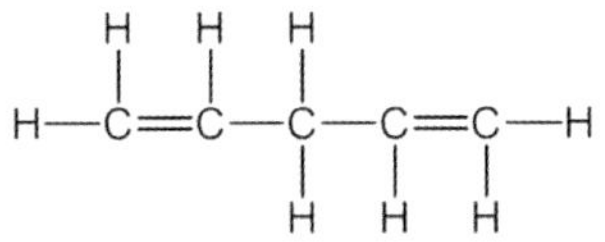

(b) $C_5H_8 + 2H_2 \rightarrow C_5H_{12}$
(c) Nickel in the form of a finely divided solid.

4. (a) $C_5H_8 + 2H_2 \rightarrow C_5H_{12}$
(b) Nickel
(c) A finely divided solid.
(d) The van der Waals attraction between molecules of isoprene is greater than the van der Waals attraction between pentane molecules.

5. (a) $C_6H_5CH{=}CHCH_2OH + H_2 \rightarrow$
$C_6H_5CH_2CH_2CH_2OH$
(b) Nickel in the form of a finely divided solid.

Exercise 2.4I

1. (a) $CH_2{=}CHCH_2CH_3 + Cl_2$
$\rightarrow CH_2ClCHClCH_2CH_3$
(b) H—C(H)(Cl)—C(H)(Cl)—C(H)(H)—C(H)(H)—H

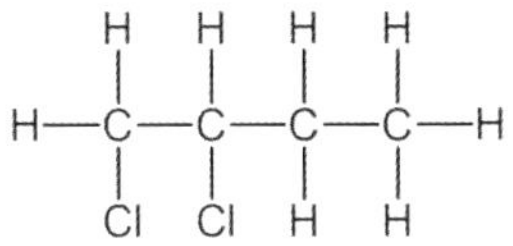

2. The bromine water is decolourised when a mixture of but-1-ene and bromine water is shaken.

3. (a) $C_2H_4 + Cl_2 \rightarrow C_2H_4Cl_2$
(b) 1,2-dichloroethane
(c) Hydrogen chloride

4. (a) $CH_2{=}CHCH_2CH{=}CH_2 + 2Br_2$
$\rightarrow CH_2BrCHBrCH_2CHBrCH_2Br$
(b) H—C(H)(Br)—C(H)(Br)—C(H)(H)—C(H)(Br)—C(H)(Br)—H

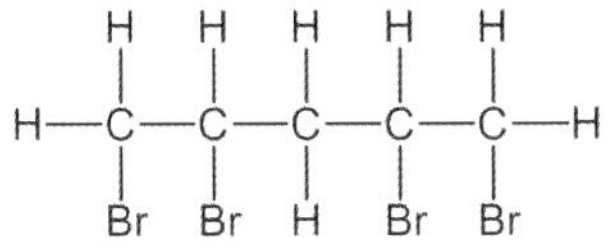

5. (a) Molecular formula: $C_{10}H_{16}$
Empirical formula: C_5H_8
(b) Bromine water is decolourised when shaken with a sample of myrcene.
(c) CH_2Cl, CCl, H_2C, $CHClCH_2Cl$, H_2C, CHCl, CCl, H_3C, CH_3

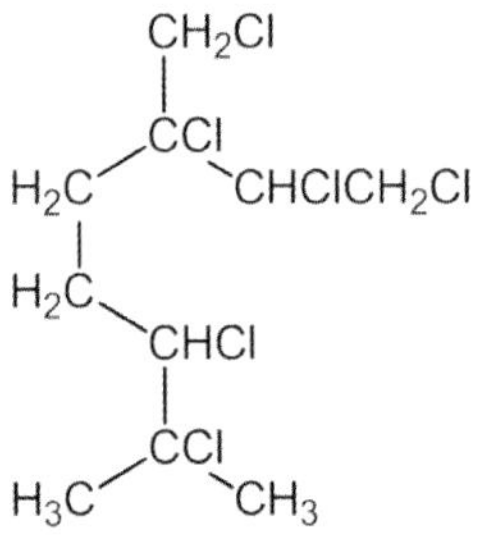

6. Moles of Br_2 used = 0.050 mol
Moles of alkene reacted = 0.050 mol

$$\text{Molar mass of alkene} = \frac{2.1\text{ g}}{0.050\text{ mol}} = 42\text{ g mol}^{-1}$$

The formula of the alkene is C_3H_6

7. Mass of Cl_2 used = 2.4 g
   Moles of Cl_2 used = 0.034 mol
   Moles of alkene reacted = 0.034 mol

$$\text{Molar mass of alkene} = \frac{1.4\text{ g}}{0.034\text{ mol}} = 41\text{ g mol}^{-1}$$

   The formula of the alkene is C_3H_6

8. $$\text{Moles of ethene} = \frac{0.120\text{ dm}^3}{24\text{ dm}^3\text{ mol}^{-1}} = 5.00 \times 10^{-3}\text{ mol}$$

   Mass of $Br_2 = 5.00 \times 10^{-3}\text{ mol} \times 160\text{ g mol}^{-1}$
   $= 0.800$ g

$$\text{Volume of } Br_2 = \frac{0.800\text{ g}}{3.2\text{ g cm}^{-3}} = 0.25\text{ cm}^3$$

Exercise 2.4J

1. The addition of an electron seeking molecule or ion across a C=C or C≡C bond.
2. Answer B
3. (a) The C=C π-bond in an alkene is electron rich and attracts electrophiles. It is also weaker than the C-C (σ) bonds in alkanes.
   (b) Electrophilic addition.
   (c)

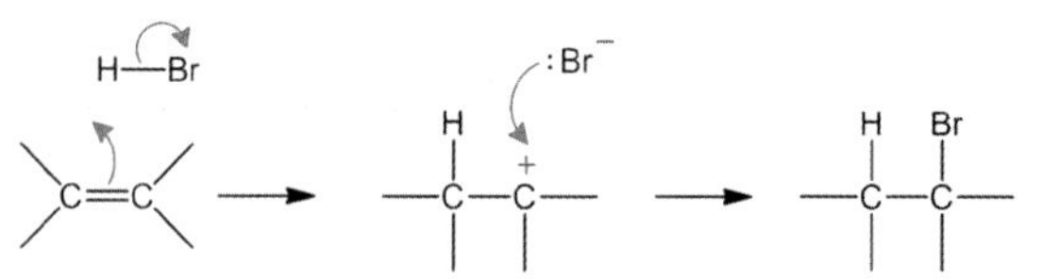

H—Br; C=C → H–C–C$^+$ (:Br$^-$) → H–C–C–Br

4.

H—Br; $H_2C=C(CH_3)CH=CH_2$ → H–C(H)(H)–C$^+$(CH_3)(CH=CH_2) (:Br$^-$) → H–C(H)(H)–C(Br)(CH_3)(CH=CH_2)

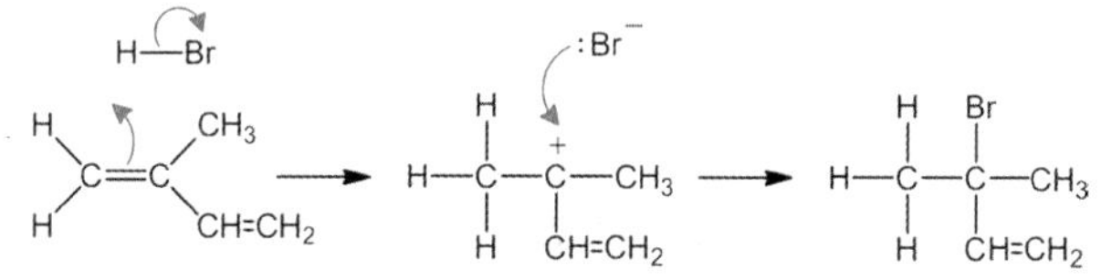

Exercise 2.4K

1. (a) The positive end of the HBr dipole is attracted to the electron rich region about the C=C bond.
   (b) Formation of 1-bromobutane:

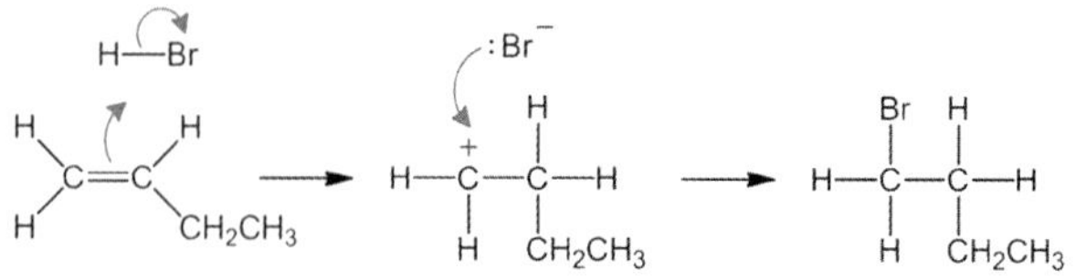

H—Br; $H_2C=CHCH_2CH_3$ → H–C$^+$(H)–C(H)(CH_2CH_3)–H (:Br$^-$) → H–C(Br)(H)–C(H)(CH_2CH_3)–H

   Formation of 2-bromobutane:

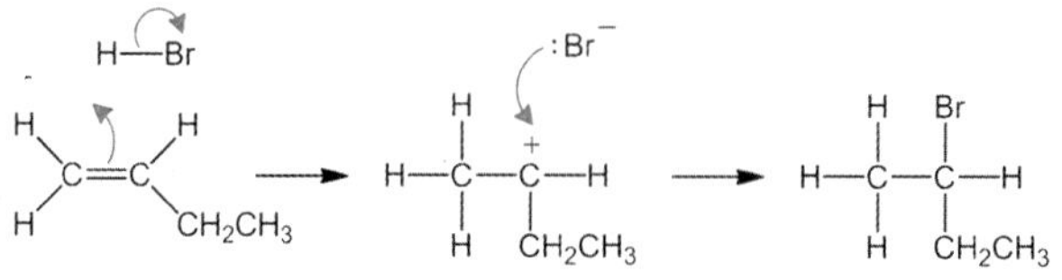

H—Br; $H_2C=CHCH_2CH_3$ → H–C(H)(H)–C$^+$(CH_2CH_3)–H (:Br$^-$) → H–C(H)(H)–C(Br)(CH_2CH_3)–H

   (c) (Fractional) distillation.
   (d) The secondary carbocation formed during the production of 2- bromobutane is more stable than the primary carbocation formed during the production of 1-bromobutane.
   (e) Electrophillic addition.

2. (a) $H_3C–C^+(CH_3)–CH(H)–CH_3$ and $H_3C–C(CH_3)(H)–C^+(H)–CH_3$
   (b) $H_3C–C(CH_3)(Cl)–C(H)(H)–CH_3$
   (c) The tertiary carbocation formed during the reaction is more stable than the secondary carbocation that would result in the other product being formed.

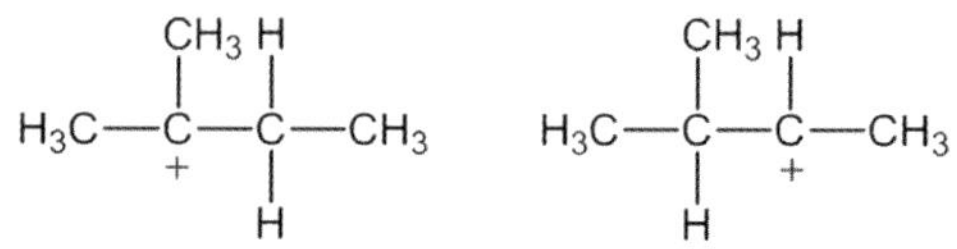

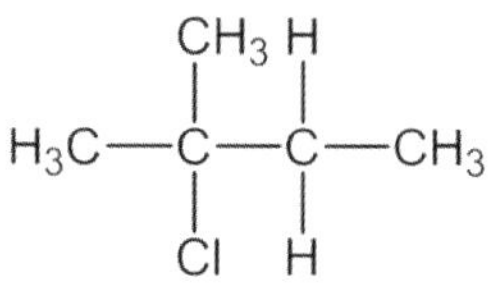

3. (a)
   (i) $H_3C–C^+(H)–C(H)(H)–CH_2CH_3$
   (ii) $H_3C–C(H)(H)–C^+(H)–CH_2CH_3$
   (b) The carbocations are equally likely to be formed as they are both secondary carbocations.

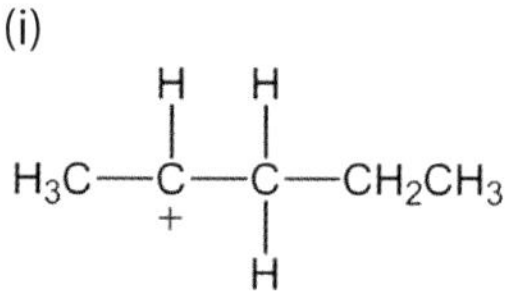

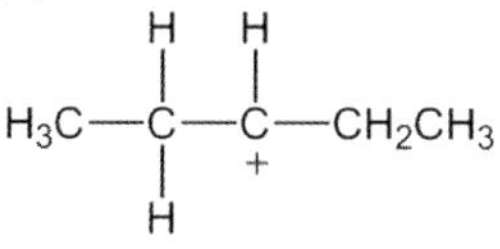

Exercise 2.4L

1. —C(H)(H)—C(CH_3)(H)—C(H)(H)—C(CH_3)(H)—C(H)(H)—C(CH_3)(H)—

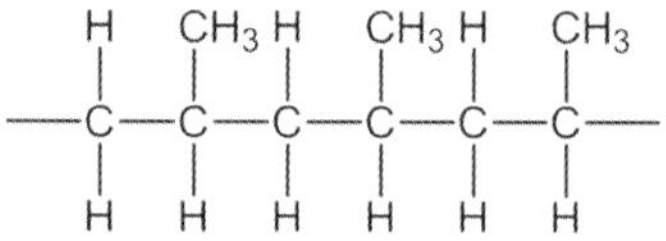

2. (a) Addition polymerisation.
   (b) —C(H)(H)—C(H)(CH_2CH_3)—C(H)(H)—C(H)(CH_2CH_3)—

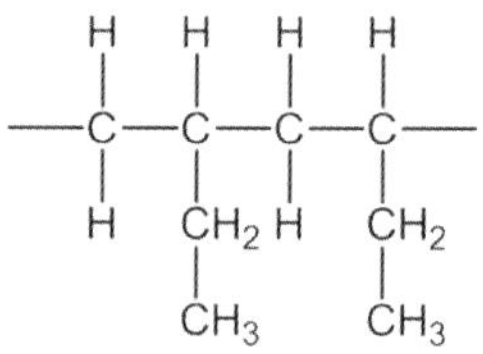

3. (a) Addition polymerisation.
   (b) —C(H)(H)—C(CH_3)(CH_3)—C(H)(H)—C(CH_3)(CH_3)—C(H)(H)—C(CH_3)(CH_3)—

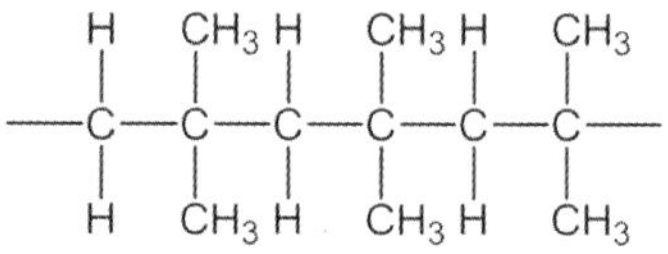

4. —C(H)(CH_3)—C(CH_3)(CH_2CH_3)—C(H)(CH_3)—C(CH_3)(CH_2CH_3)—

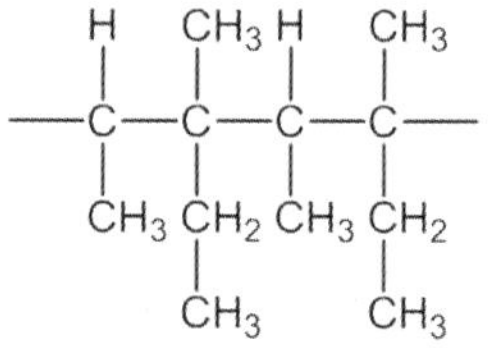

5. (a) H—C(H)=C(H)—C≡N

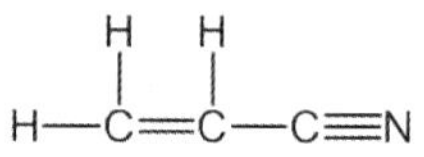

(b) Propenonitrile does not form cis-trans isomers as two hydrogen atoms are bonded to the same end of the C=C bond.

(c)

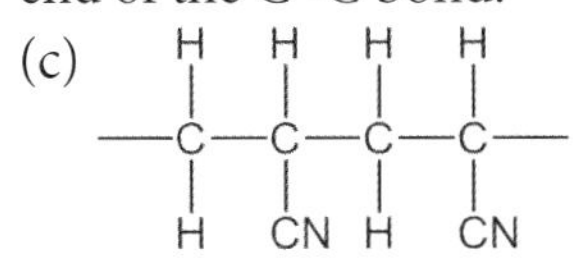

(d) Addition polymerisation.

6. (a)

(b) Tetrafluoroethene contains a C=C bond.

(c) Addition polymerisation.

7. (a) Addition polymerisation.

(b)

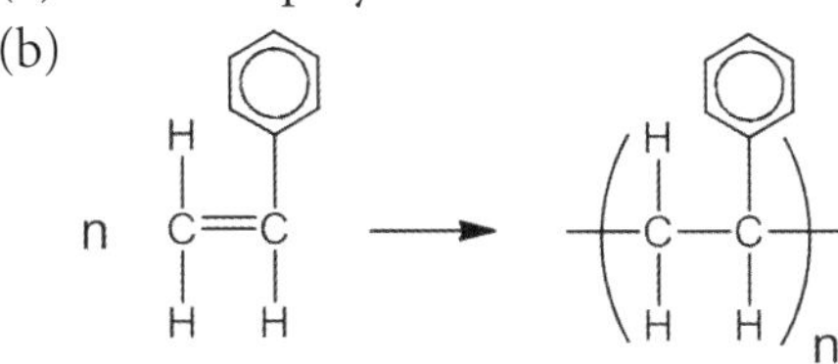

Exercise 2.5A

1. (a)

(b) Carbon and bromine have different electronegativity values.

2. (a) Structural isomers have the same molecular formula but different structural formulas.

(b) 1-bromobutane as it has a less-branched structure that gives rise to greater van der Waals attraction between molecules.

3. Boiling point increases in the order: 2-chloropropane, 1-bromopropane, 1-iodopropane as the number of electrons in the molecule increases and gives rise to greater van der Waals attraction between molecules. The boiling point of 1,1-dichloropropane is greater than the boiling point of 2-chloropropane as the second chlorine atom increases the van der Waals attraction between molecules.

4. Boiling point increases in the order: 1-chloro-2-fluoropropane, 1,2-dichloropropane, 2-bromo-1-chloropropane as the number of electrons in the molecule increases and gives rise to greater van der Waals attraction between molecules. The boiling point of 1,2-dichloropropane is greater than the boiling point of 1-chloropropane as the second chlorine atom increases the van der Waals attraction between molecules.

Exercise 2.5B

1. (a)

(b) The carbon bonded to chlorine is bonded to two other carbon atoms.

2. (a)

primary secondary

(b) Isomers have the same molecular formula but a different arrangement of atoms.

3. (a)

(b) The carbon bonded to fluorine is bonded to one other carbon atom.

Exercise 2.5C

1. CFC-12 (CCl_2F_2) is dichlorodifluoromethane and CFC-13 ($CClF_3$) is chlorotrifluoromethane.

2.

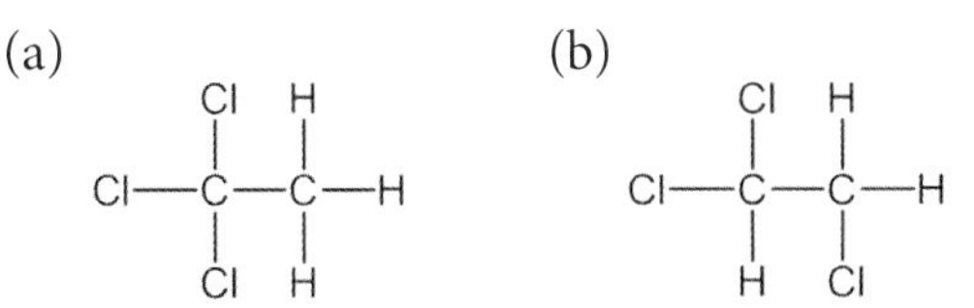

(a) 1,1,1-trichloroethane (b) 1,1,2-trichloroethane

3. (a)

1-bromo-1,1-dichloroethane

(b)

2-bromo-1,1-dichloroethane

1-bromo-1,2-dichloroethane

4.

Structure	Name	Classification
$CH_3CH_2CH_2CH_2Br$ (displayed formula)	1-bromobutane	primary
$(CH_3)_3CBr$ (displayed formula)	2-bromo-2-methylpropane	tertiary
$(CH_3)_2CHCH_2Br$ (displayed formula)	1-bromo-2-methylpropane	primary
$CH_3CHBrCH_2CH_3$ (displayed formula)	2-bromobutane	secondary

5. (a) C_4H_9Cl
   (b) $C_nH_{2n+1}Cl$
   (c) 2-chloro-2-methylpropane
   (d) secondary
   (e) The van der Waals attraction between molecules is weaker in t-butyl chloride as it has a more branched structure.

Exercise 2.5D

1. See Worked Example 2.5iv.
2. (a) To remove acidic impurities from the product.
   (b) To remove salts and other water soluble impurities from the product.
   (c) To dry the product.
   (d) To ensure smooth boiling during the distillation process.
   (e) To release carbon dioxide formed during shaking.
   (f) Moles of t-butyl alcohol used $= \frac{25\ g}{74\ g\ mol^{-1}}$
   $= 0.34$ mol
   Actual yield of t-butyl chloride $= \frac{28\ g}{92.5\ g\ mol^{-1}}$
   $= 0.30$ mol
   Percentage yield $= \frac{0.30\ mol}{0.34\ mol} \times 100 = 88\ \%$
3. Moles of butan-1-ol used $= \frac{14.8\ g}{74\ g\ mol^{-1}} = 0.200$ mol
   Actual yield of 1-bromobutane $= \frac{17.8\ g}{137\ g\ mol^{-1}}$
   $= 0.130$ mol
   Percentage yield $= \frac{0.130\ mol}{0.200\ mol} \times 100 = 65.0\ \%$
4. Answer A

Exercise 2.5E

1. Ammonia, NH_3 can use the lone pair of electrons on the nitrogen atom to form a coordinate bond. An ammonium ion, NH_4^+ does not have a lone pair of electrons and cannot form a coordinate bond.
2. Answer B

Exercise 2.5F

1. (a) $CH_3CH_2Br + KOH \rightarrow CH_3CH_2OH + KBr$
   (b) An atom or group of atoms is replaced by a nucleophile when the nucleophile uses a lone pair to form a coordinate bond with an electron deficient atom.
   (c)

$$HO^- + H_3C{-}CH_2{-}Br \rightarrow [HO\cdots CH_2(CH_3)\cdots Br]^- \rightarrow HO{-}CH_2{-}CH_3 + Br^-$$

2.

$$HO^- + H{-}CH_2{-}Cl \rightarrow [HO\cdots CH_3\cdots Cl]^- \rightarrow HO{-}CH_3 + Cl^-$$

3. Answer C
4. (a) The compounds are structural isomers as they have the same molecular formula but different structural formulas.
   (b)

$$(CH_3)_3C{-}Cl + {}^-OH \rightarrow (CH_3)_3C^+ \; Cl^- \; + \; :OH^- \rightarrow (CH_3)_3C{-}OH + Cl^-$$

5. (a) Nucleophilic substitution.
   (b) $(CH_3)_3CBr \rightarrow (CH_3)_3C^+ + Br^-$
   (c) $H_3C{-}\overset{+}{C}(CH_3){-}CH_3$
   (d) $(CH_3)_3C^+ + OH^- \rightarrow (CH_3)_3COH$

6.

```
                  H   OH
                  |   |
CH3CH2CH2CH2—C—C—H
                  |   |
                  OH  H
```

Exercise 2.5G

1. (a) $CH_3CH_2Cl + 2NH_3 \rightarrow CH_3CH_2NH_2 + NH_4Cl$
   (b) aminoethane
2. (a) (b)

```
     H   H   H   H             H   H   H   H
     |   |   |   |             |   |   |   |
  H—C—C—C—C—H           H—C—C—C—C—H
     |   |   |   |             |   |   |   |
     H  NH2  H   H             H  CN   H   H
```

3. Answer D
4. (a)

```
                  H   NH2
                  |   |
CH3CH2CH2CH2—C—C—H
                  |   |
                 NH2  H
```

   (b)

```
                  H   CN
                  |   |
CH3CH2CH2CH2—C—C—H
                  |   |
                  CN  H
```

5. (a) $Cl(CH_2)_5Cl + 4NH_3$
   $\rightarrow H_2N(CH_2)_5NH_2 + 2NH_4Cl$
   (b) Cadaverine is a base.

Exercise 2.5H

1. $CH_3Cl + D_2O \rightarrow CH_3OD + DCl$

Exercise 2.5I

1. The hydrolysis of chloroethane is slower as the C-Cl bond is stronger, and more difficult to break, than the C-Br bond.
2. Answer B
3. (a) A yellow precipitate forms on heating.
   (b) The rate of hydrolysis increases as the carbon-halogen bond becomes weaker down the group.
   (c) $CH_3CH_2CH_2CH_2Br + H_2O$
   $\rightarrow CH_3CH_2CH_2CH_2OH + HBr$
   $AgNO_3 + HBr \rightarrow AgBr + HNO_3$
4. The C-Br bond is the weakest carbon-halogen bond and the most likely to be broken when the compound is hydrolysed.
5. The compound t-butyl chloride is a tertiary halogenoalkane. In contrast, n-butyl chloride is a primary halogenoalkane. The hydrolysis of n-butyl chloride is slower as the C-Cl bond in n-butyl chloride is stronger than the C-Cl bond in t-butyl chloride.

Exercise 2.5J

1. (a) $CH_3CH_2CH_2CH_2Br + NaOH$
   $\rightarrow CH_3CH_2CH{=}CH_2 + NaBr + H_2O$
   (b) But-1-ene
   (c) Elimination
2. (a) $(CH_3)_3CBr + KOH$
   $\rightarrow H_2C{=}C(CH_3)_2 + KBr + H_2O$
   (b) 2-methylpropene
   (c) Elimination
3. (a) $CH_3CH_2Br + KOH$
   $\rightarrow CH_2{=}CH_2 + KBr + H_2O$
   (b) Ethene
   (c) Shake a sample of the gas with bromine water in a boiling tube. The orange colour of the bromine water is replaced by a colourless solution as bromine adds across the C=C bond.
   (d) None of the reactants is a gas.
4. Elimination

Exercise 2.5K

1. (a) (b)

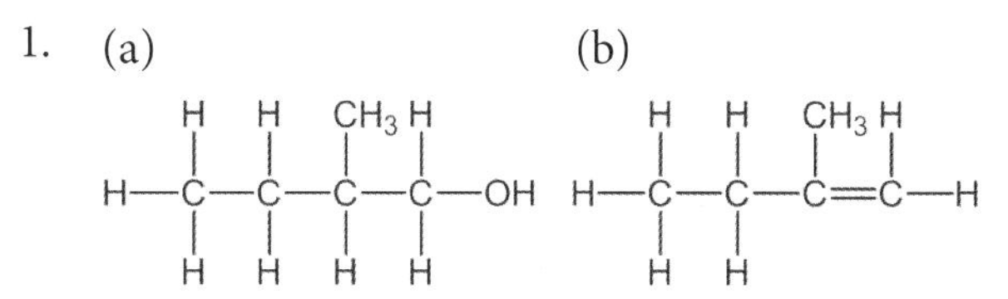

2. Answer C

Exercise 2.6A

1. (a)

```
     H   H
     |   |
  H—C—C—O—H
     |   |
     H   H
```

   (b) Ethanol can form hydrogen bonds with water.
2. Answer C
3. Ethanol can form hydrogen bonds with water. Ethene does not have a permanent dipole and cannot form hydrogen bonds with water.
4. (a) Ethanol, methanol and water can hydrogen bond with each other.
   (b) Fractional distillation.

Exercise 2.6B

1. (a) Molecule A is methanol and molecule B is water.
   (b) A hydrogen bond.
   (c) Any one of: solubility, melting point, boiling point.
2. Answer D
3. Answer B

4. Answer A
5. Answer B
6. Butan-1-ol is a primary alcohol and has a straight-chain structure. Both factors make it easier for the molecules to hydrogen bond when they pack closely together.
7. Both hydroxyl groups in ethylene glycol form hydrogen bonds with neighbouring molecules. As a result there are more hydrogen bonds between molecules, and more energy is needed to separate the ethylene glycol molecules.

Exercise 2.6C

1. (a) The carbon bonded to the hydroxyl group is bonded to one other carbon atom.
   (b) The hydroxyl groups are able to form hydrogen bonds with water.
2. Answer D
3. (a) Primary (b) Secondary
   (c) Primary (d) Tertiary

Exercise 2.6D

1.

Structure	Name	Classification
H H H H H—C—C—C—C—H H H H OH	butan-1-ol	primary
H CH_3 H H—C—C—C—H H OH H	2-methyl propan-2-ol	tertiary
H CH_3 H H—C—C—C—H H H OH	2-methyl propan-1-ol	primary
H H H H H—C—C—C—C—H H OH H H	butan-2-ol	secondary

2. Answer C
3. Answer B
4. Answer D
5. (a) ethane-1,2-diol
   (b) CH_3O
6. (a) A is ethane-1,1,2-triol
   B is propane-1,2,3-triol
   (b) A compound that contains three hydroxyl groups.
7. (a) A is 3-methylbutane-1,3-diol
   B is 2-methylbutane-1,2,4-triol
   (b) Compound A contains two hydroxyl groups.

Exercise 2.6E

1. (a) $2CH_3OH + 3O_2 \rightarrow 2CO_2 + 4H_2O$
   (b) $CH_3OH + O_2 \rightarrow CO + 2H_2O$
   (c) See Worked Example 2.6i

Exercise 2.6F

1. (a) $CH_3CHClCH_3 + KOH$
   $\rightarrow CH_3CH(OH)CH_3 + KCl$
   (b)

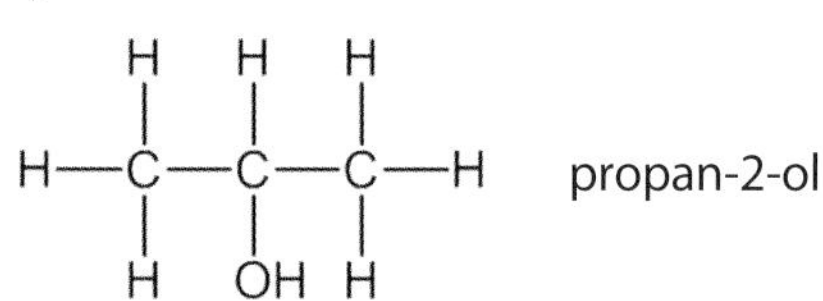

2. (a) Nucleophilic substitution.
   (b) To increase the yield of product by allowing the reactants to react at a high temperature for an extended period of time.
3. (a) $CH_2ClCH_2Cl + 2NaOH$
   $\rightarrow CH_2(OH)CH_2(OH) + 2NaCl$
   (b)

H H
H—C—C—H ethane-1,2-diol
OH OH

Exercise 2.6G

1. The sodium disappears as it reacts to form a colourless solution. Bubbles of colourless gas are produced, and the reaction mixture warms up, as the reaction mixture 'fizzes'.
2. (a) $2CH_3CH_2CH_2OH + 2Na$
   $\rightarrow 2CH_3CH_2CH_2ONa + H_2$
   (b) Sodium propoxide
3. (a) $2CH_3C(CH_3)_2OH + 2Na$
   $\rightarrow 2CH_3C(CH_3)_2ONa + H_2$
   (b) 2-methylpropan-2-ol
   (c) Tertiary

Exercise 2.6H

1. CH_3CH_2Cl (with PCl_5)
   CH_3CH_2ONa (with Na)
   CH_3CH_2Br (with HBr)
2. (a) $CH_3CH_2CH(OH)CH_3 + HBr$
   $\rightarrow CH_3CH_2CHBrCH_3 + H_2O$
   (b) Theoretical yield of C_4H_9Br
   $= 0.150 \text{ mol} \times 137 \text{ g mol}^{-1} = 20.6 \text{ g}$

(c) Actual yield = 0.85×20.6 g = 18 g

3. (a) Butan-1-ol
   (b) $CH_3CH_2CH_2CH_2OH + PCl_5$
   $\rightarrow CH_3CH_2CH_2CH_2Cl + HCl + POCl_3$
   (c) The solid disappears, the reaction mixture warms up, and steamy fumes of a pungent smelling gas are produced as the reaction mixture 'fizzes'.
4. $CH_2OHCH_2OH + 2PCl_5$
   $\rightarrow CH_2ClCH_2Cl + 2HCl + 2POCl_3$
5. Answer C

Exercise 2.6I

1. $CH_3CH_2OH + 2[O] \rightarrow CH_3COOH + H_2O$
2. Answer D

Exercise 2.6J

1.
(a)
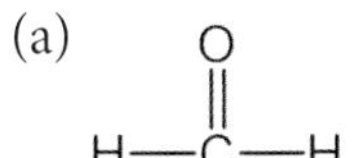

(b)
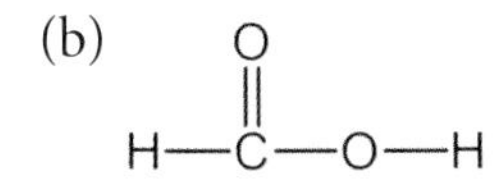

2. (a) Acidified potassium dichromate.
   (b) Cr^{3+}
   (c) Distillation
   (d) Reflux
3. (a) Orange to green.
   (b) $CH_3CH_2OH + [O] \rightarrow CH_3CHO + H_2O$
   or $CH_3CH_2OH + 2[O] \rightarrow CH_3COOH + H_2O$
4. (a) The hydroxyl group is attached to a carbon atom that has one carbon atom attached.
   (b) Oxidation
   (c) COOH
   |
   COOH
   (d) A carboxyl group.

Exercise 2.6K

1. Answer B
2. Answer B

Exercise 2.6L

1. Answer C
2. Answer D

Exercise 2.6M

1. Answer D
2. Propan-1-ol is a primary alcohol and propan-2-ol is a secondary alcohol. Both isomers will produce a colour change from orange to green when heated with acidified dichromate.
3.
4. Place samples of the alcohols in separate test tubes. Add acidified dichromate to each alcohol and warm gently. The mixture containing pentan 3-ol will turn from orange to green when heated. The mixture containing 2-methylbutan-2-ol will remain orange when heated.
5. Place a sample of Gasohol in a test tube. Add acidified dichromate and gently warm the mixture. The ethanol in the mixture will be oxidised and will turn the mixture from orange to green.

Exercise 2.7A

1. Answer D
2. Answer D

Exercise 2.7B

1. Absorption produced by the hydroxyl group in the alcohol will not be present in the IR spectrum of 2-bromo-2-methylpropane.
2. Answer D
3. Compound A is ethanoic acid as it contains an O-H bond and a C=O bond. Compound B is ethyl ethanoate as it does not contain an O-H bond. Compound C is ethanol as it does not contain a C=O bond.
4. (a) A molecule will absorb infrared radiation if the frequency of the radiation matches the frequency at which the molecule vibrates.
   (b) C-H bonds absorb at 3000 cm^{-1}, O-H bonds in an alcohol absorb at 3400 cm^{-1}, and C=O bonds absorb at 1700 cm^{-1}.
   (c) Compound B is ethanoic acid and compound C is ethanal.
5. (a) One or more bonds in a molecule will vibrate more vigorously when the molecule absorbs infrared radiation.
   (b) Spectrum A is for ethanol as it contains absorption in the range for O-H but not in the range for C=O. Spectrum B is for propanone as it contains absorption in the range for C=O but not in the range for O-H.
6. (a) COOH
   |
   COOH
   (b) Carboxyl
   (c) The IR spectrum of the product would contain absorption within the range for a C=O bond.

Exercise 2.7C

1. (a) A molecule will absorb IR radiation if bonds or groups of atoms in the molecule vibrate at the same frequency as the radiation.
   (b) A hydroxyl (-OH) group.
   (c) Match the IR spectrum of the unknown with the IR spectrum of a butanol whose identity is known.
2. Match the IR spectrum of the unknown with the IR spectrum of 1-bromopropane or 2-bromopropane.

Exercise 2.8A

1.

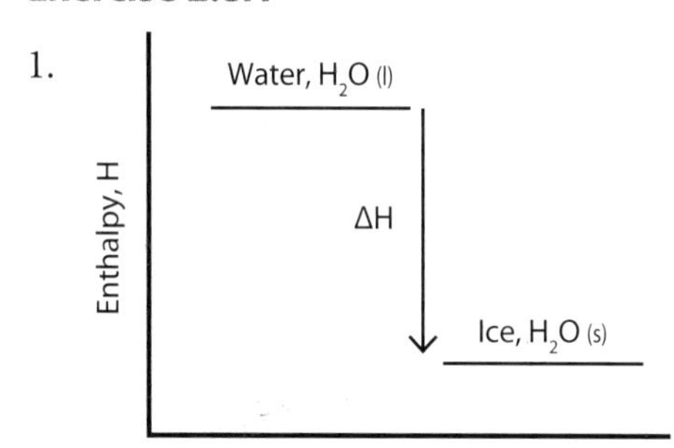

2. (a) A reaction in which the enthalpy of the products is greater than the enthalpy of the reactants.
   (b)

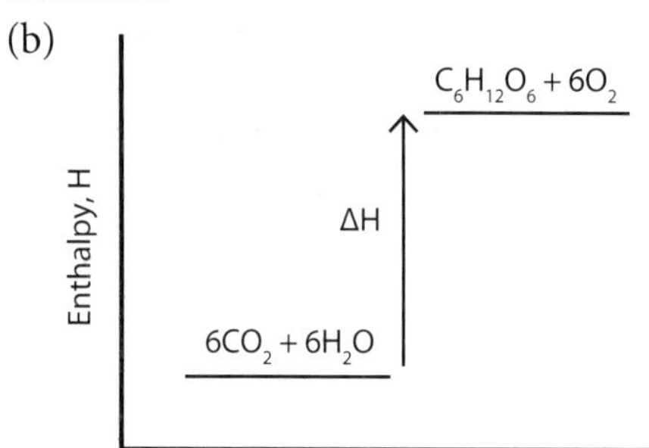

3. (a) A reaction in which the enthalpy of the products is less than the enthalpy of the reactants.
   (b)

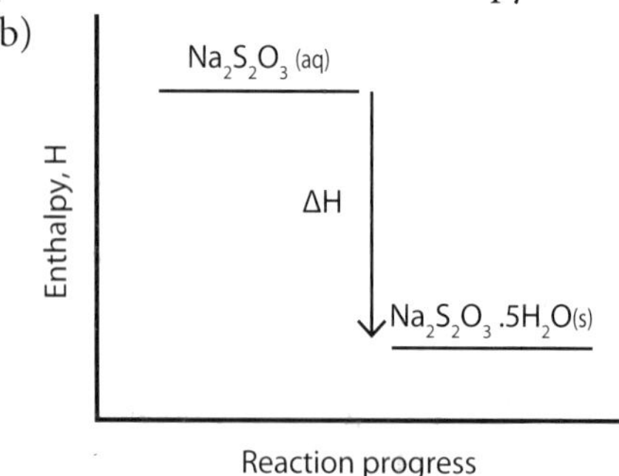

Exercise 2.8B

1. The formula $Br_2(g)$ represents a diatomic gas. The standard state of bromine is $Br_2(l)$.
2. The formula H(g) represents a monatomic gas. The standard state of hydrogen is $H_2(g)$.

Exercise 2.8C

1. (a) $q = m\,c\,\Delta T = 200.0\text{ g} \times 4.18\text{ J g}^{-1}\,{}^{o}\text{C}^{-1} \times 24.0\,{}^{o}\text{C}$
   $= 2.01 \times 10^4\text{ J}$
   (b) Mass of ethanol burnt = 1.4 g
   Moles of ethanol burnt = 0.030 mol
   $$\Delta H = \frac{-q}{\text{Moles of fuel}} = \frac{-20.1\text{ kJ}}{0.030\text{ mol}}$$
   $= -6.7 \times 10^2\text{ kJ mol}^{-1}$
2. (a) Mass of icosane burned = 0.90 g
   Temperature change = 61 °C
   (b) Moles of icosane burnt $= \dfrac{0.90\text{ g}}{282\text{ mol}^{-1}}$
   $= 3.2 \times 10^{-3}\text{ mol}$
   (c) $q = m\,c\,\Delta T = 150\text{ g} \times 4.18\text{ J g}^{-1}\,{}^{o}\text{C}^{-1} \times 61\,{}^{o}\text{C}$
   $= 3.8 \times 10^4\text{ J}$
   (d) $\Delta H = \dfrac{-q}{\text{Moles of fuel}} = \dfrac{-38\text{ kJ}}{0.0032\text{ mol}}$
   $= -1.2 \times 10^4\text{ kJ mol}^{-1}$
   (e) Any one of: heat loss from the flame to the surroundings, heat loss from the water to the surroundings, incomplete combustion of the fuel.
3. (a) $q = m\,c\,\Delta T = 50\text{ g} \times 4.2\text{ J g}^{-1}\,{}^{o}\text{C}^{-1} \times 55\,{}^{o}\text{C}$
   $= 1.2 \times 10^4\text{ J}$
   (b) Moles of butane burnt $= \dfrac{0.47\text{ g}}{58\text{ g mol}^{-1}}$
   $= 8.1 \times 10^{-3}\text{ mol}$
   $$\Delta H = \frac{-q}{\text{Moles of fuel}} = \frac{-12\text{ kJ}}{0.0081\text{ mol}}$$
   $= -1.5 \times 10^3\text{ kJ mol}^{-1}$
   (c) Heat loss from the flame to the surroundings. Heat loss from the water to the surroundings. Incomplete combustion of the fuel.
   (d) Butane produces more CO_2 and more H_2O per mole of fuel burnt. Forming bonds in CO_2 and H_2O produces heat.
4. $q = m\,c\,\Delta T = 200\text{ g} \times 4.2\text{ J g}^{-1}\,{}^{o}\text{C}^{-1} \times 28.2\,{}^{o}\text{C}$
   $= 2.4 \times 10^4\text{ J}$
   Moles of fuel burnt $= \dfrac{-q}{\Delta H} = \dfrac{-24\text{ kJ}}{-2220\text{ kJ mol}^{-1}}$
   $= 0.0011\text{ mol}$
   Molar mass $= \dfrac{0.47\text{ g}}{0.011\text{ mol}} = 43\text{ g mol}^{-1}$

Exercise 2.8D

1. (a) $q = m\,c\,\Delta T = 100\text{ g} \times 4.2\text{ J g}^{-1}\,{}^{\circ}\text{C}^{-1} \times (-0.90)\,{}^{\circ}\text{C}$
   $= -3.8 \times 10^2\text{ J}$
   (b) Molar enthalpy change, $\Delta H = \dfrac{-(-0.38)\text{ kJ}}{0.063\text{ mol}}$
   $= 6.0\text{ kJ mol}^{-1}$
   (c) The mass of salt used must increase by the ratio $\dfrac{25\,{}^{\circ}\text{C}}{0.90\,{}^{\circ}\text{C}} = 28$ and then again by the ratio $\dfrac{120\text{ g}}{100\text{ g}} = 1.2$ to $5.0\text{ g} \times 28 \times 1.2 = 1.7 \times 10^2\text{ g}$.
2. (a) Mass of TCE, $m = 100\text{ cm}^3 \times 1.33\text{ g cm}^{-3}$
   $= 133\text{ g}$
   $q = m\,c\,\Delta T = 133\text{ g} \times 1.30\text{ J g}^{-1}\,{}^{\circ}\text{C}^{-1} \times 7.2\,{}^{\circ}\text{C}$
   $= 1.2 \times 10^3\text{ J}$
   (b) Moles of hex-1-ene reacted $= \dfrac{1.4\text{ g}}{84\text{ g mol}^{-1}}$
   $= 0.017\text{ mol}$
   $\Delta H = \dfrac{-q}{\text{Moles reacted}} = \dfrac{-1.2\text{ kJ}}{0.017\text{ mol}}$
   $= -71\text{ kJ mol}^{-1}$

Exercise 2.8E

1. (a) $KOH + HCl \rightarrow KCl + H_2O$
   (b) Wear eye protection. Add a known amount of potassium hydroxide solution to an insulated container and measure the temperature of the solution with a thermometer. Add a known amount of hydrochloric acid while stirring the mixture. Measure the temperature of the solution at regular intervals, then calculate the biggest rise in temperature (ΔT) produced by the reaction. A major source of error is heat loss. Heat loss can be minimised by adding more insulation to the container, and reducing draughts around the apparatus.

Exercise 2.8F

1. (a) The energy needed to break one mole of a particular type of bond averaged over many compounds.
   (b) (i) Total bond enthalpy for products
   $= 16(750) + 18(463) = 20334\text{ kJ}$
   (ii) $\Delta H = 16548 - 20334 = -3786\text{ kJ}$
2. (a) Total bond enthalpy for the reactants
   $= 4(413) + 243 = 1895\text{ kJ}$
   Total bond enthalpy for the products
   $= 3(413) + 346 + 432 = 2017\text{ kJ}$
   $\Delta H = 1895 - 2017 = -122\text{ kJ}$
   (b)

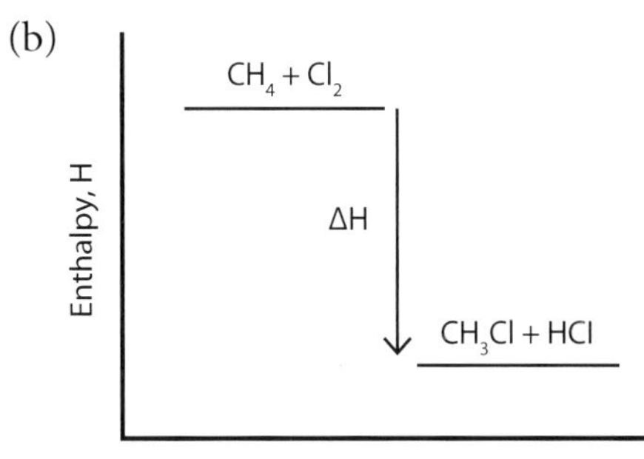

   (c) The bond enthalpies used to calculate ΔH are averaged over many compounds.
3. (a) The energy needed to break one mole of a particular type of bond averaged over many compounds.
   (b) Total bond enthalpy for the reactants
   $= 4\,E(\text{C-H}) + 2(496)\text{ kJ}$
   Total bond enthalpy for the products
   $= 2(743) + 4(463) = 3338\text{ kJ}$
   $\Delta H = 4E(\text{C-H}) + 992 - 3338 = -698$
   $E(\text{C-H}) = 412\text{ kJ}$
4. Total bond enthalpy for the reactants
   $= 612 + 4(412) + 3(497) = 3751\text{ kJ}$
   Total bond enthalpy for the products
   $= 4(803) + 4(464) = 5068\text{ kJ}$
   $\Delta H = 3751 - 5068 = -1317\text{ kJ}$
5. (a) The energy needed to break one mole of a particular type of bond averaged over many compounds.
   (b) Total bond enthalpy for the reactants
   $= 347 + 5(413) + 358 + 464 + 3(498) = 4728\text{ kJ}$
   Total bond enthalpy for the products
   $= 4(805) + 6(464) = 6004\text{ kJ}$
   $\Delta H = 4728 - 6004 = -1276\text{ kJ}$
   (c) The bond enthalpies are averaged over many compounds.
6. (a) Total bond enthalpy for the reactants
   $= 12(391) + 3(498) = 6186\text{ kJ}$
   Total bond enthalpy for the products
   $= 2(945) + 12(464) = 7458\text{ kJ}$
   Enthalpy of reaction, $\Delta H = 6186 - 7458$
   $= -1272\text{ kJ}$
   Enthalpy change per mole of ammonia
   $= -318\text{ kJ mol}^{-1}$
   (b) The energy needed to break bonds in the reactants is less than the energy produced by making bonds in the products.
7. Total bond enthalpy for the reactants
   $= 293 + 6(389) + 6(158) = 3575\text{ kJ}$

Total bond enthalpy for the products
$= 6(566) + 6(627) = 7158$ kJ
$\Delta H = 3575 - 7158 = -3583$ kJ

Exercise 2.8G

1. Answer D
2. Answer D
3. Answer D
4. Answer D
5. (a) $H_2 + F_2 \rightarrow 2HF \quad \Delta H = 2\ \Delta_f H^\circ(HF)$
   $\Delta H = 594 - 1136 = -542$ kJ
   $\Delta_f H^\circ(HF) = -271$ kJ
   (b)

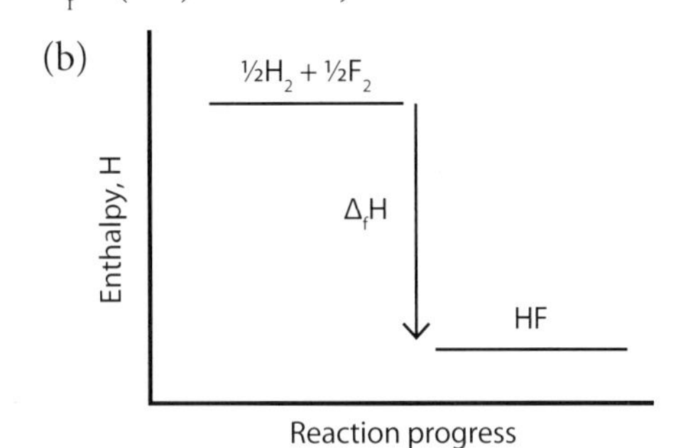

Exercise 2.8H

1. 1.0 g of water = 0.056 mol
   $\Delta H = -286\ \text{kJ mol}^{-1} \times 0.056\ \text{mol} = -16$ kJ
2. $\Delta H = 2 \times 46.2 = 92.4$ kJ

Exercise 2.8I

1. 1 kg of CO_2 contains 22.727 mol
   22.727 mol of CO_2 is formed from 5.6818 mol of butane. Energy released = 5.6818 mol × 2876.5 kJ mol^{-1} = 1.6344 kJ
2. (a) The enthalpy change when one mole of a substance is completely burnt in oxygen under standard conditions.
   (b) $\Delta_f H^\circ(\text{products}) = (-394) + 2(-286) = -966$ kJ
   $\Delta_f H^\circ(\text{reactants}) = -75 + 2(0) = -75$ kJ
   $\Delta_c H^\circ = \Delta_f H^\circ(\text{products}) - \Delta_f H^\circ(\text{reactants})$
   $= -891\ \text{kJ mol}^{-1}$
3. (a) The enthalpy change when one mole of a substance is completely burnt in oxygen under standard conditions.
   (b) $\Delta_f H^\circ(\text{products}) = 2(-394) + 3(-286)$
   $= -1646$ kJ
   $\Delta_f H^\circ(\text{reactants}) = -277 + 3(0) = -277$ kJ
   $\Delta_c H^\circ = \Delta_f H^\circ(\text{products}) - \Delta_f H^\circ(\text{reactants})$
   $= -1369\ \text{kJ mol}^{-1}$
4. (a) $\Delta_f H^\circ(\text{products}) = 4(-394) + 5(-286) = -3006$ kJ
   $\Delta_f H^\circ(\text{reactants}) = -327$ kJ
   $\Delta_c H^\circ = \Delta_f H^\circ(\text{products}) - \Delta_f H^\circ(\text{reactants})$
   $= -2679\ \text{kJ mol}^{-1}$
   (b) 25 °C (298 K) and 100 kPa
   (c) The standard enthalpy of formation for an element is zero.
5. (a) $C_{12}H_{26} + \frac{37}{2}O_2 \rightarrow 12CO_2 + 13H_2O$
   (b)

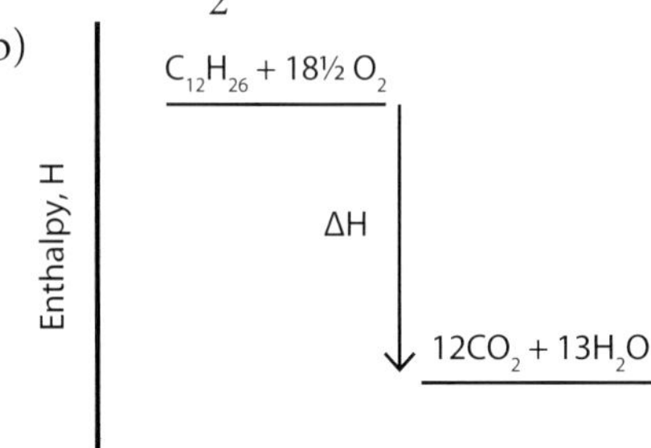

   (c) The enthalpy of the products is less than the enthalpy of the reactants.
6. (a) The enthalpy change when one mole of a substance is formed from its elements under standard conditions.
   (b) $\Delta_f H^\circ(\text{products}) = -1273 + 6(0) = -1273$ kJ
   $\Delta_f H^\circ(\text{reactants}) = 6(-394) + 6(-286)$
   $= -4080$ kJ
   $\Delta_c H^\circ = \Delta_f H^\circ(\text{products}) - \Delta_f H^\circ(\text{reactants})$
   $= 2807\ \text{kJ mol}^{-1}$

Exercise 2.8J

1.

$$\text{C (graphite)} \xrightarrow{\Delta H_1} CO_2$$
C (graphite) $\xrightarrow{\Delta H}$ C (diamond) $\xrightarrow{\Delta H_2}$ CO_2

$\Delta H = \Delta H_1 - \Delta H_2 = -393.5 - (-395.4)$
$= +1.9\ \text{kJ mol}^{-1}$

2.

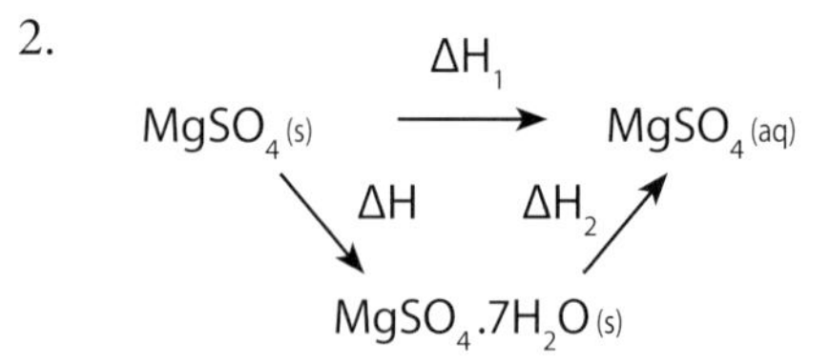

$q_1 = m\,c\,\Delta T = 100\ \text{g} \times 4.2\ \text{J g}^{-1}\ {}^\circ\text{C}^{-1} \times 9.0\ {}^\circ\text{C}$
$= 3.8 \times 10^3$ J

$$\Delta H_1 = \frac{-q_1}{\text{Moles of } MgSO_4} = \frac{-3.8\ \text{kJ}}{0.10\ \text{mol}}$$

$= -38\ \text{kJ mol}^{-1}$
$q_2 = m\,c\,\Delta T = 100\ \text{g} \times 4.2\ \text{J g}^{-1}\ {}^\circ\text{C}^{-1} \times (-3.0)\ {}^\circ\text{C}$
$= -1.3 \times 10^3$ J

$$\Delta H_2 = \frac{-q_2}{\text{Moles of } MgSO_4.7H_2O} = \frac{1.3\ \text{kJ}}{0.10\ \text{mol}}$$

$= 13 \text{ kJ mol}^{-1}$

$\Delta H = \Delta H_1 - \Delta H_2 = (-38) - (13) = -51 \text{ kJ mol}^{-1}$

3. (a) The enthalpy change when one mole of a substance is completely burnt in oxygen under standard conditions.

   (b) The enthalpy change for a reaction depends on the state of the reactants and products and does not depend on the way in which the reactants are converted to the products.

   (c)

$$C_8H_{18(l)} + 12\tfrac{1}{2}O_{2(g)} \xrightarrow{\Delta_c H^\ominus} 8CO_{2(g)} + 9H_2O_{(l)}$$

$\Delta H^\ominus_1$ (arrow from elements up to reactants) $\Delta H^\ominus_2$ (arrow from elements up to products)

Elements in their standard state: $C_{(s)}$ $H_{2(g)}$ $O_{2(g)}$

Hess's law: $\Delta H^\circ_1 + \Delta_c H^\circ = \Delta H^\circ_2$

$\Delta H^\circ_1 = -250.0$ kJ and $\Delta H^\circ_2 = 8(-393.5)$ $+ 9(-286.0) = -5722$ kJ

$\Delta_c H^\circ = \Delta H^\circ_2 - \Delta H^\circ_1 = -5472 \text{ kJ mol}^{-1}$

4.

$$C_2H_{2(g)} + 2\tfrac{1}{2}O_{2(g)} \xrightarrow{\Delta H^\ominus_2} 2CO_{2(g)} + H_2O_{(l)}$$

$\Delta H^\ominus_1$ $\Delta H^\ominus_3$

Elements in their standard state: $C_{(s)}$ $H_{2(g)}$ $O_{2(g)}$

Hess's law: $\Delta H^\circ_1 + \Delta H^\circ_2 = \Delta H^\circ_3$

$\Delta H^\circ_1 = \Delta_f H^\circ(C_2H_2)$ and $\Delta H^\circ_2 = -1300$ kJ

$\Delta H^\circ_3 = 2(-394) + (-286) = -1074$ kJ

$\Delta_f H^\circ(C_2H_2) = \Delta H^\circ_3 - \Delta H^\circ_2 = +\ 226 \text{ kJ mol}^{-1}$

5. (a) $4C_{(s)} + 5H_{2(g)} \rightarrow C_4H_{10(g)}$

   (b)

$$C_4H_{10(g)} + 6\tfrac{1}{2}O_{2(g)} \xrightarrow{\Delta_c H^\ominus} 4CO_{2(g)} + 5H_2O_{(l)}$$

$\Delta H^\ominus_1$ $\Delta H^\ominus_2$

Elements in their standard state: $C_{(s)}$ $H_{2(g)}$ $O_{2(g)}$

Hess's law: $\Delta H^\circ_1 + \Delta_c H^\circ = \Delta H^\circ_2$

$\Delta H^\circ_1 = \Delta_f H^\circ(C_4H_{10})$ and $\Delta_c H^\circ = -2876.5$ kJ

$\Delta H^\circ_2 = 4(-393.5) + 5(-285.8) = -3003$ kJ

$\Delta_f H^\circ(C_2H_2) = \Delta H^\circ_2 - \Delta H^\circ_1 = -126.5 \text{ kJ mol}^{-1}$

6.

$$C_3H_{8(g)} + 5O_{2(g)} \xrightarrow{\Delta_c H^\ominus} 3CO_{2(g)} + 4H_2O_{(l)}$$

$\Delta H^\ominus_1$ $\Delta H^\ominus_2$

Elements in their standard state: $C_{(s)}$ $H_{2(g)}$ $O_{2(g)}$

Hess's law: $\Delta H^\circ_1 + \Delta_c H^\circ = \Delta H^\circ_2$

$\Delta H^\circ_1 = \Delta_f H^\circ(C_3H_8)$ and $\Delta_c H^\circ = -2219$ kJ

$\Delta H^\circ_2 = 3(-394) + 4(-286) = -2326$ kJ

$\Delta_f H^\circ(C_2H_2) = \Delta H^\circ_2 - \Delta H^\circ_1 = -107 \text{ kJ mol}^{-1}$

7.

$$CH_3COOH_{(l)} + 2O_{2(g)} \xrightarrow{\Delta H^\ominus_2} 2CO_{2(g)} + 2H_2O_{(l)}$$

$\Delta H^\ominus_1$ $\Delta H^\ominus_3$

Elements in their standard state: $C_{(s)}$ $H_{2(g)}$ $O_{2(g)}$

Hess's law: $\Delta H^\circ_1 + \Delta H^\circ_2 = \Delta H^\circ_3$

$\Delta H^\circ_1 = \Delta_f H^\circ(CH_3COOH)$ and $\Delta H^\circ_2 = -487$ kJ

$\Delta H^\circ_3 = 2(-393) + 2(-286) = -1358$ kJ

$\Delta_f H^\circ(C_2H_2) = \Delta H^\circ_3 - \Delta H^\circ_2 = -871 \text{ kJ mol}^{-1}$

8. (a) The enthalpy change for a reaction depends on the state of the reactants and products and does not depend on the way in which the reactants are converted to the products.

   (b) Hess's law: $\Delta H_3 = \Delta H_1 + \Delta H_2$

   $\Delta H_1 = \Delta H_3 - \Delta H_2 = (2 \times 16) - (-21) = +\ 53$ kJ

Exercise 2.9A

1. The forward and reverse reactions occur at the same rate, and as a result, the amount of each reactant and product remains constant.

Exercise 2.9B

1. (a) $CH_3COONH_4 \rightarrow CH_3COOH + NH_3$

   (b) Ethanoic acid is a product of the reaction and prevents dissociation by shifting the position of the equilibrium to the left.
2. Removing HCN as the reaction proceeds increases the yield of HCN as the position of equilibrium shifts to the right.
3. Answer C
4. Answer D

Exercise 2.9C

1. The position of equilibrium shifts to the right when the pressure is decreased as there are more molecules of gas on the right side of the equilibrium.
2. The concentration of hydrogen would decrease as there are more gas molecules on the right side of the equilibrium, and the position of equilibrium shifts to the left when the pressure is increased.

Exercise 2.9D

1. Increasing the temperature increases the extent of dissociation as the forward reaction is endothermic and the position of equilibrium moves to the right.
2. Increasing the temperature decreases the yield of methanol as the reverse reaction is endothermic and the position of equilibrium moves to the left.

3. Forming ammonium cyanide is exothermic as the dissociation of ammonium cyanide occurs on heating and is therefore endothermic.
4. The forward reaction is exothermic, $\Delta H < 0$ as increasing the temperature shifts the position of equilibrium to the left, reducing the amount of methanol.
5. The forward reaction is endothermic therefore a high temperature would increase the yield of hydrogen by shifting the position of equilibrium to the right. The high temperature would also increase the rate at which hydrogen is formed.
6. (a) Ammonia is alkaline and is present at A.
   (b) Hydrogen chloride is acidic and is present at B.
   (c) To separate the ammonia vapour from the hydrogen chloride vapour.
   (d) The forward reaction is endothermic as the position of equilibrium shifts to the right on heating.

Exercise 2.9E

1. (a) Increasing the pressure shifts the position of equilibrium to the left as there are fewer gas molecules on the left side of the equilibrium.
   (b) Increasing the temperature shifts the position of equilibrium to the right as the forward reaction is endothermic.
2. (a) Increasing the pressure shifts the position of equilibrium to the left, decreasing the yield, as there are fewer gas molecules on the left side of the equilibrium.
   (b) Increasing the temperature shifts the position of equilibrium to the right, increasing the yield, as the forward reaction is endothermic.
3. (a) Increasing the pressure shifts the position of equilibrium to the right as there are fewer gas molecules on the right side of the equilibrium.
   (b) A higher temperature would shift the position of equilibrium further to the left, reducing the amount of phosgene at equilibrium.
   (c) The phosgene could be condensed to a liquid and run-off by cooling the reaction mixture to below 8 °C.
4. Answer C
5. (a) Decreasing the temperature shifts the position of equilibrium to the right as the forward reaction is exothermic.
   (b) Increasing the pressure shifts the position of equilibrium to the right as there are fewer gas molecules on the right side of the equilibrium.
6. (a) The forward and reverse reactions occur at the same rate, and as a result, the amount of each reactant and product remains constant.
   (b) Increasing the pressure shifts the position of equilibrium to the right, increasing the yield, as there are fewer gas molecules on the right side of the equilibrium.
   (c) Increasing the temperature shifts the position of equilibrium to the left, decreasing the yield, as the reverse reaction is endothermic.
   (d) A catalyst does not affect the position of equilibrium as it increases the rate of the forward and reverse reactions by an equal amount.
7. (a) Increasing the temperature shifts the position of equilibrium to the left, decreasing the yield, as the reverse reaction is endothermic.
   (b) Increasing the pressure shifts the position of equilibrium to the left, decreasing the yield, as there are fewer gas molecules on the left side of the equilibrium.
   (c) A catalyst does not affect the position of equilibrium as it increases the rate of the forward and reverse reactions by an equal amount.
8. (a) Increasing the temperature shifts the position of equilibrium to the right, increasing the yield, as the forward reaction is endothermic.
   (b) Adding nitrogen shifts the position of equilibrium to the right, increasing the yield, to reduce the amount of nitrogen.
   (c) Increasing the pressure does not affect the position of equilibrium as the number of gas molecules is the same on both sides of the equilibrium.
   (d) A catalyst does not affect the position of equilibrium as it increases the rate of the forward and reverse reactions by an equal amount.
9. (a) The silver acts as a catalyst.
   (b) Increasing the pressure shifts the position of equilibrium in Step 1 to the right, increasing the yield of ethylene oxide, as there are fewer molecules of gas on the right side of the equilibrium.
   (c) Increasing the temperature shifts the position of equilibrium in Step 1 to the right, increasing the yield of ethylene oxide, as the forward reaction is endothermic.
10. (a) The position of equilibrium shifts to the left as the reverse reaction is endothermic.

(b) Increasing the pressure has no effect on the position of equilibrium as the number of gas molecules is the same on both sides of the equilibrium.
(c) Adding a catalyst does not affect the position of equilibrium as it increases the rate of the forward and reverse reactions by an equal amount.

Exercise 2.9F

1. (a) $K_c = \dfrac{[I_3^-]}{[I_2]\,[I^-]}$
   (b) $mol^{-1}\ dm^3$
2. (a) $K_c = \dfrac{[CO_2]^2}{[CO]^2\,[O_2]}$
   (b) $mol^{-1}\ dm^3$
3. (a) A $K_c = \dfrac{[CO_2]\,[H_2]}{[CO][H_2O]}$ (no units)
   B $K_c = \dfrac{[H_2O]^2}{[H_2]^2\,[O_2]}$ $mol^{-1}\ dm^3$
   C $K_c = \dfrac{[NO]^4\,[H_2O]^6}{[NH_3]^4\,[O_2]^5}$ $mol\ dm^{-3}$
   (b) A reaction in which the reactants and products are all in the same physical state.

Exercise 2.9G

1. (a) $K_c = \dfrac{[CH_3COOC_5H_{11}][H_2O]}{[CH_3COOH]\,[C_5H_{11}OH]}$
   (b) No units
   (c) The forward reaction is endothermic as the position of equilibrium shifts to the right, and K_c increases, when the temperature increases.
2. The reverse reaction is endothermic as the position of equilibrium shifts to the left, decreasing the value of K_c, when the temperature increases. As a result the process becomes less effective as temperature increases
3. (a) Increasing the pressure shifts the position of equilibrium to the left, reducing the extent to which ozone dissociates, as there are fewer gas molecules on the left side of the equilibrium.
   (b) Increasing the pressure has no effect on the value of K_c.
4. Answer C

Exercise 2.9H

1. (a) 450 °C, 250 atm
   (b) The rate of reaction increases as the temperature increases. In contrast, the position of equilibrium shifts to the left, reducing the amount of ammonia at equilibrium, as the temperature increases.
2. (a) $N_2 + 3H_2 \rightarrow 2NH_3$
   (b) Iron
   (c) Increasing the pressure shifts the position of equilibrium to the right as there are fewer gas molecules on the right side of the equilibrium. Reducing the temperature would also shift the position of equilibrium to the right as the forward reaction is exothermic.
3. The formation of ammonia is exothermic as the reverse reaction is used to reduce the amount of ammonia in the equilibrium as the temperature is increased.

Exercise 2.9I

1. Answer C
2. (a) Increasing the pressure shifts the position of equilibrium to the right, increasing the yield, as there are fewer gas molecules on the right side of the equilibrium.
   (b) A moderate temperature is used to maintain a reasonable rate of reaction and to prevent the position of equilibrium shifting further to the left, reducing the yield, at higher temperatures.

Exercise 2.10A

1. The rate of reaction increases as the molecules are closer together and there are more collisions per second.

Exercise 2.10B

1. Answer C
2. (a) See (b) for labels. The area under the curve represents the number of molecules in the mixture.
   (b)

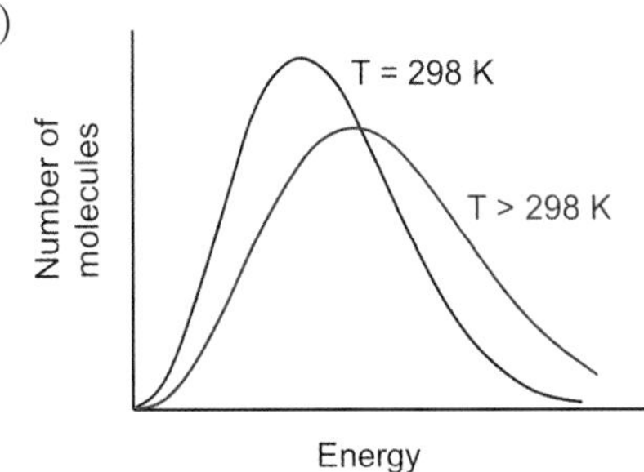

3. Answer A
4. (a) The number of particles with enough energy to react.

(b)

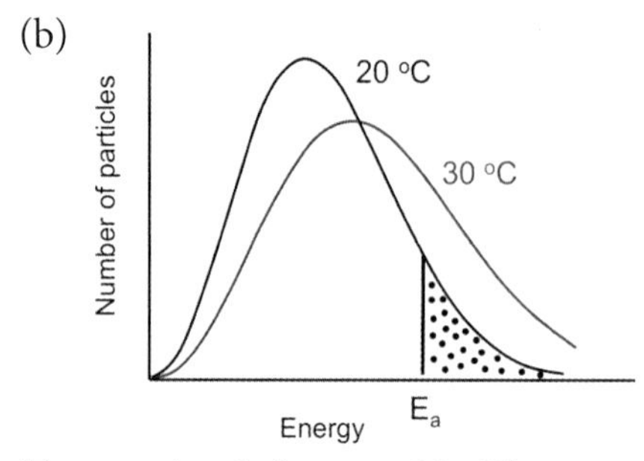

The reaction is faster at 30 °C because more of the particles in the mixture have an energy greater than the activation energy.

5. (a) See (b) for labels. The curve starts at the origin because all of the molecules have energy and are moving.

   (b)

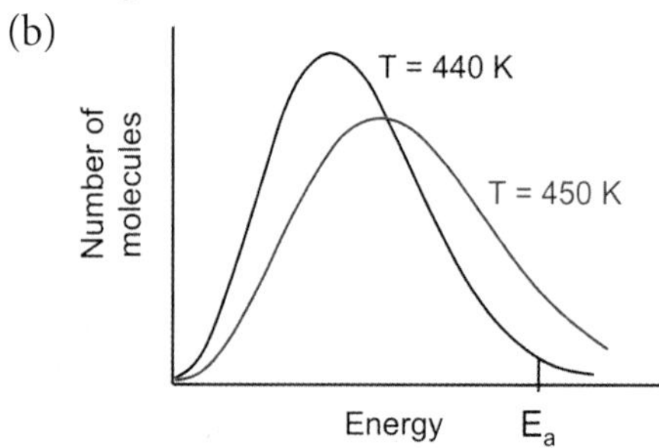

   (c) The rate of reaction decreases as the number of molecules with energy greater than the activation energy, E_a decreases.

6. (a) The minimum amount of energy needed for the reaction to occur.

   (b) Most collisions involve molecules whose energy is less than the activation energy.

   (c)

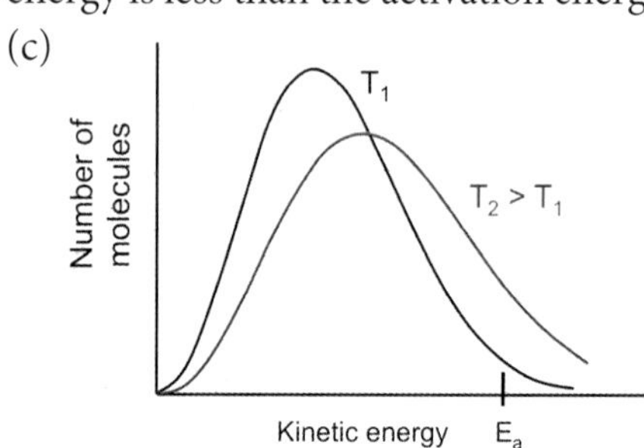

   (d) There are more successful collisions per second as a greater number of collisions involve molecules with at least the activation energy.

7. Answer D
8.

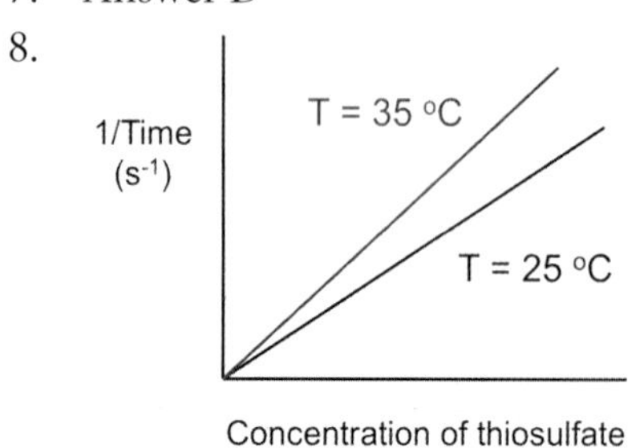

At 35 °C there are more successful collisions per second as a greater number of collisions involve molecules with at least the activation energy. As a result the average rate (=1/Time) is higher at 35 °C.

Exercise 2.10C

1. Answer A
2. Answer B
3. Answer D
4. A catalyst increases the rate of reaction by providing an alternative reaction pathway with a lower activation energy, and is not consumed by the reaction.
5. The catalyst provides an alternative reaction pathway with a lower activation energy (E_{cat}).

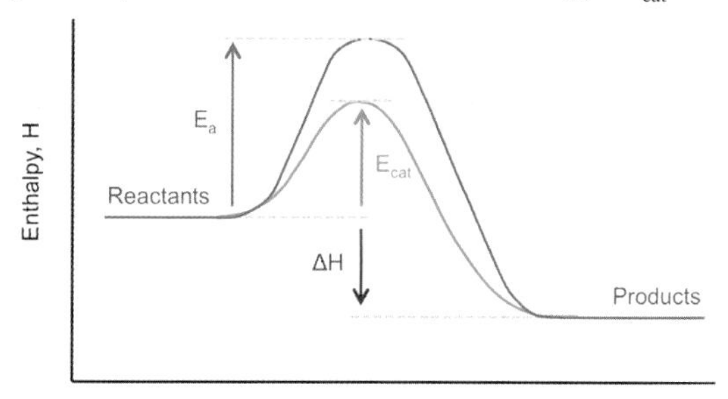

6. The spark is not a catalyst as it is not regenerated by the combustion process.
7.

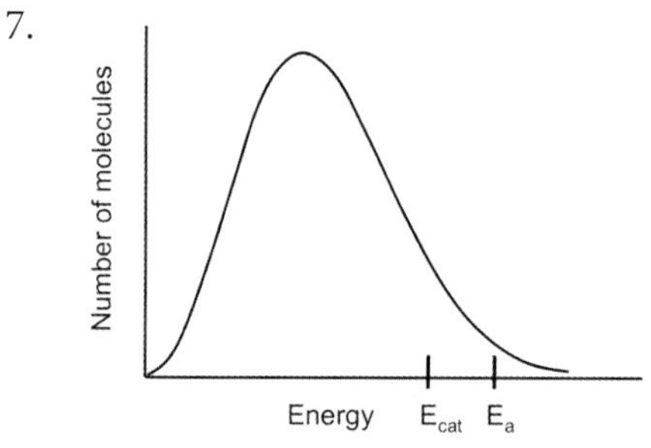

8. (a) All of the molecules are moving.

   (b) There is no limit to the amount of energy a molecule can have.

   (c) The number of molecules with energy greater than the activation energy.

   (d) A catalyst reduces the activation energy. As a result there are more successful collisions per second as a greater number of collisions involve molecules with enough energy to react.

9. A catalyst reduces the activation energy. As a result there are more successful collisions per second as a greater number of collisions involve molecules with enough energy to react.
10. (a) A catalyst lowers the activation energy,

increasing the number of molecules with energy greater than the activation energy.

(b) The number of successful collisions per second increases as a greater number of collisions involve molecules with enough energy to react.

Exercise 2.10D

1. Answer D
2. Answer C
3. Answer D
4. Answer C

Exercise 2.10E

1. The efficiency of the catalyst is higher as it has a larger surface area on which the reaction can occur.
2. (a) The platinum is a catalyst.
   (b) To increase the efficiency of the catalyst by increasing the amount of contact between the reactants and the catalyst.
   (c) A state of dynamic equilibrium cannot be established as the gases flow through the tube too quickly.

Exercise 2.11A

1. (a) Their outermost electrons are in an s-type subshell.
   (b) $Mg_{(g)} \rightarrow Mg^{+}_{(g)} + e^{-}$
   (c) The first ionisation energy decreases down the group as the outermost electrons are further from the nucleus and better shielded.
2. (a) The Group II metals are very reactive.
   (b) $1s^2\ 2s^2\ 2p^6\ 3s^2\ 3p^6\ 4s^2$
   (c) The number of filled shells increases down the group.
   (d) The Group II metals are smaller due to a greater nuclear charge.
   (e) 0.224 nm
   (f) Barium atoms have a greater atomic mass than strontium atoms and occupy a similar volume.

Exercise 2.11B

1. (a) $2Ca_{(s)} + O_{2(g)} \rightarrow 2CaO_{(s)}$
   (b) Moles of Ca = 5.0×10^{-2} mol
   Actual yield of CaO = 5.0×10^{-2} mol $\times$ 0.76
   = 3.8×10^{-2} mol
   Mass of CaO formed
   = 3.8×10^{-2} mol $\times$ 56 g mol^{-1} = 2.1 g
2. (a) Barium oxide, BaO
   (b) $2Ba + O_2 \rightarrow 2BaO$
   $Ba + O_2 \rightarrow BaO_2$

Exercise 2.11C

1. (a) Calcium sinks because it has a greater density than water.
   (b) *Any two of*: the metal sinks then rises, the metal disappears, bubbles of colourless gas, the reaction mixture warms up, fizzing, steamy fumes, a white solid forms in the solution.
   (c) $Ca + 2H_2O \rightarrow Ca(OH)_2 + H_2$
   (d)

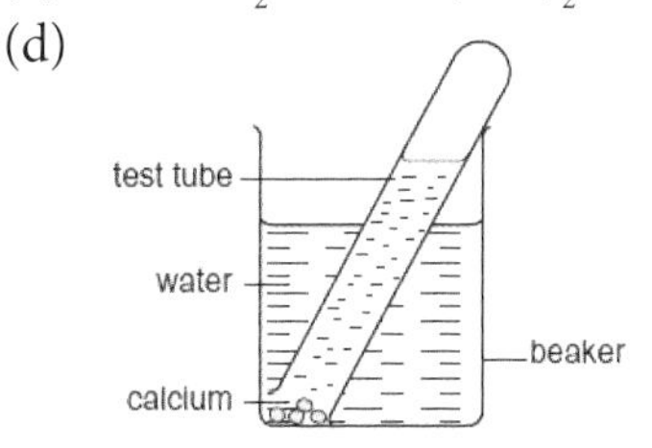

2. (a) Calcium sinks and then rises as it reacts quickly to form bubbles of colourless gas. The reaction mixture warms up and steamy fumes are produced as a white solid forms in the solution. Magnesium sinks and its surface becomes dull as it reacts very slowly to form bubbles of colourless gas and a colourless solution.
   (b) The gas produced by calcium burns with a 'pop'. The small amount of gas produced by magnesium is not sufficient to produce a 'pop'.
   (c) The solution containing calcium is strongly alkaline and turns purple on adding universal indicator. The solution containing magnesium is weakly alkaline and turns blue on adding universal indicator.
3. (a) Reactivity increases from magnesium to barium.
   (b) $Ba + 2H_2O \rightarrow Ba(OH)_2 + H_2$
4. (a) White powder formed. Intense white light produced.
   (b) $Mg_{(s)} + H_2O_{(g)} \rightarrow MgO_{(s)} + H_{2(g)}$
   (c) *Any two of*: the reaction would occur more slowly, produce less heat, emit less light.
5. Answer C

Exercise 2.11D

1. (a) $Mg + 2HCl \rightarrow MgCl_2 + H_2$
   (b) *Any two of*: fizzing, metal disappears, reaction mixture warms up.
2. (a) Both metals form a white solid when they burn in air. Burning magnesium produces intense white light. Calcium turns the flame a brick red

colour.
(b) Calcium reacts quickly, producing heat and bubbles of gas as the amount of metal decreases. A white solid forms in the solution. Magnesium reacts very slowly.
(c) Both metals react quickly. The mixture fizzes and warms up as the amount of metal decreases. A colourless solution is formed.

Exercise 2.11E

1. Answer C
2. Barium hydroxide is more soluble and produces a solution with a higher pH.
3. (a) $Ba(OH)_2 + 2HNO_3 \rightarrow Ba(NO_3)_2 + 2H_2O$
(b) The barium sulfate formed in the reaction is insoluble.
4. (a) Dissolving is endothermic as more solid dissolves, increasing the solubility, when the temperature is increased.
(b) $SrSO_4$ is more soluble than $BaSO_4$ and less soluble than $CaSO_4$.
5. Answer B
6. Answer C

Exercise 2.11F

1. (a) A compound is basic if it accepts hydrogen ions.
(b) $CaO + H_2O \rightarrow Ca(OH)_2$
(c) 5.3×10^{-3} mol $\times$ 74 g mol^{-1} = 0.39 g
(d) $Ca(OH)_2 + 2HCl \rightarrow CaCl_2 + 2H_2O$
2. (a) Add water.
(b) $Ba(OH)_2 + H_2SO_4 \rightarrow BaSO_4 + 2H_2O$
(c) White.
3. (a) $BaCO_3 + 2HCl \rightarrow BaCl_2 + H_2O + CO_2$
(b) The limewater turns milky.
(c) 3.4×10^{-3} mol $\times$ 24 dm^3 mol^{-1} = 0.082 dm^3 = 82 cm^3
4. (a) $SrO + 2HCl \rightarrow SrCl_2 + H_2O$
(b) Heat the solution in an evaporating basin until crystals begin to form. Leave the solution to cool and allow time for crystals to form. Filter the mixture then dry the crystals in a low temperature oven.
5. (a) $CaCO_3 + 2HCl \rightarrow CaCl_2 + H_2O + CO_2$
(b) Moles of $CaCO_3$ = 0.1500 mol $\times$ 0.95 = 0.14 mol
Volume of CO_2 = 0.14 mol $\times$ 24 dm^3 mol^{-1} = 3.4 dm^3
Actual yield of CO_2 = 3.60 dm^3 $\times$ 0.95 = 3.4 dm^3
6. (a) *Any one of:* fizzing, solid disappears, colourless solution formed.
(b) $CaCO_3.MgCO_3 + 4HCl \rightarrow CaCl_2 + MgCl_2 + 2H_2O + 2CO_2$
(c) Calcium ions turn the flame a brick red colour.
7. (a) $CaO.MgO + 2H_2O \rightarrow Ca(OH)_2 + Mg(OH)_2$
(b) The solubility of the hydroxides increases down the group.
(c) Molar mass of $Mg(OH)_2$ = 58 g mol^{-1}
Molar mass of $CaCl_2$ = 111 g mol^{-1}
$$\text{Atom Economy} = \frac{2 \times 58\ \text{g}}{(2 \times 58\ \text{g}) + 111\ \text{g}} \times 100 = 51\ \%$$
(d) $Mg(OH)_2 \rightarrow MgO + H_2O$

Exercise 2.11G

1. Answer C
2. (a) $BaCO_3 \rightarrow BaO + CO_2$
(b) Calcium ion (Ca^{2+}) is smaller, has a greater charge density, and is more able to polarise C-O bonds within neighbouring carbonate ions than barium ion (Ba^{2+}). As a result less heat is needed to decompose $CaCO_3$.
3. (a) $CaCO_3 \rightarrow CaO + CO_2$
(b) Magnesium ion (Mg^{2+}) is smaller, has a greater charge density, and is more able to polarise C-O bonds within neighbouring carbonate ions than calcium ion (Ca^{2+}). As a result less heat is needed to decompose $MgCO_3$ and the decomposition temperature for $MgCO_3$ would be lower.
4. (a) $Ca(OH)_2 \rightarrow CaO + H_2O$
(b) Magnesium ion (Mg^{2+}) is smaller, has a greater charge density, and is more able to polarise C-O bonds within neighbouring carbonate ions than barium ion (Ba^{2+}). As a result less heat is needed to decompose $MgCO_3$ and the decomposition temperature for $MgCO_3$ would be lower.
5. (a) $CaCO_3.MgCO_3 \rightarrow CaO.MgO + 2CO_2$
(b) $BeCO_3 < MgCO_3 < CaCO_3 < SrCO_3 < BaCO_3$
(c) As the Group II cations (M^{2+}) get bigger down the group their charge density decreases, and they are less able to polarise C-O bonds within neighbouring carbonate ions. As a result more heat is needed to decompose the carbonate, and the decomposition temperatures for the Group II carbonates increase going down the group.

6. (a) Moles of $CaSO_4.2H_2O$ = 0.200 mol
Moles of $CaSO_4$ = 0.191 mol

$$\% \text{ yield} = \frac{0.191 \text{ mol}}{0.200 \text{ mol}} \times 100 = 95.5 \%$$

(b) $CaSO_4 \rightarrow CaO + SO_3$
(c) As the Group II cations (M^{2+}) get bigger down the group their charge density decreases, and they are less able to polarise the bonds within neighbouring sulfate ions. As a result more heat is needed to decompose the sulfate, and the decomposition temperatures for the Group II sulfates increase going down the group.
7. (a) $BaCO_3 \rightarrow BaO + CO_2$
(b) Beryllium ion (Be^{2+}) is smaller, has a greater charge density, and is more able to polarise C-O bonds within neighbouring carbonate ions than barium ion (Ba^{2+}). As a result less heat is needed to decompose $BeCO_3$ and the decomposition temperature for $BeCO_3$ would be lower.
(c) $BaCO_3 + C \rightarrow BaO + 2CO$
8. (a) $2Mg(NO_3)_2 \rightarrow 2MgO + 4NO_2 + O_2$
(b) Strontium ion (Sr^{2+}) is bigger, has a lower charge density, and is less able to polarise bonds within neighbouring nitrate ions than magnesium ion (Mg^{2+}). As a result more heat is needed to decompose $Sr(NO_3)_2$ and the decomposition temperature for $Sr(NO_3)_2$ would be higher.
9. Answer C

Exercise 3A

1. (a) *Any two of:* rinse the pipette/burette, ensure there are no air bubbles in the pipette/burette after filling, gently swirl the contents of the conical flask after each addition, add solution from the burette dropwise near the end point, read the burette scale at the bottom of the meniscus.
(b) *Any two of:* repeat the titration to obtain more than one accurate titre, average the accurate titres, use only accurate titres that are in close agreement when calculating the average titre.
2. (a) Weigh 1.85 g of the acid in a small beaker. Pour the acid into a 250 cm^3 volumetric flask. Use deionised water to wash any drops of acid remaining into the volumetric flask. Add deionised water to the volumetric flask until the bottom of the meniscus lies on the fill line. Stopper and invert the flask several times to mix the contents.
(b) (i) $RCOOH + NaOH \rightarrow RCOONa + H_2O$
(ii) Colourless to pink.
(iii) Moles of NaOH added = 2.5×10^{-3} mol
Molarity of the acid = 2.5×10^{-3} mol $\times$ 40
= 0.10 mol dm^{-3}

$$\text{Molar mass of the acid} = \frac{7.4 \text{ g dm}^{-3}}{0.10 \text{ mol dm}^{-3}}$$

= 74 g mol^{-1}
RMM of the acid = 74
RMM of the R group = 74 – RMM of COOH = 29
RCOOH is CH_3CH_2COOH
3. (a) Weigh 2.90 g of the solid in a small beaker. Dissolve the solid in the minimum volume of deionised water. Pour the solution into a 250 cm^3 volumetric flask. Use deionised water to wash any drops of solution remaining into the volumetric flask. Add deionised water to the volumetric flask until the bottom of the meniscus lies on the fill line. Stopper and invert the flask several times to mix the contents.
(b) (i) $Na_2CO_3 + 2HCl \rightarrow 2NaCl + CO_2 + H_2O$
(ii) Yellow to red.
(iii) Moles of Na_2CO_3 in 25 cm^3
= 1.0×10^{-3} mol
Molarity of Na_2CO_3 = 1.0×10^{-3} mol $\times$ 40
= 0.040 mol dm^{-3}
Mass of Na_2CO_3 in 1 dm^3 = 4.2 g
Mass of water of crystallisation = 11.6 – 4.2
= 7.4 g
Moles of water of crystallisation

$$= \frac{7.4 \text{ g}}{18 \text{ g mol}^{-1}} = 0.41 \text{ mol}$$

$$x = \frac{0.41 \text{ mol}}{0.040 \text{ mol}} = 10$$

Exercise 3B

1.

Experiment	Observations	Deductions
1. Describe the appearance of A.	White solid.	Group I, II or ammonium compound.
2. Dip a wire loop in concentrated hydrochloric acid, touch sample A with the wire, then hold it in a blue Bunsen flame.	Yellow/orange flame.	Sodium ion present.
In a fume cupboard: 3. Add about 1 cm^3 of concentrated sulfuric acid to a half spatula measure of A in a test tube. Test the gas evolved using a glass rod dipped in concentrated ammonia solution.	Steamy fumes. White fumes of solid ammonium chloride.	 Hydrogen chloride gas. Chloride ion present.
4. Make up a solution of A by dissolving a half spatula-measure of A in a test tube half-full of dilute nitric acid. Put 1 cm^3 of the solution into each of two separate test tubes.	No effervescence. Colourless solution formed.	Carbonate ion not present.
(a) (i) Add a few drops of silver nitrate solution to the first test tube.	White precipitate formed.	Chloride ion present.
(ii) Add about 1 cm^3 of concentrated ammonia into the first test tube.	White precipitate dissolves.	Confirms chloride ion present.
(b) Add a few drops of barium chloride solution into the second test tube.	White precipitate formed.	Sulfate ion present.

Exercise 3C

1.

Experiment	Observations	Deductions
1. Place 10 drops of B in a test tube and add 1 cm^3 of water.	B mixes completely with water to form a colourless solution.	B can effectively hydrogen bond with water. B contains a -OH group or a -COOH group.
2. Place 10 drops of B on a watch glass placed on a heat proof mat and ignite it using a splint.	B burns with a clean blue flame.	B has a low carbon content.
3. Add approximately 10 drops of B to a test tube one quarter full of bromine water and mix well.	The orange colour of bromine remains.	B does not contain a C=C bond.
4. Add 10 drops of B to 2 cm^3 of acidified potassium dichromate solution in a test tube. Warm the mixture gently.	The mixture changes from orange to green and there is a change in smell.	B is oxidised. B is a primary or secondary alcohol.

2.

Experiment	Observations	Deductions
1. Describe the solution and place a few drops on a piece of Universal Indicator paper.	Colourless solution. Turns Universal Indicator paper green.	Neutral solution. B is not a carboxylic acid.
In a fume cupboard: 2. Shake a small volume of the solution with bromine water.	The orange colour of bromine remains.	B does not contain a C=C bond.
3. Gently heat 2 cm³ of the solution with 2 cm³ of acidified potassium dichromate solution.	The mixture changes from orange to green and there is a change in smell.	B is oxidised. B is a primary or secondary alcohol.

Exercise 3D

1.

Experiment	Observations	Deductions
1. Place 1 cm³ of Z in a test tube and add 1 cm³ of water. Add a bung and shake the test tube.	Two layers formed.	Z cannot hydrogen bond effectively with water.
2. Place 10 drops of Z on a watch glass placed on a heatproof mat and ignite it using a burning splint.	Z burns with a smoky yellow flame.	Z has a high carbon content.
3. In a fume cupboard add approximately 0.5 cm³ of Z to a test tube one quarter full of bromine water and mix well.	The bromine water is decolourised.	Z contains one or more C=C bonds.
4. Place 1 cm³ of Z in a test tube. Add 1 cm³ of ethanol and 1 cm³ of silver nitrate solution before placing the test tube in a beaker of hot water for 5 minutes.	A yellow precipitate forms.	Silver iodide precipitate. Z contains a C-I bond.

Exercise 3E

1. (a) Solid zinc sulfate that contains water of crystallisation.

(b)

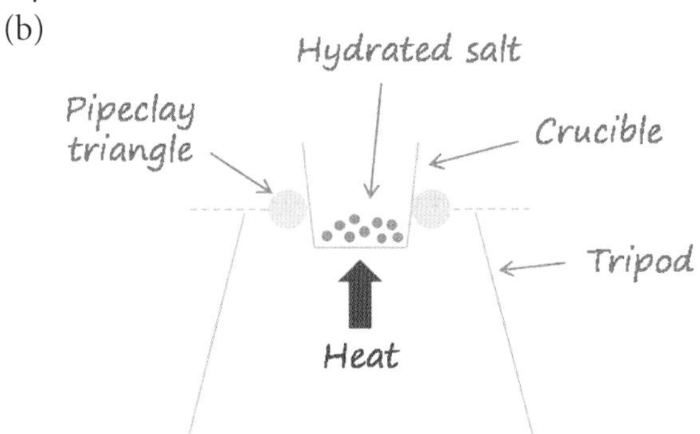

(c) The mass of the empty crucible and the mass of the crucible when it contains the sample of hydrated zinc sulfate.

(d) Heat then weigh the crucible containing the sample. Repeat until the mass of the crucible containing the sample stops decreasing.

(e) Use tongs to lift the crucible after it has been heated.

(f) (i) 8.63 - 4.85 = 3.78 g

(ii) $\dfrac{3.78 \text{ g}}{18 \text{ g mol}^{-1}} = 0.210 \text{ mol}$

(iii) $\dfrac{4.85 \text{ g}}{161 \text{ g mol}^{-1}} = 0.0301 \text{ mol}$

(iv) $x = \dfrac{0.210 \text{ mol}}{0.0301 \text{ mol}} = 7.0$

and the formula is $ZnSO_4.7H_2O$

2. (a)

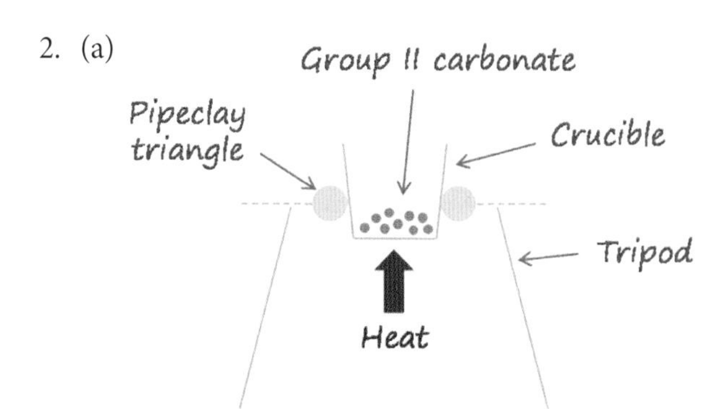

(b) Use tongs to lift the crucible after it has been heated.
(c) The mass of the empty crucible and the mass of the crucible when it contains the sample of the Group II carbonate.
(d) The crucible containing the sample is heated and weighed. The process is repeated until the mass stops decreasing.
(e) (i) 4.90 – 2.33 = 2.57 g

(ii) $\frac{2.57\text{ g}}{44\text{ g mol}^{-1}} = 0.0584\text{ mol}$

(iii) 0.0584 mol

(iv) Molar mass of Group II oxide

$= \frac{2.33\text{ g}}{0.0584\text{ mol}} = 40\text{ g mol}^{-1}$

RAM of Group II element

= 40 - RAM of O = 24

(v) The Group II element is Mg.

Exercise 3F

1. (a) The mass of the container and the mass of the container when filled with lead oxide.
   (b) Carry out the experiment in a fume cupboard as carbon monoxide is a poisonous gas.
   (c) The sample is heated in its container. The sample is then weighed in its container and the process repeated until the mass does not change.
   (d) To allow the apparatus to cool and to prevent the lead reacting with oxygen in the air.
   (e) (i) 1.39 – 1.26 = 0.13 g

   (ii) $\frac{0.13\text{ g}}{16\text{ g mol}^{-1}} = 0.0081\text{ mol}$

   (iii) $\frac{1.26\text{ g}}{207\text{ g mol}^{-1}} = 0.00609\text{ mol}$

   (iv) 1 mol of Pb reacts with $\frac{0.0081\text{ mol}}{0.00609\text{ mol}}$

   = 1.3 mol of O

   The empirical formula is Pb_3O_4

Exercise 3G

1.

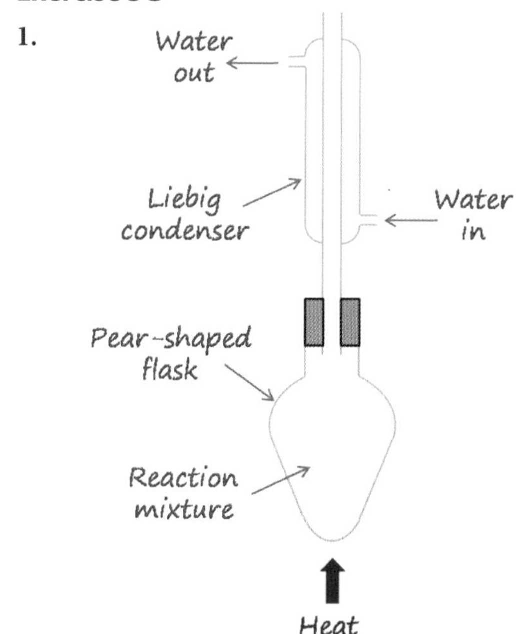

2. (a) To ensure smooth boiling of the reaction mixture.
   (b) To remove heat from the reaction mixture.
   (c) The continuous boiling and condensing of the reaction mixture using a condenser fitted in the vertical position.
   (d) To oxidise the ethanol.
   (e) Anhydrous calcium chloride/sodium sulfate/magnesium sulfate. The drying agent is removed by filtering the mixture.
   (f) Incomplete reaction and loss of product during recovery of the product from the reaction mixture.

3.

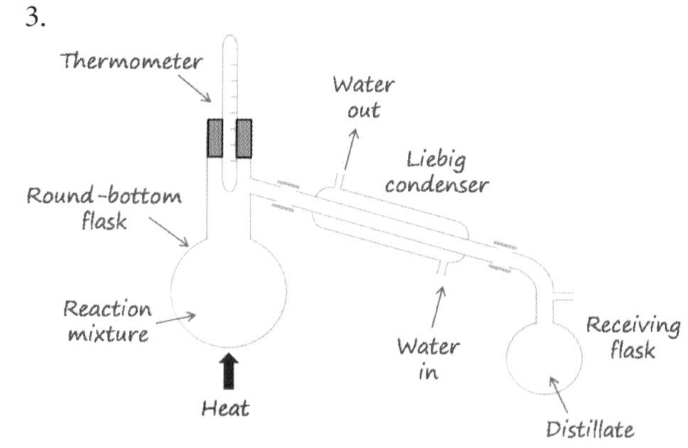

4. (a) It removes water from the reaction mixture.
   (b) To prevent evaporation of diethyl ether and reduce the risk of fire.
   (c) Ethanol.

(d) (i) To remove acidic impurities in the crude distillate.
(ii) Diethyl ether cannot hydrogen bond effectively with water.
(iii) To dry the product.
(iv) By filtering the mixture.
(e) Ethanol, water and diethyl ether have different boiling points.

5. (a) The repeated boiling and condensing of a reaction mixture using a condenser fitted in the vertical position.
(b) To remove heat from the reaction mixture.
(c) Phosphorus and phosphoric acid.
(d) Add sodium hydrogencarbonate solution to the distillate in a separating funnel. Stopper the funnel, invert and shake. Open the tap every few seconds to release any gas formed during shaking.
(e) Add anhydrous calcium chloride/sodium sulfate/magnesium sulfate to dry the distillate. The drying agent is removed by filtering the mixture.
(f) Iodoethane boils below 100 °C.
(g) Iodoethane boils within this temperature range. Ethanol does not boil within this range.

6. Mass of butan-1-ol used
$= 0.81\ \text{g cm}^{-3} \times 13.7\ \text{cm}^3 = 11\ \text{g}$

Moles of butan-1-ol used $= \dfrac{11\ \text{g}}{74\ \text{mol}^{-1}} = 0.15\ \text{mol}$

Theoretical yield of 1-bromobutane = 0.15 mol

Actual yield of 1-bromobutane $= \dfrac{10.28\ \text{g}}{137\ \text{g mol}^{-1}}$
$= 0.07504\ \text{mol}$

Percentage yield $= \dfrac{0.07504\ \text{mol}}{0.015\ \text{mol}} \times 100 = 50\%$

Glossary

1.1 Formulas, Equations and Amounts of Substance

Elements and Compounds

- An **atom** is the smallest amount of an element.
- An **element** is a pure substance that contains one type of atom.
- A **molecule** contains two or more atoms bonded together to form a single uncharged particle.
- A **molecular material** is an element, compound or mixture composed of molecules.
- A **diatomic element** is an element made up of diatomic molecules.
- A **diatomic molecule** is a molecule that contains two atoms.
- An **allotrope** is one physical form of an element.
- A **monatomic element** is an element made up of particles that each contain a single atom.
- A **compound** is a pure substance that contains two or more elements bonded together.
- An **ionic compound** is made of ions held together by attractive forces between oppositely charged ions.

Chemical Formulas

- The **chemical formula** of a substance defines the relative amount of each element in the substance and is the smallest amount of substance that can participate in a chemical reaction.
- The **valency** of an element represents the ability of the element to combine with other elements when forming compounds.

Naming Compounds

- A **cation** is an ion with a positive charge.
- An **anion** is an ion with a negative charge.
- A **molecular ion** is a positively or negatively charged ion that contains two or more atoms that are covalently bonded.

Chemical Equations

- A **chemical equation** relates the amounts of substance that react and the amounts of products formed in a chemical change.
- A **state symbol** represents the physical state of a substance under the specified conditions.

Aqueous Solutions

- An **aqueous solution** is formed by dissolving a substance in water.
- A **salt** is formed when an acid reacts with a base.
- An **acid** is a substance that forms hydrogen ions in solution.
- An **alkali** is a solution containing hydroxide ions.
- A **base** is a substance that reacts with an acid to form a salt.

Ionic Equations

- An **ionic equation** contains only those compounds and ions needed to describe the change that occurs during a chemical reaction.
- The term **precipitate** refers to an insoluble solid formed in solution by a chemical reaction.
- The term **spectator ion** refers to an ion that is present in a reaction mixture but does not participate in the reaction.
- The term **neutralisation** refers to any chemical reaction in which an acid reacts with a base to form a salt and water.

Amounts of Substance

- One **mole** of a substance contains the same number of atoms, molecules, or ions as 6.02×10^{23} formulas of the substance.
- The **molar mass** of a substance is the mass of one mole of the substance in grams.

Calculating Reacting Amounts

- A **limiting reactant** determines the amount of product formed in a reaction.
- A reactant is **in excess** if some of the reactant remains after the reaction is complete.

Salts Containing Water

- **Water of crystallisation** refers to water molecules that are chemically bonded within crystals and are part of the crystal structure.
- A **hydrated salt** is a salt that contains water of crystallisation.
- An **anhydrous salt** is a salt that does not contain water of crystallisation.
- The technique of **heating to constant mass** is used to remove water from a substance by a process of repeated heating and weighing until the mass of substance remains constant.

1.2 Atomic Structure and The Periodic Table

Atoms Ions and Isotopes

- The **atomic number** of an element is the number of protons in the nucleus of one atom of the element.
- The **mass number** of an atom is the total number

of protons and neutrons in the atom.

- **Isotopes** are atoms with the same atomic number that have different mass numbers.
- The **relative isotopic mass (RIM)** of an isotope is the mass of one atom of the isotope relative to one-twelfth of the mass of one atom of carbon-12.
- **Avogadro's constant (L)** is the number of atoms in exactly 12 g of carbon-12.
- The **relative atomic mass (RAM)** of an element is the average mass of one atom of the element relative to one-twelfth of the mass of an atom of carbon-12.
- The **mass spectrum** of a substance is the relative abundance of each ion obtained by ionising the substance.
- The **relative molecular mass (RMM)** of a molecule is the mass of the molecule relative to one-twelfth of the mass of an atom of carbon-12.
- The **relative formula mass (RFM)** of a compound is the average mass of one formula of the compound relative to one-twelfth of the mass of one atom of carbon-12.

Atomic Structure

- An **atomic orbital** is a region of space within an atom that can be occupied by up to two electrons.
- The **electron configuration** of an atom details how the electrons are distributed amongst the subshells.
- The **Aufbau principle** asserts that electrons fill the subshells in an atom in order of increasing energy.
- The **ground state** of an atom is obtained by filling the lowest energy subshells available to produce the electron configuration with the lowest energy.
- The **spin** of an electron describes the arrangement of the electron within an atomic orbital.
- The **first ionisation energy** of an element is the energy needed to remove one mole of electrons from one mole of gas phase atoms to form one mole of gas phase ions with a 1+ charge.
- The **second ionisation energy** of an element is the energy needed to remove one mole of electrons from one mole of gas phase ions with a 1+ charge to form one mole of gas phase ions with a 2+ charge.
- The **third ionisation energy** of an element is the energy needed to remove one mole of electrons from one mole of gas phase ions with a 2+ charge to form one mole of gas phase ions with a 3+ charge.
- The **shielding** experienced by an electron is the reduction in attraction to the nucleus due to electrons in subshells closer to the nucleus.

The Periodic Table

- The term **group** refers to a column of elements in the Periodic Table.
- The term **period** refers to a row of elements in the Periodic Table.
- A **periodic trend** is a variation in the properties of the elements within a group or period.
- A **semimetal** is an element with properties in common with both metals and nonmetals.
- The **metallic character** of an element refers to the extent to which the properties of the element resemble those of a metal.

1.3 Chemical Bonding and Structure

Ionic Compounds

- An **ionic lattice** is the regular structure that results from the packing together of ions in an ionic compound.
- **Ionic bonding** refers to the attractive forces between oppositely charged ions in an ionic lattice.

Covalent Bonding

- The **octet rule** states that atoms will attempt to gain, lose or share electrons to form compounds in which they have a full outer shell containing eight electrons.
- A **covalent bond** is the attractive force between the negatively charged electrons in a shared pair of electrons and the positively charged nucleus of each atom sharing the electrons.
- A **double bond** is formed when two atoms each use two electrons to share two pairs of electrons.
- A **triple bond** is formed when two atoms each use three electrons to share three pairs of electrons.
- The term **multiple bonding** refers to the formation of two or more covalent bonds between a pair of atoms.
- A **structural formula** describes the bonding in a molecule or ion by using a line between two atoms to represent a shared pair of electrons.
- A **molecular ion** is an ion containing two or more atoms held together by covalent bonds.
- A **coordinate bond** is a covalent bond that is formed when two atoms share a pair of electrons donated by one of the atoms.

- A **nonpolar bond** is a covalent bond in which the electrons forming the bond are shared equally by the atoms forming the bond.
- A **polar bond** is a covalent bond in which the electrons forming the bond are not shared equally by the atoms forming the bond.
- The **electronegativity** of an element is the extent to which an atom of the element attracts the shared electrons in a covalent bond to itself when forming covalent bonds with other atoms.
- The term **bond dipole** refers to the partial charges that result from the unequal sharing of electrons in the bond.
- The term **permanent dipole** refers to the separation of charge that gives rise to oppositely charged ends in a molecule.
- A **polar molecule** is a molecule with a permanent dipole.

Covalently Bonded Materials

- A substance has a **molecular covalent structure** if it is composed of molecules held together by weak attractive forces.
- A substance has a **giant covalent structure** if it contains a network of atoms held together by covalent bonds.
- A **delocalised electron** is an outer shell electron that is not associated with an atom or ion and is able to move freely.

Metals

- The term **metallic bond** refers to the attractive force between the metal ions and the delocalised outer shell electrons in a metal.
- A substance is **malleable** if it can be worked into different shapes.
- A substance is **ductile** if it can be drawn into wires.
- A substance is an **electrical conductor** if it allows an electric current to flow within it.

1.4 Shapes of Molecules and Ions

VSEPR Theory

- A **bonding pair** is a pair of electrons in the outer shell of an atom that is shared to form a covalent bond.
- A **lone pair** is a pair of electrons in the outer shell of an atom that is not involved in bonding.
- The term **bond angle** refers to the angle formed by two bonds to the same atom.
- The term **molecular shape** refers to the shape formed by the atoms in a molecule.

Nonpolar Molecules

- A **nonpolar molecule** does not have a permanent dipole.

1.5 Intermolecular Forces

- The term **intermolecular force** refers to any type of bonding interaction between molecules.

Van der Waals Forces

- The term **van der Waals force** refers to the attraction between an instantaneous dipole and the induced dipoles on neighbouring molecules.

Dipole Forces

- The term **dipole force** refers to the attraction between the positive end of the permanent dipole on one molecule and the negative end of the permanent dipole on neighbouring molecules.

Hydrogen Bonding

- A **hydrogen bond** is a strong dipole-like force of attraction between a lone pair on a very electronegative atom (N, O or F) and a hydrogen atom that is covalently bonded to a very electronegative atom (N, O or F) in a neighbouring molecule.

Properties of Liquids

- The term **miscible** describes liquids that mix in all proportions.
- The term **immiscible** describes liquids that do not mix.

1.6 Oxidation and Reduction

Redox Reactions

- A **redox reaction** is a chemical reaction in which oxidation and reduction occur at the same time.
- **Oxidation** is a chemical change in which electrons are lost.
- **Reduction** is a chemical change in which electrons are gained.
- A **half-equation** describes oxidation or reduction in terms of the loss or gain of electrons.
- A **reducing agent** provides the electrons needed for reduction.
- An **oxidising agent** gains the electrons produced by oxidation.
- The **oxidation state** of an element is the extent to

which the element has been oxidised.

Oxidation Numbers

- The **oxidation number** of an element indicates the number of electrons lost or gained by an element.

Disproportionation Reactions

- **Disproportionation** refers to a redox reaction in which the same element is oxidised and reduced.

1.7 Group VII: The Halogens

Reactivity

- The **oxidising ability** of a substance refers to its ability to act as an oxidising agent.

Redox Chemistry

- A **strong acid** is a solution in which all of the molecules dissociate to form hydrogen ions.
- A **weak acid** is a solution in which only a fraction of the molecules dissociate to form hydrogen ions.

1.8 Volumetric Analysis

Working with Solutions

- A **volumetric pipette** is used to accurately measure a fixed volume of solution.
- A **burette** is used to accurately dispense measured volumes of solution.
- A **volumetric flask** is used to accurately prepare a fixed volume of solution.
- The term **meniscus** refers to the surface of a liquid.

Calculations with Solutions

- The term **concentration** refers to the amount of solute, in units of mass or moles, per unit volume of solution.
- The term **molarity** refers to the concentration of a solution in units of moles per cubic decimetre (mol dm^{-3}).

Titration Experiments

- A **titration** is an experiment to accurately determine the concentration of a substance in solution by measuring the volume of a standard solution needed to react with a known volume of the solution.
- A **standard (solution)** is a stable solution whose concentration has been accurately determined.
- An **acid-base titration** is a titration experiment based on the neutralisation of an acid by an alkali or vice versa.
- The **equivalence point** is the point in a titration at which the sample has completely reacted with the standard solution.
- The **titre** is the volume of solution needed to reach the equivalence point in a titration.
- An **indicator** is a substance that changes colour over a range of pH values.
- The **end point** is the point in a titration at which the indicator changes colour.

2.1 Further Calculations

Percentage Yield

- The **percentage yield** for a chemical reaction is the actual yield expressed as a percentage of the theoretical yield.
- The **theoretical yield** of a chemical reaction is the yield assuming complete reaction and no loss of product during recovery.
- The **actual yield** of a chemical reaction is the yield obtained after recovery of the product.

Atom Economy

- The **atom economy** of a chemical process is the yield of useful products expressed as a percentage of the total yield of all products.

Calculating Gas Volumes

- **Avogadro's Law** states that the volume of one mole of any gas is exactly 24 dm^3 if measured at 20 °C and 1 atmosphere pressure.
- The **molar volume** of a gas is the volume of one mole of gas at a specified temperature and pressure.

Percentage Composition Calculations

- The **mass percentage** of an element in a substance is the mass of the element in 100 g of the substance expressed as a percentage.
- The **percentage composition** of a substance is the composition of the substance expressed in terms of mass percentages.
- The **empirical formula** of a compound is the simplest whole number ratio of atoms of each element in the compound.
- The **molecular formula** of a compound represents the number of atoms of each element in one molecule of the compound.

2.2 Organic Chemistry

Organic Compounds

- An **organic compound** is a substance whose structure is based on the element carbon.
- **Isomers** are compounds with the same molecular formula that have a different arrangement of atoms.
- A **hydrocarbon** is a compound that contains only carbon and hydrogen.
- A **general formula** is used to generate the molecular formulas for an entire family of compounds.
- A **space-filling model** represents the size of the atoms in a molecule.
- A **homologous series** is a family of compounds with similar chemical properties, whose formulas derive from the same general formula and differ by CH_2 when arranged by mass, and whose physical properties vary smoothly as the molecules get bigger.

Structural Formulas

- A **structural formula** describes the bonding in a molecule or ion by using a line between atoms to represent a shared pair of electrons.
- A **condensed formula** is obtained from a structural formula by using formulas to represent groups of atoms within the structure.

Functional Groups

- The term **functional group** refers to a group of atoms in a molecule that together determine the reactions of the compound.
- The **skeletal formula** of a molecule or ion is obtained by omitting the hydrogen atoms from the structure and using a line between atoms to represent each shared pair of electrons.
- An **alkyl group** is a functional group with the general formula C_nH_{2n+1} where n = 1, 2, 3...

2.3 Hydrocarbons: Alkanes

Structure and Properties

- A **saturated hydrocarbon** is a compound that contains only carbon and hydrogen, and does not contain any C=C or C≡C bonds.
- The term **parent alkane** refers to the straight-chain alkane on which the structure of a molecule is based.
- A **structural isomer** is used when referring to molecules with the same molecular formula that have different structural formulas.

Combustion of Alkanes

- **Complete combustion** refers to the complete reaction of a compound with an excess of oxygen.
- **Incomplete combustion** refers to the complete reaction of a compound with a limited supply of oxygen.
- A **volatile** compound is easily transformed into vapour.
- A **catalyst** is a substance that speeds up a chemical reaction without being consumed by the reaction.
- **Chemisorption** refers to the formation of a chemical bond between a molecule and a surface.
- **Heterogeneous catalysis** refers to a reaction in which the reactants are in a different physical state than the catalyst.
- The term **finely divided** is used to describe a solid that consists of small particles with a large surface area.

Halogenation of Alkanes

- A **free radical** is an atom, molecule or ion with an unpaired electron.
- **Homolytic fission** refers to the breaking of a bond in a way that each atom forming the bond gains one of the electrons shared in the bond.
- A **photochemical reaction** is a reaction that occurs when one or more of the reactants absorb radiation.
- A **reaction mechanism** details the individual reactions that together describe the change that occurs during a chemical reaction.
- A **substitution reaction** involves replacing one atom or group of atoms with a different atom or group of atoms.
- The term **free radical substitution** describes a substitution reaction which is propagated by free radicals.

2.4 Hydrocarbons: Alkenes

Structure and Properties

- An **unsaturated hydrocarbon** is is a compound that contains only carbon and hydrogen, and contains one or more C=C or C≡C bonds.
- A **diene** is a molecule or ion that contains two C=C bonds.

The C=C Functional Group

- A **sigma (σ) bond** is a covalent bond formed by the linear or 'head-on' overlap of atomic orbitals that does not restrict rotation about the bond.
- A **pi (π) bond** is a covalent bond formed by the sideways overlap of p-orbitals that restricts rotation about the bond.
- **Bond length** refers to the distance between the nuclei of two covalently bonded atoms.
- **Average bond energy** refers to the energy needed to break one mole of bonds of a specified type averaged over many compounds.
- **Geometric isomers** have the same structural formula but a different arrangement of atoms in space as the molecule cannot rotate about one or more C=C bonds.

Reactions of Alkenes

- **Hydrogenation** refers to the addition of a hydrogen molecule across a C=C or C≡C bond.
- An **addition reaction** is a reaction in which a molecule adds across a C=C or C≡C bond.
- An **electrophile** is a molecule or ion that attacks regions of high electron density.
- **Heterolytic fission** describes the breaking of a covalent bond in a way that both of the shared electrons are given to one of the atoms forming the bond.
- A **carbocation** is a cation that contains an electron deficient carbon atom (C^+).
- **Electrophilic addition** describes the addition of an electron seeking molecule or ion across a C=C or C≡C bond.
- **Markovnikov's Rule** asserts that when HX (X = Cl, Br) adds across a C=C bond, the halogen (X) attaches to the carbon with more alkyl groups bonded to it.
- A **primary carbocation** has one carbon atom bonded to the positively charged carbon atom within the carbocation.
- A **secondary carbocation** has two carbon atoms bonded to the positively charged carbon atom within the carbocation.
- A **tertiary carbocation** has three carbon atoms bonded to the positively charged carbon atom within the carbocation.
- A **polymer** is a large molecule formed by the joining together of many smaller monomers.
- The term **monomers** refers to the many small molecules that combine to form a polymer.
- The term **polymerisation** refers to a reaction in which a large polymer molecule is formed by the joining together of many monomers.
- **Addition polymerisation** refers to the formation of a polymer as a result of addition reactions between monomers.

2.5 Halogenoalkanes

Structure and Properties

- A **halogenoalkane** is a saturated hydrocarbon containing one or more halogen atoms.
- In a **primary halogenoalkane** the halogen atom is bonded to a carbon atom that has one carbon atom attached.
- In a **secondary halogenoalkane** the halogen atom is bonded to a carbon atom that has two carbon atoms attached.
- In a **tertiary halogenoalkane** the halogen atom is bonded to a carbon atom that has three carbon atoms attached.

Preparation

- **Organic synthesis** is the process of making an organic compound.
- The term **in situ** describes the formation of a reactant within the reaction mixture.
- **Reflux** refers to the continuous boiling and condensing of a reaction mixture using a condenser fitted in the vertical position.
- **Boiling chips** are small pieces of glass or other inert material that are added to a mixture to ensure smooth boiling.
- **Distillation** refers to the process of heating a mixture and condensing the vapours formed using a Liebig condenser in the horizontal position.
- The term **distillate** refers to a liquid extracted by distillation.
- A **separating funnel** is used to separate immiscible liquids.
- A **drying agent** is a substance that absorbs water.

Reactions

- In a **nucleophilic substitution** reaction an atom or group of atoms is replaced by a nucleophile when the nucleophile uses a lone pair to form a coordinate bond with an electron-deficient atom.
- A **nucleophile** is a molecule or ion that attacks

regions of low electron density by using a lone pair to form a coordinate bond with an electron-deficient atom.

- A **transition state** is an unstable structure formed during a reaction that cannot be isolated from the reaction mixture.
- An **amine** is a compound containing an amino (-NH_2) functional group.
- A **nitrile** is a compound containing a cyano (-CN) functional group.
- The term **hydrolysis** describes a reaction in which bonds are broken as the result of a compound reacting with water.
- The term **alkaline hydrolysis** refers to the hydrolysis of a compound using dilute alkali.
- The term **ethanolic** is used to describe a solution in which the solvent is ethanol.
- An **elimination reaction** is a reaction in which a small molecule is removed from a larger molecule.

2.6 Alcohols

Structure and Properties

- An **alcohol** is a compound containing a hydroxyl group.
- In a **primary alcohol** the hydroxyl group is bonded to a carbon atom that has one carbon atom attached.
- In a **secondary alcohol** the hydroxyl group is bonded to a carbon atom that has two carbon atoms attached.
- In a **tertiary alcohol** the hydroxyl group is bonded to a carbon atom that has three carbon atoms attached.
- A **diol** is a compound containing two hydroxyl groups.
- A **triol** is a compound containing three hydroxyl groups.

Reactions

- The term **halogenation** refers to a reaction in which one or more halogen atoms are added to a compound.

Oxidation of Alcohols

- A **carboxylic acid** is a compound with the structure RCOOH whose reactions are determined by the carboxyl (COOH) functional group.
- An **aldehyde** is a compound with the structure RCHO whose reactions are determined by the -CHO functional group.
- A **ketone** is a compound with the structure RCOR' whose reactions are determined by the carbonyl (C=O) functional group.

2.7 Infrared Spectroscopy

- The **wavenumber** of a molecular vibration is the reciprocal of its wavelength and is measured in units of cm^{-1}.
- The term **ground state** is used when referring to the lowest energy state of a molecular vibration.
- The **infrared (IR) spectrum** of a substance is the graph of percent transmission against frequency obtained by passing infrared radiation through the substance.
- **Infrared (IR) spectroscopy** is the procedure by which the IR spectrum of a substance is obtained.

2.8 Energetics

Enthalpy

- The **enthalpy** of a substance refers to the amount of heat energy contained within the substance.
- A chemical reaction is **exothermic** if the enthalpy of the products is less than the enthalpy of the reactants.
- The **enthalpy of reaction** is the enthalpy change when the amounts specified in the chemical equation react.
- A chemical reaction is **endothermic** if the enthalpy of the products is greater than the enthalpy of the reactants.

Measuring Enthalpy Changes

- **Standard conditions** refers to a temperature of 25 °C (298 K) and a pressure of 100 kPa.
- The **standard enthalpy change** for a process is equal to the change in heat energy that occurs when the process is carried out at constant pressure under standard conditions.
- The **standard state** of a substance refers to the physical state of the substance under standard conditions.
- A **molar enthalpy change** is the enthalpy change when one mole of substance reacts.
- The **specific heat capacity** of a substance is the energy needed to raise the temperature of 1 g of the substance by 1 °C (1 K).
- The **standard enthalpy of neutralisation** is the

enthalpy change when one mole of water is produced by a neutralisation reaction under standard conditions.

Calculating Enthalpy Changes

- An **average bond enthalpy** is the energy needed to break one mole of bonds of a specified type averaged over many compounds.
- The **standard enthalpy of formation** of a substance is the enthalpy change when one mole of the substance is formed from its elements under standard conditions.
- The **standard enthalpy of combustion** of a substance is the enthalpy change when one mole of the substance is completely burnt in oxygen under standard conditions.

Hess's Law

- **Hess's law** states that the enthalpy change for a chemical reaction depends on the state of the reactants and products and does not depend on the way in which the reactants are converted to the products.

2.9 Equilibrium

Dynamic Equilibrium

- A **reversible reaction** is a reaction that can operate in the forward direction (reactants → products) and the reverse direction (products → reactants).
- In a **dynamic equilibrium** the forward and reverse reactions occur at the same rate and, as a result, the amount of each reactant and product in the mixture remains constant.

Factors Affecting Equilibrium

- The **position of equilibrium** describes the extent to which a chemical equilibrium favours the reactants (on the left) or products (on the right).

Describing Equilibrium

- The **equilibrium constant, K_c** for a reaction is a positive number that relates the concentrations of the species in the equilibrium, and does not change unless the temperature in the equilibrium changes.
- A **homogeneous reaction** is a reaction in which the reactants and products are all in the same physical state.
- A **heterogeneous reaction** is a reaction in which the reactants and products are *not* all in the same physical state.

2.10 Chemical Kinetics

Rate of Reaction

- The **rate of reaction** is the amount of substance that reacts, or the amount of product formed, per unit time.
- The **average rate (of reaction)** refers to the average amount of a substance reacted or formed per unit time during the reaction.

Factors Affecting Rate

- A **successful collision** occurs when particles with sufficient energy and the correct orientation collide and react.
- The **Maxwell-Boltzmann distribution** describes the distribution of energy amongst the particles in a mixture of liquids or gases.
- The **activation energy** for a reaction is the minimum amount of energy needed for the reaction to occur.

Catalysis

- A **catalyst** is a substance that increases the rate of reaction by providing an alternative reaction pathway with a lower activation energy, and is not consumed by the reaction.
- A substance is **finely divided** if it has a large surface to volume ratio.

2.11 Group II: The Alkaline Earth Metals

Properties

- In an **s-block element** the outermost electrons in each atom are located in an s-type subshell.

Properties of Group II Compounds

- The **solubility** of a solid is the maximum mass of solid that will dissolve in 100 g of solvent at a given temperature.

Reactions of Group II Compounds

- A **base** is a molecule or ion that accepts a hydrogen ion (H^+) from another substance.
- **Thermal decomposition** is the breaking down of a substance into two or more simpler substances by the action of heat.

Thermal Stability of Group II Compounds

- **Thermal stability** refers to the ability of a substance to resist the effects of heating.
- The **charge density** within an ion is the charge per unit volume within the ion.

Appendix 1
Specification Mapping

This text has been written to cover the content of the Revised GCE Specification (CCEA 2016), and is presented in an order that will hopefully facilitate teaching and learning of the subject. The specific aspects of the specification addressed in each section of the text are detailed below.

AS Unit 1

Section	Specification Content
1.1 Formulas, Equations and Amounts of Substance	1.1 and 1.9.1
1.2 Atomic Structure and The Periodic Table	1.2
1.3 Chemical Bonding and Structure	1.3 (except 1.3.4) and 1.5
1.4 Shapes of Molecules and Ions	1.6 and 1.3.4
1.5 Intermolecular Forces	1.4
1.6 Oxidation and Reduction	1.7
1.7 Group VII: The Halogens	1.8
1.8 Volumetric Analysis	1.9 (except 1.9.1)
1.9 Qualitative Analysis	1.10

AS Unit 2

Section	Specification Content
2.1 Further Calculations	2.1
2.2 Organic Chemistry	Aspects of 2.2 are introduced, and then developed throughout Sections 2.3-6.
2.3 Hydrocarbons: Alkanes	2.3 and 2.5.9
2.4 Hydrocarbons: Alkenes	2.4
2.5 Halogenoalkanes	2.5 (except 2.5.9)
2.6 Alcohols	2.6
2.7 Infrared Spectroscopy	2.7
2.8 Energetics	2.8
2.9 Equilibrium	2.10
2.10 Chemical Kinetics	2.9
2.11 Group II: The Alkaline Earth Metals	2.11

Appendix 2
Working with Significant Figures

The following examples illustrate how to work with significant figures when doing calculations.

Rule 1
The number of significant figures (*sig fig*) associated with a mass, volume, temperature, or other measured quantity, depends on how it was measured.

Example 1

A solid sample with an actual mass of 5.646 g will be considered to have a mass of:

- 5.65 g (*3 sig fig*) if measured on a balance that weighs to the nearest 0.01 g,
- 5.6 g (*2 sig fig*) if measured on a balance that weighs to the nearest 0.1 g, and
- 6 g (*1 sig fig*) if measured on a balance that weighs to the nearest 1 g.

Rule 2
Only significant figures are used when doing calculations.

Example 2

Carbon, in the form of graphite, is a major component of the 'lead' in pencils. A single dot made by a pencil contains approximately 0.000**10** g of carbon. How many atoms of carbon are in the dot?

The mass of carbon has been measured to two significant figures (in bold).

$$\text{Moles of carbon in the dot} = \frac{\text{Mass}}{\text{Molar Mass}}$$

$$= \frac{\mathbf{1.0} \times 10^{-4}\ \text{g}}{12\ \text{g mol}^{-1}} = \mathbf{8.3} \times 10^{-6}\ \text{mol}$$

The molar mass is 'exact' (has unlimited significant figures) as it is obtained from RAMs on the Periodic Table, which unless otherwise instructed, are exact.

The answer is rounded to two significant figures as the accuracy of the mass (two significant figures) limits the accuracy of the answer to two significant figures.

The answer, after rounding, is used in the next step of the calculation. Retaining additional (insignificant) figures at this stage does not improve the accuracy of the final answer.

Carbon atoms in the dot = Moles × L
$= (\mathbf{8.3} \times 10^{-6})(6.02 \times 10^{23}) = \mathbf{5.0} \times 10^{18}$

The answer from the first step (8.3×10^{-6}) has two significant figures, and Avogadro's constant (6.02×10^{23}) has three significant figures.

The answer is rounded to two significant figures as the accuracy of the answer is limited to two significant figures by the accuracy of the answer from the first step.

Example 3

In industry iron is obtained from iron(III) oxide, Fe_2O_3 using a Blast Furnace. How much iron can be obtained from 1 tonne (1 tonne = 1000 kg) of iron(III) oxide?

$$\text{Moles of } Fe_2O_3 \text{ in 1 tonne} = \frac{\text{Mass}}{\text{Molar Mass}}$$

$$= \frac{1 \times 10^{6}\ \text{g}}{160\ \text{g mol}^{-1}} = 6.25 \times 10^{3}\ \text{mol}$$

In this example the mass (1 tonne) is exact, and does not limit the accuracy of the answer, as it has not been measured.

Moles of Fe in 6.25×10^{3} mol of Fe_2O_3
$= 2 \times \mathbf{6.25} \times 10^{3} = \mathbf{1.25} \times 10^{4}$ mol

The whole number (2) is exact and does not limit the accuracy of the answer.

The answer has three significant figures as the accuracy of the answer is limited by the accuracy of the answer from the previous step (6.25×10^{3}), which has three significant figures.

Mass of Fe in 1 tonne = $(\mathbf{1.25} \times 10^{4}\ \text{mol})(56\ \text{g mol}^{-1})$
$= \mathbf{7.00} \times 10^{5}\ \text{g} = 700\ \text{kg}$

The molar mass (56 g mol^{-1}) is exact and does not limit the accuracy of the answer.

Again, the answer has three significant figures as the accuracy of the answer is limited by the accuracy of the answer from the previous step (1.25×10^{4}), which has three significant figures.